# Methods in Enzymology

## Volume 349
## SUPEROXIDE DISMUTASE

# METHODS IN ENZYMOLOGY

EDITORS-IN-CHIEF

John N. Abelson     Melvin I. Simon

DIVISION OF BIOLOGY
CALIFORNIA INSTITUTE OF TECHNOLOGY
PASADENA, CALIFORNIA

FOUNDING EDITORS

Sidney P. Colowick and Nathan O. Kaplan

*Methods in Enzymology*

*Volume 349*

# Superoxide Dismutase

EDITED BY

## Lester Packer

DEPARTMENT OF MOLECULAR PHARMACOLOGY
AND TOXICOLOGY
SCHOOL OF PHARMACY
UNIVERSITY OF SOUTHERN CALIFORNIA
LOS ANGELES, CALIFORNIA

*Editorial Advisory Board*
Joe M. McCord
Irwin Fridovich

## ACADEMIC PRESS
An Elsevier Science Imprint

San Diego   San Francisco   New York   Boston   London   Sydney   Tokyo

This book is printed on acid-free paper.

Copyright © 2002, Elsevier Science (USA).

All Rights Reserved.
No part of this publication may be reproduced or transmitted in any form or by any means, electronic or mechanical, including photocopy, recording, or any information storage and retrieval system, without permission in writing from the Publisher.

The appearance of the code at the bottom of the first page of a chapter in this book indicates the Publisher's consent that copies of the chapter may be made for personal or internal use of specific clients. This consent is given on the condition, however, that the copier pay the stated per copy fee through the Copyright Clearance Center, Inc. (222 Rosewood Drive, Danvers, Massachusetts 01923), for copying beyond that permitted by Sections 107 or 108 of the U.S. Copyright Law. This consent does not extend to other kinds of copying, such as copying for general distribution, for advertising or promotional purposes, for creating new collective works, or for resale. Copy fees for pre-2002 chapters are as shown on the title pages. If no fee code appears on the title page, the copy fee is the same as for current chapters.
0076-6879/2002 $35.00

Explicit permission from Academic Press is not required to reproduce a maximum of two figures or tables from an Academic Press chapter in another scientific or research publication provided that the material has not been credited to another source and that full credit to the Academic Press chapter is given.

Academic Press
*An Elsevier Science Imprint*
525 B Street, Suite 1900, San Diego, California 92101-4495, USA
http://www.academicpress.com

Academic Press
32 Jamestown Road, London NW1 7BY, UK
http://www.academicpress.com

International Standard Book Number: 0-12-182252-4

PRINTED IN THE UNITED STATES OF AMERICA
02  03  04  05  06  07  SB  9  8  7  6  5  4  3  2  1

# Table of Contents

CONTRIBUTORS TO VOLUME 349 . . . . . . . . . . . . . . . . .  ix

PREFACE . . . . . . . . . . . . . . . . . . . . . . . . . . .  xv

VOLUMES IN SERIES . . . . . . . . . . . . . . . . . . . . .  xvii

## Section I. Superoxide Reactions and Mechanisms

1. Quantitation of Intracellular Free Iron by Electron Paramagnetic Resonance Spectroscopy — ANH N. WOODMANSEE AND JAMES A. IMLAY   3

2. Aconitase: Sensitive Target and Measure of Superoxide — PAUL R. GARDNER   9

3. Reactions of Manganese Porphyrins and Manganese-Superoxide Dismutase with Peroxynitrite — GERARDO FERRER-SUETA, CELIA QUIJANO, BEATRIZ ALVAREZ, AND RAFAEL RADI   23

4. Superoxide Dismutase Kinetics — MATTIA FALCONI, PETER O'NEILL, MARIA ELENA STROPPOLO, AND ALESSANDRO DESIDERI   38

5. Analysis of Cu,ZnSOD Conformational Stability by Differential Scanning Calorimetry — MARIA CARMELA BONACCORSI DI PATTI, ANNA GIARTOSIO, GIUSEPPE ROTILIO, AND ANDREA BATTISTONI   49

6. Catalytic Pathway of Manganese Superoxide Dismutase by Direct Observation of Superoxide — DAVID N. SILVERMAN AND HARRY S. NICK   61

7. Extracellular Superoxide Dismutase — STEFAN L. MARKLUND   74

8. Prokaryotic Manganese Superoxide Dismutases — JAMES W. WHITTAKER   80

9. Nickel-Containing Superoxide Dismutase — JIN-WON LEE, JUNG-HYE ROE, AND SA-OUK KANG   90

10. Reversible Conversion of Nitroxyl Anion to Nitric Oxide — LARS-OLIVER KLOTZ AND HELMUT SIES   101

11. Purification and Determination of Activity of Mitochondrial Cyanide-Sensitive Superoxide Dismutase in Rat Tissue Extract — PEDRO IÑARREA   106

12. Studies of Metal-Binding Properties of Cu,Zn Superoxide Dismutase by Isothermal Titration Calorimetry — PATRICIA L. BOUNDS, BARBARA SUTTER, AND WILLEM H. KOPPENOL   115

13. Superoxide Reductase from *Desulfoarculus baarsii* — VINCENT NIVIÈRE AND MURIELLE LOMBARD   123

14. Enzyme-Linked Immunosorbent Assay for Human and Rat Manganese Superoxide Dismutases — KEIICHIRO SUZUKI, TOMOMI OOKAWARA, YASUHIDE MIYAMOTO, AND NAOYUKI TANIGUCHI   129

## Section II. Mutants, Knockouts, Transgenics

15. Investigating Phenotypes Resulting from a Lack of Superoxide Dismutase in Bacterial Null Mutants — DANIÈLE TOUATI   145

16. Bacterial Superoxide Dismutase and Virulence — PAUL R. LANGFORD, ASSUNTA SANSONE, PIERA VALENTI, ANDREA BATTISTONI, AND J. SIMON KROLL   155

17. Superoxide Dismutase Null Mutants of Baker's Yeast, *Saccharomyces cerevisiae* — LORI A. STURTZ AND VALERIA CIZEWSKI CULOTTA   167

18. Measurement of "Free" or Electron Paramagnetic Resonance-Detectable Iron in Whole Yeast Cells as Indicator of Superoxide Stress — CHANDRA SRINIVASAN AND EDITH BUTLER GRALLA   173

19. Transgenic Superoxide Dismutase Overproducer: Murine — SERGE PRZEDBORSKI, VERNICE JACKSON-LEWIS, DAVID SULZER, ALI NAINI, NORMA ROMERO, CAIPING CHEN, AND JULIA ARIAS   180

20. Transgenic and Mutant Mice for Oxygen Free Radical Studies — TING-TING HUANG, INES RAINERI, FAYE EGGERDING, AND CHARLES J. EPSTEIN   191

21. Overexpression of Cu,ZnSOD and MnSOD in Transgenic *Drosophila* — ROBIN J. MOCKETT, WILLIAM C. ORR, AND RAJINDAR S. SOHAL   213

## Section III. Superoxide Dismutase Mimics

22. Manganese Porphyrins and Related Compounds as Mimics of Superoxide Dismutase — INES BATINIĆ-HABERLE   223

23. Superoxide Dismutase Mimics: Antioxidative and Adverse Effects . . . . . . . . . GIDON CZAPSKI, AMRAM SAMUNI, AND SARA GOLDSTEIN   234

24. Superoxide Reductase Activities of Neelaredoxin and Desulfoferrodoxin Metalloproteins . . . FRANK RUSNAK, CARLA ASCENSO, ISABEL MOURA, AND JOSÉ J. G. MOURA   243

25. Purification and Preparation of Prion Protein: Synaptic Superoxide Dismutase . . . MAKI DANIELS AND DAVID R. BROWN   258

## Section IV. In Vivo Sources, Cell Signaling

26. Mitochondrial Superoxide Anion Production and Release into Intermembrane Space . . . DERICK HAN, FERNANDO ANTUNES, FRANCESCA DANERI, AND ENRIQUE CADENAS   271

27. Measurement of Superoxide Radical and Hydrogen Peroxide Production in Isolated Cells and Subcellular Organelles . . . ALBERTO BOVERIS, SILVIA ALVAREZ, JUANITA BUSTAMANTE, AND LAURA VALDEZ   280

28. Biochemical Assay of Superoxide Dismutase Activity in *Drosophila* . . . ROBIN J. MOCKETT, ANNE-CÉCILE V. BAYNE, BARBARA H. SOHAL, AND RAJINDAR S. SOHAL   287

29. Transcriptional Regulation and Environmental Induction of Gene Encoding Copper- and Zinc-Containing Superoxide Dismutase . . . MUN SEOG CHANG, HAE YONG YOO, AND HYUNE MO RHO   293

30. Transcription Regulation of Human Manganese Superoxide Dismutase Gene . . . DARET K. ST. CLAIR, SUREERUT PORNTADAVITY, YONG XU, AND KELLEY KININGHAM   306

31. Assessment of Oxidants in Mitogen-Activated Protein Kinase Activation . . . LANCE S. TERADA AND RHONDA F. SOUZA   313

32. Assaying Binding Capacity of Cu,ZnSOD and MnSOD: Demonstration of Their Localization in Cells and Tissues . . . INGRID EMERIT, PAULO FILIPE, JOAO FREITAS, ALFONSO FERNANDES, FRÉDÉRIC GARBAN, AND JANY VASSY   321

## Section V. Superoxide Dismutase in Aging and Disease Therapy

33. Superoxide Dismutase in Aging and Disease: JOE M. MCCORD
    An Overview                                                           331

34. Tissue-Specific Mitochondrial Production of LINDA K. KWONG
    $H_2O_2$: Its Dependence on Substrates and Sensi-  AND RAJINDAR S. SOHAL
    tivity to Inhibitors                                                  341

35. Targeting Superoxide Dismutase to Critical Sites  MASAYASU INOUE,
    of Action                                         EISUKE SATO,
                                                      MANABU NISHIKAWA,
                                                      AH-MEE PARK,
                                                      KENSAKU MAEDA,
                                                      AND EMIKO KASAHARA  346

36. *In Vitro* Quantitation of Biological Superoxide and  KEVIN R. MESSNER
    Hydrogen Peroxide Generation                          AND JAMES A. IMLAY  354

AUTHOR INDEX . . . . . . . . . . . . . . . . . . . .   363

SUBJECT INDEX . . . . . . . . . . . . . . . . . . . .  391

# Contributors to Volume 349

Article numbers are in parentheses following the names of contributors.
Affiliations listed are current.

BEATRIZ ALVAREZ (3), *Laboratory of Enzymology, Universidad de la República, 11800 Montevideo, Uruguay*

SILVIA ALVEREZ (27), *Laboratory of Free Radical Biology, School of Pharmacy and Biochemistry, University of Buenos Aires, 1113 Buenos Aires, Argentina*

FERNANDO ANTUNES (26), *Department of Chemistry and Biochemistry, Center of Biochemistry and Physiological Studies, Universidade de Lisboa, P-1749-06 Lisbon, Portugal*

JULIA ARIAS (19), *Department of Neurology, Columbia University, New York, New York 10032*

CARLA ASCENSO (24), *Department of Chemistry and Center for Chemistry and Biotechnology, FCT, Universidade Nova de Lisboa, 2825-114 Monte de Caparica, Portugal*

INES BATINIĆ-HABERLE (22), *Department of Biochemistry, Duke University Medical Center, Durham, North Carolina 27710*

ANDREA BATTISTONI (5, 16), *Department of Biology, University of Rome "Tor Vergata," 00133 Rome, Italy*

ANNE-CÉCILE V. BAYNE (28), *Department of Molecular Pharmacology and Toxicology, University of Southern California, Los Angeles, California 90033*

MARIA CARMELA BONACCORSI DI PATTI (5), *Department of Biochemical Sciences "A. Rossi Fanelli," University of Rome "La Sapienza," 00185 Rome, Italy*

PATRICIA L. BOUNDS (12), *Laboratory of Inorganic Chemistry, Eidgenössische Technische Hochschule, CH-8093 Zurich, Switzerland*

ALBERTO BOVERIS (27), *Laboratory of Free Radical Biology, School of Pharmacy and Biochemistry, University of Buenos Aires, 1113 Buenos Aires, Argentina*

DAVID R. BROWN (25), *Department of Biology and Biochemistry, University of Bath, Bath BA2 7AY, United Kingdom*

JUANITA BUSTAMANTE (27), *Laboratory of Free Radical Biology, School of Pharmacy and Biochemistry, University of Buenos Aires, 1113 Buenos Aires, Argentina*

ENRIQUE CADENAS (26), *Department of Molecular Pharmacology and Toxicology, School of Pharmacy, University of Southern California, Los Angeles, California 90033*

MUN SEOG CHANG (29), *School of Biological Sciences, Seoul National University, Seoul 151-1742, Korea*

CAIPING CHEN (19), *Department of Neurology, Columbia University, New York, New York 10032*

VALERIA CIZEWSKI CULOTTA (17), *Division of Toxicological Sciences, Environmental Health Sciences, Bloomberg School of Public Health, Johns Hopkins University, Baltimore, Maryland 21205*

GIDON CZAPSKI (23), *Department of Physical Chemistry, Hebrew University of Jerusalem, Jerusalem 91940, Israel*

FRANCESCA DANERI (26), *Department of Molecular Pharmacology and Toxicology, School of Pharmacy, University of Southern California, Los Angeles, California 90033*

ix

MAKI DANIELS (25), *Department of Biochemistry, University of Cambridge, Cambridge CB2 1QW, United Kingdom*

ALESSANDRO DESIDERI (4), *National Institute for the Physics of the Matter (INFM) and Department of Biology, University of Rome "Tor Vergata," 00133 Rome, Italy*

FAYE EGGERDING (20), *Molecular Oncology and Cancer Genetics Laboratory, Huntington Medical Research Institutes, Pasadena, California 91105*

INGRID EMERIT (32), *Biomedical Institute of Cordeliers, Université Paris VI, Université Pierre & Marie Curie-Paris VI, 75006 Paris, France*

CHARLES J. EPSTEIN (20), *Department of Pediatrics, Genetics Division, University of California, San Francisco, California 94143*

MATTIA FALCONI (4), *National Institute for the Physics of Matter (INFM) and Department of Biology, University of Rome "Tor Vergata," 00133 Rome, Italy*

ALFONSO FERNANDES (32), *Centre of Metabolism and Endocrinology, 09726 Lisbon, Portugal*

GERADO FERRER-SUETA (3), *Department of Physical-Chemical Biology, Universidad de la República, 11800 Montevideo, Uruguay*

PAULO FILIPE (32), *Department of Dermatology, University of Lisbon, 09726 Lisbon, Portugal*

JOAO FREITAS (32), *Department of Dermatology, University of Lisbon, 09726 Lisbon, Portugal*

FRÉDÉRIC GARBAN (32), *Biomedical Institute of Cordeliers, Université Paris VI, Université Pierre & Marie Curie-Paris VI, 75006 Paris, France*

PAUL R. GARDNER (2), *Children's Hospital Medical Center, Cincinnati, Ohio 45229*

ANNA GIARTOSIO (5), *Department of Biochemical Sciences "A. Rossi Fanelli," University of Rome "La Sapienza," 00185 Rome, Italy*

SARA GOLDSTEIN (23), *Department of Physical Chemistry, Hebrew University of Jerusalem, Jerusalem 91940, Israel*

EDITH BUTLER GRALLA (18), *Department of Chemistry and Biochemistry, University of California, Los Angeles, California 90095*

DERICK HAN (26), *Department of Molecular Pharmacology and Toxicology, School of Pharmacy, University of Southern California, Los Angeles, California 90033*

TING-TING HUANG (20), *Department of Pediatrics, Genetics Division, University of California, San Francisco, California 94143*

JAMES A. IMLAY (1, 36), *Department of Microbiology, University of Illinois at Urbana-Champaign, Urbana, Illinois 61801*

PEDRO IÑARREA (11), *Department of Biochemistry and Molecular and Cellular Biology, University of Zaragoza, 50009 Zaragoza, Spain*

MASAYASU INOUE (35), *Department of Biochemistry and Molecular Pathology, Osaka City University Medical School, Osaka 545-8585, Japan*

VERNICE JACKSON-LEWIS (19), *Department of Neurology, Columbia University, New York, New York 10032*

SA-OUK KANG (9), *Laboratory of Biophysics, School of Biological Sciences, and Institute of Microbiology, Seoul National University, Seoul 151-742, Korea*

EMIKO KASAHARA (35), *Department of Biochemistry and Molecular Pathology, Osaka City University Medical School, Osaka 545-8585, Japan*

KELLEY KININGHAM (30), *Graduate School of Toxicology, University of Kentucky, Lexington, Kentucky 40536*

LARS-OLIVER KLOTZ (10), *Institute for Physiological Chemistry I, Heinrich-Heine-Universität Düsseldorf, D-40225 Düsseldorf, Germany*

WILLEM H. KOPPENOL (12), *Laboratory of Inorganic Chemistry, Eidgenössische Technische Hochschule, CH-8093 Zurich, Switzerland*

J. SIMON KROLL (16), *Molecular Infectious Diseases Group, Department of Paediatrics, Imperial College of Science, Technology and Medicine, St. Mary's Hospital Campus, London W2 1PG, United Kingdom*

LINDA K. KWONG (34), *Department of Molecular Pharmacology and Toxicology, University of Southern California, Los Angeles, California 90033*

PAUL R. LANGFORD (16), *Molecular Infectious Diseases Group, Department of Paediatrics, Imperial College of Science, Technology and Medicine, St. Mary's Hospital Campus, London W2 1PG, United Kingdom*

JIN-WON LEE (9), *Laboratory of Biophysics, School of Biological Sciences, and Institute of Microbiology, Seoul National University, Seoul 151-742, Korea*

MURIELLE LOMBARD (13), *Laboratory of Chemistry and Biochemistry of Biological Redox Centers, DBMS-CEA/ CNRS/Université Joseph Fourier, 38054 Grenoble Cedex 9, France*

KENSAKU MAEDA (35), *Department of Biochemistry and Molecular Pathology, Osaka City University Medical School, Osaka 545-8585, Japan*

STEFAN L. MARKLUND (7), *Department of Medical Biosciences, Clinical Chemistry, Umeå University Hospital, SE-901 85 Umeå, Sweden*

JOE M. MCCORD (33), *Webb-Waring Institute, University of Colorado Health Science Center, Denver, Colorado 80262*

KEVIN R. MESSNER (36), *Department of Microbiology, University of Illinois at Urbana-Champaign, Urbana, Illinois 61801*

YASUHIDE MIYAMOTO (14), *Department of Biochemistry, Osaka University Medical School, Osaka 565-0871, Japan*

ROBIN J. MOCKETT (21, 28), *Department of Molecular Pharmacology and Toxicology, University of Southern California, Los Angeles, California 90033*

ISABEL MOURA (24), *Department of Chemistry and Center for Chemistry and Biotechnology, FCT Universidade Nova de Lisboa, 2825-114 Monte de Caparica, Lisbon, Portugal*

JOSÉ J. G. MOURA (24), *Department of Chemistry and Center for Chemistry and Biotechnology, FCT Universidade Nova de Lisboa, 2825-114 Monte de Caparica, Lisbon, Portugal*

ALI NAINI (19), *Department of Neurology, Columbia University, New York, New York 10032*

HARRY S. NICK (6), *Department of Neuroscience, University of Florida College of Medicine, Gainesville, Florida 32610*

MANABU NISHIKAWA (35), *Department of Biochemistry and Molecular Pathology, Osaka City University Medical School, Osaka 545-8585, Japan*

VINCENT NIVIÈRE (13), *Laboratory of Chemistry and Biochemistry of Biological Redox Centers, DBMS-CEA/ CNRS/Université Joseph Fourier, 38054 Grenoble Cedex 9, France*

PETER O'NEILL (4), *Radiation and Genome Stability Unit, Medical Research Council, Harwell, Didcot, Oxon OX11ORD, United Kingdom*

TOMOMI OOKAWARA (14), *Department of Biochemistry, Osaka University Medical School, Osaka 565-0871, Japan*

WILLIAM C. ORR (21), *Department of Biological Sciences, Southern Methodist University, Dallas, Texas 75275*

AH-MEE PARK (35), *Department of Biochemistry and Molecular Pathology, Osaka City University Medical School, Osaka 545-8585, Japan*

SUREERUT PORNTADAVITY (30), *Graduate Center for Toxicology, University of Kentucky, Lexington, Kentucky 40536*

SERGE PRZEDBORSKI (19), *Departments of Neurology and Pathology, Columbia University, New York, New York 10032*

CELIA QUIJANO (3), *Department of Biochemistry, Universidad de la República, 11800 Montevideo, Uruguay*

RAFAEL RADI (3), *Department of Biochemistry, Universidad de la República, 11800 Montevideo, Uruguay*

INES RAINERI (20), *Department of Pediatrics, Genetics Division, University of California, San Francisco, California 94143*

HYUNE MO RHO (29), *School of Biological Sciences, Seoul National University, Seoul 151-1742, Korea*

JUNG-HYE ROE (9), *Laboratory of Molecular Microbiology, School of Biological Sciences, and Institute of Microbiology, Seoul National University, Seoul 151-742, Korea*

NORMA ROMERO (19), *Department of Neurology, Columbia University, New York, New York 10032*

GIUSEPPE ROTILIO (5), *Department of Biology, University of Rome "Tor Vergata," 00133 Roma, Italy*

FRANK RUSNAK (24), *Section of Hematology Research, Department of Biochemistry and Molecular Biology, Mayo Clinic and Foundation, Rochester, Minnesota 55905*

DARET K. ST. CLAIR (30), *Graduate Center for Toxicology, University of Kentucky, Lexington, Kentucky 40536*

AMRAM SAMUNI (23), *Department of Molecular Biology, School of Medicine, Hebrew University of Jerusalem, Jerusalem 91120, Israel*

ASSUNTA SANSONE (16), *The European Bioinformatics Institute, EMBL Outstation, Wellcome Trust Genome Campus, Cambridge CB10 1SD, United Kingdom*

EISUKE SATO (35), *Department of Biochemistry and Molecular Pathology, Osaka City University Medical School, Osaka 545-8585, Japan*

HELMUT SIES (10), *Institute of Physiological Chemistry, Heinrich-Heine-Universität Düsseldorf, D-40225 Düsseldorf, Germany*

DAVID N. SILVERMAN (6), *Department of Pharmacology, University of Florida College of Medicine, Gainesville, Florida 32610*

BARBARA H. SOHAL (28), *Department of Molecular Pharmacology and Toxicology, University of Southern California, Los Angeles, California 90033*

RAJINDAR S. SOHAL (21, 28, 34), *Department of Molecular Pharmacology and Toxicology, University of Southern California, Los Angeles, California 90033*

RHONDA F. SOUZA (31), *University of Texas Southwestern Medical Center and Dallas Veterans Administration Medical Center, Dallas, Texas 75216*

CHANDRA SRINIVASAN (18), *Department of Chemistry and Biochemistry, University of California, Los Angeles, California 90095*

MARIA ELENA STROPPOLO (4), *National Institute for the Physics of Matter (INFM) and Department of Biology, University of Rome "Tor Vergata," 00133 Rome, Italy*

LORI A. STURTZ (17), *Division of Toxicological Sciences, Environmental Health Sciences, Bloomberg School of Public Health, Johns Hopkins University, Baltimore, Maryland 21205*

DAVID SULZER (19), *Department of Neurology, Columbia University, New York, New York 10032*

BARBARA SUTTER (12), *Laboratory of Inorganic Chemistry, Eidgenössische Technische Hochschule, CH-8093 Zurich, Switzerland*

KEIICHIRO SUZUKI (14), *Department of Biochemistry, Osaka University Medical School, Osaka 565-0871, Japan*

NAOYUKI TANIGUCHI (14), *Department of Biochemistry, Osaka University Medical School, Osaka 565-0871, Japan*

LANCE S. TERADA (31), *University of Texas Southwestern Medical Center and Dallas Veterans Administration Medical Center, Dallas, Texas 75216*

DANIÈLE TOUATI (15), *Jacques Monod Institute CNRS-Universités Paris 6 et Paris 7, 75251 Paris Cedex 05, France*

LAURA VALDEZ (27), *Laboratory of Free Radical Biology, School of Pharmacy and Biology, University of Buenos Aires, 1113 Buenos Aires, Argentina*

PIERA VALENTI (16), *Institute of Microbiology, II Università di Napoli, 80138 Naples, Italy*

JANY VASSY (32), *Laboratory of Image Analysis of Cellular Pathology, Hôpital Saint Louis, 75010 Paris, France*

JAMES W. WHITTAKER (8), *Department of Biochemistry and Molecular Biology, Oregon Graduate Institute School of Science and Engineering, Oregon Health and Science University, Beaverton, Oregon 97006*

ANH N. WOODMANSEE (1), *Department of Microbiology, University of Illinois at Urbana-Champaign, Urbana, Illinois 61801*

YONG XU (30), *Graduate Center for Toxicology, University of Kentucky, Lexington, Kentucky 40536*

HAE YONG YOO (29), *School of Biological Sciences, Seoul National University, Seoul 151-742, Korea*

# Preface

The discovery of the enzyme superoxide dismutase (SOD) and its subsequent publication in 1969 by Joe McCord and Irwin Fridovich ushered in a new era in the understanding of oxidative processes in biological systems. Following their purification of the copper–zinc form of superoxide dismutase from bovine erythrocytes, SOD (or a mimic) was found to be ubiquitous in every aerobic organism from microbes to humans where the superoxide radical anion ($O_2^-$) was found to be dismutated by metal-catalyzed reactions.

SOD was the first enzyme found to use an oxygen free radical as a substrate. The removal of superoxide by SOD is a vital link in the system of proteins, enzymes, water, and lipid-soluble substances involved in antioxidant defense. A network of enzymatic and nonenzymatic reactions maintains the delicate balance between oxidants and free radicals against "oxidative stress" and susceptibility to oxidative damage. The superoxide radical anion or its protonated form causes oxidative molecular damage to lipids, proteins, and other molecules. Within cells, organelles, and in extracellular fluids different forms of SOD help maintain a lower steady state of superoxide. SOD action forms hydrogen peroxide which is a strong oxidant scavenged by catalase and peroxidases, lowering its steady-state levels.

Superoxide radicals are involved in diverse physiological and pathophysiological processes. Many enzymes producing superoxide have been characterized. Biological oxidations involved in respiratory and cytochrome P450 electron transport reactions produce superoxide as a by-product. Other enzymes can be induced to form superoxide as by the plasma membrane oxidase involved in the transient burst of superoxide produced by activated neutrophils and macrophages or by conversion of xanthine dehydrogenase to xanthine oxidase.

The vital importance of SOD in biological systems has often been demonstrated using techniques to remove the enzyme, such as culturing facultative aerobic bacteria under anaerobic conditions in which loss of the enzyme results in much decreased susceptibility of these organisms to survive when subsequently exposed to oxygen. Mutations in SOD, as in the deadly human disease familial amyotrophic lateral sclerosis, or overexpression of SOD in transgenic animals results in pathology. The demonstration of toxic effects of superoxide anion has led to efforts to find therapeutic strategies such as stabilizing the enzyme for administration *in vivo* or developing synthetic SOD mimics.

The discovery of enzymes that produce the nitric oxide free radical (NO) revealed that this free radical was also important in many physiological and pathophysiological processes. NO is generated by constitutive or endogenous nitric oxide synthase (eNOS) or an inducible nitric oxide synthase (iNOS). Low levels of NO act as a signaling molecule and regulate many normal processes such as blood flow and neurotransmission. However, iNOS producing large amounts of NO is cytotoxic. Nitrotyrosine formation interferes with protein phosphorylation reactions. However, the most damaging effects are observed when both superoxide and NO are simultaneously produced as by activated immune system cells or by endothelial cells. Very rapid reaction between NO and superoxide anion forms peroxynitrite, a very powerful oxidant capable of severely damaging proteins, DNA, and other biological molecules causing what has been called "nitrosative stress." Hence SODs are also a first line of defense against nitrosative stress.

Many of the new and updated methods and techniques needed for the investigation of SOD and the superoxide radical are described in this volume. Indeed, the advisors for this volume of *Methods in Enzymology* have been none other than Drs. Joe McCord and Irwin Fridovich, the codiscoverers of SOD, both of whom have continued their pioneering research on superoxide and SOD and related fields of study over the past thirty years. Indeed, both these individuals contributed most of the suggestions for contributors to this volume. This book is dedicated to their momumental contributions to understanding the role of superoxide and SOD in basic biological systems and the biomedical sciences.

LESTER PACKER

# METHODS IN ENZYMOLOGY

VOLUME I. Preparation and Assay of Enzymes
*Edited by* SIDNEY P. COLOWICK AND NATHAN O. KAPLAN

VOLUME II. Preparation and Assay of Enzymes
*Edited by* SIDNEY P. COLOWICK AND NATHAN O. KAPLAN

VOLUME III. Preparation and Assay of Substrates
*Edited by* SIDNEY P. COLOWICK AND NATHAN O. KAPLAN

VOLUME IV. Special Techniques for the Enzymologist
*Edited by* SIDNEY P. COLOWICK AND NATHAN O. KAPLAN

VOLUME V. Preparation and Assay of Enzymes
*Edited by* SIDNEY P. COLOWICK AND NATHAN O. KAPLAN

VOLUME VI. Preparation and Assay of Enzymes (*Continued*)
Preparation and Assay of Substrates
Special Techniques
*Edited by* SIDNEY P. COLOWICK AND NATHAN O. KAPLAN

VOLUME VII. Cumulative Subject Index
*Edited by* SIDNEY P. COLOWICK AND NATHAN O. KAPLAN

VOLUME VIII. Complex Carbohydrates
*Edited by* ELIZABETH F. NEUFELD AND VICTOR GINSBURG

VOLUME IX. Carbohydrate Metabolism
*Edited by* WILLIS A. WOOD

VOLUME X. Oxidation and Phosphorylation
*Edited by* RONALD W. ESTABROOK AND MAYNARD E. PULLMAN

VOLUME XI. Enzyme Structure
*Edited by* C. H. W. HIRS

VOLUME XII. Nucleic Acids (Parts A and B)
*Edited by* LAWRENCE GROSSMAN AND KIVIE MOLDAVE

VOLUME XIII. Citric Acid Cycle
*Edited by* J. M. LOWENSTEIN

VOLUME XIV. Lipids
*Edited by* J. M. LOWENSTEIN

VOLUME XV. Steroids and Terpenoids
*Edited by* RAYMOND B. CLAYTON

VOLUME XVI. Fast Reactions
*Edited by* KENNETH KUSTIN

VOLUME XVII. Metabolism of Amino Acids and Amines (Parts A and B)
*Edited by* HERBERT TABOR AND CELIA WHITE TABOR

VOLUME XVIII. Vitamins and Coenzymes (Parts A, B, and C)
*Edited by* DONALD B. MCCORMICK AND LEMUEL D. WRIGHT

VOLUME XIX. Proteolytic Enzymes
*Edited by* GERTRUDE E. PERLMANN AND LASZLO LORAND

VOLUME XX. Nucleic Acids and Protein Synthesis (Part C)
*Edited by* KIVIE MOLDAVE AND LAWRENCE GROSSMAN

VOLUME XXI. Nucleic Acids (Part D)
*Edited by* LAWRENCE GROSSMAN AND KIVIE MOLDAVE

VOLUME XXII. Enzyme Purification and Related Techniques
*Edited by* WILLIAM B. JAKOBY

VOLUME XXIII. Photosynthesis (Part A)
*Edited by* ANTHONY SAN PIETRO

VOLUME XXIV. Photosynthesis and Nitrogen Fixation (Part B)
*Edited by* ANTHONY SAN PIETRO

VOLUME XXV. Enzyme Structure (Part B)
*Edited by* C. H. W. HIRS AND SERGE N. TIMASHEFF

VOLUME XXVI. Enzyme Structure (Part C)
*Edited by* C. H. W. HIRS AND SERGE N. TIMASHEFF

VOLUME XXVII. Enzyme Structure (Part D)
*Edited by* C. H. W. HIRS AND SERGE N. TIMASHEFF

VOLUME XXVIII. Complex Carbohydrates (Part B)
*Edited by* VICTOR GINSBURG

VOLUME XXIX. Nucleic Acids and Protein Synthesis (Part E)
*Edited by* LAWRENCE GROSSMAN AND KIVIE MOLDAVE

VOLUME XXX. Nucleic Acids and Protein Synthesis (Part F)
*Edited by* KIVIE MOLDAVE AND LAWRENCE GROSSMAN

VOLUME XXXI. Biomembranes (Part A)
*Edited by* SIDNEY FLEISCHER AND LESTER PACKER

VOLUME XXXII. Biomembranes (Part B)
*Edited by* SIDNEY FLEISCHER AND LESTER PACKER

VOLUME XXXIII. Cumulative Subject Index Volumes I-XXX
*Edited by* MARTHA G. DENNIS AND EDWARD A. DENNIS

VOLUME XXXIV. Affinity Techniques (Enzyme Purification: Part B)
*Edited by* WILLIAM B. JAKOBY AND MEIR WILCHEK

VOLUME XXXV. Lipids (Part B)
*Edited by* JOHN M. LOWENSTEIN

VOLUME XXXVI. Hormone Action (Part A: Steroid Hormones)
*Edited by* BERT W. O'MALLEY AND JOEL G. HARDMAN

VOLUME XXXVII. Hormone Action (Part B: Peptide Hormones)
*Edited by* BERT W. O'MALLEY AND JOEL G. HARDMAN

VOLUME XXXVIII. Hormone Action (Part C: Cyclic Nucleotides)
*Edited by* JOEL G. HARDMAN AND BERT W. O'MALLEY

VOLUME XXXIX. Hormone Action (Part D: Isolated Cells, Tissues, and Organ Systems)
*Edited by* JOEL G. HARDMAN AND BERT W. O'MALLEY

VOLUME XL. Hormone Action (Part E: Nuclear Structure and Function)
*Edited by* BERT W. O'MALLEY AND JOEL G. HARDMAN

VOLUME XLI. Carbohydrate Metabolism (Part B)
*Edited by* W. A. WOOD

VOLUME XLII. Carbohydrate Metabolism (Part C)
*Edited by* W. A. WOOD

VOLUME XLIII. Antibiotics
*Edited by* JOHN H. HASH

VOLUME XLIV. Immobilized Enzymes
*Edited by* KLAUS MOSBACH

VOLUME XLV. Proteolytic Enzymes (Part B)
*Edited by* LASZLO LORAND

VOLUME XLVI. Affinity Labeling
*Edited by* WILLIAM B. JAKOBY AND MEIR WILCHEK

VOLUME XLVII. Enzyme Structure (Part E)
*Edited by* C. H. W. HIRS AND SERGE N. TIMASHEFF

VOLUME XLVIII. Enzyme Structure (Part F)
*Edited by* C. H. W. HIRS AND SERGE N. TIMASHEFF

VOLUME XLIX. Enzyme Structure (Part G)
*Edited by* C. H. W. HIRS AND SERGE N. TIMASHEFF

VOLUME L. Complex Carbohydrates (Part C)
*Edited by* VICTOR GINSBURG

VOLUME LI. Purine and Pyrimidine Nucleotide Metabolism
*Edited by* PATRICIA A. HOFFEE AND MARY ELLEN JONES

VOLUME LII. Biomembranes (Part C: Biological Oxidations)
*Edited by* SIDNEY FLEISCHER AND LESTER PACKER

VOLUME LIII. Biomembranes (Part D: Biological Oxidations)
*Edited by* SIDNEY FLEISCHER AND LESTER PACKER

VOLUME LIV. Biomembranes (Part E: Biological Oxidations)
*Edited by* SIDNEY FLEISCHER AND LESTER PACKER

VOLUME LV. Biomembranes (Part F: Bioenergetics)
*Edited by* SIDNEY FLEISCHER AND LESTER PACKER

VOLUME LVI. Biomembranes (Part G: Bioenergetics)
*Edited by* SIDNEY FLEISCHER AND LESTER PACKER

VOLUME LVII. Bioluminescence and Chemiluminescence
*Edited by* MARLENE A. DELUCA

VOLUME LVIII. Cell Culture
*Edited by* WILLIAM B. JAKOBY AND IRA PASTAN

VOLUME LIX. Nucleic Acids and Protein Synthesis (Part G)
*Edited by* KIVIE MOLDAVE AND LAWRENCE GROSSMAN

VOLUME LX. Nucleic Acids and Protein Synthesis (Part H)
*Edited by* KIVIE MOLDAVE AND LAWRENCE GROSSMAN

VOLUME 61. Enzyme Structure (Part H)
*Edited by* C. H. W. HIRS AND SERGE N. TIMASHEFF

VOLUME 62. Vitamins and Coenzymes (Part D)
*Edited by* DONALD B. MCCORMICK AND LEMUEL D. WRIGHT

VOLUME 63. Enzyme Kinetics and Mechanism (Part A: Initial Rate and Inhibitor Methods)
*Edited by* DANIEL L. PURICH

VOLUME 64. Enzyme Kinetics and Mechanism (Part B: Isotopic Probes and Complex Enzyme Systems)
*Edited by* DANIEL L. PURICH

VOLUME 65. Nucleic Acids (Part I)
*Edited by* LAWRENCE GROSSMAN AND KIVIE MOLDAVE

VOLUME 66. Vitamins and Coenzymes (Part E)
*Edited by* DONALD B. MCCORMICK AND LEMUEL D. WRIGHT

VOLUME 67. Vitamins and Coenzymes (Part F)
*Edited by* DONALD B. MCCORMICK AND LEMUEL D. WRIGHT

VOLUME 68. Recombinant DNA
*Edited by* RAY WU

VOLUME 69. Photosynthesis and Nitrogen Fixation (Part C)
*Edited by* ANTHONY SAN PIETRO

VOLUME 70. Immunochemical Techniques (Part A)
*Edited by* HELEN VAN VUNAKIS AND JOHN J. LANGONE

VOLUME 71. Lipids (Part C)
*Edited by* JOHN M. LOWENSTEIN

VOLUME 72. Lipids (Part D)
*Edited by* JOHN M. LOWENSTEIN

VOLUME 73. Immunochemical Techniques (Part B)
*Edited by* JOHN J. LANGONE AND HELEN VAN VUNAKIS

VOLUME 74. Immunochemical Techniques (Part C)
*Edited by* JOHN J. LANGONE AND HELEN VAN VUNAKIS

VOLUME 75. Cumulative Subject Index Volumes XXXI, XXXII, XXXIV–LX
*Edited by* EDWARD A. DENNIS AND MARTHA G. DENNIS

VOLUME 76. Hemoglobins
*Edited by* ERALDO ANTONINI, LUIGI ROSSI-BERNARDI, AND EMILIA CHIANCONE

VOLUME 77. Detoxication and Drug Metabolism
*Edited by* WILLIAM B. JAKOBY

VOLUME 78. Interferons (Part A)
*Edited by* SIDNEY PESTKA

VOLUME 79. Interferons (Part B)
*Edited by* SIDNEY PESTKA

VOLUME 80. Proteolytic Enzymes (Part C)
*Edited by* LASZLO LORAND

VOLUME 81. Biomembranes (Part H: Visual Pigments and Purple Membranes, I)
*Edited by* LESTER PACKER

VOLUME 82. Structural and Contractile Proteins (Part A: Extracellular Matrix)
*Edited by* LEON W. CUNNINGHAM AND DIXIE W. FREDERIKSEN

VOLUME 83. Complex Carbohydrates (Part D)
*Edited by* VICTOR GINSBURG

VOLUME 84. Immunochemical Techniques (Part D: Selected Immunoassays)
*Edited by* JOHN J. LANGONE AND HELEN VAN VUNAKIS

VOLUME 85. Structural and Contractile Proteins (Part B: The Contractile Apparatus and the Cytoskeleton)
*Edited by* DIXIE W. FREDERIKSEN AND LEON W. CUNNINGHAM

VOLUME 86. Prostaglandins and Arachidonate Metabolites
*Edited by* WILLIAM E. M. LANDS AND WILLIAM L. SMITH

VOLUME 87. Enzyme Kinetics and Mechanism (Part C: Intermediates, Stereochemistry, and Rate Studies)
*Edited by* DANIEL L. PURICH

VOLUME 88. Biomembranes (Part I: Visual Pigments and Purple Membranes, II)
*Edited by* LESTER PACKER

VOLUME 89. Carbohydrate Metabolism (Part D)
*Edited by* WILLIS A. WOOD

VOLUME 90. Carbohydrate Metabolism (Part E)
*Edited by* WILLIS A. WOOD

VOLUME 91. Enzyme Structure (Part I)
*Edited by* C. H. W. HIRS AND SERGE N. TIMASHEFF

VOLUME 92. Immunochemical Techniques (Part E: Monoclonal Antibodies and General Immunoassay Methods)
*Edited by* JOHN J. LANGONE AND HELEN VAN VUNAKIS

VOLUME 93. Immunochemical Techniques (Part F: Conventional Antibodies, Fc Receptors, and Cytotoxicity)
*Edited by* JOHN J. LANGONE AND HELEN VAN VUNAKIS

VOLUME 94. Polyamines
*Edited by* HERBERT TABOR AND CELIA WHITE TABOR

VOLUME 95. Cumulative Subject Index Volumes 61–74, 76–80
*Edited by* EDWARD A. DENNIS AND MARTHA G. DENNIS

VOLUME 96. Biomembranes [Part J: Membrane Biogenesis: Assembly and Targeting (General Methods; Eukaryotes)]
*Edited by* SIDNEY FLEISCHER AND BECCA FLEISCHER

VOLUME 97. Biomembranes [Part K: Membrane Biogenesis: Assembly and Targeting (Prokaryotes, Mitochondria, and Chloroplasts)]
*Edited by* SIDNEY FLEISCHER AND BECCA FLEISCHER

VOLUME 98. Biomembranes (Part L: Membrane Biogenesis: Processing and Recycling)
*Edited by* SIDNEY FLEISCHER AND BECCA FLEISCHER

VOLUME 99. Hormone Action (Part F: Protein Kinases)
*Edited by* JACKIE D. CORBIN AND JOEL G. HARDMAN

VOLUME 100. Recombinant DNA (Part B)
*Edited by* RAY WU, LAWRENCE GROSSMAN, AND KIVIE MOLDAVE

VOLUME 101. Recombinant DNA (Part C)
*Edited by* RAY WU, LAWRENCE GROSSMAN, AND KIVIE MOLDAVE

VOLUME 102. Hormone Action (Part G: Calmodulin and Calcium-Binding Proteins)
*Edited by* ANTHONY R. MEANS AND BERT W. O'MALLEY

VOLUME 103. Hormone Action (Part H: Neuroendocrine Peptides)
*Edited by* P. MICHAEL CONN

VOLUME 104. Enzyme Purification and Related Techniques (Part C)
*Edited by* WILLIAM B. JAKOBY

VOLUME 105. Oxygen Radicals in Biological Systems
*Edited by* LESTER PACKER

VOLUME 106. Posttranslational Modifications (Part A)
*Edited by* FINN WOLD AND KIVIE MOLDAVE

VOLUME 107. Posttranslational Modifications (Part B)
*Edited by* FINN WOLD AND KIVIE MOLDAVE

VOLUME 108. Immunochemical Techniques (Part G: Separation and Characterization of Lymphoid Cells)
*Edited by* GIOVANNI DI SABATO, JOHN J. LANGONE, AND HELEN VAN VUNAKIS

VOLUME 109. Hormone Action (Part I: Peptide Hormones)
*Edited by* LUTZ BIRNBAUMER AND BERT W. O'MALLEY

VOLUME 110. Steroids and Isoprenoids (Part A)
*Edited by* JOHN H. LAW AND HANS C. RILLING

VOLUME 111. Steroids and Isoprenoids (Part B)
*Edited by* JOHN H. LAW AND HANS C. RILLING

VOLUME 112. Drug and Enzyme Targeting (Part A)
*Edited by* KENNETH J. WIDDER AND RALPH GREEN

VOLUME 113. Glutamate, Glutamine, Glutathione, and Related Compounds
*Edited by* ALTON MEISTER

VOLUME 114. Diffraction Methods for Biological Macromolecules (Part A)
*Edited by* HAROLD W. WYCKOFF, C. H. W. HIRS, AND SERGE N. TIMASHEFF

VOLUME 115. Diffraction Methods for Biological Macromolecules (Part B)
*Edited by* HAROLD W. WYCKOFF, C. H. W. HIRS, AND SERGE N. TIMASHEFF

VOLUME 116. Immunochemical Techniques (Part H: Effectors and Mediators of Lymphoid Cell Functions)
*Edited by* GIOVANNI DI SABATO, JOHN J. LANGONE, AND HELEN VAN VUNAKIS

VOLUME 117. Enzyme Structure (Part J)
*Edited by* C. H. W. HIRS AND SERGE N. TIMASHEFF

VOLUME 118. Plant Molecular Biology
*Edited by* ARTHUR WEISSBACH AND HERBERT WEISSBACH

VOLUME 119. Interferons (Part C)
*Edited by* SIDNEY PESTKA

VOLUME 120. Cumulative Subject Index Volumes 81–94, 96–101

VOLUME 121. Immunochemical Techniques (Part I: Hybridoma Technology and Monoclonal Antibodies)
*Edited by* JOHN J. LANGONE AND HELEN VAN VUNAKIS

VOLUME 122. Vitamins and Coenzymes (Part G)
*Edited by* FRANK CHYTIL AND DONALD B. MCCORMICK

VOLUME 123. Vitamins and Coenzymes (Part H)
*Edited by* FRANK CHYTIL AND DONALD B. MCCORMICK

VOLUME 124. Hormone Action (Part J: Neuroendocrine Peptides)
*Edited by* P. MICHAEL CONN

VOLUME 125. Biomembranes (Part M: Transport in Bacteria, Mitochondria, and Chloroplasts: General Approaches and Transport Systems)
*Edited by* SIDNEY FLEISCHER AND BECCA FLEISCHER

VOLUME 126. Biomembranes (Part N: Transport in Bacteria, Mitochondria, and Chloroplasts: Protonmotive Force)
Edited by SIDNEY FLEISCHER AND BECCA FLEISCHER

VOLUME 127. Biomembranes (Part O: Protons and Water: Structure and Translocation)
Edited by LESTER PACKER

VOLUME 128. Plasma Lipoproteins (Part A: Preparation, Structure, and Molecular Biology)
Edited by JERE P. SEGREST AND JOHN J. ALBERS

VOLUME 129. Plasma Lipoproteins (Part B: Characterization, Cell Biology, and Metabolism)
Edited by JOHN J. ALBERS AND JERE P. SEGREST

VOLUME 130. Enzyme Structure (Part K)
Edited by C. H. W. HIRS AND SERGE N. TIMASHEFF

VOLUME 131. Enzyme Structure (Part L)
Edited by C. H. W. HIRS AND SERGE N. TIMASHEFF

VOLUME 132. Immunochemical Techniques (Part J: Phagocytosis and Cell-Mediated Cytotoxicity)
Edited by GIOVANNI DI SABATO AND JOHANNES EVERSE

VOLUME 133. Bioluminescence and Chemiluminescence (Part B)
Edited by MARLENE DELUCA AND WILLIAM D. MCELROY

VOLUME 134. Structural and Contractile Proteins (Part C: The Contractile Apparatus and the Cytoskeleton)
Edited by RICHARD B. VALLEE

VOLUME 135. Immobilized Enzymes and Cells (Part B)
Edited by KLAUS MOSBACH

VOLUME 136. Immobilized Enzymes and Cells (Part C)
Edited by KLAUS MOSBACH

VOLUME 137. Immobilized Enzymes and Cells (Part D)
Edited by KLAUS MOSBACH

VOLUME 138. Complex Carbohydrates (Part E)
Edited by VICTOR GINSBURG

VOLUME 139. Cellular Regulators (Part A: Calcium- and Calmodulin-Binding Proteins)
Edited by ANTHONY R. MEANS AND P. MICHAEL CONN

VOLUME 140. Cumulative Subject Index Volumes 102–119, 121–134

VOLUME 141. Cellular Regulators (Part B: Calcium and Lipids)
Edited by P. MICHAEL CONN AND ANTHONY R. MEANS

VOLUME 142. Metabolism of Aromatic Amino Acids and Amines
Edited by SEYMOUR KAUFMAN

VOLUME 143. Sulfur and Sulfur Amino Acids
*Edited by* WILLIAM B. JAKOBY AND OWEN GRIFFITH

VOLUME 144. Structural and Contractile Proteins (Part D: Extracellular Matrix)
*Edited by* LEON W. CUNNINGHAM

VOLUME 145. Structural and Contractile Proteins (Part E: Extracellular Matrix)
*Edited by* LEON W. CUNNINGHAM

VOLUME 146. Peptide Growth Factors (Part A)
*Edited by* DAVID BARNES AND DAVID A. SIRBASKU

VOLUME 147. Peptide Growth Factors (Part B)
*Edited by* DAVID BARNES AND DAVID A. SIRBASKU

VOLUME 148. Plant Cell Membranes
*Edited by* LESTER PACKER AND ROLAND DOUCE

VOLUME 149. Drug and Enzyme Targeting (Part B)
*Edited by* RALPH GREEN AND KENNETH J. WIDDER

VOLUME 150. Immunochemical Techniques (Part K: *In Vitro* Models of B and T Cell Functions and Lymphoid Cell Receptors)
*Edited by* GIOVANNI DI SABATO

VOLUME 151. Molecular Genetics of Mammalian Cells
*Edited by* MICHAEL M. GOTTESMAN

VOLUME 152. Guide to Molecular Cloning Techniques
*Edited by* SHELBY L. BERGER AND ALAN R. KIMMEL

VOLUME 153. Recombinant DNA (Part D)
*Edited by* RAY WU AND LAWRENCE GROSSMAN

VOLUME 154. Recombinant DNA (Part E)
*Edited by* RAY WU AND LAWRENCE GROSSMAN

VOLUME 155. Recombinant DNA (Part F)
*Edited by* RAY WU

VOLUME 156. Biomembranes (Part P: ATP-Driven Pumps and Related Transport: The Na, K-Pump)
*Edited by* SIDNEY FLEISCHER AND BECCA FLEISCHER

VOLUME 157. Biomembranes (Part Q: ATP-Driven Pumps and Related Transport: Calcium, Proton, and Potassium Pumps)
*Edited by* SIDNEY FLEISCHER AND BECCA FLEISCHER

VOLUME 158. Metalloproteins (Part A)
*Edited by* JAMES F. RIORDAN AND BERT L. VALLEE

VOLUME 159. Initiation and Termination of Cyclic Nucleotide Action
*Edited by* JACKIE D. CORBIN AND ROGER A. JOHNSON

VOLUME 160. Biomass (Part A: Cellulose and Hemicellulose)
*Edited by* WILLIS A. WOOD AND SCOTT T. KELLOGG

VOLUME 161. Biomass (Part B: Lignin, Pectin, and Chitin)
*Edited by* WILLIS A. WOOD AND SCOTT T. KELLOGG

VOLUME 162. Immunochemical Techniques (Part L: Chemotaxis and Inflammation)
*Edited by* GIOVANNI DI SABATO

VOLUME 163. Immunochemical Techniques (Part M: Chemotaxis and Inflammation)
*Edited by* GIOVANNI DI SABATO

VOLUME 164. Ribosomes
*Edited by* HARRY F. NOLLER, JR., AND KIVIE MOLDAVE

VOLUME 165. Microbial Toxins: Tools for Enzymology
*Edited by* SIDNEY HARSHMAN

VOLUME 166. Branched-Chain Amino Acids
*Edited by* ROBERT HARRIS AND JOHN R. SOKATCH

VOLUME 167. Cyanobacteria
*Edited by* LESTER PACKER AND ALEXANDER N. GLAZER

VOLUME 168. Hormone Action (Part K: Neuroendocrine Peptides)
*Edited by* P. MICHAEL CONN

VOLUME 169. Platelets: Receptors, Adhesion, Secretion (Part A)
*Edited by* JACEK HAWIGER

VOLUME 170. Nucleosomes
*Edited by* PAUL M. WASSARMAN AND ROGER D. KORNBERG

VOLUME 171. Biomembranes (Part R: Transport Theory: Cells and Model Membranes)
*Edited by* SIDNEY FLEISCHER AND BECCA FLEISCHER

VOLUME 172. Biomembranes (Part S: Transport: Membrane Isolation and Characterization)
*Edited by* SIDNEY FLEISCHER AND BECCA FLEISCHER

VOLUME 173. Biomembranes [Part T: Cellular and Subcellular Transport: Eukaryotic (Nonepithelial) Cells]
*Edited by* SIDNEY FLEISCHER AND BECCA FLEISCHER

VOLUME 174. Biomembranes [Part U: Cellular and Subcellular Transport: Eukaryotic (Nonepithelial) Cells]
*Edited by* SIDNEY FLEISCHER AND BECCA FLEISCHER

VOLUME 175. Cumulative Subject Index Volumes 135–139, 141–167

VOLUME 176. Nuclear Magnetic Resonance (Part A: Spectral Techniques and Dynamics)
*Edited by* NORMAN J. OPPENHEIMER AND THOMAS L. JAMES

VOLUME 177. Nuclear Magnetic Resonance (Part B: Structure and Mechanism)
*Edited by* NORMAN J. OPPENHEIMER AND THOMAS L. JAMES

VOLUME 178. Antibodies, Antigens, and Molecular Mimicry
*Edited by* JOHN J. LANGONE

VOLUME 179. Complex Carbohydrates (Part F)
*Edited by* VICTOR GINSBURG

VOLUME 180. RNA Processing (Part A: General Methods)
*Edited by* JAMES E. DAHLBERG AND JOHN N. ABELSON

VOLUME 181. RNA Processing (Part B: Specific Methods)
*Edited by* JAMES E. DAHLBERG AND JOHN N. ABELSON

VOLUME 182. Guide to Protein Purification
*Edited by* MURRAY P. DEUTSCHER

VOLUME 183. Molecular Evolution: Computer Analysis of Protein and Nucleic Acid Sequences
*Edited by* RUSSELL F. DOOLITTLE

VOLUME 184. Avidin-Biotin Technology
*Edited by* MEIR WILCHEK AND EDWARD A. BAYER

VOLUME 185. Gene Expression Technology
*Edited by* DAVID V. GOEDDEL

VOLUME 186. Oxygen Radicals in Biological Systems (Part B: Oxygen Radicals and Antioxidants)
*Edited by* LESTER PACKER AND ALEXANDER N. GLAZER

VOLUME 187. Arachidonate Related Lipid Mediators
*Edited by* ROBERT C. MURPHY AND FRANK A. FITZPATRICK

VOLUME 188. Hydrocarbons and Methylotrophy
*Edited by* MARY E. LIDSTROM

VOLUME 189. Retinoids (Part A: Molecular and Metabolic Aspects)
*Edited by* LESTER PACKER

VOLUME 190. Retinoids (Part B: Cell Differentiation and Clinical Applications)
*Edited by* LESTER PACKER

VOLUME 191. Biomembranes (Part V: Cellular and Subcellular Transport: Epithelial Cells)
*Edited by* SIDNEY FLEISCHER AND BECCA FLEISCHER

VOLUME 192. Biomembranes (Part W: Cellular and Subcellular Transport: Epithelial Cells)
*Edited by* SIDNEY FLEISCHER AND BECCA FLEISCHER

VOLUME 193. Mass Spectrometry
*Edited by* JAMES A. MCCLOSKEY

VOLUME 194. Guide to Yeast Genetics and Molecular Biology (Part A)
*Edited by* CHRISTINE GUTHRIE AND GERALD R. FINK

VOLUME 195. Adenylyl Cyclase, G Proteins, and Guanylyl Cyclase
*Edited by* ROGER A. JOHNSON AND JACKIE D. CORBIN

VOLUME 196. Molecular Motors and the Cytoskeleton
*Edited by* RICHARD B. VALLEE

VOLUME 197. Phospholipases
*Edited by* EDWARD A. DENNIS

VOLUME 198. Peptide Growth Factors (Part C)
*Edited by* DAVID BARNES, J. P. MATHER, AND GORDON H. SATO

VOLUME 199. Cumulative Subject Index Volumes 168–174, 176–194

VOLUME 200. Protein Phosphorylation (Part A: Protein Kinases: Assays, Purification, Antibodies, Functional Analysis, Cloning, and Expression)
*Edited by* TONY HUNTER AND BARTHOLOMEW M. SEFTON

VOLUME 201. Protein Phosphorylation (Part B: Analysis of Protein Phosphorylation, Protein Kinase Inhibitors, and Protein Phosphatases)
*Edited by* TONY HUNTER AND BARTHOLOMEW M. SEFTON

VOLUME 202. Molecular Design and Modeling: Concepts and Applications (Part A: Proteins, Peptides, and Enzymes)
*Edited by* JOHN J. LANGONE

VOLUME 203. Molecular Design and Modeling: Concepts and Applications (Part B: Antibodies and Antigens, Nucleic Acids, Polysaccharides, and Drugs)
*Edited by* JOHN J. LANGONE

VOLUME 204. Bacterial Genetic Systems
*Edited by* JEFFREY H. MILLER

VOLUME 205. Metallobiochemistry (Part B: Metallothionein and Related Molecules)
*Edited by* JAMES F. RIORDAN AND BERT L. VALLEE

VOLUME 206. Cytochrome P450
*Edited by* MICHAEL R. WATERMAN AND ERIC F. JOHNSON

VOLUME 207. Ion Channels
*Edited by* BERNARDO RUDY AND LINDA E. IVERSON

VOLUME 208. Protein–DNA Interactions
*Edited by* ROBERT T. SAUER

VOLUME 209. Phospholipid Biosynthesis
*Edited by* EDWARD A. DENNIS AND DENNIS E. VANCE

VOLUME 210. Numerical Computer Methods
*Edited by* LUDWIG BRAND AND MICHAEL L. JOHNSON

VOLUME 211. DNA Structures (Part A: Synthesis and Physical Analysis of DNA)
*Edited by* DAVID M. J. LILLEY AND JAMES E. DAHLBERG

VOLUME 212. DNA Structures (Part B: Chemical and Electrophoretic Analysis of DNA)
*Edited by* DAVID M. J. LILLEY AND JAMES E. DAHLBERG

VOLUME 213. Carotenoids (Part A: Chemistry, Separation, Quantitation, and Antioxidation)
*Edited by* LESTER PACKER

VOLUME 214. Carotenoids (Part B: Metabolism, Genetics, and Biosynthesis)
*Edited by* LESTER PACKER

VOLUME 215. Platelets: Receptors, Adhesion, Secretion (Part B)
*Edited by* JACEK J. HAWIGER

VOLUME 216. Recombinant DNA (Part G)
*Edited by* RAY WU

VOLUME 217. Recombinant DNA (Part H)
*Edited by* RAY WU

VOLUME 218. Recombinant DNA (Part I)
*Edited by* RAY WU

VOLUME 219. Reconstitution of Intracellular Transport
*Edited by* JAMES E. ROTHMAN

VOLUME 220. Membrane Fusion Techniques (Part A)
*Edited by* NEJAT DÜZGUÜNES

VOLUME 221. Membrane Fusion Techniques (Part B)
*Edited by* NEJAT DÜZGÜNES

VOLUME 222. Proteolytic Enzymes in Coagulation, Fibrinolysis, and Complement Activation (Part A: Mammalian Blood Coagulation Factors and Inhibitors)
*Edited by* LASZLO LORAND AND KENNETH G. MANN

VOLUME 223. Proteolytic Enzymes in Coagulation, Fibrinolysis, and Complement Activation (Part B: Complement Activation, Fibrinolysis, and Nonmammalian Blood Coagulation Factors)
*Edited by* LASZLO LORAND AND KENNETH G. MANN

VOLUME 224. Molecular Evolution: Producing the Biochemical Data
*Edited by* ELIZABETH ANNE ZIMMER, THOMAS J. WHITE, REBECCA L. CANN, AND ALLAN C. WILSON

VOLUME 225. Guide to Techniques in Mouse Development
*Edited by* PAUL M. WASSARMAN AND MELVIN L. DEPAMPHILIS

VOLUME 226. Metallobiochemistry (Part C: Spectroscopic and Physical Methods for Probing Metal Ion Environments in Metalloenzymes and Metalloproteins)
*Edited by* JAMES F. RIORDAN AND BERT L. VALLEE

VOLUME 227. Metallobiochemistry (Part D: Physical and Spectroscopic Methods for Probing Metal Ion Environments in Metalloproteins)
*Edited by* JAMES F. RIORDAN AND BERT L. VALLEE

VOLUME 228. Aqueous Two-Phase Systems
*Edited by* HARRY WALTER AND GÖTE JOHANSSON

VOLUME 229. Cumulative Subject Index Volumes 195–198, 200–227

VOLUME 230. Guide to Techniques in Glycobiology
*Edited by* WILLIAM J. LENNARZ AND GERALD W. HART

VOLUME 231. Hemoglobins (Part B: Biochemical and Analytical Methods)
*Edited by* JOHANNES EVERSE, KIM D. VANDEGRIFF, AND ROBERT M. WINSLOW

VOLUME 232. Hemoglobins (Part C: Biophysical Methods)
*Edited by* JOHANNES EVERSE, KIM D. VANDEGRIFF, AND ROBERT M. WINSLOW

VOLUME 233. Oxygen Radicals in Biological Systems (Part C)
*Edited by* LESTER PACKER

VOLUME 234. Oxygen Radicals in Biological Systems (Part D)
*Edited by* LESTER PACKER

VOLUME 235. Bacterial Pathogenesis (Part A: Identification and Regulation of Virulence Factors)
*Edited by* VIRGINIA L. CLARK AND PATRIK M. BAVOIL

VOLUME 236. Bacterial Pathogenesis (Part B: Integration of Pathogenic Bacteria with Host Cells)
*Edited by* VIRGINIA L. CLARK AND PATRIK M. BAVOIL

VOLUME 237. Heterotrimeric G Proteins
*Edited by* RAVI IYENGAR

VOLUME 238. Heterotrimeric G-Protein Effectors
*Edited by* RAVI IYENGAR

VOLUME 239. Nuclear Magnetic Resonance (Part C)
*Edited by* THOMAS L. JAMES AND NORMAN J. OPPENHEIMER

VOLUME 240. Numerical Computer Methods (Part B)
*Edited by* MICHAEL L. JOHNSON AND LUDWIG BRAND

VOLUME 241. Retroviral Proteases
*Edited by* LAWRENCE C. KUO AND JULES A. SHAFER

VOLUME 242. Neoglycoconjugates (Part A)
*Edited by* Y. C. LEE AND REIKO T. LEE

VOLUME 243. Inorganic Microbial Sulfur Metabolism
*Edited by* HARRY D. PECK, JR., AND JEAN LEGALL

VOLUME 244. Proteolytic Enzymes: Serine and Cysteine Peptidases
*Edited by* ALAN J. BARRETT

VOLUME 245. Extracellular Matrix Components
*Edited by* E. RUOSLAHTI AND E. ENGVALL

VOLUME 246. Biochemical Spectroscopy
*Edited by* KENNETH SAUER

VOLUME 247. Neoglycoconjugates (Part B: Biomedical Applications)
*Edited by* Y. C. LEE AND REIKO T. LEE

VOLUME 248. Proteolytic Enzymes: Aspartic and Metallo Peptidases
*Edited by* ALAN J. BARRETT

VOLUME 249. Enzyme Kinetics and Mechanism (Part D: Developments in Enzyme Dynamics)
*Edited by* DANIEL L. PURICH

VOLUME 250. Lipid Modifications of Proteins
*Edited by* PATRICK J. CASEY AND JANICE E. BUSS

VOLUME 251. Biothiols (Part A: Monothiols and Dithiols, Protein Thiols, and Thiyl Radicals)
*Edited by* LESTER PACKER

VOLUME 252. Biothiols (Part B: Glutathione and Thioredoxin; Thiols in Signal Transduction and Gene Regulation)
*Edited by* LESTER PACKER

VOLUME 253. Adhesion of Microbial Pathogens
*Edited by* RON J. DOYLE AND ITZHAK OFEK

VOLUME 254. Oncogene Techniques
*Edited by* PETER K. VOGT AND INDER M. VERMA

VOLUME 255. Small GTPases and Their Regulators (Part A: Ras Family)
*Edited by* W. E. BALCH, CHANNING J. DER, AND ALAN HALL

VOLUME 256. Small GTPases and Their Regulators (Part B: Rho Family)
*Edited by* W. E. BALCH, CHANNING J. DER, AND ALAN HALL

VOLUME 257. Small GTPases and Their Regulators (Part C: Proteins Involved in Transport)
*Edited by* W. E. BALCH, CHANNING J. DER, AND ALAN HALL

VOLUME 258. Redox-Active Amino Acids in Biology
*Edited by* JUDITH P. KLINMAN

VOLUME 259. Energetics of Biological Macromolecules
*Edited by* MICHAEL L. JOHNSON AND GARY K. ACKERS

VOLUME 260. Mitochondrial Biogenesis and Genetics (Part A)
*Edited by* GIUSEPPE M. ATTARDI AND ANNE CHOMYN

VOLUME 261. Nuclear Magnetic Resonance and Nucleic Acids
*Edited by* THOMAS L. JAMES

VOLUME 262. DNA Replication
*Edited by* JUDITH L. CAMPBELL

VOLUME 263. Plasma Lipoproteins (Part C: Quantitation)
*Edited by* WILLIAM A. BRADLEY, SANDRA H. GIANTURCO, AND JERE P. SEGREST

VOLUME 264. Mitochondrial Biogenesis and Genetics (Part B)
*Edited by* GIUSEPPE M. ATTARDI AND ANNE CHOMYN

VOLUME 265. Cumulative Subject Index Volumes 228, 230–262

VOLUME 266. Computer Methods for Macromolecular Sequence Analysis
*Edited by* RUSSELL F. DOOLITTLE

VOLUME 267. Combinatorial Chemistry
*Edited by* JOHN N. ABELSON

VOLUME 268. Nitric Oxide (Part A: Sources and Detection of NO; NO Synthase)
*Edited by* LESTER PACKER

VOLUME 269. Nitric Oxide (Part B: Physiological and Pathological Processes)
*Edited by* LESTER PACKER

VOLUME 270. High Resolution Separation and Analysis of Biological Macromolecules (Part A: Fundamentals)
*Edited by* BARRY L. KARGER AND WILLIAM S. HANCOCK

VOLUME 271. High Resolution Separation and Analysis of Biological Macromolecules (Part B: Applications)
*Edited by* BARRY L. KARGER AND WILLIAM S. HANCOCK

VOLUME 272. Cytochrome P450 (Part B)
*Edited by* ERIC F. JOHNSON AND MICHAEL R. WATERMAN

VOLUME 273. RNA Polymerase and Associated Factors (Part A)
*Edited by* SANKAR ADHYA

VOLUME 274. RNA Polymerase and Associated Factors (Part B)
*Edited by* SANKAR ADHYA

VOLUME 275. Viral Polymerases and Related Proteins
*Edited by* LAWRENCE C. KUO, DAVID B. OLSEN, AND STEVEN S. CARROLL

VOLUME 276. Macromolecular Crystallography (Part A)
*Edited by* CHARLES W. CARTER, JR., AND ROBERT M. SWEET

VOLUME 277. Macromolecular Crystallography (Part B)
*Edited by* CHARLES W. CARTER, JR., AND ROBERT M. SWEET

VOLUME 278. Fluorescence Spectroscopy
*Edited by* LUDWIG BRAND AND MICHAEL L. JOHNSON

VOLUME 279. Vitamins and Coenzymes (Part I)
*Edited by* DONALD B. MCCORMICK, JOHN W. SUTTIE, AND CONRAD WAGNER

VOLUME 280. Vitamins and Coenzymes (Part J)
*Edited by* DONALD B. MCCORMICK, JOHN W. SUTTIE, AND CONRAD WAGNER

VOLUME 281. Vitamins and Coenzymes (Part K)
*Edited by* DONALD B. MCCORMICK, JOHN W. SUTTIE, AND CONRAD WAGNER

VOLUME 282. Vitamins and Coenzymes (Part L)
*Edited by* DONALD B. MCCORMICK, JOHN W. SUTTIE, AND CONRAD WAGNER

VOLUME 283. Cell Cycle Control
*Edited by* WILLIAM G. DUNPHY

VOLUME 284. Lipases (Part A: Biotechnology)
*Edited by* BYRON RUBIN AND EDWARD A. DENNIS

VOLUME 285. Cumulative Subject Index Volumes 263, 264, 266–284, 286–289

VOLUME 286. Lipases (Part B: Enzyme Characterization and Utilization)
*Edited by* BYRON RUBIN AND EDWARD A. DENNIS

VOLUME 287. Chemokines
*Edited by* RICHARD HORUK

VOLUME 288. Chemokine Receptors
*Edited by* RICHARD HORUK

VOLUME 289. Solid Phase Peptide Synthesis
*Edited by* GREGG B. FIELDS

VOLUME 290. Molecular Chaperones
*Edited by* GEORGE H. LORIMER AND THOMAS BALDWIN

VOLUME 291. Caged Compounds
*Edited by* GERARD MARRIOTT

VOLUME 292. ABC Transporters: Biochemical, Cellular, and Molecular Aspects
*Edited by* SURESH V. AMBUDKAR AND MICHAEL M. GOTTESMAN

VOLUME 293. Ion Channels (Part B)
*Edited by* P. MICHAEL CONN

VOLUME 294. Ion Channels (Part C)
*Edited by* P. MICHAEL CONN

VOLUME 295. Energetics of Biological Macromolecules (Part B)
*Edited by* GARY K. ACKERS AND MICHAEL L. JOHNSON

VOLUME 296. Neurotransmitter Transporters
*Edited by* SUSAN G. AMARA

VOLUME 297. Photosynthesis: Molecular Biology of Energy Capture
*Edited by* LEE MCINTOSH

VOLUME 298. Molecular Motors and the Cytoskeleton (Part B)
*Edited by* RICHARD B. VALLEE

VOLUME 299. Oxidants and Antioxidants (Part A)
*Edited by* LESTER PACKER

VOLUME 300. Oxidants and Antioxidants (Part B)
*Edited by* LESTER PACKER

VOLUME 301. Nitric Oxide: Biological and Antioxidant Activities (Part C)
*Edited by* LESTER PACKER

VOLUME 302. Green Fluorescent Protein
*Edited by* P. MICHAEL CONN

VOLUME 303. cDNA Preparation and Display
*Edited by* SHERMAN M. WEISSMAN

VOLUME 304. Chromatin
*Edited by* PAUL M. WASSARMAN AND ALAN P. WOLFFE

VOLUME 305. Bioluminescence and Chemiluminescence (Part C)
*Edited by* THOMAS O. BALDWIN AND MIRIAM M. ZIEGLER

VOLUME 306. Expression of Recombinant Genes in Eukaryotic Systems
*Edited by* JOSEPH C. GLORIOSO AND MARTIN C. SCHMIDT

VOLUME 307. Confocal Microscopy
*Edited by* P. MICHAEL CONN

VOLUME 308. Enzyme Kinetics and Mechanism (Part E: Energetics of Enzyme Catalysis)
*Edited by* DANIEL L. PURICH AND VERN L. SCHRAMM

VOLUME 309. Amyloid, Prions, and Other Protein Aggregates
*Edited by* RONALD WETZEL

VOLUME 310. Biofilms
*Edited by* RON J. DOYLE

VOLUME 311. Sphingolipid Metabolism and Cell Signaling (Part A)
*Edited by* ALFRED H. MERRILL, JR., AND YUSUF A. HANNUN

VOLUME 312. Sphingolipid Metabolism and Cell Signaling (Part B)
*Edited by* ALFRED H. MERRILL, JR., AND YUSUF A. HANNUN

VOLUME 313. Antisense Technology (Part A: General Methods, Methods of Delivery, and RNA Studies)
*Edited by* M. IAN PHILLIPS

VOLUME 314. Antisense Technology (Part B: Applications)
*Edited by* M. IAN PHILLIPS

VOLUME 315. Vertebrate Phototransduction and the Visual Cycle (Part A)
*Edited by* KRZYSZTOF PALCZEWSKI

VOLUME 316. Vertebrate Phototransduction and the Visual Cycle (Part B)
*Edited by* KRZYSZTOF PALCZEWSKI

VOLUME 317. RNA–Ligand Interactions (Part A: Structural Biology Methods)
*Edited by* DANIEL W. CELANDER AND JOHN N. ABELSON

VOLUME 318. RNA–Ligand Interactions (Part B: Molecular Biology Methods)
*Edited by* DANIEL W. CELANDER AND JOHN N. ABELSON

VOLUME 319. Singlet Oxygen, UV-A, and Ozone
*Edited by* LESTER PACKER AND HELMUT SIES

VOLUME 320. Cumulative Subject Index Volumes 290–319

VOLUME 321. Numerical Computer Methods (Part C)
*Edited by* MICHAEL L. JOHNSON AND LUDWIG BRAND

VOLUME 322. Apoptosis
*Edited by* JOHN C. REED

VOLUME 323. Energetics of Biological Macromolecules (Part C)
*Edited by* MICHAEL L. JOHNSON AND GARY K. ACKERS

VOLUME 324. Branched-Chain Amino Acids (Part B)
*Edited by* ROBERT A. HARRIS AND JOHN R. SOKATCH

VOLUME 325. Regulators and Effectors of Small GTPases (Part D: Rho Family)
*Edited by* W. E. BALCH, CHANNING J. DER, AND ALAN HALL

VOLUME 326. Applications of Chimeric Genes and Hybrid Proteins (Part A: Gene Expression and Protein Purification)
*Edited by* JEREMY THORNER, SCOTT D. EMR, AND JOHN N. ABELSON

VOLUME 327. Applications of Chimeric Genes and Hybrid Proteins (Part B: Cell Biology and Physiology)
*Edited by* JEREMY THORNER, SCOTT D. EMR, AND JOHN N. ABELSON

VOLUME 328. Applications of Chimeric Genes and Hybrid Proteins (Part C: Protein-Protein Interactions and Genomics)
*Edited by* JEREMY THORNER, SCOTT D. EMR, AND JOHN N. ABELSON

VOLUME 329. Regulators and Effectors of Small GTPases (Part E: GTPases Involved in Vesicular Traffic)
*Edited by* W. E. BALCH, CHANNING J. DER, AND ALAN HALL

VOLUME 330. Hyperthermophilic Enzymes (Part A)
*Edited by* MICHAEL W. W. ADAMS AND ROBERT M. KELLY

VOLUME 331. Hyperthermophilic Enzymes (Part B)
*Edited by* MICHAEL W. W. ADAMS AND ROBERT M. KELLY

VOLUME 332. Regulators and Effectors of Small GTPases (Part F: Ras Family I)
*Edited by* W. E. BALCH, CHANNING J. DER, AND ALAN HALL

VOLUME 333. Regulators and Effectors of Small GTPases (Part G: Ras Family II)
*Edited by* W. E. BALCH, CHANNING J. DER, AND ALAN HALL

VOLUME 334. Hyperthermophilic Enzymes (Part C)
*Edited by* MICHAEL W. W. ADAMS AND ROBERT M. KELLY

VOLUME 335. Flavonoids and Other Polyphenols
*Edited by* LESTER PACKER

VOLUME 336. Microbial Growth in Biofilms (Part A: Developmental and Molecular Biological Aspects)
*Edited by* RON J. DOYLE

VOLUME 337. Microbial Growth in Biofilms (Part B: Special Environments and Physicochemical Aspects)
*Edited by* RON J. DOYLE

VOLUME 338. Nuclear Magnetic Resonance of Biological Macromolecules (Part A)
*Edited by* THOMAS L. JAMES, VOLKER DÖTSCH, AND ULI SCHMITZ

VOLUME 339. Nuclear Magnetic Resonance of Biological Macromolecules (Part B)
*Edited by* THOMAS L. JAMES, VOLKER DÖTSCH, AND ULI SCHMITZ

VOLUME 340. Drug–Nucleic Acid Interactions
*Edited by* JONATHAN B. CHAIRES AND MICHAEL J. WARING

VOLUME 341. Ribonucleases (Part A)
*Edited by* ALLEN W. NICHOLSON

VOLUME 342. Ribonucleases (Part B)
*Edited by* ALLEN W. NICHOLSON

VOLUME 343. G Protein Pathways (Part A: Receptors)
*Edited by* RAVI IYENGAR AND JOHN D. HILDEBRANDT

VOLUME 344. G Protein Pathways (Part B: G Proteins and Their Regulators)
*Edited by* RAVI IYENGAR AND JOHN D. HILDEBRANDT

VOLUME 345. G Protein Pathways (Part C: Effector Mechanisms)
*Edited by* RAVI IYENGAR AND JOHN D. HILDEBRANDT

VOLUME 346. Gene Therapy Methods
*Edited by* M. IAN PHILLIPS

VOLUME 347. Protein Sensors and Reactive Oxygen Species (Part A: Selenoproteins and Thioredoxin)
*Edited by* HELMUT SIES AND LESTER PACKER

VOLUME 348. Protein Sensors and Reactive Oxygen Species (Part B: Thiol Enzymes and Proteins)
*Edited by* HELMUT SIES AND LESTER PACKER

VOLUME 349. Superoxide Dismutase
*Edited by* LESTER PACKER

VOLUME 350. Guide to Yeast Genetics and Molecular and Cell Biology (Part B) (in preparation)
*Edited by* CHRISTINE GUTHRIE AND GERALD R. FINK

VOLUME 351. Guide to Yeast Genetics and Molecular and Cell Biology (Part C) (in preparation)
*Edited by* CHRISTINE GUTHRIE AND GERALD R. FINK

VOLUME 352. Redox Cell Biology and Genetics (Part A) (in preparation)
*Edited by* CHANDAN K. SEN AND LESTER PACKER

VOLUME 353. Redox Cell Biology and Genetics (Part B) (in preparation)
*Edited by* CHANDAN K. SEN AND LESTER PACKER

VOLUME 354. Enzyme Kinetics and Mechanism (Part F-Detection and Characterization of Enzyme Reaction Intermediates) (in preparation)
*Edited by* DANIEL L. PURICH

VOLUME 355. Cumulative Subject Index Volumes 321–354

VOLUME 356. Laser Capture Microscopy and Microdissection (in preparation)
*Edited by* P. MICHAEL CONN

# Section I

# Superoxide Reactions and Mechanisms

# [1] Quantitation of Intracellular Free Iron by Electron Paramagnetic Resonance Spectroscopy

By ANH N. WOODMANSEE and JAMES A. IMLAY

## Background

Cells contain "free" iron, which is distinct from the tightly bound iron found in the iron–sulfur clusters and hemes of proteins.[1-6] It has been suggested that this pool performs several productive roles, including supplying iron for the synthesis of iron-containing enzymes,[1,2] functioning in the process of cellular iron transport and storage,[7-10] and contributing to the expression or repression of iron-responsive genes.[11-13] This iron pool has not been chemically defined. Although iron binds avidly *in vitro* to numerous metabolites, including citrate, peptides, and nucleotides, its predominant physiological ligands remain unknown. Instead, free iron is simply conceived of as the intracellular iron that is not stably integrated into enzymes or iron-storage proteins.

Intracellular free iron is of additional interest because of its role in both oxidative DNA damage and lipid peroxidation.[14,15] Changes in its concentration directly affect the rate of both processes. The primary experimental evidence of this assertion is that cell-permeable iron chelators block these forms of oxidative damage in cells of all types.[16,17] Conversely, circumstances that

---

[1] R. R. Crichton and M. Charloteaux-Waters, *Eur. J. Biochem.* **164**, 485 (1987).
[2] A. Jacobs, *Blood* **50**, 433 (1970).
[3] R. J. Rothman, A. Serroni, and J. C. Faber, *Mol. Pharmacol.* **42**, 703 (1992).
[4] P. Ponka, R. W. Grady, A. Wilczynska, and H. M. Schulman, *Biochim. Biophys. Acta* **802**, 477 (1984).
[5] D. Y. Yegorov, A. V. Kozlov, O. A. Azizova, and Y. A. Vladimirov, *Free Radic. Biol. Med.* **15**, 565 (1993).
[6] R. R. Crichton and R. J. Ward, in "Iron and Human Diseases" (R. B. Laufer, ed.), p. 23. CRC Press, Boca Raton, Florida, 1992.
[7] T. Ramasarma, *Free Radic. Res. Commun.* **2**, 153 (1986).
[8] S. P. Young, S. Roberts, and A. Bomford, *Biochem J.* **232**, 819 (1985).
[9] C. G. D. Morley, K. Rewers, and A. Bezkorovainy, in "The Biochemistry and Physiology of Iron" (P. Saltman and J. Hegenauer, eds.), p. 171. Elsevier, Amsterdam, 1982.
[10] A. Jacobs, *Ciba Found. Symp.* **51**, 91 (1977).
[11] R. D. Klausner, T. A. Rouault, and J. B. Harford, *Cell* **72**, 19 (1993).
[12] A. Bagg and J. B. Neiland, *Microbiol. Rev.* **51**, 509 (1987).
[13] T. A. Rouault and R. D. Klausner, *Trends Biochem. Sci.* **21**, 174 (1996).
[14] K. Keyer and J. A. Imlay, *Proc. Natl Acad. Sci. U.S.A.* **93**, 13635 (1996).
[15] D. M. Miller, N. H. Spear, and S. D. Aust, *Arch. Biochem. Biophys.* **295**, 240 (1992).
[16] M. Larramendy, A. C. Mello-Filho, E. A. Leme Martins, and R. Meneghini, *Mutat. Res.* **178**, 57 (1983).
[17] J. A. Imlay, S. M. Chin, and S. Linn, *Science* **240**, 640 (1988).

increase the amount of intracellular free iron—including mutations that accelerate iron transport or diminish storage—raise the rates of membrane and DNA injury.[18] Of particular interest to investigators of oxidative stress is the fact that free iron pools enlarge when superoxide destroys the iron–sulfur clusters of a subclass of dehydratases,[19] thereby promoting DNA and lipid damage as an indirect consequence.[14,20]

The involvement of free iron pools in oxidative injury is spurring the development of methods to quantitate them. One difficulty is that free iron comprises only about 5% of total cellular iron, so that even substantial changes in the iron pool cannot be accurately calculated from measurement of the total iron content. For that reason, methods must be designed to specifically detect loosely bound iron. A second problem is that intracellular free iron concentrations cannot be inferred from measurements made on cell lysates. The disruption of cells causes, either by chemical oxidation or dilution, sufficient damage to metalloenzymes and storage proteins that they release abundant iron. This artifactual iron flux dominates any subsequent measurement of soluble iron.

Several different spectroscopic techniques have appeared that circumvent these difficulties, albeit with different sensitivities and ease of application. The first is the use of calcein, a cell-penetrating iron chelator whose fluorescence is quenched when it binds iron. By determining its specific fluorescence *in vivo*, it is possible to obtain information about the concentration of iron available to bind it. A complication of the calcein method is that in order to preserve a workable signal-to-noise ratio, the amount of calcein added to the sample must be limited. Because more abundant intracellular metabolites, such as citrate, effectively compete with calcein for iron, the data from such experiments do not directly report the amount of total free iron. Investigators have addressed this issue by infusing cells with standard amounts of free iron. The intracellular iron concentrations that have been arrived at by this method—0.01 $\mu M$ in plant cells[21] and 0.5 $\mu M$ in human erythroleukemia cells[22]—are far lower than the typically $10^{-5}$ $M$ values reported for bacteria and yeast, using the electron paramagnetic resonance (EPR) technique described below.[14,23] The calcein method is most facile in comparing the relative amounts of free iron in well-matched samples for which the concentrations of competing intracellular chelators are unlikely to vary.

---

[18] D. Touati, M. Jacques, B. Tardat, L. Bouchard, and S. Despied, *J. Bacteriol.* **177**, 2305 (1995).
[19] D. H. Flint, J. F. Tuminello, and M. H. Emptage, *J. Biol. Chem.* **268**, 22369 (1993).
[20] S. I. Liochev and I. Fridovich, *Free Radic. Biol. Med.* **16**, 29 (1994).
[21] S. Epsztejn, O. Kakhlon, H. Glicktein, W. Breuer, and Z. I. Cabantchik, *Anal. Biochem.* **248**, 31 (1997).
[22] W. Breuer, S. Epsztejn, and Z. I. Cabantchik, *J. Biol. Chem.* **270**, 24209 (1995).
[23] C. Srinivasa, A. Liba, J. A. Imlay, J. S. Valentine, and E. B. Gralla, *J. Biol. Chem.* **275**, 29187 (2000).

Free intracellular ferrous iron has also apparently been visualized *in vivo* by a second noninvasive technique, Mössbauer spectroscopy.[24–26] This method has not been used to monitor iron homeostasis during oxidative stress, but it may yet prove useful, particularly because it can provide information about the ligand field of the "free" iron.

A third method, whole-cell EPR, has been used to quantitate free iron levels in both bacteria and yeast[14,23] and is described in detail. The technique is relatively fast and simple, has a detection limit of about 1 $\mu M$ iron, and can be adapted to discriminate between ferrous and ferric forms. In principle, this method could be adapted for application to mammalian cells, but the authors are unaware of any attempts to do so.

## Theory

Because they are paramagnetic, both $Fe^{2+}$ and $Fe^{3+}$ present EPR signals. However, the signal for $Fe^{2+}$ is much broader and weaker than that for $Fe^{3+}$; thus, free $Fe^{3+}$ can be measured directly by EPR, whereas $Fe^{2+}$ can be measured easily only after being oxidized to $Fe^{3+}$. Ferrous iron can be converted to the ferric form *in vivo* by incubating cells with the penetrating iron chelator desferrioxamine mesylate (DF). Because DF binds $Fe^{3+}$ more strongly than $Fe^{2+}$, DF lowers the redox potential of the latter species, thereby facilitating its oxidation by the molecular oxygen present in the sample. The consequence is that ferrous iron is quantitatively converted to a ferric : DF chelate. This chelate exhibits a sharp, easily detectable EPR signal at $g = 4.3$. Prosthetic iron in enzymes does not resonate at this $g$ value.

A variety of controls have validated the use of DF to identify free iron.[14] Because DF completely blocks iron-mediated DNA damage in bacteria, it appears that the DF : $Fe^{3+}$ signal includes all the iron that participates in adventitious oxidative chemistry in the cell. DF failed to extract iron from a variety of iron-containing enzymes, which indicates that the signal does not represent iron that had been integrated into proteins. Further, the exposure of bacterial cells to millimolar amounts of $H_2O_2$ yielded an EPR signal of the same intensity as when cells were exposed to DF, indicating that the sole effect of DF is to stimulate the oxidation of the pool of free ferrous iron.[27] $H_2O_2$ therefore provides an alternative method of oxidizing intracellular iron in those organisms that DF cannot penetrate easily.

[24] T. H. Abdul, A. J. Hudson, Y. S. Chang, A. R. Timms, C. Hawkins, J. M. Williams, P. M. Harrison, J. R. Guest, and S. C. Andrews, *J. Bacteriol.* **181**, 1415 (1999).
[25] H. Abdul-Tehrani, A. J. Hudson, Y. Chang, A. R. Timms, C. Hawkins, J. M. Williams, P. M. Harrison, J. R. Guest, and S. C. Andrews, *J. Bacteriol.* **181**, 1415 (1999).
[26] B. F. Matzanke, G. I. Muller, E. Bill, and A. X. Trantwein, *Eur. J. Biochem.* **183**, 371 (1989).
[27] A. L. Nguyen and J. A. Imlay, unpublished results (2001).

In the absence of DF or $H_2O_2$ treatment, the ferric iron signal in *Escherichia coli* is small, indicating that the great majority of free iron is in the reduced form. The opposite is true in yeast.[23]

## Preparation of Reagents

All chemicals are from Sigma (St. Louis, MO). Laboratory deionized water is further purified with using a Labconco (Kansas City, MO) Water Pro PS system. All reagents should be prepared fresh for each set of experiments. $Fe_2(SO_4)_3$ for iron standards is prepared as a 100 m$M$ stock in water. It is diluted to the appropriate concentrations immediately before use, because iron at lower concentrations quickly forms ferric hydroxides and precipitates out of solution. A stock of 0.2 $M$ desferrioxamine mesylate is dissolved in water, and the pH is adjusted to 8.0 with 6 $M$ stock of KOH. Cold 20 m$M$ Tris-HCl, pH 7.4, is maintained on ice, as is the same buffer supplemented with 10% (v/v) glycerol.

## Cell Growth and Sample Preparation

Cell are grown in either rich or defined medium with vigorous shaking. Low-density cultures require large culture volumes, so that the cell sample ultimately contains sufficient biomass, approximately 50 mg of protein. Cultures are centrifuged at 10,000 rpm for 5 min at 4°, and the pellet is resuspended in 9 ml of the same medium. For measurement of total free iron, 1 ml of 0.2 $M$ DF stock is added; for measurement of only ferric iron, the DF is omitted. The resuspended cells are then incubated with shaking at 37° for 10 min. The cells are then centrifuged again, washed with ice-cold 20 m$M$ Tris-HCl buffer at pH 7.4, and resuspended in 0.2–0.4 ml of the same buffer containing 10% (v/v) glycerol. The volume of the sample is measured, and 0.2 ml of the sample is loaded into a 3-mm quartz EPR tube, with care to avoid the introduction of air bubbles. The sample is immediately frozen in dry ice and stored at −70°. It is essential that all steps be performed as rapidly as possible to ensure the most accurate measurements. The remaining sample can be diluted and the absorbance determined, or lysed and the protein content assayed, in order to provide a measure of cell density for comparison with other samples.

A standard curve is generated using solutions of $Fe_2(SO_4)_3 \cdot 7H_2O$. A stock solution of ferric sulfate dissolved in deionized water is precisely quantitated by dilution into 20 m$M$ Tris-HCl–1 m$M$ DF, using $\epsilon_{420} = 2.865$ m$M^{-1}$ cm$^{-1}$ for the $Fe^{3+}$ : Df complex.[28] Ferric sulfate is then diluted to concentrations that are similar to that which will be found in the cell sample tubes—approximately 10 to 50 $\mu M$—in 20 m$M$ Tris buffer and 10% (v/v) glycerol. The standard samples are

---

[28] D. Y. Yegorev, A. V. Kozlov, O. A. Azizova, and Y. A. Vladimirov, *Free Radic. Biol. Med.* **15**, 565 (1993).

loaded into EPR tubes at the same volume used for the samples, and they are immediately frozen at $-70°$. Samples and standards may be stored indefinitely at $-70°$ and reused as long as they remain frozen. Total sample preparation time (exclusive of cell growth) is approximately 40 min.

### Electron Paramagnetic Resonance Parameters

The signal is measured with a Varian (Palo Alto, CA) Century E-112 X-band spectrophotometer equipped with a Varian TE102 cavity and a Varian temperature controller. The temperature is measured with a T-type thermocouple fixed next to the sample and an Omega Engineering (Stamford, CT) 670/680 microprocessor-based thermocouple meter. Data are analyzed with software from Scientific Software Services (Bloomington, IL). The EPR spectrometer settings are as follows: field center, 1570 G; receiver gain, 2500; field sweep, 400 G; modulation amplitude, 1.25 G; T, $-125°$; power, 30 mW.

### Calculation of Free Iron Concentration

If the EPR signals in paired samples are similar in shape, then the free iron concentration is most accurately determined by comparing the amplitude of the first derivative peak amplitude of the sample with that of a standard that contains a similar iron concentration. The signal shape may vary in samples that do not contain DF, because the ligand sphere influences the spectrum. In these cases the free iron is quantitated by double integration. This procedure provides the amount of iron in the sample. The actual intracellular concentration is calculated by dividing the iron amount by the intracellular volume of the sample. The latter can be determined by establishing the relationship between intracellular volume and the biomass of the harvested culture. For example, 1 liter of E. coli at 1.0 OD$_{600}$ (containing 190 mg of protein) represents a cytoplasmic volume of 0.5 ml. Ferric iron is directly determined from the EPR spectra of samples that have not been treated with either DF or $H_2O_2$; total iron, from the treated samples; and ferrous iron, by subtraction of the former from the latter.

### Performance and Application of Assay

Whole cell EPR allows for the measurement of non-protein-bound, chelatable iron inside whole cells. This chelatable "free" iron is distinct from the forms of iron found in Fe–S clusters, in heme, or in enzymes that use iron as a prosthetic metal for catalysis. A limitation of this method is its potential inability to distinguish free $Fe^{3+}$ from the ferric iron core of iron storage proteins, which also present a signal at $g = 4.3$. However, studies with E. coli mutants that lack storage proteins showed that in this organism the EPR iron signals arise predominantly from free rather than stored iron.

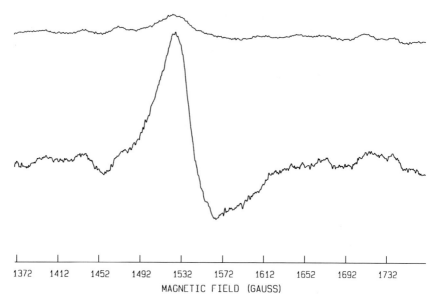

FIG. 1. Iron EPR signals from whole cell preparations of *E. coli*. Two thousand data points were collected per scan and each spectrum was the average of 50 scans per sample. The peak intensity of the scans was normalized to cell concentration. *Top scan:* AB1157, wild type, containing approximately 12 $\mu M$ intracellular free iron. *Bottom scan:* JI132, the isogenic superoxide dismutase-deficient strain [K. Keyer and J. A. Imlay, *Proc. Natl. Acad. Sci. U.S.A.* **93**, 13635 (1996)], containing 86 $\mu M$ free iron.

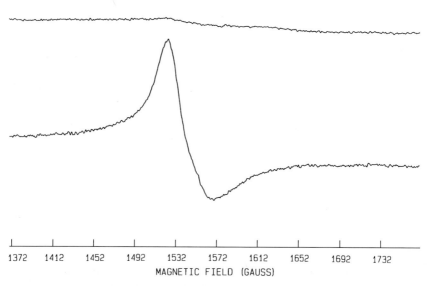

FIG. 2. *Top scan:* AN387, wild type, containing 20 $\mu M$ intracellular free iron. *Bottom scan:* SM1095, an isogenic dipicolinate-accumulating strain [S. Maringanti and J. A. Imlay, *J. Bacteriol.* **181**, 3792 (1999)], containing 320 $\mu M$ free iron.

In log-phase, aerobic, wild-type *E. coli* that are grown in standard laboratory media, the basal level of free iron ranges from 15 to 30 $\mu M$, depending on growth conditions. With the methods described here intracellular iron concentrations as low as 1 to 10 $\mu M$ can be measured, albeit with significant (1 $\mu M$) variance. Some improvement in the signal-to-noise ratio can be achieved by increasing the number of scans per sample (to reduce the sample noise) or by increasing the concentration of cells in the EPR tube (to increase the signal). Nevertheless, because normal intracellular iron concentrations approach the detection limits, increases in free iron concentration are more easily detected than are decreases.

Typical results with superoxide-stressed cells are shown in Fig. 1.[14] Figure 2 demonstrates that a strain that accumulates dipicolinate, an iron-binding metabolite, also contains a high amount of "free" iron. In this example the iron was almost certainly bound to the metabolite before the addition of DF. EPR analysis with such strains succeeds because DF binds iron more tightly than do most intracellular metabolites and therefore extracts the iron from them.

### Acknowledgments

This work was supported by Grant GM59030 from the National Institutes of Health. We gratefully acknowledge the contributions of Kay Keyer, Amy Gort, and Sujatha Maringanti, previous members of this laboratory, in the development and application of these methods.

# [2] Aconitase: Sensitive Target and Measure of Superoxide

*By* PAUL R. GARDNER

### Introduction

$O_2^{\cdot-}$ is abundantly produced in aerobic organisms, and the superoxide dismutases (SODs) appear ample for the task of scavenging $O_2^{\cdot-}$. The SODs maintain the steady-state $[O_2^{\cdot-}]$ at a remarkably low level of $\sim 10^{-10}$ $M$ inside diverse cells and organelles.[1] This $[O_2^{\cdot-}]$ is roughly equivalent to 1 $O_2^{\cdot-}$ molecule in 20 bacteria or 5000 molecules in a much larger mammalian cell. These low, steady-state intracellular $O_2^{\cdot-}$ levels and high levels of SOD expression are essential because of the extreme reactivity and toxicity of $O_2^{\cdot-}$.[2,3]

---

[1] J. A. Imlay and I. Fridovich, *J. Biol. Chem.* **266**, 6957 (1991).
[2] J. M. McCord, B. B. J. Keele, and I. Fridovich, *Proc. Natl. Acad. Sci. U.S.A.* **68**, 1024 (1971).
[3] I. Fridovich, *Science* **201**, 875 (1978).

A myriad of biomolecules are oxidized or reduced by $O_2^{·-}$; however, it is often unclear whether or how these reactions cause toxic (or beneficial) actions to cells under normal or pathological conditions. I. Fridovich has provided comprehensive reviews of the reactions of $O_2^{·-}$ with biomolecules,[4-6] and there have been some notable additions to the literature.

More recent attention has focused on the reactions of $O_2^{·-}$ with members of the growing family of [4Fe–4S]-containing (de)hydratases, including the critical citric acid cycle enzyme aconitase,[6-8] and with the gaseous signaling molecule and toxin nitric oxide (NO).[9] Both reactions occur rapidly, with estimated second-order rate constants in the range of $10^7$ and $10^{10}\ M^{-1}\ \text{sec}^{-1}$, respectively.[10-12] Moreover, evidence indicates that these targets have a critical role in $O_2^{·-}$ toxicity in various cells under a variety of conditions. Attack of either of these targets also has the potential to generate more lethal secondary damage to DNA, membranes, and proteins. The reaction of $O_2^{·-}$ with NO forms the more reactive and indiscriminant oxidant and nitrating agent peroxynitrite,[9,13] and $O_2^{·-}$ leaches highly reactive, catalytic, and damaging iron from these labile iron–sulfur centers.[14-17] The aconitases have provided a paradigm for understanding $O_2^{·-}$ toxicity and SOD function in bacteria, yeast, plants, and mammals and have yielded a sensitive method for measuring changes in the picomolar steady-state levels of $O_2^{·-}$ inside diverse cells and within mitochondria. This article describes the reaction of $O_2^{·-}$ with aconitase and the principles, practice, applications, advantages, and limitations of the "aconitase method."

## Principles of Aconitase Method

Members of the family of [4Fe–4S]-containing (de)hydratases are particularly susceptible to attack and oxidative inactivation by $O_2^{·-}$.[6-8,18,19] The mammalian

[4] I. Fridovich, *Arch. Biochem. Biophys.* **247**, 1 (1986).
[5] I. Fridovich, *J. Biol. Chem.* **264**, 7761 (1989).
[6] I. Fridovich, *Annu. Rev. Biochem.* **64**, 97 (1995).
[7] P. R. Gardner, *Biosci. Rep.* **17**, 33 (1997).
[8] D. H. Flint and R. M. Allen, *Chem. Rev.* **96**, 2315 (1996).
[9] W. A. Pryor and G. L. Squadrito, *Am. J. Physiol.* **268**, L699 (1995).
[10] A. Hausladen and I. Fridovich, *J. Biol. Chem.* **269**, 29405 (1994).
[11] D. H. Flint, J. F. Tuminello, and M. H. Emptage, *J. Biol. Chem.* **268**, 22369 (1993).
[12] R. E. Huie and S. Padmaja, *Free Radic. Res. Commun.* **18**, 195 (1993).
[13] J. S. Beckman, T. W. Beckman, J. Chen, P. A. Marshall, and B. A. Freeman, *Proc. Natl. Acad. Sci. U.S.A.* **87**, 1620 (1990).
[14] S. I. Liochev and I. Fridovich, *Free Radic. Biol. Med.* **16**, 29 (1994).
[15] K. Keyer and J. A. Imlay, *Proc. Natl. Acad. Sci. U.S.A.* **93**, 13635 (1996).
[16] C. Srinivasan, A. Liba, J. A. Imlay, J. S. Valentine, and E. B. Gralla, *J. Biol. Chem.* **275**, 29187 (2000).
[17] J. Vásquez-Vivar, B. Kalyanaraman, and M. C. Kennedy, *J. Biol. Chem.* **275**, 14064 (2000).
[18] P. R. Gardner, G. Costantino, C. Szabó, and A. L. Salzman, *J. Biol. Chem.* **272**, 25071 (1997).
[19] J.-C. Drapier and J. B. J. Hibbs, *Methods Enzymol.* **269**, 26 (1996).

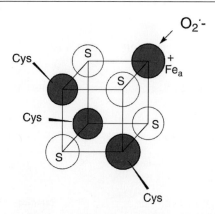

FIG. 1. Reaction of $O_2^{·-}$ with the aconitase [4Fe–4S] center. The avid binding and exceptional reactivity of $O_2^{·-}$ with the $[4Fe–4S]^{2+}$ center in aconitase and related (de)hydratases are attributed to the electrophilic character of the solvent-accessible iron atom ($Fe_a$), the extreme nucleophilicity of $O_2^{·-}$, the increased oxidizing potential of $O_2^{·-}$ in the iron-liganded state, and the sensitivity of these $Fe_a$–S bonds to oxidation and dissolution.

mitochondrial citric acid cycle enzyme aconitase and the mammalian cytosolic iron-regulatory protein and aconitase are perhaps the best studied and most understood members of this family. The aconitases contain cubane [4Fe–4S] centers with three irons liganded to cysteine residues. The fourth iron, $Fe_a$, is exposed to the solvent and to attack by $O_2^{·-}$ (Fig. 1). $Fe_a$ is essential for the catalytic dehydration of citrate to form the intermediate cis-aconitate and the hydration of cis-aconitate to form isocitrate (Scheme 1). The $Fe_a$ atom is thought to function as a Lewis acid in these reactions.[8] In the absence of substrates, $Fe_a$ ligands water or hydroxide. The Michaelis constants ($K_m$ values) for citrate, cis-aconitate, and isocitrate vary with assay conditions and enzyme source, but $K_m$ values are generally less than 0.5 m$M$, with the $K_m$ values for cis-aconitate and isocitrate being lower than that for citrate. Given the relatively high concentrations of aconitase substrates measured in cells and mitochondria, aconitase would appear to be near saturation in vivo.

$O_2^{·-}$ attacks the electrophilic solvent-exposed catalytic iron atom in the $[4Fe–4S]^{2+}$ center, causing one electron oxidation of the iron–sulfur cluster, the release of the exposed iron atom in the ferrous oxidation state, and the loss of

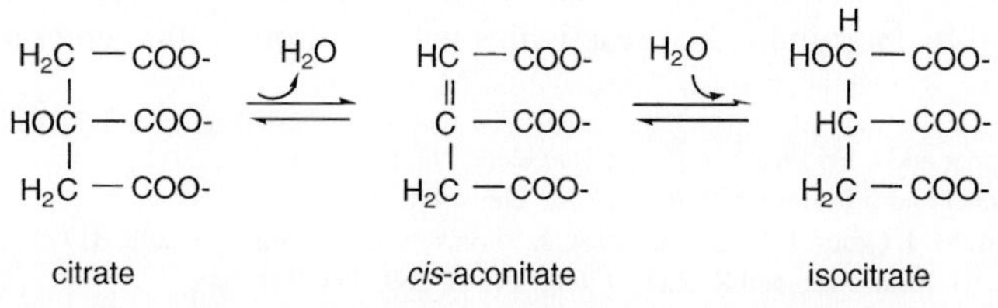

citrate          cis-aconitate          isocitrate

SCHEME 1

*Inactivation*

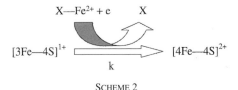

*Reactivation*

[3Fe—4S]$^{1+}$ $\xrightarrow{\text{X—Fe}^{2+}+e \quad X}_{k}$ [4Fe—4S]$^{2+}$

SCHEME 2

aconitase activity (Scheme 2). Second-order rate constants for the reaction of $O_2^{·-}$ with various aconitases are estimated in the range of $3 \times 10^6$ to $3 \times 10^7 M^{-1}$ sec$^{-1}$ at 25°.[10,11] The avid reactivity of $O_2^{·-}$ with (de)hydratase [4Fe–4S] centers has been attributed to the electrophilic character of Fe$_a$. Hausladen and Fridovich have reported a complete protection of the purified *Escherichia coli* and mitochondrial aconitases by citrate[10] and other (de)hydratases are partly protected by virtual substrates.[20,21] Substrates and inhibitors decrease the reactivity of $O_2^{·-}$ with aconitase and other (de)hydratases by sterically hindering $O_2^{·-}$ access. Electronic changes in the [4Fe–4S]$^{2+}$ center during substrate binding have been reported[22] and may also influence the avidity of $O_2^{·-}$ for Fe$_a$ and the susceptibility of the cluster to oxidation and destabilization. The demonstration of the reaction of NO with aconitases during catalytic turnover,[18] but not in the substrate-free resting state,[10,23] has emphasized the need for further investigation of the reactivity of $O_2^{·-}$ and NO with intermediates of the aconitase [4Fe–4S]$^{2+}$-catalyzed reaction. It is becoming clear that the reactions of the purified substrate-free enzyme are not fully representative of the *in vivo* reactions.

In cells, aconitase and other [4Fe–4S] enzymes are in a dynamic state of inactivation and reactivation. The iron–sulfur center is under continuous threat from attack by physiological oxidants including $O_2^{·-}$, $H_2O_2$, $O_2$, NO, and perhaps even ONOO$^-$. The roles of individual oxidants in aconitase inactivation *in vivo* are determined by their individual reactivities with the iron–sulfur center as well as

---

[20] P. R. Gardner and I. Fridovich, *J. Biol. Chem.* **266**, 19328 (1991).
[21] P. R. Gardner and I. Fridovich, *J. Biol. Chem.* **266**, 1478 (1991).
[22] B. Bennett, M. J. Gruer, J. R. Guest, and A. J. Thomson, *Eur. J. Biochem.* **233**, 317 (1995).
[23] L. Castro, M. Rodriguez, and R. Radi, *J. Biol. Chem.* **269**, 29409 (1994).

their steady-state concentrations. The second-order rate constants for the reactions of $O_2^{\cdot-}$, $H_2O_2$, $O_2$, and $ONOO^-$ with substrate-free aconitase are estimated to be on the order of $10^7$, $10^3$, $10^2$, and $10^5$ $M^{-1}$ sec$^{-1}$, respectively.[10,11,23] The second-order rate constant for the reaction of NO with pig mitochondrial aconitase during catalytic turnover with cis-aconitate at pH 6.8 is estimated to be on the order of 30 $M^{-1}$ sec$^{-1}$, and this rate constant is lower at higher pH and with citrate as the substrate.[18] In cells, reactions of physiological oxidants with aconitase are minimized by substrates and by competitive reactions with the cellular antioxidant systems. $O_2^{\cdot-}$ is by far the most reactive and selective physiological oxidant for aconitase that has been studied in bacteria and mammalian cells under normal conditions and in the presence of normal antioxidant defenses.[24,25] Nevertheless, $H_2O_2$, $O_2$, $ONOO^-$, and NO also inactivate aconitases in cells that are exposed to sufficiently high amounts of these agents for sufficient durations. Significant inactivations of aconitase have been reported after exposures of various cells to 100 $\mu M$ $H_2O_2$, 1 m$M$ $O_2$, 1 m$M$ $ONOO^-$, and 200 n$M$ NO.[18,25–27]

With the exception of NO-inactivated aconitases,[18] oxidatively inactivated aconitases are rapidly reactivated in vitro and in vivo. Reactivation requires cluster reduction and ferrous ion reinsertion.[7,28,29] The physiological mechanisms for reduction of the [3Fe–4S]$^{1+}$ center and for Fe$^{2+}$ insertion into the [3Fe–4S]$^0$ center are currently unknown. Reactivation may be facilitated by glutathione[30] and by a putative Fe$^{2+}$ carrier–donor designated as factor X in Scheme 2. Notwithstanding, pseudo-first-order rate constants ($k$) for the multistep reaction can be measured for the $O_2^{\cdot-}$-inactivated enzymes. In E. coli and human cells, the $t_{1/2}$ values for reactivation are $\sim$3 and $\sim$15 min, respectively, and the calculated $k$ values are 0.006 and 0.001 sec$^{-1}$, respectively.[24,25] Furthermore, an inactive fraction of aconitase can be measured under normal or stress conditions. Under normal aerobic conditions, 10–15% of aconitase in bacteria and mammalian cells is inactive and this fraction increases with increasing $[O_2^{\cdot-}]$.[24,25]

Measurements of the balance of inactive and active aconitase provide a sensitive measure of the steady-state $[O_2^{\cdot-}]$ in cells or mitochondria. In principle, knowledge of the second-order rate constant for inactivation ($k_2$), the pseudo-first-order rate constant for reactivation ($k$), and the inactive fraction of aconitase allows a determination of steady-state $[O_2^{\cdot-}]$. At a steady-state balance of inactivation and reactivation of aconitase under a given condition, we can apply the equation

---

[24] P. R. Gardner and I. Fridovich, J. Biol. Chem. **267**, 8757 (1992).
[25] P. R. Gardner, I. Raineri, L. B. Epstein, and C. W. White, J. Biol. Chem. **270**, 13399 (1995).
[26] P. R. Gardner, D.-D. H. Nguyen, and C. W. White, Proc. Natl. Acad. Sci. U.S.A. **91**, 12248 (1994).
[27] L. A. Castro, R. L. Robalinho, A. Cayota, R. Meneghini, and R. Radi, Arch. Biochem. Biophys. **359**, 215 (1998).
[28] M. C. Kennedy, M. H. Emptage, J.-C. Dreyer, and H. Beinert, J. Biol. Chem. **258**, 11098 (1983).
[29] M. H. Emptage, J.-L. Dreyer, M. C. Kennedy, and H. Beinert, J. Biol. Chem. **258**, 11106 (1983).
[30] P. R. Gardner and I. Fridovich, Arch. Biochem. Biophys. **301**, 98 (1993).

representing this balance, $k_2[O_2^{·-}]$[active aconitase] = $k$[inactive aconitase], to the determination of $[O_2^{·-}]$. Using 15% inactive aconitase, the respective $k$ values, and $k_2 = 10^7 M^{-1}$ sec$^{-1}$, we can calculate that $[O_2^{·-}] = \sim 2.5 \times 10^{-11} M$ in *E. coli* and $\sim 1.0 \times 10^{-10} M$ in human cells under normal aerobic growth conditions. These experimental values approximate the theoretical value for steady-state $[O_2^{·-}]$ of $10^{-10} M$ derived from the total cellular rate of $O_2^{·-}$ production and rate of removal by SODs.[1]

In the absence of precise values for the rate constant for inactivation, $k_2$, relative changes in intracellular $[O_2^{·-}]$ are readily calculated by measuring changes in the fraction of inactive aconitase as described by Eq. (1), where $X$ and $Y$ represent the two different conditions.[25]

$$\frac{[XO_2^{·-}]}{[YO_2^{·-}]} = \frac{[X \text{ inactive aconitase}]/[X \text{ active aconitase}]}{[Y \text{ inactive aconitase}]/[Y \text{ active aconitase}]} \qquad (1)$$

## Practice of Aconitase Method

*Preparation of Samples*

To measure the steady-state concentration of $[O_2^{·-}]$ in bacteria, yeast, plants, and mammalian cells and tissues, it is important that the cells, animals, and plants not be perturbed in such a way as to unwittingly alter $O_2^{·-}$ or aconitase activity. Cells, organisms, and tissues are rapidly harvested and quickly frozen because aconitase rapidly loses activity as $O_2^{·-}$ increases, and aconitase rapidly regains activity as $O_2^{·-}$ decreases. Anoxia produced during the handling of respiring cells and tissues allows aconitase reactivation by decreasing $[O_2^{·-}]$. Aconitase loses activity in extracts because of the reactions of $O_2$, $O_2^{·-}$, of $H_2O_2$ and because of freezing–thawing.

To minimize changes in aconitase activity, cells are rapidly collected and washed by centrifugation, and cell pellets are frozen on dry ice or in liquid $N_2$. Tissues and organs are rapidly excised from animals and frozen in liquid nitrogen. Cell-free extracts are prepared by rapid homogenization of tissue and sonication of cells on ice in a buffer containing stabilizing agents, and extracts are prepared at sufficiently high protein concentrations (>2 mg of protein per milliliter) to avert the loss of aconitase activity during freezing, storage, and thawing. A minimal (<5 : 1) volume ratio of 50 m$M$ Tris-HCl, pH 7.4, buffer containing 2 m$M$ sodium citrate or 20 $\mu M$ sodium fluorocitrate and 0.6 m$M$ MnCl$_2$ is used for extract preparation.

MnCl$_2$ and fluorocitrate or citrate are added to the buffer to protect the aconitase from oxidative inactivation. MnCl$_2$ scavenges $O_2^{·-}$ and is also added as a cofactor for the isocitrate dehydrogenase in the aconitase activity assay. Fluorocitrate is a competitive inhibitor of aconitases and can be sufficiently diluted to avoid interference in assays of the bacterial aconitase activity. However, citrate is preferred over fluorocitrate as a stabilizing agent for most purposes because fluorocitrate inhibition of the mammalian aconitases appears to contribute to undesirable lags in aconitase activity assays. On the other hand, citrate can generate sufficient isocitrate in extracts to produce a rapid initial rate of NADPH formation in the aconitase assay. These effects are minimized by delaying rate measurements in the aconitase assay. Sodium fluorocitrate is prepared by titrating 20 m$M$ barium ($dl$) fluorocitrate (Sigma, St. Louis, MO) dissolved in water with a slight molar excess of sulfuric acid, centrifuging to remove insoluble barium sulfate, and then neutralizing the supernatant with NaOH. The addition of reductants to cell extracts is not necessary, nor is it advised, because reductants can cause the inactivation of aconitase by liberating redox-active metals and generating $O_2^{\cdot-}$.

Cell lysates are rapidly and immediately frozen on dry ice and stored at $\leq -70°$. Aconitase activity in frozen bacteria and mammalian cell extracts is stable for more than 2 weeks. Frozen cell and tissue lysates are thawed in a 25° water bath, chilled on ice, clarified of membranes and other debris by centrifugation, and immediately assayed for aconitase activity. It is imperative that all samples be handled quickly and similarly. A 2-min delay between the thawing of extract and the assay of aconitase activity is considered normal.

Experiments designed to test the effects of various agents on aconitase activity in cells or mitochondria are carried out in the presence of protein synthesis inhibitors to block nascent aconitase synthesis and the induction of SODs or other protective proteins. Chloramphenicol (200 $\mu$g/ml) is added to bacteria $\geq$12 min before the exposure condition, and cycloheximide (100 $\mu$g/ml) is added to mammalian cells for this purpose. Controls are prepared to determine the effects of protein synthesis inhibitors on aconitase activity. Loss of activity due to proteolysis does not appear to be a significant problem in bacteria or mammalian cells, and the effects of various agents can be monitored over the course of minutes or hours.

*Assay of Aconitase*

*Reagents*

Tris-HCl (50 m$M$), pH 7.4, plus $MnCl_2$ (0.6 m$M$)
$NADP^+$ (20 m$M$) prepared in distilled water and stored at $-20°$
Sodium citrate (250 m$M$), pH 7.0
Isocitrate dehydrogenase (0.5 units/$\mu$l) prepared in Tris-HCl (50 m$M$), pH 7.4, plus $MnCl_2$ (0.6 m$M$) and stored at $-20°$

Aconitase activity is measured by monitoring absorbance at 340 nm in a freshly prepared reaction mix containing 0.2 m$M$ $NADP^+$, 5 m$M$ sodium citrate, and a 1-unit/ml concentration of pig heart isocitrate dehydrogenase in Tris-HCl plus $MnCl_2$ buffer. To start the assay, freshly thawed aliquots of clarified extracts ($\leq$30 $\mu$l and 10–100 $\mu$g of protein) are added to the 1.0-ml reaction mix preequilibrated at 25°. Measurements at 340 nm are recorded in a 1-cm cuvette at intervals of 3 min or less, and aconitase activity is calculated from the linear increase in absorbance at 340 nm during the latter half of a 60-min assay. Activity is calculated by using an extinction coefficient for NADPH of $6.22 \times 10^3 M^{-1}$ cm$^{-1}$ and assuming the conversion of one molecule of citrate to one molecule of NADPH via isocitrate dehydrogenase. One milliunit of aconitase activity is the amount catalyzing the conversion of 1 nmol of citrate to isocitrate per minute. Aconitase-specific activity is expressed relative to extract protein concentration or a stable enzyme marker such as fumarase. For extracts containing low aconitase activity such as most mammalian cell lines (<10 mU/mg), an initial lag of ~10 min is observed for NADPH formation, which may be due to the slow accumulation of the cis-aconitate intermediate. Given the long assay times required, the efficiency of aconitase activity measurements is improved by employing a thermostatted oscillating multicuvette holder.

Interference in the assay of aconitase is possible when assaying high concentrations of extract protein. NADP$^+$ can be reduced by enzymes and substrates independent of the aconitase activity in the extracts. The contribution of other enzymes and substrates to NADP$^+$ reduction can be assessed by inactivating aconitase with hydrophobic iron chelators such as $o$-phenanthroline or dipyridyl (0.5 m$M$) during a 60-min preincubation or by inhibiting aconitase with the competitive inhibitor fluorocitrate ($\geq 100$ $\mu M$). Further, it is important to demonstrate the linear dependence of aconitase activity on extract protein in the activity range being studied. The author prefers the coupled reaction to assays that monitor the formation or disappearance of cis-aconitate at 240 nm[19,31] because of the larger background absorbances of cell extracts at 240 nm.

A similar assay is used for extracts containing higher aconitase activity (>10 mU/mg protein) except that the assay time can be reduced to as little as 5 min and no lag in NADPH formation is observed. In addition, reaction volumes can be scaled down to 0.2 ml and measurements of aconitase activity can be made in 96-well microplates, using a 25° thermostatted microplate spectrophotometer equipped for measurements at 340 nm at 15-sec intervals. The assays are set up by quickly loading clarified cell or tissue extracts in less than a 10-$\mu$l volume to wells in triplicate and then immediately adding 190 $\mu$l of aconitase assay reaction mix preequilibrated to 25°, using a multichannel pipette. When using a microplate assay, it is important to standardize the measurements to millunits of aconitase activity as defined by the 1-cm cuvette assay method.

*Fractionation of Mitochondrial and Cytosolic Aconitases*

Cells are washed in 5 volumes of ice-cold fractionation buffer containing 0.25 $M$ sucrose, 10 m$M$ Tris-HCl (pH 7.4), 0.1 m$M$ EDTA, 2 m$M$ sodium citrate,

[31] A. Hausladen and I. Fridovich, *Methods Enzymol.* **269**, 37 (1996).

and 1 m$M$ sodium succinate. Citrate and succinate are added to maintain the pools of substrates in mitochondria and cytosol that normally protect the aconitase from inactivation. Cells are then centrifuged at 200$g$ for 5 min at 4° and the supernatant is removed. The cell pellet is homogenized in an equal volume of fractionation buffer on ice in a Teflon–glass Duall tissue homogenizer (Kontes Glass, Vineland, NJ) by making seven passes at a maximum speed of 500 rpm. Cell lysates are then centrifuged at 1000$g$ for 5 min at 4° to remove intact cells, debris, cell membranes, and nuclei from the supernatant containing cytoplasm, mitochondria, and other small membrane-bound organelles. Mitochondria are then separated from the cytosolic components by centrifuging at 10,000$g$ for 10 min at 4°. The supernatant is then immediately measured for cytosolic aconitase activity because delay or freezing of extracts under these conditions leads to a significant loss of aconitase activity. The mitochondrial pellet is then gently overlaid with a small volume of ice-cold fractionation buffer and centrifuged at 10,000$g$ for 2 min to remove residual cytosolic components. The mitochondrial pellet is then sonicated for 10 sec in 0.1 ml of ice-cold fractionation buffer and clarified by centrifugation for 30 sec at 14,000$g$. Mitochondrial extracts are immediately assayed for aconitase activity.

Cytosolic lactate dehydrogenase and mitochondrial glutamate dehydrogenase are also assayed to evaluate the quality of the fractionations. Lactate dehydrogenase and glutamate dehydrogenase activities are assayed at 25° by monitoring the change in absorbance at 340 nm in a 1-ml reaction mixture containing 50 m$M$ potassium phosphate buffer, pH 7.4, plus 0.2 m$M$ NADH with either 1 m$M$ sodium pyruvate, or 2 m$M$ $\alpha$-ketoglutarate and 50 m$M$ NH$_4$Cl, added as substrate, respectively.

Mitochondrial and cytosolic aconitase activities can also be differentiated semi-quantitatively, using a gel activity stain. Electrophoretic separation and staining of aconitases are carried out according to the procedure of Koen and Goodman,[32] which uses isocitrate dehydrogenase-catalyzed phenazine methosulfate reduction and redox-cycling to generate $O_2^{\cdot -}$ and to reduce nitroblue tetrazolium dye to the insoluble purple formazan. An inherent difficulty of this method is the loss of aconitase activity staining due to $O_2^{\cdot -}$-mediated inactivation. Sufficient nitroblue tetrazolium must be included to scavenge the $O_2^{\cdot -}$. Excess nitroblue tetrazolium protects the aconitase from inactivation and increases formazan formation.

*Determination of Rate Constants for Aconitase Inactivation*

Rate constants for aconitase inactivation are determined by measuring the competitive effects of SOD or other $O_2^{\cdot -}$ scavengers on the rate of inactivation of isolated, reconstituted, and fully active aconitases.[10,11,31] Measurements of aconitase reactivity with $O_2^{\cdot -}$ under various substrate, salt, and buffer conditions provide

---

[32] A. L. Koen and M. Goodman, *Biochim. Biophys. Acta* **191**, 698 (1969).

a range of values for the second-order rate constant ($k_2$) for the reaction of $O_2^{\cdot-}$ with a given aconitase that can be applied to the estimation of intracellular [$O_2^{\cdot-}$].

*Measurement of Rate Constants for Aconitase Reactivation*

In vivo reactivation rate constants for aconitases are determined by measuring the anaerobic recovery of aconitase activity after aerobic inactivation by an $O_2^{\cdot-}$-generating agent such as phenazine methosulfate in the presence of a protein synthesis inhibitor.[24,25] Briefly, cells are exposed to a concentration of phenazine methosulfate (1–100 $\mu M$) for a sufficient duration (15–60 min) to inactivate 50–80% of the aconitase activity. Cells are then harvested and distributed to several 2.4-ml septum-sealed Vacutainer tubes (47 × 10 mm; Becton Dickinson Labware, Franklin Lakes, NJ) filled with a medium that supports respiration. In this procedure, sufficient cells are needed to rapidly consume the $O_2$ in the medium, and cell respiration rates are determined with a Clark-type $O_2$ electrode.[25] Otherwise, deoxygenation of the medium is achieved by first sparging with $N_2$ and then adding cells resuspended in a small volume of medium. After various times of incubation at the appropriate temperature, cells are harvested, and extracts are prepared and assayed for aconitase activity. To harvest cells, Vacutainer tubes are centrifuged in cushioned tube holders at 7000$g$ for <2 min. The rubber septum is removed, the medium is aspirated, lysis buffer is added, and the cells are sonicated for 10 sec on ice, and the lysate is immediately frozen on dry ice. The extracts are then quickly thawed, clarified by centrifugation, and assayed for aconitase activity. The half-time for reactivation is approximated from plots of aconitase activity versus time, and the pseudo first-order rate constant for aconitase reactivation ($k$) is calculated. The half-time for reactivation of the *E. coli* and human aconitases are 3 and 15 min, respectively. The intervals for harvest need to be determined for specific organisms and conditions, but it is likely that many half-times for reactivation will fall within this range. Slower reactivation rates are expected in iron-limited and older unhealthy cells, and in such a case the intervals for harvest need to be extended.

*Measurement of the Inactive Fraction of Aconitase*

The inactive fraction of aconitase is determined by measuring the specific activity of aconitase in cells incubated under anaerobic conditions in Vacutainer tubes for a period exceeding five reactivation half-times and by comparing this activity with that of identical cells incubated under aerobic or stressed conditions. Cells are lysed, and extracts are clarified and assayed for aconitase activity as described for the measurement of aconitase reactivation rate constants. In practice, it is possible to estimate the inactive fraction under various stress conditions by measuring the percent aconitase activity in stressed cells relative to control cells and by correcting for the inactive fraction measured under normal unstressed conditions. For example, 50% aconitase activity in stressed cells normally containing a 15% inactive fraction of aconitase represents a total 67.5% inactive fraction of aconitase and an ~12-fold increase in steady-state $[O_2^{·-}]$ [Eq. (1)]. The inactive fraction of aconitase can also be determined in tissues by anaerobically activating extract aconitase with iron and thiols when whole-cell reactivation measurements are not feasible.[31,33]

## Applications of Aconitase Method

*Effects of Superoxide Dismutase Expression on Aconitase and $[O_2^{·-}]$*

The effects of SOD levels on aconitase activity have been measured in bacteria, yeast, *Drosophila*, mice, and cultured human cells.[24,25,33–36] In all cases, a lower level of SOD activity causes a lower aconitase activity. In the case of *E. coli* and cultured human cells,[24,25] quantitative relationships between SOD activity levels and active and inactive aconitase fractions have been established that support the steady-state approximations of $[O_2^{·-}]$ described by Eq. (1). In general, a 2-fold increase in $[O_2^{·-}]$ due to a 50% reduction in SOD activity increases the inactive fraction of aconitase from ~15 to ~30%.[24,33] An increase in SOD activity diminishes the inactive fraction of aconitase as approximated by Eq. (1), and the magnitude of the effect of elevated SOD levels on the inactive fraction of aconitase is greatest at elevated $O_2^{·-}$ concentrations. A 10-fold increase in SOD activity protects aconitase against up to ~50% inactivation under stress conditions as anticipated from Eq. (1) and from the ability of SOD to competitively scavenge $O_2^{·-}$.[24,25] In SOD-deficient cells, a residual 10–30% active fraction of aconitase is measured,[20,24,35] and this residual activity is presumably due to the rapid Fe–S reactivation pathway and to its contribution to cellular $O_2^{·-}$ removal and decomposition.

*Measurements of Diffusion of $O_2^{·-}$ through Cell Membranes*

Measurements of the diffusion of $O_2^{·-}$ through erythrocyte membranes resealed with the $O_2^{·-}$ sensor cytochrome *c* have demonstrated that $O_2^{·-}$ readily

---

[33] M. D. Williams, H. Van Remmen, C. C. Conrad, T. T. Huang, C. J. Epstein, and A. Richardson, *J. Biol. Chem.* **273**, 28510 (1998).

[34] Y. Li, T.-T. Huang, E. J. Carlson, S. Melov, P. C. Ursell, J. L. Olson, L. J. Noble, M. P. Yoshimura, C. Berger, P. H. Chan, D. C. Wallace, and C. J. Epstein, *Nat. Genet.* **11**, 376 (1995).

[35] S. Melov, P. Coskun, M. Patel, R. Tuinstra, B. Cottrell, A. S. Jun, T. H. Zastawny, M. Dizdaroglu, S. I. Goodman, T.-T. Huang, H. Miziorko, C. J. Epstein, and D. C. Wallace, *Proc. Natl. Acad. Sci. U.S.A.* **96**, 846 (1999).

[36] V. D. Longo, L.-L. Liou, J. S. Valentine, and E. B. Gralla, *Arch. Biochem. Biophys.* **365**, 131 (1999).

passes through anion channels and that the hydrophobic lipid bilayer of cell membranes otherwise shields cells against the charged $O_2^{·-}$ molecule.[37] Measurements of the effects of high extracellular fluxes on aconitase activity in intact human lung epithelial cells have demonstrated that this cell membrane is also impervious to xanthine oxidase-generated $O_2^{·-}$.[25] In these experiments, it is important to include catalase to prevent inactivation of cellular aconitase by the high levels of $H_2O_2$ formed directly by xanthine oxidase and indirectly through $O_2^{·-}$ dismutation.

*Effects of Redox-Cycling Drugs on Aconitase and $[O_2^{·-}]$*

Aconitase has been used as a sensitive measure of $[O_2^{·-}]$ in cells exposed to potential $O_2^{·-}$-generating agents.[24,25,38] Aconitase measurements have facilitated the demonstration of the potent $O_2^{·-}$-generating ability of the mycobacterial and pseudomonad quinoid compounds phthiocol and pyocyanine inside human lung cells.[38] The aconitase method has also proved to be more sensitive and reliable than measurements of $O_2^{·-}$ from cyanide-resistant respiration. This is understandable because cyanide-resistant respiration rates represent only a small fraction of the total respiration rate of cells and because basal cyanide-resistant respiration rates do not represent $O_2^{·-}$ production rates.

*Effects of Inhibitors of Mitochondrial Electron Transfer on Aconitase and $[O_2^{·-}]$*

Mitochondrial respiration is an important source of intracellular and intramitochondrial $O_2^{·-}$. Critical sites of $O_2^{·-}$ production have been previously determined by using isolated mitochondria and submitochondrial particles. The aconitase method provides a unique method for measuring $[O_2^{·-}]$ in the mitochondria of intact cells. Inhibitors of mitochondrial respiration increase $O_2^{·-}$ production in intact cells in roughly similar proportions to those measured with isolated mitochondria and membranes.[25] The aconitase method also allows the measurement of the effects of receptors and cellular agonists on mitochondrial $O_2^{·-}$ production in natural and minimally perturbed intracellular environments.

*Effects of Receptor Ligands and Peptides on Aconitase and $[O_2^{·-}]$*

Tumor necrosis factor (TNF) induces the mitochondrial manganese-containing SOD in mammalian cells and has been proposed to act as an inducer and cytotoxin through mitochondrial $O_2^{·-}$ production. However, measurements of aconitase demonstrate no increase in mitochondrial $O_2^{·-}$ with TNF exposure before SOD induction, loss of cell respiration, or cell death.[39] The neurotoxins kainate and

---

[37] R. E. Lynch and I. Fridovich, *J. Biol. Chem.* **253**, 4697 (1978).
[38] P. R. Gardner, *Arch. Biochem. Biophys.* **333**, 267 (1996).
[39] P. R. Gardner and C. W. White, *Arch. Biochem. Biophys.* **334**, 158 (1996).

$N$-methyl-D-aspartate acting through the membrane-bound glutamate receptor have been shown to inactivate aconitase and stimulate $O_2^{\cdot -}$ production in neurons.[40,41] $\beta$-Amyloid peptide, associated with Alzheimer's disease, also increases $O_2^{\cdot -}$ levels in neurons as measured by its effects on the fraction of active and inactive aconitase.[42]

*Effects of Superoxide Dismutase Mimetics on Aconitase and $[O_2^{\cdot -}]$*

The aconitase method has been used to test the effectiveness of SOD mimetics in scavenging intracellular $O_2^{\cdot -}$. Manganoporphyrins act as effective SOD mimetics against the excitatory neurotoxins kainate and $N$-methyl-D-aspartate in rat hippocampal or cortical cells[40,41] and the redox-cycling agents pyocyanine[43] or paraquat.[40] In the case of pyocyanine, the SOD mimetic manganese tetrakis-(1-methyl-4-pyridyl)porphyrin bolstered the total $O_2^{\cdot -}$-scavenging ability of cultured human lung A549 cells 3-fold as approximated from aconitase activity measurements.

*Effects of Specific Oxidases on Aconitase Activity and $[O_2^{\cdot -}]$*

The aconitases have been used to demonstrate increased production of $O_2^{\cdot -}$ by specific oxidases. Fibroblasts transformed with the phagocytic NADPH oxidase homolog Mox1 (Nox-1) were shown to have decreased aconitase-specific activity and an increased growth rate.[44] The authors did not, however, demonstrate an increase in the inactive fraction of aconitase. Measurements of the inactive fraction of aconitase are required when examining independently derived clones of the same cell line or when comparing aconitase-specific activities in different cell lines because different cells and independent clones have been found to express different amounts of aconitase.

*New and Old Problems*

The dynamic state of aconitase in cells suggests mechanisms for iron handling and delivery because ferrous ion released during the $O_2^{\cdot -}$-mediate-inactivation of [4Fe–4S] (de)hydratases poses a normal, but potentially lethal, threat to cells.[14–16] Moreover, in the absence of a carrier for iron, iron would be difficult to recover for the reactivation process especially under conditions of cellular iron limitation. Measurements of the inactivation–reactivation of aconitase offer a means to assess

---

[40] M. Patel, B. J. Day, J. D. Crapo, I. Fridovich, and J. O. McNamara, *Neuron* **16**, 345 (1996).
[41] L. P. Liang, Y. S. Ho, and M. Patel, *Neuroscience* **10**, 275 (2000).
[42] V. D. Longo, K. L. Viola, W. L. Klein, and C. E. Finch, *J. Neurochem.* **75**, 1977 (2000).
[43] P. R. Gardner, D.-D. H. Nguyen, and C. W. White, *Arch. Biochem. Biophys.* **325**, 20 (1996).
[44] Y.-A. Suh, R. S. Arnold, B. Lassegue, J. Shi, X. Xu, D. Sorescu, A. B. Chung, K. K. Griendling, and J. D. Lambeth, *Nature (London)* **401**, 79 (1999).

the role of various factors in Fe–S stability. In addition, the methods used to investigate $O_2^{·-}$ can be applied to investigations of other physiological inactivators of aconitase and protective strategies. For example, widely distributed, inducible, and protective NO-detoxifying NO dioxygenases have been identified by applying the methods established for $O_2^{·-}$ and the protective SODs.[45,46]

## Advantages and Limitations of Aconitase Method

An overriding advantage of the aconitase method is that cells and tissues are minimally perturbed. There is no need for high millimolar concentrations of spin traps that are frequently employed in electron paramagnetic methods. These agents can alter the steady-state balance of metabolic pathways, especially those involving $O_2^{·-}$. The aconitase method measures instantaneous changes in steady-state $[O_2^{·-}]$. The method is also readily accessible to most investigators and does not require expensive specialized equipment. Moreover, the method is superior in specificity to methods utilizing oxidizable chemiluminescent or fluorescent probes, such as lucigenin or dihydrorhodamine, that can also generate $O_2^{·-}$.[47] Further, because of the *in situ* localization of aconitase, the method can be exploited to differentially monitor $O_2^{·-}$ levels in cellular compartments and organelles such as mitochondria or chloroplasts.

One important limitation of the aconitase method is the necessity of determining precise rate constants for inactivation and reactivation for an absolute measure of steady-state $[O_2^{·-}]$. In the absence of these values, one is limited to the measurement of relative changes in $[O_2^{·-}]$ as described by Eq. (1). Factors that may influence aconitase and $[O_2^{·-}]$ measurements include changes in the availability of iron or altered citrate, *cis*-aconitate, and isocitrate concentrations. In addition, other Fe–S enzyme inactivators including NO,[18,19] ONOO$^-$,[10,19,23,27] $H_2O_2$,[25,48] and $O_2$,[26]; other oxidants; and manganese,[49] zinc, or other reactive heavy metals may confound approximations of intracellular $[O_2^{·-}]$ from measurements of aconitase activity. In such cases, it is imperative to demonstrate the role of $O_2^{·-}$ in the mechanism of enzyme inactivation by altering the $O_2^{·-}$-scavenging activity in cells or organelles by using SOD mimetics or cells altered in SOD expression. Another limitation of the aconitase method is that not all cellular compartments will be accessible to measurements because aconitase expression is limited in its distribution.

---

[45] P. R. Gardner, G. Costantino, and A. L. Salzman, *J. Biol. Chem.* **273,** 26528 (1998).
[46] P. R. Gardner, A. M. Gardner, L. A. Martin, and A. L. Salzman, *Proc. Natl. Acad. Sci. U.S.A.* **95,** 10378 (1998).
[47] S. I. Liochev and I. Fridovich, *Free Radic. Biol. Med.* **25,** 926 (1998).
[48] F. Verniquet, J. Gaillard, M. Neuburger, and R. Douce, *Biochem. J.* **276,** 643 (1991).
[49] W. Zheng, S. Ren, and J. H. Graziano, *Brain Res.* **799,** 334 (1998).

## Conclusions

Aconitase and other [4Fe–4S]-containing (de)hydratases are continuously attacked and oxidized by $O_2{}^{\cdot-}$ in cells. Aconitase is in a dynamic cycle of inactivation and reactivation. $O_2{}^{\cdot-}$ drives this cycle, and the balance of inactive and active aconitase provides a senstive measure of steady-state $[O_2{}^{\cdot-}]$ in cells and mitochondria. Investigations of the reactions of $O_2{}^{\cdot-}$, NO and other oxidants with this ubiquitous "Achilles heel" of intermediary metabolism and iron metabolism continue to reveal important mechanisms for the protection of labile Fe–S enzymes. Studies of aconitase offer a quantitative and mechanistic description of the role of $O_2{}^{\cdot-}$ in the etiology of various pathological states. Moreover, aconitase measurements provide a measure of the efficacy of SOD mimetics.

## Acknowledgment

The author gratefully acknowledges the support of American Heart Association Scientist Development Grant 9730193N.

# [3] Reactions of Manganese Porphyrins and Manganese-Superoxide Dismutase with Peroxynitrite

*By* GERARDO FERRER-SUETA, CELIA QUIJANO, BEATRIZ ALVAREZ, and RAFAEL RADI

## Introduction

Peroxynitrite anion ($ONOO^-$) is a strong oxidant and cytotoxic species formed *in vivo*, which readily reacts with transition metal-containing centers in redox reactions that yield secondary reactive intermediates. In this article we describe methodology and results relating to the reactions of peroxynitrite with manganese porphyrins and manganese-containing superoxide dismutase (MnSOD) and discuss its importance for nitric oxide-dependent oxidative damage and pharmacological control.

*Superoxide, Nitric Oxide, and Peroxynitrite*

The discovery of nitric oxide ($\cdot$NO) in biology led to the elaboration of a hypothesis in the early 1990s, proposing that the pathways of $\cdot$NO and superoxide ($O_2{}^{\cdot-}$) toxicity can merge into a common route involving the formation of peroxynitrite.[1,2] Indeed, the reaction between ·NO and $O_2^{·-}$ to form peroxynitrite anion ($ONOO^-$) is close to the diffusion-control limit, with a rate constant of $\sim 10^{10}\,M^{-1}\,\sec^{-1}$ [Eq. (1)].[3]

$$\cdot NO + O_2^{·-} \rightarrow ONOO^- \qquad (1)$$

Because this rate constant is 5 to 10 times higher than that of $O_2^{·-}$ with MnSOD or Cu,ZnSOD ($1-2 \times 10^9\,M^{-1}\,\sec^{-1}$), it can be estimated that in the presence of physiological concentrations of SOD of 10 to 20 $\mu M$, ·NO at micromolar concentrations will be able to trap a significant fraction of the $O_2^{·-}$ formed. Peroxynitrite formation has been demonstrated in different cell, organ, and animal systems[4-7] and proposed to contribute to oxidative stress-mediated pathology. Indeed, peroxynitrite is more reactive than its precursors and will be able to damage cell components such as proteins, nucleic acids, and lipids.[8]

*Peroxynitrite Diffusion and Reactivity*

Peroxynitrite is a weak base with a p$K_a$ of 6.8,[2] which means that at a pH of 7.4, 80% will be in the basic form. Diffusion considerations imply that peroxynitrite will be formed close to the sites of $O_2^{·-}$ formation, because ·NO can easily traverse biological membranes. Once formed, peroxynitrite can reach other compartments crossing membranes by passive diffusion as ONOOH or by passage through anion channels as $ONOO^-$.[9]

Peroxynitrite has high one- and two-electron reduction potentials of 1.6–1.7 and 1.3–1.37 V, respectively,[10] but still is a relatively selective biological oxidant.[8] Great effort has been devoted to determine the rate constants of peroxynitrite

---

[1] J. S. Beckman, T. W. Beckman, J. Chen, P. A. Marshall, and B. A. Freeman, *Proc. Natl. Acad. Sci. U.S.A.* **87**, 1620 (1990); R. Radi, J. S. Beckman, K. M. Bush, and B. A. Freeman, *Arch. Biochem. Biophys.* **288**, 481 (1991).

[2] R. Radi, J. S. Beckman, K. M. Bush, and B. A. Freeman, *J. Biol. Chem.* **266**, 4244 (1991).

[3] R. E. Huie and S. Padmaja, *Free Radic. Res. Commun.* **18**, 195 (1993); S. Goldstein, and G. Czapski, *Free Radic. Biol. Med.* **19**, 505 (1995); R. Kissner, T. Nauser, P. Bugnon, P. G. Lye, and W. H. Koppenol, *Chem. Res. Toxicol.* **10**, 1285 (1997).

[4] H. Ischiropoulos, L. Zhu, and J. S. Beckman, *Arch. Biochem. Biophys.* **298**, 446 (1992).

[5] A. G. Estevez, N. Spear, S. M. Manuel, L. Barbeito, R. Radi, and J. S. Beckman, *Prog. Brain Res.* **118**, 269 (1998); A. G. Estevez, J. P. Crow, J. B. Sampson, C. Reiter, Y. Zhuang, G. J. Richardson, M. M. Tarpey, L. Barbeito, and J. S. Beckman, *Science* **286**, 2498 (1999).

[6] C. Brito, M. Naviliat, B. C. Tiscornia, F. Vuillier, G. Gualco, G. Dighiero, R. Radi, and A. M. Cayota, *J. Immunol.* **162**, 3356 (1999).

[7] J. Ara, S. Przedborski, A. B. Naini, V. Jackson-Leuis, R. R. Trifiletti, J. Horwitz, and H. Ischiropoulos, *Proc. Natl. Acad. Sci. U.S.A.* **95**, 7659 (1998).

[8] R. Radi, A. Denicola, B. Alvarez, G. Ferrer-Sueta, and H. Rubbo, in "Nitric Oxide: Biology and Pathobiology" (L. Ignarro, ed.), pp. 57–82. Academic Press, San Diego, California, 2000.

[9] A. Denicola, J. M. Souza, and R. Radi, *Proc. Natl. Acad. Sci. U.S.A.* **95**, 3566 (1998).

[10] G. Merenyi and J. Lind, *Chem. Res. Toxicol.* **10**, 1216 (1997); W. H. Koppenol, and R. Kissner, *Chem. Res. Toxicol.* **11**, 87 (1998).

reactions. Because at pH 7.4 and 37° peroxynitrite decomposes in the absence of targets in less than 10 sec, owing to the homolysis of ONOOH,[8] the kinetics of its reactions must be studied by fast techniques such as stopped-flow spectrophotometry.[11]

To date, tens of rate constants of peroxynitrite reactions have been determined.[8] Overall they show that the preferential biotargets of peroxynitrite will be metal centers, sulfur and selenium compounds, and carbon dioxide. Concerning metal centers relevant to this article, it has been shown that some metal centers can be oxidized rapidly by peroxynitrite with rate constants exceeding $10^5 M^{-1}$ sec$^{-1}$.[12,13] Peroxynitrite will also react with superoxide dismutase mimetics such as Mn-porphyrins with rate constants ranging from $10^5$ to $10^7 M^{-1}$ sec$^{-1}$.[14,15] Because manganese-containing molecules have an important role in detoxifying $O_2^{·-}$, their reaction with peroxynitrite will undoubtedly affect the outcome and are described in detail below.

Targets that do no react directly with peroxynitrite may nevertheless be modified by radicals formed from it. This is the case for tyrosine nitration, a particularly relevant reaction because it has been put forward as a footprint of peroxynitrite formation *in vivo,* although it has been shown that nitration can also be caused by reactions not involving peroxynitrite.[16] Peroxynitrous acid-dependent nitration has low yields (∼6%) and is a relatively slow process, because its rate is limited by the rate of formation of ·OH and ·NO$_2$. On the other hand, direct reactions of peroxynitrite with target molecules are much faster. Thus, kinetic considerations suggest that in biological systems, with high concentrations of fast-reacting targets, less than 1% of peroxynitrite will be able to homolize to the free radicals. Thus, in those cases in which nitration has been detected *in vivo* and pharmacological evidence confirms that it is mediated by peroxynitrite, a different mechanism needs to be proposed in order to explain nitration.

Indeed, it can be generalized that a Lewis acid (LA) is needed for nitration to occur, and CO$_2$ or certain metal centers can account for peroxynitrite-mediated nitration at biologically relevant rates. The reaction of ONOO$^-$ with the LA proceeds to the Lewis adduct, which in turn homolyzes to yield ·NO$_2$ and the

---

[11] R. Radi, *Methods Enzymol.* **269,** 354 (1996).
[12] L. Castro, M. Rodriguez, and R. Radi, *J. Biol. Chem.* **269,** 29409 (1994); R. Floris, S. R. Piersma, G. Yang, P. Jones, and R. Wever, *Eur. J. Biochem.* **215,** 767 (1993); J. P. Crow, J. S. Beckman, and J. M. McCord, *Biochemistry* **34,** 3544 (1995).
[13] C. Quijano, D. Hernandez-Saavedra, L. Castro, J. M. McCord, B. A. Freeman, and R. Radi, *J. Biol. Chem.* **276,** 11631 (2001).
[14] J. Lee, J. A. Hunt, and J. T. Groves, *J. Am. Chem. Soc.* **120,** 6053 (1998).
[15] G. Ferrer-Sueta, I. Batinic-Haberle, I. Spasojevic, I. Fridovich, and R. Radi, *Chem. Res. Toxicol.* **12,** 442 (1999).
[16] J. P. Eiserich, M. Hristova, C. E. Cross, A. D. Jones, B. A. Freeman, B. Halliwell, and A. van der Vliet, *Nature (London)* **391,** 393 (1998); M. R. Gunther, L. C. Hsi, J. F. Curtis, J. K. Gierse, L. J. Marnett, T. E. Eling, and R. P. Mason, *J. Biol. Chem.* **272,** 17086 (1997).

corresponding oxy radical:

$$ONOO^- + LA \rightarrow ONOO-LA^- \rightarrow \cdot NO_2 + \cdot O-LA^-(O=LA^{\cdot-}) \quad (2)$$

For example,

$$ONOO^- + H^+ \rightarrow ONOO-H \rightarrow \cdot NO_2 + \cdot O-H \quad (3)$$

$$ONOO^- + CO_2 \rightarrow ONOO-CO_2^- \rightarrow \cdot NO_2 + \cdot O-CO_2^- \quad (4)$$

$$ONOO^- + Mn^{III}P \rightarrow ONOO-Mn^{III}P \rightarrow \cdot NO_2 + \cdot O-Mn^{III}P$$

$$\rightarrow \cdot NO_2 + O=Mn^{IV}P \quad (5)$$

The overall reaction can be regarded as an oxene ($O^{\cdot-}$) transfer from peroxynitrite to the LA. In some cases the oxy radical rearranges to yield the corresponding radical of the oxo compound ($O=LA^{\cdot-}$) via oxidation of the LA as in Eq. (5). The rate-limiting step in the overall reaction and the yield of radicals diffusing out of the solvent cage depend on the LA involved. For instance, for $H^+$ homolysis is rate limiting and the radical yield is $\sim$30%, whereas for many manganese complexes adduct formation is the slow step and the radical yield is close to 100%.

*Redox Properties of Manganese*

The redox chemistry of manganese is rich and varied; being at the center of the $d$ block it can either gain or lose electrons with comparative ease. Oxidation states have been reported ranging from $-3$ to $+7$ for inorganic and organometallic compounds.[17] In biology, manganese-containing enzymes generally employ $Mn^{II}$ ion in hydrolases such as arginase and probably serine/threonine protein phosphatase 1, or the $Mn^{II}/Mn^{III}$ couple in redox enzymes such as Mn-catalase and MnSOD.[18] The appearance of a $Mn^{IV}$ intermediate has been reported for the oxygen-evolving complex of photosystem II.[19] Both $Mn^{2+}$ and $Mn^{3+}$ ions are "hard" Lewis acids, that is, they show preference for interaction with hard Lewis bases, that is, highly charged and rather small nucleophiles such as carboxylates and other oxyanions. The redox potentials of manganese are dependent on the ligands that surround the metal. For instance, aqueous $Mn^{II}/Mn^{III}$ couple has a $E^{\circ\prime}$ of 1.51 V, whereas $E^{\circ\prime}$ for the same couple in MnSOD is only 0.31 $V^{20}$ (see Scheme 1[20,21,21a,21b]).

---

[17] N. N. Greenwood and A. Earnshow, "Chemistry of the Elements," 2nd Ed. Butterworth-Heinemann, Woburn, Massachusetts, 1997.

[18] D. W. Christianson, *Prog. Biophys. Mol. Biol.* **67,** 217 (1997).

[19] C. F. Yocum and V. L. Pecoraro, *Curr. Opin. Chem. Biol.* **3,** 182 (1999).

[20] W. H. Koppenol, *Bioelectrochem. Bioenerget.* **18,** 3 (1987).

[21] I. Batinic-Haberle, I. Spasojevic, P. Hambright, L. Benov, A. L. Crumbliss, and I. Fridovich, *Inorg. Chem.* **38,** 4011 (1999).

[21a] W. H. Koppenol, J. J. Moreno, W. A. Pryor, H. Ischiropoulos, and J. S. Beckman, *Chem. Res. Toxicol.* **5,** 834 (1992).

[21b] N. Jin, J. L. Bourassa, S. C. Tizio, and J. T. Groves, *Angew. Chem. Int. Ed.* **39,** 3849 (2000); F. C. Chen, S. H. Cheng, C. H. Yu, M. H. Liu, and Y. O. Su, *J. Electroanal. Chem.* **474,** 52 (1999).

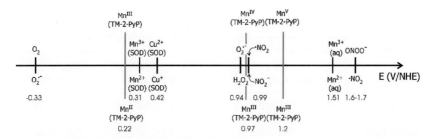

SCHEME 1. Redox potentials at pH 7 of some manganese ions and reactive oxygen and nitrogen species. In the scheme, oxidants are located above the line and ordered according to oxidizing strength from left to right; reductants are below the line, with reducing strength increasing from right to left. In this kind of scheme, thermodynamically favored reactions are those of an oxidant with any reductant placed at its left. Redox potentials are obtained for SOD, $O_2{}^{·-}$, and $Mn^{3+}$(aq) [Koppenol, *Bioelectrochem. Bioenerget.* **18**, 3 (1987)]; ONOO⁻ and ·NO$_2$ [Koppenol *et al.*, *Chem. Res. Toxicol.* **5**, 834 (1992)]; and MnTM-2-PyP [Batinic-Haberle *et al.*, *Inorg. Chem.* **38**, 4011 (1999); Jin *et al.*, *Angew. Chem. Int. Ed.* **39**, 3849 (2000); Chen *et al.*, *J. Electroanal. Chem.* **474**, 52 (1999)].

## Mn-Porphyrins and Peroxynitrite

Porphyrin ligands have been designed and synthesized to alter the redox potential of manganese complexes and thus enhance their SOD activity[21] (and see [22] in this volume [22]). This approach has proved successful to the point of obtaining functional mimics with $k_{cat}$ values of $5 \times 10^7 M^{-1}$ sec$^{-1}$ [23] and even higher at the expense of decreased complex stability.[24]

In these complexes the rate-limiting step of the catalytic cycle is the reduction of Mn$^{III}$ by $O_2{}^{·-}$. Thus, even with slow catalysts, the $O_2{}^{·-}$ can be scavenged rapidly if the reduction is performed by reductants such as NADPH or glutathione, which can keep manganese in oxidation state II. In this respect some complexes with modest SOD activity have been proposed to act as "reductant : $O_2{}^{·-}$ oxidoreductases."[25]

SOD activity, which involves Mn$^{III}$/Mn$^{II}$ transitions, and reactivity toward peroxynitrite, involving the initial oxidation of Mn$^{III}$ to Mn$^{IV}$ [Eq. (5)], follow a similar trend in rate constants (see Table I). Nevertheless, the oxidation of Mn$^{III}$-porphyrins with ONOO⁻ [Eq. (5)] is not catalytic as the O=Mn$^{IV}$ complex remains oxidized unless a reductant is present. Mn-porphyrins can be regarded as catalytic scavengers of peroxynitrite only if the O=Mn$^{IV}$ to Mn$^{III}$ reduction is performed efficiently by some other molecule. This scavenging is not complete because the products of this reaction are powerful oxidants. Both Mn$^{IV}$ complexes and ·NO$_2$ have redox potentials of about 1 V/NHE (normal hydrogen electrode).

---

[22] I. Batinic-Haberle, *Methods Enzymol.* **349**, [22], 2002 (this volume); I. Spasojevic, and I. Batinic-Haberle, *Inorg. Chim. Acta* **317**, 230 (2001).
[23] I. Batinic-Haberle, L. Benov, I. Spasojevic, and I. Fridovich, *J. Biol. Chem.* **273**, 24521 (1998).
[24] I. Batinic-Haberle, S. I. Liochev, I. Spasojevic, and I. Fridovich, *Arch. Biochem. Biophys.* **343**, 225 (1997); R. Kachadourian, I. Batinic-Haberle, and I. Fridovich, *Inorg. Chem.* **38**, 391 (1999).
[25] K. M. Faulkner, S. I. Liochev, and I. Fridovich, *J. Biol. Chem.* **269**, 23471 (1994).

TABLE I
RATES OF Mn$^{III}$-PORPHYRIN REACTIONS WITH PEROXYNITRITE AND SUPEROXIDE
AND ITS RELATION TO REDOX POTENTIALS OF Mn$^{II}$/Mn$^{III}$ PAIR

| Complex[a] | $k_{ONOO^-}$ (pH 7.4, $M^{-1}$ sec$^{-1}$) | $k_{cat}$ SOD ($M^{-1}$ sec$^{-1}$)[b] | $E_{1/2}$(Mn$^{II}$/Mn$^{III}$) V/NHE[b] |
|---|---|---|---|
| MnTM-2-PyP | $1.85 \times 10^{7c}$ | $6.2 \times 10^7$ | 0.228 |
| MnTE-2-PyP | $3.06 \times 10^{7d}$ | $5 \times 10^7$ | 0.22 |
| MnTM-3-PyP | $3.82 \times 10^{6c}$ | $4.1 \times 10^6$ | 0.052 |
| MnTM-4-PyP | $4.33 \times 10^{6c}$ | $3.8 \times 10^6$ | 0.06 |
| MnTSPP | $2.1 \times 10^{5d}$ | ND[e] | −0.16 |
| MnTCPP | $1.0 \times 10^{5c}$ | $3.2 \times 10^4$ | −0.194 |

[a] TM(E)-2(3,4)-PyP, 5,10,15,20-tetrakis[N-methyl(ethyl)pyridinium-2(3,4)-yl]porphyrin; TSPP, 5,10,15,20-tetrakis(4-sulfonatophenyl)porphyrin; TCPP, 5,10,15,20-tetrakis(4-carboxylatophenyl)porphyrin.
[b] Batinic-Haberle et al., Inorg. Chem. **38**, 4011 (1999).
[c] Ferrer-Sueta et al., Chem. Res. Toxicol. **12**, 442 (1999).
[d] Unpublished results.
[e] ND, No data available.

In some systems (e.g., DNA oxidation and nitration reactions) peroxynitrite decomposition can be accompanied by an increase in oxidation.[26,27]

The reactions with Mn$^{III}$-porphyrins are among the fastest of peroxynitrite chemistry. With second-order rate constants above $10^7 M^{-1}$ sec$^{-1}$, Mn$^{III}$TM-2-PyP and Mn$^{III}$TE-2-PyP are unique in that they can compete, at low micromolar concentrations, with physiological levels of $CO_2$ for reaction with peroxynitrite (see Fig. 3) and may scavenge carbonate radical ($CO_2{\cdot}^-$) as well.[15]

However, catalytic scavenging of peroxynitrite (represented by reactions a and b in Scheme 2) and $O_2{\cdot}^-$ dismutation (reaction e in Scheme 2) are not the only possible interactions of Mn-porphyrins with reactive nitrogen and oxygen species. It has been shown that Mn$^{II}$TE-2-PyP forms an addition complex with ·NO,[28] the reactivity and biological relevance of which remain to be investigated. Finally, it is known that Mn-porphyrins can catalyze the autooxidation of reductants such as ascorbate and glutathione (reactions d and c in Scheme 2).[29] This usually happens at high oxygen concentrations such as those used for in vitro experiments and may lead to waste of reducing equivalents and to radical formation.

One possibility to avoid enhanced oxidation and nitration by Mn$^{IV}$ and ·NO$_2$ is to reduce peroxynitrite by two electrons with Mn$^{II}$ complexes to yield O=Mn$^{IV}$ and NO$_2^-$ (reaction f in Scheme 2). O=Mn$^{IV}$ can in turn be reduced by intracellular

[26] J. T. Groves and S. S. Marla, J. Am. Chem. Soc. **117**, 9578 (1995).
[27] G. Ferrer-Sueta, L. Ruiz-Ramirez, and R. Radi, Chem. Res. Toxicol. **10**, 1338 (1997).
[28] I. Spasojevic, I. Batinic-Haberle, and I. Fridovich, Nitric Oxide **4**, 526 (2000).
[29] P. R. Gardner, D. D. Nguyen, and C. W. White, Arch. Biochem. Biophys. **325**, 20 (1996).

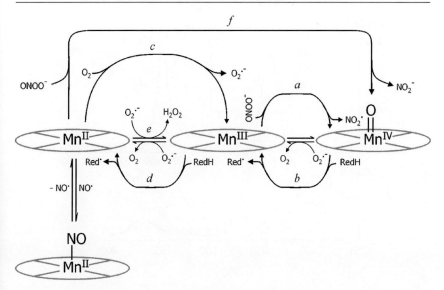

SCHEME 2. Some of the possible reactions of Mn-porphyrins with reactive nitrogen and oxygen species and reductants.

reductants to complete the cycle. Our laboratory is currently exploring this possibility.

## Experimental Approach and Results

Mn-porphyrin reactions are conveniently studied by ultraviolet/visible (UV/vis) spectrophotometry (Fig. 1). Each oxidation state displays unique spectral properties and redox reactions can be followed at different wavelengths. In reactions in which only two oxidation states are involved, the maxima of either reactant or product are usually the wavelengths of choice. In reactions with more than two species involved, several wavelengths must be explored and isosbestic points are particularly informative about the rate of the individual redox reactions that take place. Modern stopped-flow instrumentation has allowed the identification and spectral characterization of transient species either by diode array or rapid-scanning spectrophotometry. Spectroelectrochemical techniques have also been used to characterize the spectra and redox behavior of species too reactive to remain in solution.

Oxidation of $Mn^{III}$-porphyrins with peroxynitrite was studied under pseudo-first-order conditions with peroxynitrite in excess over the Mn-porphyrin in a stopped-flow spectrophotometer (SF.17MV; Applied Photophysics, Leatherhead, England). In all cases, peroxynitrite (dissolved in a known solution of NaOH) was mixed with the Mn-porphyrin dissolved in an acidic phosphate buffer from

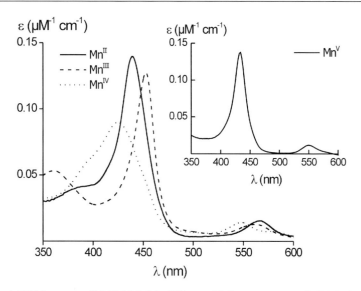

FIG. 1. UV/vis spectra of MnTM-2-PyP in different oxidation states at neutral pH, Maxima of the Soret bands ($\varepsilon$): Mn$^{II}$, 439 nm ($1.4 \times 10^5 M^{-1}$ cm$^{-1}$); Mn$^{III}$, 453.4 nm ($1.29 \times 10^5 M^{-1}$ cm$^{-1}$); Mn$^{IV}$, 425 nm ($9 \times 10^4 M^{-1}$ cm$^{-1}$); Mn$^{V}$ at pH 14, 433 nm ($1.38 \times 10^5 M^{-1}$ cm$^{-1}$).

which $CO_2$ was eliminated by boiling and purging with argon. The Mn-porphyrin concentration was between 1 and 2 $\mu M$ and peroxynitrite was in at least 15-fold excess. In all experiments the final phosphate concentration was 50 m$M$, $T = 37 \pm 0.1°$, and pH was measured at the outlet of the stopped-flow apparatus. Conditions were set so that the observed rate constant of Mn-porphyrin oxidation was always more than 10 times larger than the known rate constant of the proton-catalyzed decay of peroxynitrite at that pH (Fig. 2). The second-order rate constants were also determined under pseudo-first-order conditions with Mn-porphyrin (3–18 $\mu M$) in 10-fold excess over peroxynitrite at pH 7.4, and essentially identical rate constants were obtained.

For the oxidation of Mn$^{III}$-porphyrins with peroxynitrite in the presence of $CO_2$, the complexes were dissolved in buffer containing a known concentration of total carbonate,[29b] this solution was mixed in the stopped-flow apparatus with ONOO$^-$ dissolved in dilute NaOH. Initial concentrations of both ONOO$^-$ and Mn-porphyrins were 5 $\mu M$ and the initial $CO_2$ concentration ranged from 0 to 2 m$M$ (Fig. 3).

Reduction of O=Mn$^{IV}$-porphyrins (1 to 7 $\mu M$) is carried out under pseudo-first-order conditions with the reductants in excess over the complex. Mn$^{IV}$-porphyrins

[29b] R. Radi, A. Denicola, and B. A. Freeman, *Methods Enzymol.* **301**, 353 (1999).

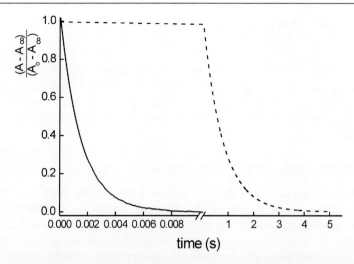

FIG. 2. Time course of the reaction of MnTM-4-PyP (1.9 $\mu M$) with peroxynitrite (150 $\mu M$), which was mixed with MnTM-4-PyP at 37°, pH 7.4. The reaction was monitored at 462.2 nm (solid line, $Mn^{III}$ oxidation) and 302 nm (dotted line, peroxynitrite decay [Radi, *Methods Enzymol.* **269**, 354 (1996)]).

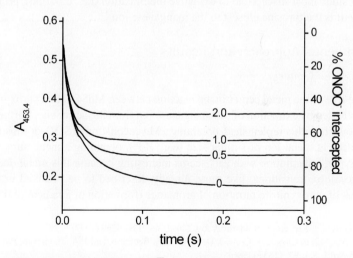

FIG. 3. Kinetic competition between $Mn^{III}$TM-2-PyP and $CO_2$ for peroxynitrite; $ONOO^-$ and $Mn^{III}$TM-2-PyP (5 $\mu M$ each) were mixed in the stopped-flow apparatus with $CO_2$ present. The millimolar concentrations of $CO_2$ are indicated by the numbers, pH 7.4, 37°. Absorbance at 453.4 nm serves to assess the fraction of peroxynitrite reduced by $Mn^{III}$. This is shown on the right y axis.

were prepared immediately before the stopped-flow experiment by mixing with approximately 0.9 equivalents of potassium peroxymonosulfate. Under these conditions all peroxymonosulfate is consumed to yield initially $Mn^V$, which decays rapidly to $Mn^{IV}$.[30] This $Mn^{IV}$-porphyrin solution decays with $t_{1/2} > 10$ min. In the case of MnTM-2-PyP and MnTE-2-PyP, where the $Mn^V$ species is unusually stable, an excess (2- or 3-fold) of $NO_2^-$ reduces all $Mn^V$ to $Mn^{IV}$ rapidly but can also reduce $Mn^{IV}$ to $Mn^{III}$ more slowly, so that the study must be done immediately after $NO_2^-$ is added.

### MnSOD Reactions with Peroxynitrite

Peroxynitrite reacts with MnSOD *in vitro*, leading to the nitration of Tyr-34 and thus to enzyme inactivation.[4,13,31–33] Nitration and inactivation of MnSOD has been reported to occur in chronic rejection of human renal allografts,[31] strongly supporting the relevance of this reaction *in vivo*. MnSOD along with prostacyclin synthase and tyrosine hydroxylase are representative enzymes reported to be inactivated solely by tyrosine nitration *in vivo*.[7,34]

Metalloproteins are kinetically preferential targets of peroxynitrite, with the metal center prevailing over amino acids.[35] Thus, it is probable that amino acid modifications in metalloproteins exposed to peroxynitrite, arise from their reaction between secondary species formed by the reaction of peroxynitrite with the metal center. This is what occurs in peroxynitrite reaction with MnSOD, where Tyr-34 is the residue most susceptible to oxidative modification (i.e., nitration), probably because it is the tyrosine closest to the manganese ion.[13]

### Experimental Approach and Results

*Role of Metal Center*

The role of the metal center in the reaction between MnSOD and peroxynitrite was evaluated by extraction of the manganese ion from the active site. Further, the $Mn^{III}$ could be either replenished, obtaining a Mn-reconstituted SOD, or substituted by a different metal ion of similar size (e.g., $Zn^{II}$), obtaining a metal-substituted apoenzyme. Substitution with $Zn^{II}$ seems interesting because this metal does not undergo redox transitions, like those already described to be involved in metal catalysis of peroxynitrite nitration of aromatics (formation of a radical of the oxo

---

[30] J. T. Groves, J. Lee, and S. S. Marla, *J. Am. Chem. Soc.* **119**, 6269 (1997).
[31] L. A. MacMillan-Crow, J. P. Crow, J. D. Kerby, J. S. Beckman, and J. A. Thompson, *Proc. Natl. Acad. Sci. U.S.A.* **93**, 11853 (1996).
[32] F. Yamakura, H. Taka, T. Fujimura, and K. Murayama, *J. Biol. Chem.* **273**, 14085 (1998).
[33] L. A. MacMillan-Crow, J. P. Crow, and J. A. Thompson, *Biochemistry* **37**, 1613 (1998).
[34] M. Zou, A. Yesilkaya, and V. Ullrich, *Drug. Metab. Rev.* **31**, 343 (1999).
[35] B. Alvarez, G. Ferrer-Sueta, B. A. Freeman, and R. Radi, *J. Biol. Chem.* **274**, 842 (1999).

compound, O=LA·⁻). The metal extraction and replacement can be performed in *Escherichia coli* MnSOD as previously described.[36] Nevertheless, for human MnSOD the procedure needs to be performed in the presence of 0.5 $M$ sucrose, because otherwise the enzyme largely precipitates.

Metal content after these procedures was determined by atomic absorption with a graphite furnace atomic absorption spectrometer (Spectra 20; Varian Instruments, Victoria, Australia), and SOD activity was measured as described previously.[37] The manganese content of the enzyme was proportional to both enzyme activity and reactivity with peroxynitrite, as is shown below.

*Kinetic Studies*

In studies of the peroxynitrite reaction with MnSOD it is not possible to achieve pseudo-first-order conditions with excess target over peroxynitrite, which are usually of choice when studying peroxynitrite kinetics.[11] This is due to the limited availability of MnSOD and the fact that peroxynitrite concentrations should not be much less than 0.1 m$M$ in order to detect a variation in absorbance of at least 0.1 AU. So the kinetics of this reaction can be studied by an initial rate approach, with enzyme concentrations similar or lower to those of peroxynitrite.[35]

Because the absorbance at 302 nm is proportional to peroxynitrite concentration and the change in MnSOD absorbance is negligible at this wavelength, the initial rate of peroxynitrite decomposition follows Eq. (6):

$$\frac{d[ONOO^-]_0}{dt} = -[ONOO^-]_0 \frac{dA/dt}{(A_0 - A_f)} = k_{obs}[ONOO^-]_0$$

$$= k[MnSOD]_0[ONOO^-]_0 \quad (6)$$

$dA/dt$ is the slope of the initial change in absorbance at 302 nm (first 0.1–0.2 sec, ~10–20% of peroxynitrite decay), and $A_0$ and $A_f$ are the initial and final absorbance measurements. After the initial rate assumption that the concentration of reagents remains constant during this period, the observed rate constant could be determined as[35] shown in Eq. (7):

$$k_{obs} = \frac{-dA/dt}{(A_0 - A_f)} \quad (7)$$

To assure the accuracy of the rate constant determinations, 200 absorbance measurements were acquired during the initial part of the reaction (0.1–0.2 sec) and 200 further points were acquired until more than 99.9% of peroxynitrite had decomposed.

---

[36] D. E. Ose and I. Fridovich, *J. Biol. Chem.* **251,** 1217 (1976); W. F. Beyer, J. A. J. Reynolds, and I. Fridovich, *Biochemistry* **28,** 4403 (1989).

[37] L. Flohé and F. Otting, *Methods Enzymol.* **105,** 93 (1987).

By performing this experiment at various MnSOD concentrations a second-order rate constant ($k$) for the reaction between MnSOD and peroxynitrite can be determined from the slope of the plot of $k_{obs}$ versus [MnSOD]$_0$ (Fig. 4A, inset). The experiments performed with $E.\ coli$ MnSOD, holoenzyme and apoenzyme, resulted in $k$ values of $1.4 \times 10^5$ and $<10^4\ M^{-1}\ sec^{-1}$, respectively, confirming that the manganese ion is a preferential target in the peroxynitrite reaction with MnSOD.

Concentrations of MnSOD were lower than those of peroxynitrite, and as observed in Fig. 4A the kinetic traces did not follow a single exponential function. The decomposition of peroxynitrite in the presence of MnSOD was initially fast but it declined with time, suggesting that the enzyme was being consumed.

Thus, simulations of the kinetics of the reaction between MnSOD and peroxynitrite were performed, using the simulation program Gepasi.[38] In this simulation we considered the proton-catalyzed decomposition of peroxynitrite [Eq. (8)[8]] and a series of bimolecular reactions of peroxynitrite with MnSOD that ultimately led to a modified form of the enzyme (MnSOD$_{n+1ox}$) that can no longer react with peroxynitrite.

$$ONOO^- \rightarrow NO_3^- \qquad k_{H^+} = 1.2\ sec^{-1} \qquad (8)$$
$$ONOO^- + MnSOD \rightarrow MnSOD_{1ox} \qquad k_{1ox} = 1.4 \times 10^5 M^{-1}\ sec^{-1} \qquad (9)$$
$$ONOO^- + MnSOD_{1ox} \rightarrow MnSOD_{2ox} \qquad k_{2ox} = 1.4 \times 10^5 M^{-1}\ sec^{-1} \qquad (10)$$
$$\vdots$$
$$ONOO^- + MnSOD_{nox} \rightarrow MnSOD_{n+1ox} \qquad k_{n+1ox} = 1.4 \times 10^5 M^{-1}\ sec^{-1} \qquad (11)$$

The simulation indicated that the experimental data fit to a model in which $n + 1 = 5$ (Fig. 4B). Because the first oxidative modification promoted by peroxynitrite on MnSOD is the nitration of Tyr-34,[13,32] which in turn leads to enzyme inactivation, it can be suggested that, on average, five molecules of peroxynitrite are required for nitration of the critical tyrosine, which represents a 20% nitration yield (see also below).

*Nitration of Self and Low Molecular Weight Aromatics*

The reactions of peroxynitrite with transition metal centers typically lead to the formation of nitrating species and therefore we evaluated whether the manganese ion in MnSOD participated in the formation of nitrating species.

MnSOD holoenzyme, apoenzyme (Apo), manganese-reconstituted SOD (Apo/Mn), and zinc-substituted apoenzyme (Apo/Zn) were exposed to peroxynitrite and tyrosine nitration was assessed by Western blot, using a highly specific anti-nitrotyrosine antibody produced in our laboratory[6] (Fig. 5A). Quantitation of nitrated tyrosine residues was achieved by densitometry of the Western blots and

---

[38] P. Mendes, *Comput. Appl. Biosci.* **9**, 563 (1993); P. Mendes, *Trends Biochem. Sci.* **22**, 361 (1997).

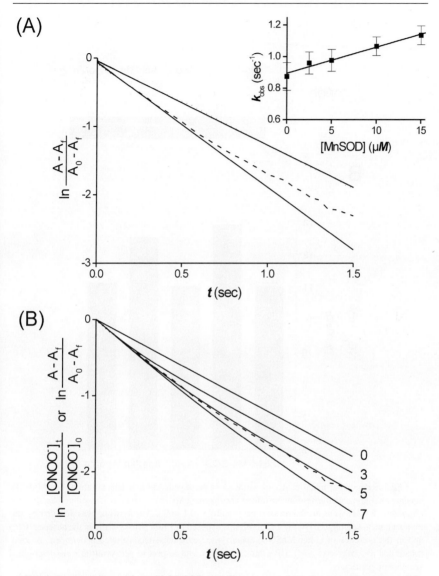

FIG. 4. Peroxynitrite reaction with MnSOD. (A) Time course of the decomposition peroxynitrite (0.1 m$M$) in the absence (dotted line) and in the presence (dashed line) of MnSOD (5 $\mu M$) at 37° and pH 7.5. The reaction was monitored by measuring the absorbance ($A$) at 302 nm and the linear fit of the data (solid lines) was performed in the first 100 msec. *Inset:* Second-order plot for the reaction of peroxynitrite (0.2 m$M$) with MnSOD. (B) The solid lines represent simulations of peroxynitrite decay in the absence (0) and in the presence (3–7) of MnSOD [same conditions as in (A)]; the numbers to the right represent $n + 1$. The reaction was monitored as $[ONOO^-]/[ONOO^-]_0$ versus $t$. The dashed line represents the experimental data of peroxynitrite decay in the presence of MnSOD [same as (A)].

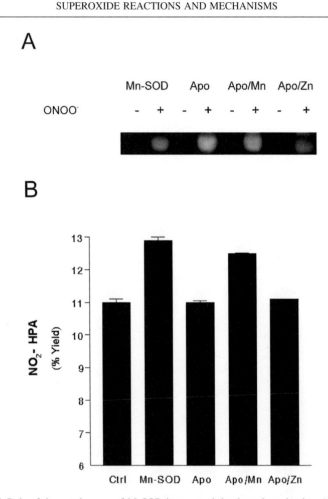

FIG. 5. Role of the metal center of MnSOD in peroxynitrite-dependent nitration. (A) MnSOD, apoenzyme (Apo), manganese-reconstituted apoenzyme (Apo/Mn), and zinc-substituted apoenzyme (Apo/Zn) (5 $\mu M$) were incubated with peroxynitrite (0.1 m$M$). Nitrotyrosine was detected by immunochemical techniques. (B) HPA (5 m$M$) was exposed to 1 m$M$ peroxynitrite in the absence (Ctrl) and in the presence of either MnSOD, apoenzyme, manganese-reconstituted apoenzyme, or zinc-substituted apoenzyme (5 $\mu M$). HPA nitration yields with respect to peroxynitrite were determined spectophotometrically.

electrospray mass spectrometry. The results show more nitration in the manganese-reconstituted enzyme than in the zinc-substituted enzyme, suggesting that the metal center would be involved in protein-tyrosine nitration. Because zinc does not promote nitration[8,13] the formation of nitrotyrosine observed in the zinc-containing enzyme is due to the reactions of $\cdot$OH and $\cdot$NO$_2$ radicals arising from the proton-catalyzed decomposition of peroxynitrite, which is in large excess over SOD. The greater nitration observed in the apoenzyme by comparison to native MnSOD is consistent with a larger surface exposure of tyrosines in the former. In the case of MnSOD, the nitrating species formed from the reaction of peroxynitrite preferentially reacts with the adjacent Tyr-34 that is located only 5 Å away from the active site, resulting in "site-specific" nitration.

To further confirm the formation of secondary nitrating species by MnSOD, the nitration of a water-soluble analog of tyrosine, hydroxyphenylacetic acid (HPA), was assessed. HPA was exposed to peroxynitrite in the absence and presence of MnSOD, apoenzyme, manganese-reconstituted apoenzyme, and zinc-substituted apoenzyme. After the reaction had taken place, 4-nitro-HPA was determined spectrophotometrically at 430 nm[39] and by high-performance liquid chromatography (HPLC)-based techniques.[13,40,41] MnSOD and manganese-reconstituted SOD increased the yield of HPA nitration by peroxynitrite, whereas the apoenzyme and the zinc-substituted apoenzyme did not (Fig. 5B), confirming that the manganese metal center reacts with peroxynitrite, leading to the formation of nitrating species.

## Concluding Remarks

This article underscores the relevance of peroxynitrite reactions with mangnese-containing SOD and SOD mimetics. The reactions are fast and lead to the formation of secondary reactive species. The interactions of peroxynitrite with MnSOD leading to enzyme nitration and inactivation may play an important role during · NO-dependent alterations of mitochondrial homeostasis. Protection of Tyr-34 from nitration appears to be a critical issue to maintain SOD activity during periods of excess mitochondrial reactive nitrogen species formation. On the other hand, our increasing knowledge of the reactions of peroxynitrite with transition metal centers allows for the development and evaluation of compounds such as the Mn-porphyrins, which can operate as "peroxynitrite reductases." Although more research is needed to study and define the predominant redox mechanisms by which the Mn-porphyrins exert biological effects, they represent a class of compounds of potential use for attenuating the toxic effects of peroxynitrite *in vivo*.

## Acknowledgments

This work was supported by the International Centre for Genetic Engineering and Biotechnology (Trieste, Italy), the Swedish Agency for Research Cooperation (Sweden), the Howard Hughes Medical Institute, the Programa de Desarrollo de Ciencias Básicas (Uruguay), Comisión Sectorial de Investigación Científica (Universidad de la República, Uruguay), and the Third World Academy of Sciences. R.R. is an International Research Scholar of the Howard Hughes Medical Institute.

[39] J. S. Beckman, H. Ischiropoulos, L. Zhu, M. van der Woerd, C. Smith, J. Chen, J. Harrison, J. C. Martin, and M. Tsai, *Arch. Biochem. Biophys.* **298,** 438 (1992).
[40] J. P. Crow and H. Ischiropoulos, *Methods Enzymol.* **269,** 185 (1996).
[41] J. P. Crow, *Methods Enzymol.* **301,** 151 (1999).

## [4] Superoxide Dismutase Kinetics

By MATTIA FALCONI, PETER O'NEILL, MARIA ELENA STROPPOLO, and ALESSANDRO DESIDERI

### Introduction

Cu,Zn superoxide dismutases (Cu,ZnSODs) constitute a class of ubiquitous metalloenzymes that catalyze the dismutation of the superoxide radical anion into oxygen and hydrogen peroxide[1,2] according to the schematic reaction

$$2O_2^{\cdot -} + 2H^+ \longrightarrow H_2O_2 + O_2$$

The dismutation reaction involves successive encounters of two distinct superoxide anions with the enzyme catalytic center [in the Cu(II) and Cu(I) oxidation states, respectively], which is hosted at the surface of the enzyme $\beta$ barrel at the dead end of a narrow protein channel. The reaction is diffusion limited and the substrate is highly unstable, two characteristics that make difficult the use of direct assay to determine its activity.

In this article we describe three methods to determine Cu,ZnSOD activity: the main, direct experimental assay based on generation of the superoxide anion by water pulse radiolysis; a fast, indirect method based on pyrogalol autoxidation; and a computational method based on the calculation of the enzyme–substrate association rate.

### Direct Experimental Method

*Determination of Activity of Superoxide Dismutases by Pulse Radiolysis*

The technique of pulse radiolysis has proved effective in determining the activity (rate constants) for the reaction of superoxide dismutase with its substrate, superoxide ($O_2^{\cdot -}$).[3] Pulse radiolysis provides a clean method of producing $O_2^{\cdot -}$ in aqueous solution at various pH values and ionic strengths and allows direct observation of $O_2^{\cdot -}$ through its well-characterized optical absorption centered at 250 nm. Because the technique of pulse radiolysis is time resolved, reactions may

---

[1] J. V. Bannister, W. H. Bannister, and G. Rotilio, *CRC Crit. Rev. Biochem.* **22**, 111 (1987).
[2] D. Bordo, A. Pesce, M. Bolognesi, M. E. Stroppolo, M. Falconi, and A. Desideri, in "Handbook of Metalloproteins" (K. Wieghardt, R. Huber, T. Poulos, and A. Messerschmidt, eds.). John Wiley & Sons, London, 2001.
[3] E. M. Fielden, P. B. Roberts, R. Bray, D. Lowe, G. Mautner, G. Rotilio, and L. Calabrese, *Biochem. J.* **139**, 49 (1974).

## TABLE I
### CATALYTIC RATES ($k_{cat}/K_m$) FOR NINE DIFFERENT Cu,Zn SUPEROXIDE DISMUTASES

| | $k_{cat}/K_m (\times 10^{-10} M^{-1} sec^{-1})$ |
|---|---|
| Ox | $0.39^a$ |
| Human | $0.25^a$ |
| Yeast | $0.33^a$ |
| Sheep | $0.32^a$ |
| Spinach | $0.39^a$ |
| Shark | $0.39^a$ |
| E. coli | $0.20^b$ |
| P. leiognathi | $0.85^b$ |
| S. typhimurium | $1.30^b$ |

[a] Catalytic rate constants determined at pH 8 and $\mu = 0.02\ M$.
[b] Catalytic rate constants determined at pH 7.5 and $\mu = 0.02\ M$.

be followed on the time scale of 1 $\mu$s to several seconds, ideal for monitoring the reaction of SOD with $O_2^{\cdot-}$. This method has been used to determine the activity of a variety of SODs, a representative selection of which are presented in Table I. The "ping–pong" mechanism for interaction of Cu,ZnSOD with $O_2^{\cdot-}$, whereby the copper center undergoes successive reduction–oxidation processes [reactions (1) and (2)], was first identified by the technique of pulse radiolysis.[4] The mechanism was established[3] by monitoring the quantitative redox cycling of the copper center through changes in the characteristic blue color of $Cu^{2+}$SOD at 610 nm.

$$O_2^{\cdot-} + Cu^{2+}SOD \longrightarrow O_2 + Cu^+SOD \tag{1}$$

$$O_2^{\cdot-} + Cu^{2+}SOD \xrightarrow{+2H^+} O_2 + Cu^{2+}SOD \tag{2}$$

*Principle of Using Pulse Radiolysis to Generate $O_2^{\cdot-}$*

Irradiation of buffered, aqueous solution at pH 6–12 with high-energy electrons, as produced by an accelerator used for pulse radiolysis, produces the following water radicals: hydroxyl radical ($\cdot$OH), hydrated electrons ($e_{aq}^-$), and hydrogen atoms ($\cdot$H).

$$H_2O \longrightarrow\!\!\!\!\!\sim\!\!\!\!\!\longrightarrow \cdot OH,\ e_{aq}^-,\ \cdot H$$

---

[4] J. Butler and E. J. Land, in "Free Radicals: A Pratical Approach" (N. A. Punchard and F. J. Kelly, eds.), p. 47. Oxford University Press, Oxford, 1996.

The majority of the energy is deposited in the water. If the solution is oxygenated and the buffer does not react efficiently with $e_{aq}^-$ or ˙H, then $e_{aq}^-$ and ˙H are converted into $O_2^{˙-}$ by reaction with oxygen.

$$e_{aq}^- + O_2 \longrightarrow O_2^{˙-} \tag{3}$$

$$˙H + O_2 \longrightarrow HO_2^{˙-} \tag{4}$$

The $pK_a$ of $HO_2^{˙-}$ is[5] 4.8, so that at the pH values generally used to determine the activity of SOD, the major species is $O_2^{˙-}$.

To convert ˙OH into $O_2^{˙-}$, the aqueous solution contains a high concentration of an ˙OH scavenger, which produces $O_2^{˙-}$ on reaction with ˙OH. Ethanol (0.1 $M$) or 0.1 $M$ sodium formate, as the ˙OH scavenger, is generally included in the solution.

$$˙OH + CH_3CH_2OH \longrightarrow CH_3˙CHOH \xrightarrow{+O_2} CH_3CHOH(O_2)˙ \tag{5}$$

$$\downarrow$$

$$CH_3CHO + H^+ + O_2^{˙-} \tag{6}$$

$$HCO_2^- + ˙OH \longrightarrow CO_2^{˙-} + H_2O \tag{7}$$

$$CO_2^{˙-} + O_2 \longrightarrow CO_2 + O_2^{˙-} \tag{8}$$

Formate is a cleaner system but as it is ionic, it contributes to the ionic strength of the solution. Ethanol is therefore preferred if ionic strength dependencies are to be undertaken, although the slow formation of $O_2^{˙-}$ in reactions (5) and (6)[6] needs to be considered when assaying at pH < 7. Using these systems, the majority of the reactions that produce $O_2^{˙-}$ are complete within 100 $\mu$sec. Because pulse radiolysis generally delivers the high-energy electrons within 1–2 $\mu$sec, the reactions of $O_2^{˙-}$ with SOD may be followed readily over time. Assay of SOD is generally undertaken at pH 8–8.5.

*Assay Solutions*

Purified water obtained from a water purification system such as Milli-Q (Millipore, Bedford, MA) is used to prepare all solutions.
The following stock solutions are prepared.

Tris, 0.5 $M$
Morpholine propane sulfonic acid (MOPS), 0.5 $M$
EDTA, 0.1 $M$
NaOH, 0.5 $M$
Ethanol, 17 $M$
Sodium borate, 0.2 $M$

---

[5] B. H. J. Bielski and D. E. Cabelli, *Int. J. Radiat. Biol.* **59**, 291 (1991).
[6] C. von Sonntag, "The Chemical Basis of Radiation Biology." Taylor & Francis, Philadelphia, 1987.

The assay buffers used are as follows:

pH 7–9.5
Tris–MOPS
EDTA, 0.1 m$M$
Ethanol (or sodium formate), 0.1 $M$

pH 9.5–12
Sodium borate
EDTA, 0.1 m$M$
Ethanol, 0.1 $M$

The concentration of the buffers (Tris–MOPS or sodium borate) is adjusted so that the ionic strength of the solution is constant at 0.02. To increase the ionic strength of the solution, if required, a known concentration of sodium perchlorate is added. These assay buffers and perchlorate are used to avoid interference from binding of anions such as phosphate or chloride to the metal ion sites of SOD. EDTA is used to chelate any inadventitious metal ions present in the assay buffer to minimize unwanted interactions with $O_2{}^{\cdot-}$. The concentration of SOD dissolved in these assay buffers is generally in the range 0.1–2 $\mu M$. Before pulse irradiation and determination of the rate of reaction of $O_2{}^{\cdot-}$ with a known concentration of SOD, the solution is saturated with oxygen by bubbling 100% oxygen through the solution for ~20 min.

*Pulse Radiolysis Formation of $O_2{}^{\cdot-}$*

The principle of pulse radiolysis has been well documented.[4] A short pulse of high-energy electrons of typically 1–2 $\mu$sec in duration impinges on a quartz cell containing the solution containing SOD. The pulse of ionizing radiation produces $O_2{}^{\cdot-}$ as described above. For assay conditions, a radiation dose is used to produce >10 $\mu M$ $O_2{}^{\cdot-}$ so that the $[O_2{}^{\cdot-}] \gg [SOD]$ under turnover conditions. The optical absorbance of $O_2{}^{\cdot-}$ may be readily monitored at 250 nm, a wavelength where $O_2{}^{\cdot-}$ absorbs, by passing an analyzing light beam through the solution contained in the quartz cell of known path length. The analyzing light beam is orthogonal to the electron beam. The light beam then passes through a monochromator to select the desired wavelength and finally reaches the photodetector (e.g., photomultiplier). The detector changes the analyzing light intensity into an electrical signal, which is digitized, displayed, stored, and processed by a microprocessor. The dependence of the change in optical absorbance with time is measured in the presence and absence of SOD, as shown in Fig. 1. In the absence of SOD the lifetime of $O_2{}^{\cdot-}$ at the given pH is determined and must be compared with its lifetime in the presence of a known concentration of SOD. Changes in absorbance as low as $10^{-3}$ may

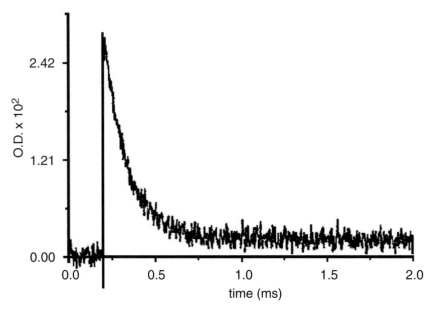

FIG. 1. Typical time dependence for the loss of optical absorbance of superoxide in the presence of 0.9 mM *Photobacterium leiognathi* Cu,Zn SOD at pH 7.0.

readily be determined with the detection systems used for pulse radiolysis. This absorbance change at 250 nm would represent a concentration of $O_2^{\cdot-}$ of 0.5 $\mu M$ if determined in a cell with a 1-cm path length.

The kinetics of the reaction of $O_2^{\cdot-}$ with SOD are determined on the basis of reactions (9)–(11).

$$\Delta A_{O_2^{\cdot-}} = \varepsilon \cdot (\text{concentration of } O_2^{\cdot-}) \cdot (\text{path length of cell}) \quad (9)$$

where $\varepsilon$ is the extinction coefficient[7] of $O_2^{\cdot-}$ at the monitoring wavelength ($\varepsilon_{250\,nm} = 2000\ M^{-1}\ cm^{-1}$). Knowing $\Delta A_{O_2^{\cdot-}}$, determined at given times after the irradiation pulse, the concentration of $O_2^{\cdot-}$ is obtained. Assuming reactions (1) and (2) occur, then the kinetics for the reaction of $O_2^{\cdot-}$ with SOD may be determined by reaction (10):

$$-\frac{d[O_2^{\cdot-}]}{dt} = k_1[E_A][O_2^{\cdot-}] + k_2[E_B][O_2^{\cdot-}] \quad (10)$$

---

[7] J. Rabani and S. O. Nielson, *J. Phys. Chem.* **73**, 3736 (1969).

where $[E_A]$ and $[E_B]$ are the concentrations of the oxidized and reduced forms of SOD. If $k_1 = k_2 = k$, then

$$-\frac{d[O_2{}^{\cdot -}]}{dt} = k[E_A + E_B][O_2{}^{\cdot -}] \qquad (11)$$

Using $[E_A] + [E_B] = [E_0]$, where $[E_0]$ is the total concentration of SOD, then

$$-\frac{d[O_2{}^{\cdot -}]}{dt} = k[E_0][O_2{}^{\cdot -}] \qquad (12)$$

Under turnover conditions in which $[O_2{}^{\cdot -}] \gg [SOD]$, a plot of $\ln[O_2{}^{\cdot -}]$ versus time ($t$) gives a straight line with a slope equal to $k[E_0]$, the rate of reaction of $O_2{}^{\cdot -}$ with SOD at a given [SOD]. By determining the dependence of the first-order rate constant at various $[E_0]$, the value of $k$, the activity of SOD, is obtained.

*Determination of $k_1$ and $k_2$*

The rate constants $k_1$ and $k_2$ may be determined as described above, except that the following condition applies: $[O_2{}^{\cdot -}] \ll [SOD]$.

## Indirect Experimental Method

*Determination of Activity of Superoxide Dismutase by Pyrogallol Method*

A simple and efficient assay to determine Cu,ZnSOD activity involves the reaction of the superoxide anion with an indicator such as pyrogallol (1,2,3-benzenetriol),[8] which can be monitored spectrophotometrically. Superoxide, produced by an *in vitro* reaction, reacts with the indicator, giving absorbance at a specific wavelength that increases with time. Addition of the enzyme, which induces superoxide disproportion, decreases the superoxide concentration itself, slowing down the reaction of superoxide with the indicator. The correlation between the Cu,ZnSOD concentration and the inhibition rates of the superoxide reaction with the indicator allows determination of enzyme activity.

Pyrogallol autoxides rapidly in alkaline aqueous solution. Its autoxidation rate is measured from the linear increase in absorbance at 420 nm for some minutes (usually from 1 to 3 min for each assay). The increment of the band is linear within 7–10 min, and then the color turns brown and green.[8] Addition of Cu,ZnSOD inhibits pyrogallol autoxidation, dismutating the superoxide in hydrogen peroxide and molecular oxygen and blocking the absorbance increase at 420 nm. At pH 7.9 the autoxidation is inhibited up to 99% by Cu,ZnSOD,[8] as the enzyme activity is

---

[8] S. Marklund and G. Marklund, *Eur. J. Biochem.* **47**, 469 (1974).

independent of pH in the range of pH 5.5–9.5 and 7.0–8.4 for the eukaryotic[9,10] and the prokaryotic[11] enzyme, respectively. The assay is usually carried out in 50 m$M$ Tris-HCl buffer (pH 8.2)–1 m$M$ DPTA (diethylenetriaminepentaacetic acid) in the presence of 0.2 m$M$ pyrogallol. Cu,ZnSOD activity is calculated in relation to the percentage of inhibition of the pyrogallol autoxidation rate.

This method is fast, inexpensive, simple, sensitive, and requires only common diagnostic tools such as a commercial spectrophotometer operating in the visible light range, and only a small amount of enzyme.

It is useful (1) to check for the presence of SOD by measuring its activity *in vitro*, from tissue extracts or cells, or after a purification step; (2) to roughly determine the concentration of SOD, knowing the activity/concentration relation of the specific SOD investigated; (3) to compare two or more SOD species, or wild type and mutants of the same enzyme, on measuring the relative activity; and (4) to follow the activity trend under various conditions that can accelerate or slow the enzyme catalytic rate.

## Computational Method

### Measurement of Superoxide Dismutase–Superoxide Association Rate through Brownian Dynamics Simulation

Cu,Zn superoxide dismutases are usually described as perfect enzymes, because their second-order catalytic rate ($k_{cat}/K_m$) is on the order of $10^9 \, M^{-1} \, \text{sec}^{-1}$. These values are at the upper range of the observed rate and are typical for reactions limited by the frequency with which enzyme and substrate encounter each other. The dismutation reaction of SOD can be described by the following scheme[12]:

$$\text{SOD} + \text{O}_2^{\cdot -} \underset{k_{-1}}{\overset{k_1}{\longleftrightarrow}} \text{O}_2^{\cdot -}\text{–SOD} \overset{k_2}{\rightarrow} \text{products}$$

where $k_1$ is the second-order association rate, $k_{-1}$ is the first-order enzyme–substrate dissociation constant, and $k_2$ is the first-order catalytic rate constant. These quantities are related in the Michaelis–Menten model of enzyme kinetics by the following equation:

$$k_f = k_2/K_m = k_1 k_2/(k_{-1} + k_2)$$

---

[9] P. O'Neill, S. Davies, L. Calabrese, C. Capo, F. Marmocchi, G. Natoli, and G. Rotilio, *Biochem. J.* **251**, 41 (1988).
[10] F. Polticelli, A. Battistoni, P. O'Neill, G. Rotilio, and A. Desideri, *Protein Sci.* **5**, 248 (1996).
[11] M. E. Stroppolo, M. Sette, P. O'Neill, F. Polizio, M. T. Cambria, and A. Desideri, *Biochemistry* **37**, 12287 (1998).
[12] Y. Wong, T. W. Clark, J. Shen, and J. A. McCammon, *Mol. Simul.* **10**, 277 (1993).

where $k_f$ is the second-order enzyme rate constant and $K_m$ is the Michaelis–Menten constant. In the case of SOD the catalytic process is diffusion limited, that is, $k_f = k_1$, because $k_2 \gg k_{-1}$.

The SOD–$O_2^{\cdot-}$ bimolecular reaction rate is increased by the electrostatic attraction between the reactants,[13] and this "electrostatically driven" enzyme–substrate recognition process has been suggested to be characteristic of all Cu,ZnSODs on the basis of the similarity of their electrostatic potential distribution[14] and because of the catalytic rate constant value.[9] A brownian dynamics (BD) simulation method[15] has confirmed that electrostatic forces play an important role in steering superoxide into the active site.[16,17]

In this section we discuss the BD simulation method that permits calculation of the association rate constant between $O_2^{\cdot-}$ and Cu,ZnSODs and has been shown to provide values comparable to the experimentally determined catalytic rate.[9,18]

*Protein Structure and Electrostatic Potential Calculation*

The availability of the three-dimensional structure of the protein under investigation is essential to start a BD simulation study. Structures of proteins can be retrieved from the RCSB Protein Data Bank (PDB)[19] (Web address, *http://www.rcsb.org/pdb*), or can be built by homology modeling.[20]

Because substrate recognition in SOD is electrostatically enhanced, the second step concerns the calculation of the electrostatic potential around the protein structure. This is usually done through a continuum macroscopic approach by solving the finite difference Poisson–Boltzmann equation (FDPB) around a macromolecule with arbitrary shape and charge distribution. The protein is usually mapped into five different cubic grids of $81 \times 81 \times 81$ site points, spaced by 1.0, 1.5, 2.0, 4.0, and 8.0 Å, respectively. The FDPB is then solved at each lattice point, using the results obtained with the bigger grids to accurately determine the boundary conditions of the finer grids. The five different potential grids are used to generate a $81 \times 81 \times 81 \times 5$ map from which forces are calculated. Values of 4 and 78.5 are typically used for the dielectric constant of the protein interior and

---

[13] I. Klapper, R. Hagstrom, R. Fine, K. Sharp, and B. Honig, *Proteins* **1**, 47 (1986).
[14] A. Desideri, M. Falconi, F. Polticelli, M. Bolognesi, K. Djinovic, and G. Rotilio, *J. Mol. Biol.* **223**, 337 (1992).
[15] S. H. Northrup, S. A. Allison, and J. A. McCammon, *J. Chem. Phys.* **80**, 1517 (1984).
[16] S. A. Allison and J. A. McCammon, *J. Phys. Chem.* **89**, 1072 (1985).
[17] K. Sharp, R. Fine, K. Schulten, and B. Honig, *J. Phys. Chem.* **91**, 3624 (1987).
[18] A. Sergi, M. Ferrario, F. Polticelli, P. O'Neill, and A. Desideri, *J. Phys. Chem.* **98**, 10554 (1994).
[19] H. M. Berman, J. Westbrook, Z. Feng, G. Gilliland, T. N. Bhat, H. Weissig, I. N. Shindyalov, and P. E. Bourne, *Nucleic Acids Res.* **28**, 235 (2000).
[20] M. Falconi and A. Desideri, *in* "Handbook of Copper Pharmacology and Toxicology" (E. J. Massaro, ed.). Humana Press, Totowa, New Jersey, 2001 (in press).

the solvent, respectively. The proteins are analyzed at pH 7, considering all lysines and arginines protonated and all carboxyl groups dissociated.

This type of calculation can be efficiently done with commercial programs such as DelPhi[21,22] (Web address, *http://trantor.bioc.columbia.edu/delphi*).

*Brownian Dynamics*

The diffusion of superoxide relative to SOD is typically evaluated on an ensemble of 10,000 different trajectories. The trajectory propagation is described by the following equation:

$$\mathbf{r}(t + \Delta t_g) = \mathbf{r}(t) + \Delta t_g D\mathbf{F}(t)/k_B T + \mathbf{S}$$

where $\mathbf{r}$ is the position vector, $D$ is the diffusion constant of superoxide relative to SOD (0.1283 Å/psec),[23,24] $k_B$ is the Boltzmann constant, $T = 298$ K, $\Delta t_g$ is the time step, $\mathbf{F}(t)$ is the electrostatic force applied to superoxide, and $\mathbf{S}$ is a vector whose components are independent Gaussian numbers determined by the properties $< S_i > = 0$ and $< S_i^2 > = 2D\, \Delta t_g$ ($i = x, y, z$).

The time step value is directly related to the distance of the superoxide from the protein center; in particular, a different time step is used in each grid, its value being 0.125 psec for the finer grid (grid index $g = 1$) and increasing proportionally with the increase in grid spacing.

*Trajectory Propagation and Rate Calculation*

A trajectory is typically initiated at $R_1 = 61.5$ Å and continues until it reacts by entering a sphere of radius 4 Å centered on the copper atom,[23] or escapes reaching a sphere of $R_2 = 150$ Å centered on the protein center. When the diffusing particle collides with a nonreactive area of the protein surface (i.e., outside the metal active site) the particle is "reflected," starting from the position occupied before collision occurred and translated with an appropriate random vector.[17] The aim of this procedure is to prevent a particle close to the protein surface from colliding with the protein because of a strong attractive force, and remaining trapped in that position without reacting with the copper or escaping.

The association rate $k_a$ between superoxide and superoxide dismutase is calculated by multiplying the probability $P$ that a trajectory starting at $R_1$ hits the target by the rate $k_d(R_1)$ of diffusion of particles to the sphere of radius $R_1$[15]:

$$k_a = k_d(R_1)P$$

---

[21] B. Honig and A. Nicholls, *Science* **268**, 1144 (1995).
[22] M. K. Gilson, K. A. Sharp, and B. Honig, *J. Comp. Chem.* **9**, 327 (1987).
[23] S. A. Allison, R. J. Bacquet, and J. A. McCammon, *Biopolymers* **27**, 251 (1988).
[24] J. J. Sines, S. A. Allison, and J. A. McCammon, *Biochemistry* **29**, 9403 (1990).

However, the probability $P$ that the diffusing particle hits the target must be corrected because calculations are carried out in a limited space and there is a finite probability that a trajectory that escaped at $R_2$ reenters the simulation space at $R_1$.[23] This correction can be analytically calculated only if the flux at the starting radius $R_1$ has no angular dependence or, in other words, if the potential for $r > R_1$ either vanishes or can be considered centrosymmetric.

### Applications of Brownian Dynamics Simulation

Cu,ZnSOD has been the subject of several computational studies concerning the quantitative evaluation of the enzyme–substrate association rate. BD simulations have been shown to be able to reproduce the trend of the ionic strength dependence experimentally observed with the native and chemically modified bovine enzymes,[23,25] and to reproduce the experimental catalytic rate in natural variants.[18] The BD approach has been shown to be a useful tool with which to estimate the role exerted by each charged residue in the enzyme–substrate encounter and to engineer enzymes having an efficiency higher than that displayed by the native enzyme.[24]

BD simulation and mutagenesis were used to predict, and then express, mutants of the human enzyme that are significantly more active than the wild-type enzyme.[26] In particular, substitution of the negative Glu-131 and/or Glu-130 with glutamine creates faster reaction rates and increased ionic strength dependence, matching brownian dynamics simulations incorporating electrostatic terms.

To quantitatively define the electrostatic and mechanistic contributions of sequence-invariant Arg-143 in human Cu,ZnSOD, single-site mutants at this position were investigated experimentally and computationally.[27] Rate constants for several Arg-143 mutants were determined at different pH values and ionic strengths, using pulse radiolytic methods, and were compared with results from BD simulations. The calculated and experimental ionic strength profiles gave similar slopes for all but the glutamate mutant. Differences between the calculated and experimental rates for the glutamate and lysine mutants reflect the structural contribution of Arg-143 to the dismutation process.

The association rate, calculated by brownian dynamics simulations incorporating electrostatic terms, and the enzyme reaction rate, evaluated by the pulse radiolysis method, were measured for both bovine and shark SODs, native and chemically modified at lysine residues by carbamoylation.[28] The experimental

---

[25] K. Sharp, R. Fine, and B. Honig, *Science* **236**, 1460 (1987).
[26] E. D. Getzoff, D. E. Cabelli, C. L. Fisher, H. E. Parge, M. S. Viezzoli, L. Banci, and R. A. Hallewell, *Nature (London)* **358**, 347 (1992).
[27] C. L. Fisher, D. E. Cabelli, J. A. Tainer, R. A. Hallewell, and E. D. Getzoff, *Proteins* **19**, 24 (1994).
[28] F. Polticelli, M. Falconi, P. O'Neill, R. Petruzzelli, A. Galtieri, A. Lania, L. Calabrese, G. Rotilio, and A. Desideri, *Arch. Biochem. Biophys.* **312**, 22 (1994).

and calculated rates indicated that residue 134, which is an arginine and a lysine in the shark and bovine enzyme, respectively, contributes 19% to the guidance of superoxide toward the copper ion.

In another work,[29] the role of Lys-120 and Lys-134, located at the edge of the active site channel in most Cu,Zn superoxide dismutases, in steering the anionic substrate toward the catalytic copper ion has been investigated by mutating these residues in *Xenopus laevis* Cu,ZnSOD to leucine and threonine, respectively. Experimental and BD-evaluated rates indicate that neutralization of each of these residues decreases the catalytic rate by about 40%.

BD simulations have also been shown to be able to reproduce the pH dependence of the catalytic rate of native, singly (Lys-120→Leu, or Lys-134→Thr), and doubly (Lys-120→Leu, and Lys-134→Thr) mutated *Xenopus laevis* B Cu,Zn superoxide dismutase.[10]

Diffusion enhancement of the substrate toward the active site is maintained in the prokaryotic Cu,ZnSODs despite modifications in their different tertiary and quaternary structures that displace the location of the electrostatic loop as compared with the eukaryotic enzymes.[30] In fact, experimental and computational data show that, at low ionic strength, neutralization of Lys-60 of the Cu,ZnSOD from *Escherichia coli* strongly reduces the catalytic activity of the enzyme (~50%), indicating that this residue plays a primary role in the electrostatic attraction of the substrate, where as neutralization of Lys-63 does not significantly influence the catalytic rate.[31] In the case of the prokaryotic enzyme from *Photobacterium leiognathi*, it was possible to engineer a superefficient enzyme by introducing a Glu-59→Gln mutation.[32] At neutral pH this mutant was found to have a $k_{cat}/K_m$ value of $1 \times 10^{10}\ M^{-1}\ sec^{-1}$, the highest value ever found for any superoxide dismutase, likely caused by a series of combined effects (such as enhanced substrate attraction by the modified electrostatic field distribution, a large solvent-accessible active site, and a specific intersubunit interaction that may influence the active site arrangement). Remarkably, it has been observed that the catalytic efficiency of the prokaryotic enzyme can be modulated by mutation of a neutral residue located at the intersubunit interface, about 18 Å away from the active center.[33] BD simulation is not able to reproduce the experimentally detected functional changes.[33]

---

[29] F. Polticelli, A. Battistoni, G. Bottaro, M. T. Carrì, P. O'Neill, A. Desideri, and G. Rotilio, *FEBS Lett.* **352,** 76 (1994).

[30] D. Bordo, D. Matak, K. Djinovic-Carugo, C. Rosano, A. Pesce, M. Bolognesi, M. E. Stroppolo, M. Falconi, A. Battistoni, and A. Desideri, *J. Mol. Biol.* **285,** 283 (1999).

[31] S. Folcarelli, A. Battistoni, M. Falconi, P. O'Neill, G. Rotilio, and A. Desideri, *Biochem. Biophys. Res. Commun.* **244,** 908 (1998).

[32] S. Folcarelli, F. Venerini, A. Battistoni, P. O'Neill, G. Rotilio, and A. Desideri, *Biochem. Biophys. Res. Commun.* **256,** 425 (1999).

[33] M. E. Stroppolo, A. Pesce, M. Falconi, P. O'Neill, M. Bolognesi, and A. Desideri, *FEBS Lett.* **483,** 17 (2000).

This result indicates that the enhanced catalytic behavior of an SOD mutant is a consequence of altered mechanical properties of the enzyme and is not directly related to a different distribution of charges in the static structure. A comparison of experimental and BD calculated rates may then be a useful approach to discriminate when the modulated functional properties due to mutation/perturbation are of electrostatic or of structural/dynamic origin.

# [5] Analysis of Cu,ZnSOD Conformational Stability by Differential Scanning Calorimetry

*By* MARIA CARMELA BONACCORSI DI PATTI, ANNA GIARTOSIO, GIUSEPPE ROTILIO, and ANDREA BATTISTONI

## Introduction

Since the discovery of its activity in 1969, Cu,Zn superoxide dismutase (Cu,ZnSOD) has been shown to be a stable enzyme whose properties are not altered by isolation procedures utilizing organic solvents.[1] Shortly afterward, the exceptional stability of bovine Cu,ZnSOD (the first Cu,ZnSOD to be characterized) was further highlighted by studies showing that this enzyme retains its native structure in 8 $M$ urea[2,3] and 4% (w/v) sodium dodecyl sulfate (SDS),[2] is denatured slowly in the presence of high concentrations of guanidinium hydrochloride,[4,5] and unfolds at high temperatures.[6,7] On the whole, these properties indicate that bovine Cu,ZnSOD thermal stability is comparable to or higher than that typical of many proteins from thermophilic organisms.[8] Further studies have shown that the Cu,ZnSODs isolated from other sources are all stable[7,9–12] and that these enzymes constitute one of the most stable globular protein families studied so far. Structural

[1] J. M. McCord and I. Fridovich, *J. Biol. Chem.* **244**, 6049 (1969).
[2] H. S. Forman and I. Fridovich, *J. Biol. Chem.* **248**, 2645 (1973).
[3] D. P. Malinowski and I. Fridovich, *Biochemistry* **18**, 5055 (1979).
[4] J. L. Abernethy, H. M. Steinman, and R. L. Hill, *J. Biol. Chem.* **249**, 7339 (1974).
[5] H. Mach, Z. Dong, C. R. Middaugh, and R. V. Lewis, *Arch. Biochem. Biophys.* **287**, 41 (1991).
[6] J. R. Lepock, L. D. Arnold, B. H. Torrie, B. Andrews, and J. Kruuv, *Arch. Biochem. Biophys.* **241**, 243 (1985).
[7] J. A. Roe, A. Butler, D. M. Scholler, J. S. Valentine, L. Marky, and K. J. Breslauer, *Biochemistry* **27**, 950 (1988).
[8] R. Scandurra, V. Consalvi, R. Chiaraluce, L. Politi, and P. C. Engel, *Biochimie* **80**, 933 (1988).
[9] J. R. Lepock, H. E. Frey, and R. A. Hallewell, *J. Biol. Chem.* **265**, 21612 (1990).
[10] M. C. Bonaccorsi di Patti, A. Giartosio, G. Musci, P. Carlini, and L. Calabrese, *in* "Frontiers of Reactive Oxygen Species in Biology and Medicine" (K. Asada and T. Yoshikawa, eds.), p. 129. Elsevier Science, Amsterdam, 1994.

investigations have established that the enzyme stability is due to a combination of different factors, including the intrinsic stability of the eight-stranded $\beta$-barrel fold, the close packing of the hydrophobic interfaces between the subunits,[13] the presence of an intrachain disulfide bond,[4,14] and the active site stabilization induced by the prosthetic metal ions.[2,7,12] Interestingly, although enzymes isolated from different natural sources maintain a similar three-dimensional structure,[15] they often exhibit substantial differences in their heat stability.[7,10,16] Analysis of molecular determinants modulating the stability of Cu,ZnSOD is therefore a promising strategy to understand factors governing protein stability.

## Experimental Approaches to Analysis of Cu,ZnSOD Stability

A survey of methods used to study the stability of Cu,ZnSOD is outside the scope of this work, but it is worthwhile to recall that the most widely used approach to compare the heat stability of different Cu,ZnSOD samples is that of measuring the residual activity of the enzyme after incubation at a high temperature (e.g., 70°) for a defined time interval. This method is simple and has the advantage that it does not require any complex instrumentation, but its major limitation is that it does not discriminate between differences due to variations in conformational stability or in the reversibility of protein unfolding. An example of the difficulties in the interpretation of results of thermal inactivation assays is provided by a comparison of the heat stability at 70° of the *Escherichia coli* monomeric enzyme with that of the bovine and *Xenopus laevis* dimeric enzymes, at a monomer concentration of $4 \times 10^{-4}$ $M$.[17] Under these conditions the activity of the *E. coli* enzyme is scarcely affected by incubation at high temperatures, whereas both of the eukaryotic enzymes undergo a progressive loss of activity on incubation. The *E. coli* enzyme unfolds at a temperature distinctly lower than the temperature of bovine and *X. laevis* Cu,ZnSOD unfolding.[12] This result is essentially due to the high reversibility of denaturation of the *E. coli* enzyme whereas that of the two eukaryotic Cu,ZnSODs is completely irreversible. Nonetheless, this method has

---

[11] Y. Bourne, S. M. Redford, H. M. Steinman, J. R. Lepock, J. A. Tainer, and E. D. Getzoff, *Proc. Natl. Acad. Sci. U.S.A.* **93,** 12774 (1996).
[12] A. Battistoni, S. Folcarelli, L. Cervoni, F. Polizio, A. Desideri, A. Giartosio, and G. Rotilio, *J. Biol. Chem.* **273,** 5655 (1998).
[13] J. A. Tainer, E. D. Getzoff, K. M. Beem, J. S. Richardson, and D. C. Richardson, *J. Mol. Biol.* **160,** 287 (1982).
[14] A. Battistoni, A. P. Mazzetti, and G. Rotilio, *FEBS Lett.* **413,** 313 (1999).
[15] D. Bordo, K. Djinovic-Carugo, and M. Bolognesi, *J. Mol. Biol.* **238,** 366 (1994).
[16] M. C. Bonaccorsi di Patti, M. T. Carri, R. Gabbianelli, R. Da Gai, C. Volpe, A. Giartosio, G. Rotilio, and A. Battistoni, *Arch. Biochem. Biophys.* **377,** 284 (2000).
[17] A. Battistoni and G. Rotilio, *FEBS Lett.* **374,** 199 (1995).

proved useful to test the effect of site-specific mutations on enzyme stability,[18,19] as well as to reveal that dimeric Cu,ZnSODs show a concentration-dependent inactivation rate (the inactivation being faster at low protein concentration), which suggests that, under these experimental conditions, the heat-induced denaturation of the enzyme occurs on monomerization.[17]

A more informative approach to the problem of understanding the structural basis of Cu,ZnSOD stability has been that of analyzing the conformational transitions occurring during the unfolding of different Cu,ZnSODs by steady-state and dynamic fluorescence spectroscopy, usually as a function of guanidinium hydrochloride concentration.[19-22] These studies rely on the presence of aromatic residues (usually a single trypthophan or tyrosine residue) buried inside the protein or at the dimer interface. Alterations of the environment surrounding these intrinsic probes, such as those due to protein denaturation or subunit dissociation, may be monitored by changes in the emission spectra or in the rotational correlation times of the fluorophore. Unfortunately, most eukaryotic and some prokaryotic Cu,ZnSODs lack tryptophan residues, making this approach amenable only for a small subset of enzyme variants.

The most useful information about the factors that affect the thermal stability of Cu,ZnSOD have undoubtedly been obtained by differential scanning calorimetry (DSC), the technique of choice for the quantitative measurements of heat effects on protein stability. Detailed descriptions of the principles underlying DSC may be found elsewhere,[23,24] but we would like to note that the calorimetric parameters that are obtained by DSC are the $T_m$ (temperature of maximum heat capacity), $\Delta H_{cal}$ (calorimetric enthalpy of denaturation), and $\Delta C_p$ (difference in heat capacity between the denatured and the native state). $T_m$ values are related to protein conformational stability and analysis of DSC traces allows determination of the thermodynamic parameters $\Delta G$, $\Delta H$, and $\Delta S$ of unfolding. Furthermore, it is also possible by DSC to evaluate how different factors affect the reversibility of unfolding by repeated scans of the same protein sample.

[18] R. A. Hallewell, K. C. Imlay, P. Lee, N. M. Fong, C. Gallegos, E. D. Getzoff, J. A. Tainer, D. E. Cabelli, P. Tekamp-Olson, G. T. Mullenbach, and L. S. Cousens, *Biochem. Biophys. Res. Commun.* **181**, 474 (1991).
[19] S. Folcarelli, A. Battistoni, M. T. Carrì, F. Polticelli, M. Falconi, L. Stella, N. Rosato, G. Rotilio, and A. Desideri, *Protein Eng.* **9**, 323 (1996).
[20] G. Mei, N. Rosato, N. Silva, R. Rusch, E. Gratton, I. Savini, and A. Finazzi Agró, *Biochemistry* **31**, 7724 (1992).
[21] F. Malvezzi-Campeggi, M. E. Stroppolo, G. Mei, N. Rosato, and A. Desideri, *Arch. Biochem. Biophys.* **370**, 201 (1999).
[22] M. E. Stroppolo, F. Malvezzi-Campeggi, G. Mei, N. Rosato, and A. Desideri, *Arch. Biochem. Biophys.* **377**, 215 (2000).
[23] J. M. Sturtevant, *Annu. Rev. Phys. Chem.* **38**, 463 (1987).
[24] P. L. Privalov, *Adv. Protein Chem.* **33**, 167 (1979).

Despite some difficulties in extrapolating accurate thermodynamic data from the thermograms of Cu,ZnSOD (see below), calorimetric studies have provided a wealth of information on the factors that stabilize the enzyme and have allowed analysis of the effect of specific residues on protein stability.

### Setting up a Differential Scanning Calorimetry Experiment

The parameter that provides a first indication of conformational stability is the $T_m$ value. However, the experimenter should bear in mind that $T_m$ values obtained by DSC are not absolute values, but depend on the experimental conditions employed for DSC, such as pH and buffer composition, scan rate, and, in the case of SOD, also the oxidation state of copper. Therefore, comparative analyses of the thermal stability of different samples based on the $T_m$ value can be carried out only if the DSC experiments are performed under the same experimental conditions. Examples of $T_m$ values obtained for various enzymes under different conditions are reported in Table I.[7,9–12,16,25]

When examining a new protein it is always advisable to perform a pH dependence study of thermal stability. As a matter of fact, the $T_m$ of the bovine enzyme is nearly constant in the pH 6.0–8.0 range,[7] whereas that of the monomeric *E. coli* Cu,ZnSOD is greatly affected by small pH variations.[12] Different buffers have been used for studying the stability of Cu,ZnSOD, but the most commonly used buffer is 100 m$M$ potassium phosphate, as this buffer allows measurement of protein stability in a physiologically relevant pH range (between pH 6.0 and 8.0) and because its pH is minimally affected by temperature and is characterized by a low enthalpy of ionization. It should also be noted that buffers with an affinity for metal ions (e.g., Tris or pyrophosphate) should be avoided when investigating the properties of a metalloprotein such as Cu,ZnSOD.

Before DSC analysis, highly purified Cu,ZnSOD samples are extensively dialyzed into the buffer of choice. The protein solution and the buffer (which is used to fill the reference cell) are then degassed under mild vacuum for 5–10 min and immediately scanned. It is important to check for scan rate dependence of $T_m$ and/or of the shape of the DSC peak by employing different scan rates (0.2–2°/min), because if such a dependence is found, either the scan rates are too high to allow the denaturing protein to reach equilibrium at each temperature of the DSC scan or denaturation of the protein is under kinetic control. The scan rate routinely employed for the analysis of Cu,ZnSODs is 1°/min.

All traces are corrected for the instrumental baseline obtained by filling both cells with the buffer. Reversibility of thermal transitions can be evaluated by a second heating of the same sample immediately after ending the scan and cooling the sample, and by partial scans to midtransition.

We have used a MicroCal (Northampton, MA) MC-2 calorimeter, which requires 1–2 mg of protein for each scan and used the software package (Origin)

## TABLE I
### THERMAL STABILITY OF VARIOUS Cu,ZnSODs

| Cu,ZnSOD | pH | Scan rate (°C/min) | $T_m$ (°C) | Ref. | Ref. no. |
|---|---|---|---|---|---|
| Eukaryotic | | | | | |
| Human | 7.0 (PBS) | 10 | 101 | Biliaderis et al. (1987) | 25 |
| | 7.8 (phosphate) | 1 | 80.1 | Lepock et al. (1990) | 9 |
| Bovine | 7.0 (PBS) | 10 | 96–104 | Biliaderis et al. (1987) | 25 |
| | 5.5 (acetate) | 0.82 | 89–96 | Roe et al. (1988) | 7 |
| | 7.8 (phosphate) | 1 | 88.3 | Lepock et al. (1990) | 9 |
| | 7.8 (glycine) | 1 | 86.3 | M. C. Bonaccorsi and A. Giartosio, unpublished (2001) | |
| | 7.5 (HEPES) | 1 | 91.5 | M. C. Bonaccorsi and A. Giartosio, unpublished (2001) | |
| | 8.4 (HEPES) | 1 | 91.5 | M. C. Bonaccorsi and A. Giartosio, unpublished (2001) | |
| Sheep | 7.8 (phosphate) | 1 | 87.1 | Bonaccorsi et al. (1994) | 10 |
| | 7.5 (HEPES) | 1 | 95.4 | M. C. Bonaccorsi and A. Giartosio, unpublished (2001) | |
| Shark | 7.8 (phosphate) | 1 | 84.1 | Bonaccorsi et al. (1994) | 10 |
| Xenopus laevis A | 7.8 (phosphate) | 1 | 68.75 | Bonaccorsi et al. (2000) | 16 |
| Xenopus laevis B | 7.8 (phosphate) | 1 | 77.16 | Bonaccorsi et al. (2000) | 16 |
| | 7.4 (HEPES) | 1 | 84.65 | M. C. Bonaccorsi and A. Giartosio, unpublished (2001) | |
| Yeast | 5.5 (acetate) | 0.82 | 82 | Roe et al. (1988) | 7 |
| | 7.8 (phosphate) | 1 | 73.1 | Bonaccorsi et al. (1994) | 10 |
| Prokaryotic | | | | | |
| Escherichia coli | 7.8 (phosphate) | 1 | 65.9 | Battistoni et al. (1998) | 12 |
| Photobacterium leiognathi | 7.8 (phosphate) | 1 | 71 | Bourne et al. (1996) | 11 |

provided by MicroCal for data acquisition and analysis. However, protein availability might not be a problem any longer, as new-generation instruments allow the use of decreased amounts of sample, down to 0.15–0.3 mg.

## Thermostability of Cu,ZnSOD by Differential Scanning Calorimetry

DSC traces of Cu,ZnSOD generally exhibit a single asymmetric endotherm due to unfolding of the protein. The denaturation of eukaryotic enzymes is usually highly irreversible, as demonstrated by rescanning experiments. In contrast, bacterial Cu,ZnSODs refold with high efficiency. Irrespective of the method used (see below), deconvolutions of experimental traces show the presence of two components underlying the single DSC peak in most Cu,ZnSODs. Because no dependence of $T_m$ on protein concentration (at least in the range of concentrations used in DSC experiments) has been reported, it has been assumed that Cu,ZnSOD denatures as a dimer. This assumption is in apparent contradiction with circular dichroism and fluorescence spectroscopy studies, which have suggested the presence of a monomeric intermediate in the Cu,ZnSOD-unfolding process.[20,22] However, such a monomeric intermediate may be observed only at low protein concentrations (in the submicromolar range) in the presence of 2–3 $M$ guanidinium hydrochloride. Probably, under the conditions used in DSC experiments, the monomerization process, if it occurs, is strictly associated with denaturation. The two components of the bovine and yeast Cu,ZnSOD thermograms have been attributed to the presence of two forms of SOD, one with copper in the oxidized state (showing lower thermostability) and the other with copper in the reduced state (this one being more thermostable).[7] By analogy with the results obtained with these enzymes, identification of two components underlying the DSC profile of other Cu,ZnSODs has been attributed to the presence of two populations of enzyme molecules differing in the oxidation state of the copper ion.[9,16,26] Some caution must be exercised, however, in extrapolating this interpretation to all cases as there could also be other possible sources of heterogeneity in the protein sample, such as the lack of a full complement of metal ions. Metal occupancy is one of the most important determinants of the extremely high thermal stability of Cu,ZnSOD, and a dramatic decrease in the thermal stability of the protein has been observed in different Cu,ZnSODs on removal of copper and zinc (in the case of the bovine enzyme the difference in $T_m$ approximates 40° at pH 7.8).[7,12] The zinc ion provides the most important contribution to the enzyme stability, but also removal of copper influences the $T_m$. The presence of a fraction of copper-devoid enzyme is not uncommon in Cu,ZnSOD preparations, and copper content should be accurately determined by analytical techniques in order to exclude such a possible cause of heterogeneity. Moreover, although purified eukaryotic Cu,ZnSODs usually possess a complete zinc complement, some mutant Cu,ZnSODs associated with the onset of the familial form of amyotrophic lateral sclerosis have been shown to have a decreased affinity for the zinc ion,[27] and the presence of a consistent amount of zinc-free protein may be found in recombinant prokaryotic enzymes.[12]

[25] C. G. Biliaderis, R. J. Weselake, A. Petkau, and A. D. Friesen, *Biochem. J.* **248**, 981 (1987).
[26] D. E. McRee, S. M. Redford, E. D. Getzoff, J. R. Lepock, R. A. Hallewell, and J. A. Tainer, *J. Biol. Chem.* **265**, 14234 (1990).
[27] A. G. Estevez, J. P. Crow, J. B. Sampson, C. Reiter, Y. Zhuang, G. J. Richardson, M. M. Tarpey, L. Barbeito, and J. S. Beckman, *Science* **286**, 2498 (1999).

The effect of metal ions on Cu,ZnSOD stability may be studied by DSC analyses of enzyme derivatives lacking one or both of the prosthetic metals, or enzymes remetalated under controlled conditions. An example of the contribution of metal ions to the stability of a Cu,ZnSOD is provided by the DSC analysis of *E. coli* Cu,ZnSOD reported in Fig. 1.[12] The unfolding of this enzyme is highly reversible, and deconvolution of the DSC traces shows the presence of two independent two-state transitions with significantly different $T_m$ values. In agreement with metal content analysis by atomic absorption, we have found that the $T_m$ of the species with lower thermal stability is lower than that of the copper-free enzyme and identical to that of an enzyme derivative deprived of both copper and zinc.[12]

## Preparation of Cu,ZnSOD Derivatives

### Preparation of Apo-SOD

Cu,ZnSOD lacking both the copper and zinc ion can be obtained by extensive dialysis of the enzyme (at a concentration of 15–30 mg/ml) against 100 m$M$ acetate buffer, pH 3.8, containing 5 m$M$ EDTA. The time required to remove metals varies depending on the specific enzyme, but the complete removal of copper may require up to 3 days at room temperature The use of concentrated solutions of the enzyme allows monitoring copper removal either by visual inspection (the protein loses its green color) or spectrophotometrically at 680 nm. Shorter incubation times and lower EDTA concentrations may be used to remove metals from most bacterial Cu,ZnSODs.

The enzyme is then dialyzed overnight against the same buffer, containing 100 m$M$ NaClO$_4$ to remove excess EDTA. As zinc is an abundant contaminant of distilled water, from this point on special attention must be paid to using metal-free glassware (obtained by treatment with diluted solutions of nitric acid) and high purity distilled water, further treated with Chelex-100 (Bio-Rad, Hercules, CA).

Apo-Cu,ZnSOD is dialyzed against 100 m$M$ acetate buffer, pH 3.8, to remove NaClO$_4$ and, as a last step, against the buffer of choice for subsequent analyses or reconstitution.

A comprehensive review of the methods described to reconstitute metal-devoid Cu,ZnSOD may be found elsewhere.[28] To avoid occupancy of the wrong site, the experimenter should pay attention to the protein–metal stoichiometry, to the pH at which reconstitution is carried out, and to the choice of the buffer used. Useful remetalated forms for DSC analyses may be obtained by reconstitution of the apoprotein at pH 5.5–6.0 with one or two zinc ions per subunit, followed by a 12-hr incubation at 4°. This procedure allows the obtaining of derivatives with the zinc site or both the metal sites occupied by zinc.

[28] J. S. Valentine and M. W. Pantoliano, *Met. Ions Biol. Syst.* **3**, 291 (1981).

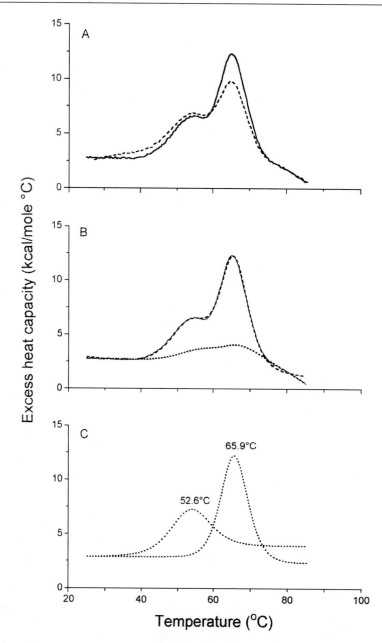

FIG. 1. Thermal stability of *E. coli* Cu,ZnSOD. All calorimetric traces were corrected by a buffer–buffer baseline. (A) Solid line, first heating scan; dotted line, reheating. (B) Solid line, first scan; dotted line, progress baseline; dashed line, theoretical curve obtained using a reversible two-state model. (C) Fitted transition curves. [Reprinted from Battistoni *et al.*, *J. Biol. Chem.* **273**, 5655 (1998).]

## Preparation of Copper-Free and Zinc-Free Superoxide Dismutase

As a first step in the preparation of Cu,ZnSOD lacking the copper ion, potassium ferrocyanide is added to the protein solution (15–30 mg/ml) until the green color disappears (due to copper reduction). The enzyme is subsequently dialyzed for 12 hr against 100 m$M$ potassium phosphate buffer, 50 m$M$ KCN, pH 6.0. The sample is then further dialyzed for 24 hr at 4° (twice) against phosphate buffer to remove KCN. The final copper content may be verified by atomic absorption spectroscopy, but it is generally much lower than 2% of the initial value.

Zinc-free SOD can be obtained by a procedure similar to that described to prepare the apoenzyme by decreasing the pH of the acetate buffer to pH 3.6 and by avoiding the addition of EDTA. However, under these conditions extensive copper loss has been observed in prokaryotic Cu,ZnSODs and some copper loss may also occur in the eukaryotic enzymes. Moreover, the copper bound to a zinc-free enzyme may migrate to the zinc site as the pH is raised close to neutrality, making this derivative scarcely useful for DSC analysis.[28]

## Thermodynamic Analysis of Differential Scanning Calorimetry Traces

DSC traces of all Cu,ZnSODs so far anlyzed show a sharp drop in heat capacity ($C_p$) above the $T_m$. The occurrence of an exothermic process that overlaps the denaturational endotherm distorts the postdenaturation baseline and prevents an accurate measurement of $\Delta H$ and $\Delta C_p$ values. Interestingly, such a drop in heat capacity is not observed in Cu,ZnSOD derivatives lacking copper and zinc (see Fig. 1), suggesting that protonation of metal ligands on protein unfolding is involved in the exothermic process. Moreover, protein aggregation may also contribute to this phenomenon.

Uncertainties in determination of $\Delta H_{cal}$ and $\Delta C_p$ and irreversibility of unfolding in the case of eukaryotic enzymes hamper rigorous thermodynamic analysis of DSC curves obtained for Cu,ZnSOD. Nevertheless, if the denaturation process is apparently irreversible, a thermodynamic analysis of DSC traces can be attempted by assuming that protein unfolding occurs through a two-step mechanism, N↔D→I, where the irreversible step is slower than the reversible unfolding over the temperature range used for the analysis.[23,29] Thermodynamic parameters can be obtained directly by fitting the DSC profile by using the van't Hoff equation, as long as N and D are in equilibrium. van't Hoff enthalpy is related to calorimetric enthalpy by

$$\Delta H_{vH} = 4RT^2 C_p / \Delta H_{cal} \qquad (1)$$

Comparison of calculated $\Delta H_{vH}$ and experimental $\Delta H_{cal}$ values allows determination of whether the transition is two-state and, if not, deconvolution of the

---

[29] J. M. Sanchez-Ruiz, *Biophys. J.* **61**, 921 (1992).

experimental DSC trace into two-state components can be performed. Deconvolution can be carried out excluding the high-temperature tail of the peak (beyond the peak half-height on the down slope), where distortion due to aggregation takes place. The thermodynamic parameters ($T_m$ and $\Delta H_{cal}$) obtained by deconvolution of the DSC traces according to the two-state reversible model can be employed to calculate $\Delta \Delta G$ values for mutant proteins at the appropriate $T_m$ of the wild type according to the Gibbs–Helmholtz equation [Eq. (2)]:

$$\Delta G = \Delta H_{cal}[1 - (T/T_m)] - \Delta C_p[(T_m - T) + T \ln(T/T_m)] \quad (2)$$

Because of difficulty in obtaining reasonable $\Delta C_p$ values for most Cu,ZnSODs, calculation of $\Delta G$ values requires the assumption that either $\Delta H$ is temperature independent ($\Delta C_p = 0$) or, alternatively, the temperature dependence of $\Delta H$ can be estimated by using a $\Delta C_p$ value of 0.12 cal/g, which is the average $\Delta C_p$ experimentally determined for globular proteins of size comparable to Cu,ZnSOD.[24]

Alternatively, a deconvolution according to an irreversible model can be carried out by assuming that each peak represents an irreversible one-step transition of the form N→D, obeying pseudo-first-order kinetics, and that the temperature dependence of the rate constant for denaturation follows the Arrhenius relation.[9,26] According to this model, the heat capacity during the transition is given by Eq. (3):

$$C_p(T) = \Delta H_{cal}(df_D/dT) \quad (3)$$

where $f_D$ is the fraction of protein denatured, which can be calculated as

$$df_D[T(t)]/dt = e^{A-E_A/R(T_0+vt)}(1 - f_D[T(t)]) \quad (4)$$

where $T_0$ is the initial temperature and $v$ is the DSC scan rate. The variable parameters are $\Delta H_{cal}$ and the Arrhenius constants $A$ and $E_A$, which can then be used to calculate the rates of inactivation for Cu,ZnSOD, according to the Arrhenius relation [Eq. (5)]:

$$k(T) = e^{A-E_A/RT} \quad (5)$$

## Differential Scanning Calorimetry Investigations on Effects of Point Mutations on Cu,ZnSOD Stability

Examples of DSC applications are provided by studies aimed at evaluating the effect of point mutations on Cu,ZnSOD thermal stability. An open question concerns factors responsible for the irreversibility of denaturation of Cu,ZnSODs. Several eukaryotic Cu,ZnSODs possess cysteine residues not involved in disulfide bond formation (free cysteines), which could mediate formation of incorrect disulfide bonds on thermally induced protein unfolding, thus leading to protein aggregation. In support of this hypothesis it has been observed that the only natural eukaryotic Cu,ZnSOD that displays partial reversibility of unfolding after a DSC scan is the yeast enzyme, which does not contain free cysteines.[6] Several investigations have, therefore, been focused on the contribution of free cysteines to the irreversibility of protein unfolding at high temperatures (assayed by repeated scans on the same sample)[9,26] and to the conformational stability of the enzyme (evaluated as changes in $T_m$ values).[16] Substitution of cysteine residues was shown to notably increase the reversibility of unfolding in bovine and human Cu,ZnSODs, with only minor effects on conformational stability.[9,26] Denaturation of bovine Cu,ZnSOD Ala-6 was 52% reversible after heating to 84°, compared with 32% reversibility for the wild-type enzyme. For human Cu,ZnSOD Ala-6, Ser-111 reversibility was 89% after heating to 80°, compared with only 16% for the wild type. These studies clearly demonstrate that free cysteine residues are a major cause of irreversibility of unfolding. Nonetheless, the mutant enzymes lacking free cysteines also exhibit significant irreversibility of unfolding when heated at temperatures well above the $T_m$, suggesting that other factors also contribute to eukaryotic Cu,ZnSOD aggregation. Bacterial Cu,ZnSODs do not possess free cysteines. We have found that the thermal unfolding of *E. coli* Cu,ZnSOD, which is monomeric, is highly reversible,[12] whereas the unfolding of *Photobacterium leiognathi* Cu,ZnSOD, which is dimeric, has been reported to be irreversible.[11] These findings suggest that exposure to the solvent of the hydrophobic subunit interfaces greatly enhances the irreversibility of unfolding. However, we have observed that for other dimeric bacterial enzymes scanned under conditions identical to those employed for analyzing *P. leiognathi* Cu,ZnSOD, significant (close to 80%) reversibility may be attained (A. Battistoni and A. Giartosio, unpublished observations, 2001). Besides the absence of free cysteines, differences in the organization of the subunit interface, which in the bacterial Cu,ZnSODs is characterized by the presence of a water-filled intermolecular cavity and by looser intermolecular contacts with respect to the eukaryotic enzymes,[11,30,31] could favor the reversibility of unfolding of bacterial dimeric Cu,ZnSODs. Other factors that may influence the reversibility of unfolding in DSC experiments are the incorrect rebinding of metal ions after unfolding (bovine apo-SOD shows significant refolding at pH 3.8)[7] and amino acid damage at high temperatures.

Work has outlined that in some cases free cysteines may also affect conformational stability. *Xenopus laevis* possesses two Cu,ZnSOD variants, XSOD A and XSOD B, which present different heat stabilities as measured by conventional inactivation assays.[32] A serine-to-cysteine substitution at position 150 near the

---

[30] D. Bordo, D. Matak, K. Djinovic-Carugo, C. Rosano, A. Pesce, M. Bolognesi, M. E. Stroppolo, M. Falconi, A. Battistoni, and A. Desideri, *J. Mol. Biol.* **285**, 283 (1999).

[31] A. Pesce, A. Battistoni, M. E. Stroppolo, F. Polizio, M. Nardini, J. S. Kroll, P. R. Langford, P. O'Neill, M. Sette, A. Desideri, and M. Bolognesi, *J. Mol. Biol.* **302**, 465 (2000).

[32] C. Capo, F. Polticelli, L. Calabrese, M. E. Schininà, M. T. Carrì, and G. Rotilio, *Biochem. Biophys. Res. Commun.* **173**, 1186 (1990).

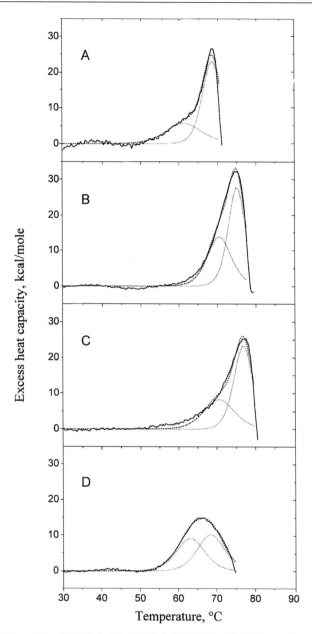

FIG. 2. DSC profiles of XSOD A (A), XSOD A Cys150Ala (B), XSOD B (C), and XSOD B Ser150Cys (D). All the experimental calorimetric traces (solid lines) were corrected by a buffer–buffer baseline. Dashed lines: theoretical curves obtained by deconvolution according to the two-state reversible model. Dotted lines: best two-component fit. [Reprinted from Bonaccorsi di Patti *et al.*, *Arch. Biochem. Biophys.* **377,** 284 (2000).]

dimer interface was identified as a possible candidate for the lower heat stability of XSOD A compared with XSOD B. In this case, DSC has shown the dramatic effect of the removal of Cys-150 from the A variant by site-directed mutagenesis and insertion of cysteine at position 150 in the B variant.[16] Apart from the expected effect on reversibility of unfolding, this substitution was sufficient to completely reverse the conformational stability of the two enzymes as evidenced by the $T_m$ values obtained by DSC for mutant and wild-type proteins (Fig. 2).[16] Thus, DSC analysis has allowed the pinpointing of a single amino acid residue as the major molecular determinant of the different thermal stability of two Cu,ZnSOD natural variants.

## Concluding Remarks

With the expanding potential of site-directed mutagenesis and recombinant protein production methods, it might be anticipated that in future DSC will become increasingly important in clarifying the role of different residues in the mechanisms and forces that govern protein stability. As a matter of fact, despite some limitations, this technique has already proved valuable in studies of Cu,ZnSOD. In this respect, bacterial SODs appear to be more promising than eukaryotic SODs, because of their distinctive structural features and their high(er) reversibility of unfolding. Comparison of various SODs will undoubtly help in defining the molecular features responsible for the exceptionally high thermal stability of this protein.

## Acknowledgment

This work was partially supported by the CNR Target Project on Biotechnology.

## [6] Catalytic Pathway of Manganese Superoxide Dismutase by Direct Observation of Superoxide

*By* DAVID N. SILVERMAN and HARRY S. NICK

The UV absorption of superoxide[1] provides a means of measuring superoxide dismutase (SOD) catalysis by direct observation of the substrate. However, few researchers measure catalysis of superoxide dismutases in this manner because of several inherent difficulties, mostly related to the short lifetime of this radical anion. Much more practicable are indirect methods that, for example, generate

[1] J. Rabani and S. O. Nielson, *J. Phys. Chem.* **73**, 3736 (1969).

$O_2^{\cdot-}$ by the action of xanthine oxidase on xanthine and indirectly measure $O_2^{\cdot-}$ by its ability to reduce cytochrome $c$.[2,3] Yet direct measurements of $O_2^{\cdot-}$ are useful in studies of the kinetic mechanisms and catalytic pathways of the SODs because they allow initial velocity experiments as well as progress curves of catalysis to be measured. These are essential for the analysis of catalysis in terms of well-known steady-state parameters or Michaelis–Menten constants that are the basis for our understanding of the kinetic mechanisms of enzymes. Two major methods utilize this approach: stopped-flow spectrophotometry and pulse radiolysis. When applied to catalysis by MnSOD, these methods reveal diffusion-controlled catalysis at small $O_2^{\cdot-}$ concentrations, and a proton-transfer limited catalysis near maximal velocities. MnSOD appears unique among the superoxide dismutases in forming a peroxide-bound inhibited complex.

## Methods That Use Direct Observation of Superoxide

*Stopped-Flow Spectrophotometry*

Despite the short lifetime of superoxide in aqueous solution,[4] it still has been possible to adapt the useful and versatile methods of stopped-flow spectrophotometry to measure catalysis by superoxide dismutases. This method rapidly mixes superoxide into aqueous solutions containing the enzyme, and follows the subsequent decay of superoxide by detecting its UV absorbance near 250 nm ($\epsilon_{250} = 2000$[1]). The basic adaptation that allows stopped-flow instrumentation to be used is the stabilization of superoxide in an organic solvent,[5–7] in most cases dimethyl sulfoxide (DMSO) with $N,N$-dimethylformamide added to depress the freezing point because DMSO solidifies at 18°. The solubility of $KO_2$ in DMSO can be enhanced by addition of nearly equimolar amounts of 18-crown-6 ether, which serves to complex the potassium ion. In this manner stock solutions containing 0.1 $M$ superoxide in DMSO, prepared under dry nitrogen, can be attained. These stock solutions maintain superoxide concentrations for hours at room temperature, and if frozen at $-20°$, for months.[8] Before the stopped-flow experiment, the dilution of the stock solution with DMSO adjusts for the concentration of superoxide needed.

---

[2] C. Beauchamp and I. Fridovich, *Anal. Biochem.* **44**, 276 (1971).
[3] I. Fridovich, *Adv. Enzymol.* **58**, 61 (1986).
[4] D. Behar, G. Czapski, J. Rabani, L. M. Dorfman, and H. A. Schwarz, *J. Phys. Chem.* **74**, 3209 (1970).
[5] G. J. McClune and J. A. Fee, *Biophys. J.* **24**, 65 (1978).
[6] G. J. McClune and J. A. Fee, *FEBS Lett.* **67**, 294 (1976).
[7] D. P. Riley, W. J. Rivers, and R. H. Weiss, *Anal. Biochem.* **196**, 344 (1991).
[8] C. Bull, G. J. McClune, and J. A. Fee, *J. Am. Chem. Soc.* **105**, 5290 (1983).

A second problem then becomes to achieve efficient and rapid mixing in which the solution of superoxide in DMSO is greatly diluted by an aqueous solution containing superoxide dismutase. Although DMSO and water are miscible, they have different refractive indices and densities, and it is difficult to achieve efficient mixing because the combined solutions tend to form layers. This has been overcome by mixing chambers, specifically designed for this purpose, that disperse the solutions into small droplets by the turbulence of forcing them under pressure into the mixing chamber, where the droplets from the two solutions homogenize. McClune and Fee[6] have described the design of such a mixer. With such an apparatus, dilutions of the superoxide stock yield concentrations of superoxide as high as 5 m$M$.[9] At such concentrations the oxygen resulting from the catalysis exceeds its solubility limit unless pressurized in the stop syringe. In a two-syringe stopped-flow instrument, it is usual to dilute the DMSO solution 25-fold with the aqueous solution containing enzyme, buffer, and metal chelator. A 2-msec deadtime can be achieved. In a three-syringe variation of this procedure, the stock solution is first diluted 10-fold with a buffered aqueous solution in the first mixer, and this solution is then diluted 10-fold again 3 msec later with a second solution containing enzyme. This method has the advantage that the 3-msec interval promotes more efficient mixing of the DMSO and aqueous solution. This procedure is easily adapted to modern, sequential mixing, four-syringe stopped-flow spectrophotometers.

Concentrations of initial superoxide immediately after mixing are obtained by extrapolation of the absorbance of superoxide to the time of mixing. The averaging of independent shots taken under identical conditions improves the signal-to-noise ratio in the progress curve. Hearn et al.[10] have used these procedures with a scanning stopped-flow spectrophotometer to observe the transient, product-inhibited human MnSOD that is formed during catalysis and has an absorbance in the visible region.

*Pulse Radiolysis*

Pulse radiolysis avoids the problems of mixing and using nonaqueous solvents but has its own limitations related to the relatively low amounts of superoxide that can be produced (less than about 30 $\mu M$). Pulse radiolysis experiments utilize a van de Graaff accelerator (generally about 2 MeV) to form the solvated electron $e_{aq}^-$. The reaction with water to form hydrogen atoms and hydroxyl radicals limits the lifetime of $e_{aq}^-$. Superoxide radical anions are generated from oxygen in aqueous solutions containing 10–30 m$M$ sodium formate as OH· scavenger.[11]

[9] C. Bull and J. A. Fee, *J. Am. Chem. Soc.* **107**, 3295 (1985).
[10] A. S. Hearn, C. K. Tu, H. S. Nick, and D. N. Silverman, *J. Biol. Chem.* **274**, 24457 (1999).
[11] H. A. Schwarz, *J. Chem. Ed.* **58**, 101 (1981).

Under these conditions the formation of superoxide radicals is more than 90% complete by the first microsecond after the pulse. Decay of $O_2{^{\cdot-}}$ is monitored spectrophotometrically as in stopped-flow spectrophotometry.

### Kinetic Mechanism of Human Manganese Superoxide Dismutase

Human MnSOD is a homotetramer of 22-kDa subunits, each of which contains one manganese atom.[12] The crystal structure of human MnSOD has been determined to 2.2-Å resolution,[13] and it is similar to that of the bacterial MnSODs, from *Bacillus stearothermophilus* at 2.4 Å,[14] and from *Thermus thermophilus* at 1.8 Å.[15] The structures of FeSOD from *Pseudomonas ovalis* at 2.9 Å[16] and from *Escherichia coli* at 1.85 Å[17] are also similar, including the same ligands to and approximate geometry about the metal. In human MnSOD, the geometry about the metal is trigonal bipyramidal with three histidine ligands (His-26, -74, and -163) and one aspartate (Asp-159) with a solvent molecule as a fifth ligand. These active-site ligands and geometry are conserved, with small variations, in the bacterial Mn and FeSODs. In addition, the tetrameric structure formed by the subunits of human MnSOD forms a ring of positive electrostatic charge surrounding an active site; this is suggested to enhance attraction of substrate[13] as has been shown to occur in Cu,ZnSOD.[18]

*Efficiency of Catalysis*

The superoxide dismutases function through sequential redox processes in which the metal cycles between oxidized and reduced states, a rough and perhaps inaccurate representation of which is given in Eqs. (1) and (2).

$$P-Mn^{3+} + O_2{^{\cdot-}} \leftrightharpoons [P-Mn^{3+}-O_2{^{\cdot-}}] \to P-Mn^{2+} + O_2 \quad (1)$$

$$P-Mn^{2+} + O_2{^{\cdot-}} + 2H^+ \overset{k_3}{\leftrightharpoons} [P-Mn^{2+}-O_2{^{\cdot-}}] + 2H^+ \overset{k_4}{\leftrightharpoons} P-Mn^{3+} + H_2O_2 \quad (2)$$

$$\phantom{P-Mn^{2+} + O_2{^{\cdot-}} + 2H^+ \overset{k_3}{\leftrightharpoons} [}k_5 \updownarrow k_{-5}$$

$$P-Mn^{3+}-X$$

---

[12] J. M. McCord, J. A. Boyle, E. D. Bay, L. J. Rizzolo, and M. L. Salin, *in* "Superoxide and Superoxide Dismutase" (A. M. Michelson, J. M. McCord, and I. Fridovich, eds.), p. 129. Academic Press, London, 1977.

[13] G. E. O. Borgstahl, H. E. Parge, M. J. Hickey, W. F. Beyer, R. A. Hallewell, and J. A. Tainer, *Cell* **71**, 107 (1992).

[14] M. W. Parker and C. C. F. Blake, *J. Mol. Biol.* **199**, 649 (1988).

[15] M. L. Ludwig, A. L. Metzger, K. A. Pattridge, and W. C. Stallings, *J. Mol. Biol.* **219**, 335 (1991).

[16] B. L. Stoddard, D. Ringe, and G. A. Petsko, *Protein Eng.* **4**, 113 (1990).

[17] M. S. Lah, M. M. Dixon, K. A. Pattridge, W. C. Stallings, J. A. Fee, and M. L. Ludwig, *Biochemistry* **34**, 1646 (1995).

[18] E. D. Getzoff, D. E. Cabelli, C. L. Fisher, H. E. Parge, M. S. Viezzoli, L. Banci, and R. A. Hallewell, *Nature (London)* **358**, 347 (1992).

TABLE I
STEADY-STATE CONSTANTS FOR DISMUTATION OF SUPEROXIDE
BY VARIOUS FORMS OF SOD

| Enzyme | $k_{cat}/K_m$ ($\mu M^{-1}$ sec$^{-1}$) | $k_{cat}$ (msec$^{-1}$) | Ref. |
|---|---|---|---|
| Human MnSOD | 800 | 40 | 21 |
| *Thermus thermophilus* MnSOD | 310 | 13 | 19 |
| *Escherichia coli* FeSOD | 330 | 26 | 9 |
| Bovine Cu,ZnSOD | ~2000 | >5000 | 9 |

Here P–Mn$^{3+}$ represents the metal bound to protein and P–Mn$^{3+}$–X is a product-inhibited complex. This scheme is the same as that used by Bull *et al.*[19] to fit stopped-flow data on catalysis by MnSOD from *T. thermophilus*. A scheme simpler than Eqs. (1) and (2) was found by McAdam *et al.*[20] to fit pulse radiolysis data from *B. stearothermophilus* MnSOD. Their scheme used only irreversible steps and the inhibited complex was formed directly in the bimolecular encounter of enzyme and $O_2^{\cdot-}$ and dissociated directly to Mn$^{3+}$SOD. Steady-state constants for catalysis of superoxide decay by MnSOD from *T. thermophilus* and for the human MnSOD are similar, with $k_{cat}$ near $10^4$ sec$^{-1}$ and $k_{cat}/K_m$ between $10^8$ and $10^9$ $M^{-1}$ sec$^{-1}$. These constants are compared with iron- and copper-containing SODs in Table I. In general, there is no pH dependence of the kinetic constants $k_{cat}$ and $k_{cat}/K_m$ for MnSOD in the range of pH 7.0 to 9.5. There is some decrease in $k_{cat}/K_m$ above pH 9.5 for wild type, perhaps because of the addition of a hydroxide ligand to the metal.[21] Similar observations were made with FeSOD.[9]

*Redox Properties of MnSOD.* The redox potentials of the SODs must be finely tuned to lie midway between that of the standard potentials for the reactions to be catalyzed: $O_2 + e^- \rightarrow O_2^{\cdot-}$ (−0.16 V) and $O_2^{\cdot-} + 2H^+ + e^- \rightarrow H_2O_2$ (+0.87 V).[22] Site-specific mutants with midpoint potentials outside this range will catalyze efficiently only one of the half-reactions and become a superoxide reductase or a superoxide oxidase.[23] Reduction midpoint potentials ($E°$ values) for the MnSODs and FeSODs have been difficult to measure because these enzymes do not equilibrate directly with electrodes and it is difficult to find mediators that equilibrate over reasonable times. The most definitive measurement of $E°$ among the bacterial Mn- and FeSODs is that of Vance and Miller,[24] who have

[19] C. Bull, E. C. Niederhoffer, T. Yoshida, and J. A. Fee, *J. Am. Chem. Soc.* **113**, 4069 (1991).
[20] M. E. McAdam, R. A. Fox, F. Lavelle, and E. M. Fielden, *Biochem. J.* **165**, 71 (1977).
[21] C. A. Ramilo, V. Leveque, Y. Guan, J. R. Lepock, J. A. Tainer, H. S. Nick, and D. N. Silverman, *J. Biol. Chem.* **274**, 27711 (1999).
[22] R. H. Holm, P. Kennepohl, and E. I. Solomon, *Chem. Rev.* **96**, 2239 (1996).
[23] S. I. Liochev and I. Fridovich, *J. Biol. Chem.* **275**, 38482 (2000).
[24] C. Vance and A.-F. Miller, *J. Am. Chem. Soc.* **120**, 461 (1998).

taken the precautions of using long-lived mediators that equilibrate properly in measurements extending up to 10 hr. They have obtained the same $E°$ values for SOD by using mediators that themselves have different $E°$ values, and have measured the same $E°$ value for both the oxidative and reductive directions. The value of $E°$ for FeSOD from *E. coli* is 223 ± 6 mV[24] and for human MnSOD it is 407 ± 21 mV.[25] For iron-substituted MnSOD $E°$ is −240 mV,[24] nearly 500 mV less than that for FeSOD, showing how finely tuned the active sites of MnSOD and FeSOD are to achieve the necessary redox potentials. This is such a sensitively determined system that it is not surprising that there is evidence of a significant effect on the values of $E°$ for MnSOD with certain active-site mutants of MnSOD.[26,27]

*Proton Transfer in Pathway*

Observation of solvent hydrogen isotope effects near 2.1 on the maximal velocity of catalysis[19] is consistent with proton transfer as a rate-contributing step in the pathway of a bacterial MnSOD. Enhancement of the maximal velocity of catalysis with increasing concentrations of primary amines as exogenous proton donors has been observed by Bull and Fee[9] for FeSOD. (Enhancement of $k_{cat}$ for MnSOD is more difficult to measure because of product inhibition, discussed below.) These experiments confirm the rate-contributing role of proton transfer from solution to FeSOD. This means that when the enzyme is working under maximal velocity or $k_{cat}$ conditions, proton transfer becomes rate contributing. The solvent hydrogen isotope effects on $k_{cat}/K_m$ are near unity. That is, when catalysis proceeds at low substrate concentration compared with $K_m$, under $k_{cat}/K_m$ conditions, it does not have limiting proton transfer steps. This understanding of solvent hydrogen isotope effects and mechanism is consistent with the kinetic equations derived for various mechanisms of MnSOD including Eqs. (1) and (2). A significant literature has accumulated on similar results in catalysis by carbonic anhydrase.[28,29]

There are many complications and unknown features for the proton transfer requirement in MnSOD and FeSOD. One is based on the important observation that the reduction of FeSOD is accompanied by the uptake of a proton in the pH range 7–10, although there is no evidence that this proton is involved in the catalytic pathway.[9] This is possibly the protonation of $Fe^{3+}OH^-$ to form $Fe^{2+}OH_2$.

---

[25] V. J.-P. Leveque, K. Carrie, C. Vance, H. S. Nick, and D. N. Silverman, *Biochemistry* **40**, 10586 (2001).
[26] Y. Hsieh, Y. Guan, C. K. Tu, P. J. Bratt, A. Angerhofer, J. R. Lepock, M. J. Hickey, J. A. Tainer, H. S. Nick, and D. N. Silverman, *Biochemistry* **37**, 4731 (1998).
[27] V. J.-P. Leveque, M. E. Stroupe, J. R. Lepock, D. E. Cabelli, J. A. Tainer, H. S. Nick, and D. N. Silverman, *Biochemistry* **39**, 7131 (2000).
[28] D. N. Silverman, C. K. Tu, X. Chen, S. M. Tanhauser, A. J. Kresge, and P. J. Laipis, *Biochemistry* **32**, 10757 (1993).
[29] H. Steiner, B.-H. Jonsson, and S. Lindskog, *Eur. J. Biochem.* **59**, 253 (1975).

Hence by analogy, Eqs. (1) and (2) might be written as follows:

$$P-Mn^{3+} + O_2^{\cdot-} + H^+ \rightarrow H^+-P-Mn^{2+} + O_2 \quad (3)$$

$$H^+-P-Mn^{2+} + O_2^{\cdot-} + H^+ \rightleftharpoons P-Mn^{3+} + H_2O_2 \quad (4)$$

Here P represents the enzyme and $H^+$ placed before P indicates a protonated group of the enzyme or possibly metal-bound water. These equations are different from Eqs. (1) and (2) in that proton transfer is also associated with the first stage of catalysis. The activity and azide inhibition of FeSOD imply that the reduced state $Fe^{2+}SOD$ contains a group of $pK_a$ 8.6–9.0.[9] Observation of this $pK_a$ in reduced FeSOD from *E. coli* by proton nuclear magnetic resonance (NMR) showed it to influence the chemical shifts of many active-site residues, perhaps another manifestation of a hydrogen bond network.[30] Further NMR studies of Y34F FeSOD associated this $pK_a$ with Tyr-34, supporting evidence of its role in this network.[31] An excellent discussion of the role of proton transfer in the cataytic pathway of FeSOD is presented by Lah et al.[17]

*Product Inhibition*

Initial studies of MnSOD from *B. stearothermophilus* used pulse radiolysis to determine that there is an initial burst of activity followed by an extended region of zero-order decay of superoxide. The superoxide consumed in this initial transient is not stoichiometric and greatly exceeds the enzyme concentration. The zero-order region was explained as being due to the presence of a reversibly inhibited form of the enzyme that can interconvert to an active form.[20] A solvent hydrogen isotope effect on the zero-order rate constant during catalysis by human MnSOD was measured at $3.1 \pm 0.2$,[32] a value significantly large enough to indicate involvement of a proton transfer. One interpretation is that active enzyme is regenerated by a rate-contributing proton transfer allowing the dissociation of peroxide (or, possibly, peroxoyl anion) from the enzyme, and that this rate-contributing proton transfer is from water or a residue of the enzyme itself. Direct observation of an absorbance at 420 nm attributed to this inhibited complex provided further information on its rate of formation and decay.[10]

Bull et al.[19] observed this absorption spectrum of inhibited MnSOD from *T. thermophilus* during the zero-order phase of catalysis and, on the basis of comparison with visible absorption spectra of inorganic complexes, suggested product inhibition by peroxide. More specifically, they suggested a side-on peroxo complex of Mn(III)SOD resulting from the oxidative addition of $O_2^{\cdot-}$ to Mn(II)SOD. This inactive complex of the enzyme is represented as $P-Mn^{3+}-X$ in Eq. (2). The solvent hydrogen isotope effect on the zero-order region of catalysis[32] as well as

---

[30] D. L. Sorkin and A.-F. Miller, *Biochemistry* **36**, 4916 (1997).
[31] D. L. Sorkin, D. K. Duong, and A.-F. Miller, *Biochemistry* **36**, 8202 (1997).
[32] J. L. Hsu, Y. Hsieh, C. K. Tu, D. O'Connor, H. S. Nick, and D. N. Silverman, *J. Biol. Chem.* **271**, 17687 (1996).

more detailed examination of step 5 of Eq. (2) by pulse radiolysis[33] suggest that protonation of the side-on peroxo complex to an Mn(III)-hydroperoxide intermediate is involved in its conversion to free enzyme. It is interesting that FeSOD does not exhibit evidence of a side-on inhibited complex,[9] although such side-on complexes have been observed with inorganic ferric complexes.[34] Both Cu,ZnSOD and FeSOD are irreversibly inactivated by hydroxyl radicals produced by the Fenton reactions of $H_2O_2$ with their metal ions; MnSOD is not irreversibly inactivated in this manner. Thus, in this review, product inhibition refers to the transitory inhibited complex of the zero-order phase observed during catalysis by MnSOD, and not to the irreversible type of inactivation that occurs in FeSOD and Cu,ZnSOD.

## Identification of Catalytic Residues in Human MnSOD

*Crystal Structures*

The crystal structure of human MnSOD shows a network of hydrogen-bonded side chains and water in the active-site cavity (Fig. 1).[13,34a] This network comprises the following continuous sequence of hydrogen-bonded side chains and solvent molecules: Mn–solvent–Q143–Y34–$H_2O$–H30–Y166, this latter residue

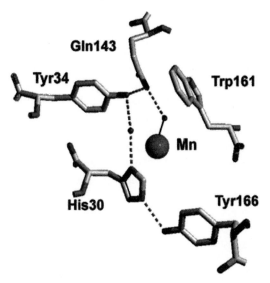

FIG. 1. The hydrogen bond network in the active-site region of human MnSOD from the data of Borgstahl *et al.*, *Cell* **71,** 107 (1992). The ligands of the manganese are His-26, His-74, His-163, and Asp-159 (not shown), and a water molecule, which is indicated as a small sphere. A second water is shown hydrogen bonded between Tyr-34 and His-30. Tyr-166 is from an adjacent subunit of the homotetramer. Another possibility for the inhibited enzyme involves an inner sphere complex of peroxide anion with the manganese, with catalysis in Eq. (2) occurring in an outer sphere complex [Cabelli *et al.*, in "Biomimetic Oxidations Catalyzed by Transition Metal Complexes" (B. Meunier, ed.), p. 461. Imperial College Press].

## TABLE II
### STEADY-STATE KINETIC CONSTANTS FOR DECAY OF SUPEROXIDE CATALYZED BY HUMAN MnSOD AND SITE-SPECIFIC MUTANTS AT pH 9.4–9.6 AND 20°

| Enzyme | $k_{cat}/K_m$ ($\mu M^{-1}$ sec$^{-1}$) | $k_{cat}$ (msec$^{-1}$) | Ref. |
|---|---|---|---|
| Wild type | 800 | 40 | 21 |
| Q143N | 0.8 | 0.3 | 26 |
| Y34F | 870 | 3 | 36 |
| H30N | 130 | 4 | 21 |
| Y166F | 95 | 1 | 21 |

originating on an adjacent subunit. This hydrogen-bonded network has been altered or interrupted by amino acid replacements at each of these sites; the catalytic properties of some of these site-specific mutants are shown in Table II.

Crystal structures are available for each of the mutants of Table II except Y166F MnSOD. In the vicinity of the manganese, the positions of the inner shell ligands are shifted to a negligible extent compared with the wild-type enzyme in each case. The conservative replacement Gln-143 → Asn in human MnSOD, corresponding to shortening the length of the side chain by one methylene group, increased the distance between the amide nitrogen and the manganese by 1.7 Å, whereas the side-chain carbonyl oxygen distance remained unchanged.[26] Although this mutation could potentially leave a cavity in the active site, in the crystal structure a water molecule (which is not a ligand of the metal) bridges Asn-143 and the hydroxyl/water molecule which is a ligand of the manganese. Because of the shift of the Asn-143 side chain and the altered position of Tyr-34, the hydrogen bond between these two residues can no longer be maintained in Q143N. This mutation interrupts the hydrogen bond network in the active site. Crystal structures are also reported and analyzed for Y34F, Q146L, and Q146H MnSOD from *E. coli,* in which residue 146 is the equivalent of residue 143 in the human enzyme.[35]

In the wild-type MnSOD, the hydroxyl group of Tyr-34 forms a hydrogen bond to the amide side chain of Gln-143 (2.9 Å in distance; Fig. 1); this interaction is absent in Y34F because of the missing hydroxyl group. Many more details are provided by Guan *et al.*[36] for human Y34F MnSOD and by Edwards *et al.*[35]

---

[33] A. S. Hearn, M. E. Stroupe, D. E. Cabelli, J. R. Lepock, J. A. Tainer, H. S. Nick, and D. N. Silverman, *Biochemistry* **40,** 12051 (2001).

[34] F. Neese and E. I. Solomon, *J. Am. Chem. Soc.* **120,** 12829 (1998).

[34a] D. E. Cabelli, D. Riley, J. A. Rodriguez, J. S. Valentine, and H. Zhu, *in* "Biomimetic Oxidations Catalyzed by Transition Metal Complexes" (B. Meunier, ed.), p. 461. Imperial College Press, 2000.

[35] R. A. Edwards, M. M. Whittaker, J. W. Whittaker, E. N. Baker, and G. B. Jameson, *Biochemistry* **40,** 15 (2001).

[36] Y. Guan, M. J. Hickey, G. E. O. Borgstahl, R. A. Hallewell, J. A., Lepock, D. O'Connor, Y. Hsieh, H. S. Nick, D. N. Silverman, and J. A. Tainer, *Biochemistry* **37,** 4722 (1998).

for *E. coli* Y34F MnSOD. Therefore, the most significant structural change in the Y34F mutant compared with the wild type is the breaking of the hydrogen-bonded network in the active site. The crystal structure of the mutant H30N human MnSOD shows minimal changes in the orientation of residues in the active-site cavity, with Asn-30 having the same dihedral angle about the $C_\alpha$–$C_\beta$ bond as His-30 in the wild-type enzyme.[21] The primary structural changes in the H30N mutant involve altered local hydrogen bonds to solvent and side chains. A water molecule between Tyr-34 and Asn-30 was also found in the H30N mutant as in the wild type; however, in the H30N mutant it lost its hydrogen bonding to the side chain of Asn-30. Therefore, the hydrogen-bonded network is also broken in the H30N mutant. In addition, Tyr-166 from the adjacent subunit of the dimer, which in the wild type is hydrogen bonded to the side chain of His-30, is no longer hydrogen bonded to Asn-30 in the H30N mutant because of the shorter length and orientation of the side chain of Asn-30. This hydrogen bond is also removed in Y166F, although there is no crystal structure of this mutant.

*Catalytic Activity of Site-Specific Mutants*

The steady-state kinetic constants for the catalysis of superoxide decay by mutants at several active-site positions of human MnSOD are given in Table II. These steady-state constants were obtained by measurements of initial velocity of catalysis by stopped-flow spectrophotometry or by pulse radiolysis. The activity that remains for these mutants is substantial; thus, these residues are not absolutely essential for catalysis. This conclusion was also reached for H30A MnSOD from *Saccharomyces cerevisiae* by Borders *et al.*,[37] for Y34F MnSOD from *E. coli*,[38,39] and for Y34F FeSOD from *E. coli*.[31] These reports used indirect kinetic methods, such as a pyrogallol autooxidation assay or the xanthine oxidase and cytochrome *c* assay, which are not amenable to the determination of $k_{cat}/K_m$ and $k_{cat}$.

The substitutions at residues 30, 34, and 166 in Table II reduce $k_{cat}$ for MnSOD by about an order of magnitude and the resulting values of $k_{cat}$ have a solvent hydrogen isotope effect near 2, facts that are consistent with a reduced effectiveness of the proton delivery network in each of these variants. This is perhaps another manifestation of the hydrogen bond network observed in the crystal structure. It is interesting that this is approximately the magnitude by which $k_{cat}$ for $CO_2$ hydration catalyzed by carbonic anhydrase II is reduced when the intramolecular proton shuttle His-64 is replaced by alanine.[40] Moreover, the observation that

---

[37] C. L. Borders, M. J. Bjerrum, M. A. Schirmer, and S. G. Oliver, *Biochemistry* **37**, 11323 (1988).
[38] T. Hunter, K. Ikebukuro, W. H. Bannister, J. V. Bannister, and G. J. Hunter, *Biochemistry* **36**, 4925 (1997).
[39] M. M. Whittaker and J. W. Whittaker, *Biochemistry* **36**, 8923 (1997).
[40] C. K. Tu, D. N. Silverman, C. Forsman, B.-H. Jonsson, and S. Lindskog, *Biochemistry* **28**, 7913 (1989).

many mutations at position 30 result in about the same reduced value of $k_{cat}$ compared with wild type[21] signifies that no residue used to replace His-30 participates in a proton transfer network as effective in catalysis as the network in the wild type. Mutation of Gln-143, which is hydrogen bonded with the manganese-bound solvent, causes the additional effect of altering the redox potential significantly and is discussed separately below. Mutations at residues 30, 34, and 166, which are farther from the metal, probably also alter the redox potential but to a smaller extent than replacements at position 143.

For the wild-type MnSOD, $k_{cat}/K_m$ at $8 \times 10^8 \, M^{-1} \, sec^{-1}$ is nearly encounter controlled. The 6-fold lower value for the mutant H30N indicates a change in the rate-limiting step, in the sense that diffusion is less limiting. Catalysis by the mutant Y34F MnSOD showed $k_{cat}$ to be decreased but $k_{cat}/K_m$ remained unchanged compared with wild type (Table II). This was interpreted as an effect of Tyr-34 on the proton transfer processes that limit $k_{cat}$, but that have no effect on $k_{cat}/K_m$, which involves no rate-contributing proton transfers. In this property, Y34F MnSOD is like carbonic anhydrase, in which replacement of the proton shuttle residue His-64 causes a decrease in $k_{cat}$ by approximately 10-fold with no change in $k_{cat}/K_m$.[40]

One significant change caused by the Gln → Asn replacement is the apparent shift of the reduction potential of manganese to a more positive value that renders Q143N MnSOD stable in the Mn(II) state under purification conditions, whereas the wild-type MnSOD is clearly and predominantly in the Mn(III) state.[26] Support for this conclusion comes first from the absence of a visible absorption spectrum of the mutant Q143N, whereas wild-type Mn(III)SOD has a broad absorbance with a maximum at 480 nm. A second result indicating the predominance of Mn(II) in the resting state of Q143N MnSOD is the presence of a complex electron paramagnetic resonance (EPR) spectrum.[26] Wild-type MnSOD from *E. coli* reduced to the Mn(II) state by dithionite also showed spectra with many resonances in this region; SOD containing Mn(III) is EPR silent.[41]

The replacement of Gln-143 by asparagine in human MnSOD causes a decrease in the value of $k_{cat}/K_m$ for $O_2^{--}$ decay by three orders of magnitude and a decrease in $k_{cat}$ by two orders of magnitude compared with the wild type at pH 9.6 (Table II). The Q146L and Q146H mutants of *E. coli* MnSOD retain 5–10% of the activity of wild type.[35] Thus, the mutant Q143N MnSOD exhibits catalysis that is not diffusion controlled. It is unlikely that the replacement of Gln-143 by asparagine has significantly impeded the access of superoxide to the active site; rather, this mutation has changed the rate-limiting step for $k_{cat}/K_m$ to a step other than diffusion of superoxide. A new rate-limiting step or steps for Q143N is also consistent with the appearance of an apparent value of the $pK_a$ for $k_{cat}/K_m$ to a value near or below 8.5.[26] This $pK_a$ for $k_{cat}/K_m$ is not observed in the wild-type MnSOD.[32] The altered positions of Tyr-34 and Asn-143 in Q143N may impair the proton transfer to the

[41] J. A. Fee, E. R. Shapiro, and T. H. Moss, *J. Biol. Chem.* **251**, 6157 (1976).

active site, resulting in a reduced value of $k_{cat}$. Kinetic and structural studies of site-directed mutants replacing Gln-143 with several other residues show catalysis similar to that of Q143N, with additional evidence of changes in the redox potential and the breaking of the hydrogen bond network.[27]

Trp-161 is a highly conserved residue that forms a hydrophobic side of the active-site cavity of MnSOD, with its indole ring adjacent to and about 5 Å from the manganese (Fig. 1). In the structure of W161F MnSOD the side chain of Phe-161 superimposes on the indole ring of Trp-161 in the wild type.[42] However, in the mutant, the hydroxyl side chain of Tyr-34 is 3.9 Å from the manganese, closer by 1.2 Å than in the wild type. Moreover, there appear to be no significant changes in the backbone positions between the mutant and wild type. What has occurred in the mutant is a general tightening of the hydrogen-bonded chain from the metal-bound aqueous ligand to His-30, which is characterized by the shortening of several hydrogen bonds. The conservative replacement of Trp-161 with phenylalanine resulted in an enzyme that enhanced only the oxidative stage of catalysis [Eq. (1)] with the apparent $k_{cat}/K_m$ reduced by only 3-fold compared with wild type. The catalysis in the half-cycle of Eq. (2) could not be measured because of the efficient formation of product-inhibited complex described in the next section.

## Product Inhibition of Human MnSOD

The catalysis of superoxide decay by MnSOD is unique among all the SODs in its susceptibility to product inhibition. This is different from the inhibition due to Fenton chemistry that is characterized for FeSOD[43] and Cu,ZnSOD,[44] but is most likely due to peroxo complexes at the manganese. In the reverse direction, the rapid mixing of $H_2O_2$ with human Mn(III)SOD and subsequent scanning stopped-flow spectrophotometry resulted in the appearance of a visible absorption spectrum with a maximum at 420 nm and an extinction coefficient $\epsilon_{420}$ of $\sim 500\ M^{-1}\ cm^{-1}$.[10] This is similar to that obtained for the inhibited complex in the dismutation direction after mixing $O_2^{·-}$ and MnSOD. This study shows that the inhibited complex can be reached through both the forward ($O_2^{·-}$) and reverse ($H_2O_2$) reactions, supporting a mechanism in which the zero-order phase results from product inhibition. This result renders much less likely the attribution of this zero-order phase to a conformational change not related to product binding.

Site-specific mutagenesis has revealed a surprising array of residues in the active-site cavity that influence the formation of the product-inhibited complex. Cabelli et al.[42] show a typical pulse radiolysis measurement of catalysis by W161F

---

[42] D. E. Cabelli, Y. Guan, V. Leveque, A. S. Hearn, J. A. Tainer, H. S. Nick, and D. N. Silverman, *Biochemistry* **36**, 11686 (1999).
[43] W. F. Beyer and I. Fridovich, *Biochemistry* **26**, 1251 (1987).
[44] R. C. Bray, S. A. Cockle, E. M. Fielden, P. B. Roberts, G. Rotilio, and L. Calabrese, *Biochem. J.* **139**, 43 (1974).

MnSOD extending to long times (>10 msec). The rate constant of the zero-order region of $O_2^{\cdot-}$ decay, $k_0/[E_0]$, was observed to be 50 $\text{sec}^{-1}$; this is about 10-fold smaller than for wild type, indicating greater product inhibition by W161F MnSOD.[42] The mutant Y34F human MnSOD also showed greater product inhibition than wild type, with $k_0/[E_0] = 80\ \text{sec}^{-1}$.[36] On the other hand, several variants of human MnSOD show much less product inhibition than wild type. Generally, this class of variants has lower overall activity; that is, there is less product inhibition when the rate of product formation is lower; examples are Q143N[26] and H30N human MnSOD.[21] It should be pointed out that the changes in product inhibition among these mutants may be a result of different values of the rate constants of steps 3 and 4 rather than of $k_5$ and $k_{-5}$ of Eq. (2).

Cabelli et al.[42] showed that the extent of inhibition of W161F Mn(II)SOD is nearly complete with the consumption of a concentration of superoxide equivalent to the concentration of this reduced enzyme. This is important in showing that the inhibited complex is formed by reaction of superoxide with the reduced form of enzyme. So favorable is the formation of this inhibited state that, for the most part, the reactants W161F Mn(II)SOD and $O_2^{\cdot-}$ proceed directly to the product-inhibited form; that is, a catalytic burst is attributed to the first half-cycle [Eq. (1)] and the subsequent consumption of $O_2^{\cdot-}$ yields the inhibited complex. The oxidation of the manganese in W161F MnSOD by superoxide appears mainly to proceed through the inhibited complex [$k_5 > k_4$ of Eq. (2)]; however, in the wild-type human MnSOD, complex formation appears more competitive with direct oxidation ($k_4 \approx k_5$). In contrast, for *T. thermophilus*, $k_4$ is approximately 40 times faster than $k_5$.[19]

## Summary

Measurement of catalysis by MnSOD using direct observation of the UV absorbance of superoxide allows determination of steady-state catalytic constants. Stabilizing superoxide in aprotic solvents such as dimethyl sulfoxide permits the use of stopped-flow spectrophotometry, although significant information is lost in the 2- to 4-msec mixing time; generating superoxide by pulse radiolysis requires no mixing time. Studies show that $k_{cat}/K_m$ for the decay of superoxide catalyzed by MnSOD proceeds at diffusion control. Investigations using solvent hydrogen isotope effects and enhancement of catalysis by exogenous proton donors show that $k_{cat}$ near $10^4\ \text{sec}^{-1}$ contains a significant contribution from proton transfer steps. The active site of MnSOD is dominated by a hydrogen bond network comprising the manganese-bound aqueous ligand, the side chains of four residues (Gln-143, Tyr-34, His-30, and Tyr-166 from an adjacent subunit), as well as other water molecules. Interrupting this hydrogen bond network by conservative replacement of residues 30, 34, and 166 causes a 10- to 40-fold decrease in maximal velocity, interpreted as an effect on proton transport to the active site, with smaller effects on $k_{cat}/K_m$. Replacement of Gln-143 causes a much greater decrease in catalytic activity, by two to three orders of magnitude, and causes significant changes to the redox potential as well. During catalysis, MnSOD is inhibited by a peroxide complex of the metal in the active site, different from the inhibition of FeSOD and Cu,ZnSOD by Fenton chemistry. Site-specific mutagenesis of active-site residues alters the extent of product inhibition of MnSOD as well, indicating that this is not only a property of the metal. The replacement of Trp-161 with phenylalanine results in a variant that is completely blocked in catalysis by product inhibition.

# [7] Extracellular Superoxide Dismutase

*By* STEFAN L. MARKLUND

## Introduction

Mammalian extracellular superoxide dismutase (EC 1.15.1.1, EC-SOD) is a secreted, tetrameric, copper- and zinc-containing glycoprotein with a molecular mass of the subunit protein of about 25,000 Da.[1-3] EC-SOD is the major SOD isoenzyme in extracellular fluids such as plasma, lymph,[4] and synovial fluid.[5] It also occurs in cerebrospinal fluid,[6] and seminal plasma,[7] but here the normally cytosolic Cu,ZnSOD shows higher activity. In mammals 90–99% of the EC-SOD is located in the tissues. Most of this tissue enzyme is probably bound to heparan sulfate proteoglycans on cell surfaces, in basal membranes, and in the connective tissue matrix.[8-10] However, in some cases, for example, placenta,[11] and lung during development,[12] in mouse brain,[13] and mouse aorta (S. L. Marklund, unpublished data, 2001), a large portion of the enzyme stains intracellularly. It may be stored in

---

[1] S. L. Marklund, *Proc. Natl. Acad. Sci. U.S.A.* **79**, 7634 (1982).
[2] L. Tibell, K. Hjalmarsson, T. Edlund, G. Skogman, Å. Engström, and S. L. Marklund, *Proc. Natl. Acad. Sci. U.S.A.* **84**, 6634 (1987).
[3] L. M. Carlsson, S. L. Marklund, and T. Edlund, *Proc. Natl. Acad. Sci. U.S.A.* **93**, 5219 (1996).
[4] S. L. Marklund, E. Holme, and L. Hellner, *Clin. Chim. Acta* **126**, 41 (1982).
[5] S. L. Marklund, A. Bjelle, and L.-G. Elmqvist, *Ann. Rheum. Dis.* **45**, 847 (1986).
[6] J. Jacobsson, P. A. Jonsson, P. M. Andersen, L. Forsgren, and S. L. Marklund, *Brain* **124**, 1461 (2001).
[7] R. Peeker, L. Abramsson, and S. L. Marklund, *Hum. Mol. Reprod.* **3**, 1061 (1997).
[8] K. Karlsson and S. L. Marklund, *Lab. Invest.* **60**, 659 (1989).
[9] K. Karlsson, J. Sandström, A. Edlund, T. Edlund, and S. L. Marklund, *Free Radic. Biol. Med.* **14**, 185 (1993).
[10] K. Karlsson, J. Sandström, A. Edlund, and S. L. Marklund, *Lab. Invest.* **70**, 705 (1994).
[11] K. A. Boggess, H. H. Kay, J. D. Crapo, W. F. Moore, H. B. Suliman, and T. D. Oury, *Am. J. Obstet, Gynecol.* **183**, 199 (2000).

secretory vesicles, and whether it there contributes to protection against the superoxide radical is unknown. Consequently, the total content of EC-SOD measured in a tissue extract does not necessarily reflect the SOD activity in the extracellular space.

A distinguishing feature of EC-SOD is its affinity for heparin and other sulfated glycosaminoglycans, primarily heparan sulfate proteoglycan,[14] which determines the distribution and retention of the enzyme *in vivo*.[8-10] The affinity is conferred by a carboxy-terminally located positively charged heparin-binding domain.[15] On chromatography on heparin-Sepharose, plasma EC-SOD from most mammals is divided into at least three fractions: A, which lacks affinity; B, which shows intermediate affinity; and C, which has high affinity.[16,17] The fractions with reduced affinity are composed of heterotetramers containing subunits with proteolytically truncated carboxy-terminal ends.[15,18] Tissue EC-SOD is mainly composed of homotetrameric high-affinity type C.[18] The rat lacks the high-affinity type C.[17] The heparin-binding domain of the rat EC-SOD is identical to those of the mouse and human enzymes. However, owing to a difference in one critical amino acid residue in a subunit interaction area, rat EC-SOD is dimeric.[3] High heparin affinity thus requires the cooperative action of four heparin-binding domains.

The middle portion of the EC-SOD sequence shows high similarity with that part of the CuZn-SOD sequence that defines the active site. All the ligands to the prosthetic metals as well as several others for the function of important residues can be identified in EC-SOD.[2] The spectral properties of EC-SOD are similar to those of the Cu,ZnSODs,[19] and the enzyme behaves like Cu,ZnSOD in the Tsuchihashi chloroform–methanol procedure.[1] The consequence of this is that all the inhibitors and treatments commonly used to identify the cytosolic Cu,ZnSODs, affect EC-SOD almost identically. Thus, EC-SOD is reversibly inhibited by cyanide and azide, and inactivated by diethyldithiocarbamate, phenylglyoxal, and hydrogen peroxide.[20] In general, the EC-SODs are more sensitive to these treatments than the Cu,ZnSODs.[20] We have so far failed to identify any inhibitor that efficiently distinguishes between the mammalian copper-containing SODs; EC-SOD and Cu,ZnSOD.

---

[12] E. Nozik-Grayk, C. S. Dieterle, C. A. Piantadosi, J. J. Enghild, and T. D. Oury, *Am. J. Physiol. Lung. Cell. Mol. Physiol.* **279**, L997 (2000).
[13] T. D. Oury, J. P. Card, and E. Klann, *Brain Res.* **850**, 96 (1999).
[14] K. Karlsson, U. Lindahl, and S. L. Marklund, *Biochem. J.* **256**, 29 (1988).
[15] J. Sandström, L. Carlsson, S. L. Marklund, and T. Edlund, *J. Biol. Chem.* **267**, 18205 (1992).
[16] K. Karlsson and S. L. Marklund, *Biochem. J.* **242**, 55 (1987).
[17] K. Karlsson and S. L. Marklund, *Biochem. J.* **255**, 223 (1988).
[18] J. Sandström, K. Karlsson, T. Edlund, and S. L. Marklund, *Biochem. J.* **294**, 853 (1993).
[19] L. Tibell, R. Aasa, and S. L. Marklund, *Arch. Biochem. Biophys.* **304**, 429 (1993).
[20] S. L. Marklund, *Biochem. J.* **220**, 269 (1984).

Here we describe some methods that can be used to probe the physical properties of EC-SOD and that can aid in distinguishing the enzyme from the Cu,ZnSODs.

## Experimental Procedures

*Preparation of Samples*

Tissues are homogenized with a mixer [Ultra-Turrax (IKA Labortechnik, Staufen, Germany) or similar apparatus] in 10–25 volumes of ice-cold potassium phosphate, pH 7.4, with 0.3 $M$ KBr (chaotropic salt that may increase the extraction severalfold from some tissues[21]), 3 m$M$ diethylenetriaminepentaacetic acid, 100 kIU of aprotinin, and 0.5 m$M$ phenylmethylsulfonyl fluoride [the latter three additions to inhibit proteases; commercial cocktails such as Complete (Boehringer, Mannheim, Germany) can also be used]. The tissue extract may then be subjected to ultrasonication followed by centrifugation (20,000$g$, 20 min, 4°). The supernatants are used for the separations.

For subsequent separation on concanavalin A (ConA)-Sepharose, by gel chromatography, or with immobilized antibodies, samples can be processed fresh or after storage below $-70°$. The affinity for heparin is, however, sensitive to treatment of the samples and is easily lost owing to proteolytic truncation of the carboxy-terminal heparin-binding domain.[18] To avoid this, tissues should preferably be processed fresh. If the tissues are allowed to thaw after freezing, we have noted a loss of heparin affinity. We interpret the effect to be due to proteolytic enzymes released by freezing-induced cellular and subcellular rupture. Frozen tissues can, however, be pulverized at liquid $N_2$ temperature, followed by addition of ice-cold extraction buffer and subsequent sonication without loss of heparin affinity.[18] In the final extract, the antiproteolytic measures seem to prevent further degradation.

The stability problem is apparently not so great in plasma. The plasma EC-SOD heparin affinity pattern was not influenced by freezing and thawing or by a 3-day storage in a refrigerator.[16] Apparently the large amounts of protein and the numerous antiproteases in plasma confer protection. EDTA (or citrate) should be used as anticoagulant. Heparin interferes with the heparin-Sepharose procedure and also induces a large increase in the apparent molecular mass of EC-SOD fraction C.[17]

*Highly Diluted Samples*

Sometimes little material is available, which is why larger dilutions than suggested above must be used. This will lead to markedly reduced recovery of EC-SOD by the various procedures suggested, probably because of unspecific adsorption of the enzyme to surfaces. We have found that addition of 0.2% bovine serum albumin to the extracts and media used for the separations generally prevents the problem.

[21] S. L. Marklund, *J. Clin. Invest.* **74,** 1398 (1984).

## Analysis of Superoxide Dismutase Activity

The EC-SOD activity is low both in extracellular fluids and in most tissue extracts. The amount of EC-SOD in the fractions collected by the suggested procedures is consequently low and varies between about 4 and 200 ng/ml. The corresponding activities are difficult to analyze with the more common SOD assays. Highly sensitive assays are necessary, and we use only the direct spectrophotometric assay with superoxide obtained from $KO_2$.[22,23]

## Separation by Gel Chromatography

The principle behind the procedure is that most EC-SODs show an apparent molecular mass of about 150,000 Da on gel chromatography, whereas Cu,ZnSOD elutes at a position corresponding to 30,000 Da. The high apparent molecular mass is dependent on the hydrodynamic effects of the single carbohydrate chains on the subunits; mutant nonglycosylated human EC-SOD elutes close to the expected molecular mass of 97,000 Da.[24] EC-SOD in plasma from human, cat, pig, sheep, mouse, rabbit, guinea pig, and rat can easily be separated from the relatively small Cu,ZnSOD peaks.[17] The rat EC-SOD, which is dimeric,[3] elutes at an apparent molecular mass of 90,000 Da. Such atypical molecular masses might also occur in EC-SODs from other phyla. In tissue extracts EC-SOD is mostly a minor isoenzyme, but it is still usually possible to assess the amount of EC-SOD from the gel chromatography pattern. This was not possible in rat tissue homogenates, however, where the EC-SOD peak was hidden behind the large Cu,ZnSOD and MnSOD peaks.[25]

*Procedure.* Any gel chromatography system with good resolution in the range of 30–150 kDa can be used. We typically use the following procedure. The sample (1–5 ml) is applied to a column (1.6 × 90 cm) of Sephacryl S-300 (AP Biotech, Uppsala, Sweden), eluted at 20 ml/hr with 10 m$M$ potassium phosphate (pH 7.4)–0.15 $M$ NaCl, and collected in 3-ml fractions. The absorbance at 280 nm and the SOD activity are determined in collected fractions.

## Separation on Concanavalin A-Sepharose

EC-SOD is, unlike the other SOD isoenzymes, a glycoprotein and has been found to bind to lectins such as concanavalin A, lentil lectin, and wheat germ lectin.[1,2] Chromatography of samples on concanavalin A-substituted Sepharose

---

[22] S. L. Marklund, *J. Biol. Chem.* **251**, 7504 (1976).
[23] S. L. Marklund, *in* "Handbook of Methods for Oxygen Radical Research" (R. Greenwald, ed.), p. 249. CRC Press, Boca Raton, Florida, 1985.
[24] A. Edlund, T. Edlund, K. Hjalmarsson, S. L. Marklund, J. Sandström, M. Strömkvist, and L. Tibell, *Biochem. J.* **288**, 451 (1992).
[25] S. L. Marklund, *Biochem. J.* **222**, 649 (1984).

(ConA-Sepharose; AP Biotech) has proved to be a useful and reliable procedure for distinguishing EC-SOD from other SOD isoenzymes. The EC-SOD is bound and can then be eluted with $\alpha$-methylmannoside. The recovery of EC-SOD in the suggested procedure is about 70–80%. As tested with pure enzyme or by analysis of extracts and collected fractions with an enzyme-linked immunosorbent assay (ELISA) for EC-SOD, the recovery from human extracts is regularly close to 75%. We always compensate our results for that.

*Procedure.* The chromatography is carried out manually in a stepwise fashion. The tissue extract (1–2 ml) is applied to a 1-ml ConA-Sepharose column equilibrated with 50 m$M$ Na-HEPES (pH 7.0)–0.25$M$ NaCl. The sample is applied in 0.5-ml portions at 5-min intervals. After 5 min, 3 ml of equilibration buffer is added. The eluting fluid from the tissue extract and buffer additions is collected and contains the SOD activity that lacks ConA affinity. The column is then washed with 10 ml of equilibration buffer. EC-SOD is finally eluted with 5 ml of 0.5 $M$ $\alpha$-methylmannoside added in 1-ml portions at 5-min intervals. The column is regenerated with 5 ml of $\alpha$-methylmannoside followed by 10 ml of equilibration buffer.

*Chromatography on Heparin-Sepharose*

*Procedure.* The chromatography is carried out on a 2-ml heparin-Sepharose column equilibrated with 15 m$M$ sodium cacodylate, pH 6.5, with 50 m$M$ NaCl, eluted at 5 ml/hr. The tissue extract (1–10 ml) or plasma (up to 2 ml) is applied. Many plasma proteins bind to heparin, and if more than 2 ml is applied there is a risk of column saturation.[16] The samples should be dialyzed against the equilibration buffer. After application of samples, the column is eluted with 15 ml of the buffer. Thereafter, bound proteins are eluted with a linear gradient of NaCl in the buffer (0–1 $M$; total volume, 50 ml). The eluent is collected in 1.5-ml fractions, and the SOD activity and absorbance at 280 nm are determined.

By definition EC-SOD A is the fraction that elutes without binding, B is the fraction that elutes early in the gradient, and C is the fraction that elutes relatively late. In human, cat, pig, mouse, rabbit, and guinea pig plasma, the B fractions eluted between 0.17 and 0.30 $M$ NaCl, and the C fractions eluted between 0.42 and 0.62 $M$ NaCl.[16,17] In all these species Cu,ZnSOD and MnSOD eluted without binding, together with EC-SOD A. To distinguish between EC-SOD A and the other isoenzymes in the nonbinding fraction, the ConA-Sepharose procedure or immobilized antibodies can be used. All SOD activity in the gradient in these species was given by EC-SOD. Note, however, that it cannot be taken for granted that all SOD activity in the gradient represents EC-SOD. The strongly negatively charged heparin-Sepharose gel will also function as a cation-exchange chromatography column. The net charge of Cu,ZnSODs and MnSODs from other taxa might be such that they bind to the column.

## Analyses with Antibodies

Despite some sequence similarities we have so far not found any antigenic cross-reactivity between human EC-SOD and Cu,ZnSODs from a few mammalian species. Apparently immunological methods can safely be used for specific identification of the two isoenzymes.

Western blots and ELISAs function well for both, and are standard methods. Regarding immunohistochemistry we generally find the staining for Cu,ZnSODs to be more robust than that for EC-SOD. The antigenic reactivity of EC-SOD appears sensitive to strongly denaturing or cross-linking fixatives such as formaldehyde and glutaraldehyde, and better staining is often seen with frozen sections or with alcohol fixation. However, proteolytic digestion (proteinase 1; Ventana, Tucson, AZ) has markedly improved the staining of formalin-fixed paraffin-embedded samples.

A convenient means of distinguishing between EC-SOD and Cu,ZnSOD in extracts is to use antibodies against the enzymes immobilized on particles such as Sepharose. The extracts are then incubated with the immobilized antibodies or control particles, and after centrifugation the (cyanide-sensitive) SOD activities of the supernatants are analyzed. The lost activity corresponds to the SOD isoenzyme against which the antibody is directed, and the remaining activity corresponds to the other isoenzyme. At this time, commercial antibodies versus EC-SOD are not available, whereas antibodies versus Cu,ZnSOD are easier to find. Using antibodies versus Cu,ZnSOD, distinction between the activities of the two isoenzymes can be achieved.

*Procedure.* The antibody is incubated with CNBr-activated Sepharose (AP Biotech) at a concentration of 10 mg of protein per milliliter of swollen gel, followed by blocking and washing according to the instructions of the manufacturer. The capacity of the immobilized antibody is then tested by incubation of the gel with a solution containing SOD. Isolated enzyme can be used, or a tissue extract containing only the isoenzyme of interest. If, for example, an anti-Cu,ZnSOD antibody is to be tested, different dilutions of a hemolysate can be used. Typically we add 50 $\mu$l of antibody–gel to a series of tubes containing 1 ml of enzyme/tissue extract/hemolysate with stepwise increasing (doubling) concentrations. The tubes are gently shaken at 4° for a couple of hours or overnight. After centrifugation, the SOD activity is determined in the supernatants. The antibodies typically have bound about 100 $\mu$g of SOD per milliliter of wet gel. For the assay we never use more than 25% of the maximal capacity of the gel, which ensures almost complete adsorption of the SOD isoenzyme.

Tissue extracts diluted to suit the capacity of the gel (see above) in neutral buffer, for example, phosphate-buffered saline, are incubated with antibody–gel (typically 50 $\mu$l of gel per milliliter of extract) or Sepharose 4B (blank). The suspensions are shaken at about 4° for 2hr or overnight. After centrifugation, the cyanide-sensitive SOD activities of the supernatants are analyzed.

## General Comments

It should be possible to demonstrate and quantify EC-SOD in samples from most mammalian species, using the procedures outlined here. Secreted copper-containing SODs have been found in several other phyla. Whether some or all of these are on the same evolutionary branch as the mammalian EC-SODs or represent separate branches has not yet been comprehensively analyzed. Even less is known about specific properties such as heparin binding, glycosylation, and tetrameric versus dimeric state. The present procedures might help in probing for such properties.

## [8] Prokaryotic Manganese Superoxide Dismutases

*By* JAMES W. WHITTAKER

### Introduction

Manganese superoxide dismutases (MnSODs)[1-3] are the front-line antioxidant defense in many prokaryotes, protecting against oxidative challenges resulting from environmental or biological interactions. The enzymes are typically multimers of small subunits (~23-kDa molecular mass), and are usually localized in the cytoplasm,[4] where they may be associated with DNA.[5,6] Enzymes from mesophilic organisms are generally homodimeric proteins, whereas thermophilic and hyperthermophilic enzymes are often tetramers.[2] Each subunit contains a mononuclear manganese complex (Fig. 1) that forms the catalytic active site. The metal ion is redox active, shuttling between Mn(II) and Mn(III) oxidation states during turnover.

Superoxide dismutases are widespread among bacterial and archaeal life, occurring in both prokaryotic domains of the phylogenetic tree.[2] Even some organisms normally identified as strict anaerobes have been found to contain a superoxide dismutase, and the exceptional aerobes that lack the enzyme (i.e., the lactic acid bacteria) generally have a fermentative metabolism and may contain manganese salts that mimic SOD activity.[7] The following sections provide a basic guide to the properties of the prokaryotic MnSODs.

---

[1] J. M. McCord, *New Horiz.* **1,** 70 (1993).
[2] I. Fridovich, *J. Biol. Chem.* **272,** 18515 (1997).
[3] J. W. Whittaker, *Metals Biol. Syst.* **37,** 587 (2000).
[4] H. M. Steinman, L. Weinstein, and M. Brenowitz, *J. Biol. Chem.* **269,** 28629 (1994).
[5] K. A. Hopkin, M. A. Papazian, and H. M. Steinman, *J. Biol. Chem.* **267,** 24253 (1992).
[6] R. A. Edwards, H. M. Baker, M. M. Whittaker, J. W. Whittaker, G. B. Jameson, and E. N. Baker, *J. Biol. Inorg. Chem.* **3,** 161 (1998).

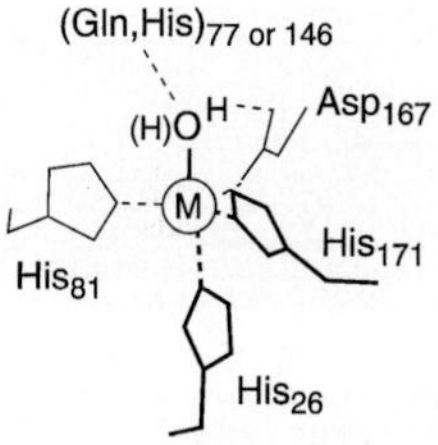

FIG. 1. Consensus metal-binding site of prokaryotic Mn- and FeSODs. M, metal (manganese or iron); *E. coli* MnSOD sequence numbering.

## Genomic Analysis

The explosion in genomics that has occurred has generated a wealth of information about the genetic contents of a broad spectrum of organisms. Identifying structural genes and assigning functions are two of the principal goals of this bioinformatics revolution, and superoxide dismutases have proved to be particularly easy to identify because many structural elements are highly conserved.[8–10] SOD gene studies have an additional significance in the context of clinical research, where superoxide dismutases are potentially important as virulence factors for pathogens.

This conservation of SOD structure have made it possible to design oligonucleotide probes that may be used to detect and amplify unique SOD sequences in genomic DNA isolates from a wide variety of organisms. The approach has been used to identify SOD genes in a range of gram-positive bacteria, including *Clostridiun, Enterococcus, Lactococcus, Staphylococcus,* and *Streptococcus*.[11] For these organisms, degenerate polymerase chain reaction (PCR) primers (forward primer, 5'-CCITAYICITAYGAYGCIYTIGARCC-3'; reverse primer, 5'-ARRTARTAIGC RTGYTCCCAIACRTC-3') incorporating ambiguous nucleotides have been designed by back-translation of two highly conserved protein sequence motifs, one occurring near the amino terminal of the protein [PY(PAT)YDALEP] and the other occurring near the carboxyl terminus (DVWEHAYYL) (Fig. 2). SOD gene sequences have been successfully amplified with 2 units of *Taq* DNA polymerase in a reaction containing 50 ng of genomic DNA, a 0.1 $\mu M$ concentration of each primer, a 200 $\mu M$ concentration of each dNTP. The PCR mixture is subjected

---

[7] F. S. Archibald, *Crit. Rev. Microbiol.* **13,** 63 (1986).
[8] M. W. Parker and C. C. F. Blake, *FEBS Lett.* **229,** 377 (1988).
[9] M. W. Parker, C. C. F. Blake, D. Barra, F. Bossa, M. E. Schinina, W. H. Bannister, and J. V. Bannister, *Protein Eng.* **1,** 393 (1987).
[10] T. Hunter, K. Ikeburkuro, W. H. Bannister, J. V. Bannister, and G. J. Hunter, *Biochemistry* **36,** 4925 (1997).
[11] C. Poyart, P. Berche, and P. Trieu-Cuot, *FEMS Microbiol. Lett.* **131,** 41 (1995).

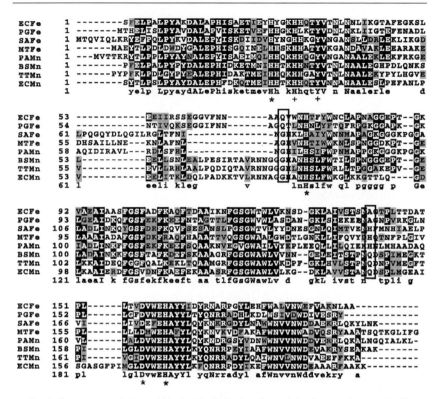

FIG. 2. Sequence correlations within the Mn, Fe family of bacterial and archaeal superoxide dismutases. Residues that serve as metal ligands are marked with an asterisk (*). Boxed residues correlate with metal specificity. *Key:* ECFe, *Escherichia coli* FeSOD (GenBank accession number 147842); PGFe, *Porphyromonas gingivalis* FeSOD (GenBank accession number 97324); SAFe, *Sulfolobus acidocaldarius* FeSOD (GenBank accession number 396203); MTFe, *Mycobacterium tuberculosis* FeSOD (GenBank accession number 98822); PAMn, *Pyrobaculum aerophilum* MnSOD (GenBank accession number 7290015); BSMn, *Bacillus stearothermophilus* MnSOD (GenBank accession number 143552); TTMn, *Thermus thermophilus* MnSOD (GenBank accession number 494880); ECMn, *Escherichia coli* MnSOD (GenBank accession number 147594). Sequences were aligned by the BLAST local sequence similarity search program [Altschul *et al.*, *Nucleic Acids Res.* **25**, 3389 (1997); Tatusova *et al.*, *FEMS Microbiol. Lett.* **174**, 247 (1999)] and the alignment was displayed by using CLUSTAL W with Boxshade [Thompson *et al.*, *Nucleic Acids Res.* **22**, 4673 (1994)].

to denaturation (3 min at 95°), followed by 35 cycles of amplification (30 sec of denaturation at 95°, 2 min of annealing at 37°, and 90 sec of elongation at 72°). The PCR product (comprising approximately 480 bp, nearly 85% of a typical SOD structural gene) is then available for cloning and subsequent sequence analysis.[11] The PCR product may also be tagged (end-labeled by kinase reaction or condensation with a biotin or streptavidin tag, or internally labeled by an additional stage of PCR using radiolabeled or biotinylated nucleotides) and used as a probe in

Southern hybridization to select DNA fragments containing the intact SOD gene from a genomic library. PCR amplification of SOD gene fragments can also be used for clinical analysis of pathogenic bacteria, and has been successfully used to detect as few as 10 cells of the obligate intracellular parasite *Coxiella burnetii* (the causative agent of Q fever) in clinical samples.[12]

## Identification of Superoxide Dismutase Type from Genomic Data

As indicated above, the availability of extensive sequence information from both genomic databases and fragment analysis has allowed the rapid and relatively easy identification of genes encoding the Mn, FeSOD family of enzymes in a wide range of organisms. Distinguishing between manganese and iron homologs tends to be more difficult, because the two subfamilies share extensive structural and sequence similarity. However, the correlation of a large number of SOD protein sequences for biochemically well-characterized enzymes has revealed subtle differences between the Fe- and MnSODs that may be useful for predicting metal specificity.[8-10] An alignment of amino acid sequences for eight SOD enzymes representing three distinct subgroups (iron only, cambialistic, and manganese only) is shown in Fig. 2[13-15] to illustrate this point. The shaded areas correspond to strictly conserved regions of the amino acid sequence, including metal ligands (identified by an asterisk in Fig. 2). The metal-binding sites are identical for both manganese and iron enzymes, with three histidine residues and an aspartate comprising the inner coordination sphere, and X-ray crystallographic studies have shown that the structural similarities extend even to the outer sphere of the metal complexes.[6,16-18]

In spite of the remarkable similarity of the Mn- and FeSOD structures, empirical correlations permit metal specificity to be confidently predicted for certain sequences. The crucial factor is the identification of a conserved outer sphere residue in the polypeptide sequence.[8-10] This residue forms an essential hydrogen bond to the coordinated solvent molecule in the active site, and is either glutamine or histidine in all known SOD structures (Fig. 1). For MnSODs, the residue is a glutamine arising from the C-terminal end of the $\beta_2$ sheet in domain II (residue 146), whereas for most FeSODs, the corresponding residue arises from helix $\alpha_2$ in domain I (residue 77 in *Escherichia coli* MnSOD sequence numbering). The

---

[12] A. Stein and D. Raoult, *J. Clin. Microbiol.* **30**, 2462 (1992).
[13] S. F. Altschul, T. L. Madden, A. A. Schaffer, J. Zhang, Z. Zhang, W. Miller, and D. J. Lipman, *Nucleic Acids Res.* **25**, 3389 (1997).
[14] T. A. Tatusova and T. L. Madden, *FEMS Microbiol. Lett.* **174**, 247 (1999).
[15] J. D. Thompson, D. G. Higgins, and T. J. Gibson, *Nucleic Acids Res.* **22**, 4673 (1994).
[16] M. S. Lah, M. M. Dixon, K. A. Pattridge, W. C. Stallings, J. A. Fee, and M. L. Ludwig, *Biochemistry* **34**, 1646 (1995).
[17] W. C. Stallings, K. A. Pattridge, R. K. Strong, and M. L. Ludwig, *J. Biol. Chem.* **259**, 10695 (1984).
[18] B. L. Stoddard, D. Ringe, and G. A. Petsko, *Protein Eng.* **4**, 113 (1990).

side chains of residues 77 and 146 lie close together in the conserved packing of the protein interior, imposing a mutual steric constraint on residues occupying these two positions in the sequence. Thus, the presence of glutamine (or histidine) at one position requires that the second position be occupied by a residue with a more compact side chain (e.g., glycine or alanine). The two regions are enclosed in boxes in Fig. 2.

Simple inspection of a given sequence can usually determine which of these limiting cases applies. Most iron-only SODs (like *E. coli* FeSOD) have glutamine near position 77 in the primary structure. The presence of a glutamine in this position does not demand iron specificity, however, because *Porphyromonas gingivalis* SOD has this feature and exhibits relatively low metal specificity, functioning nearly equally well with either iron or manganese bound to the protein, characteristic of cambialistic SODs.[19-22] On the other hand, a subgroup of iron-only SODs (including *Sulfolobus acidocaldarius* and *Mycobacterium tuberculosis* enzymes) has been found experimentally to be iron specific but to have a histidine residue at position 146 (*E. coli* MnSOD numbering) substituting for glutamine in the outer sphere. Complicating matters further, cambialistic enzymes may show a slight preference for one metal or the other, leading to further distinction of iron- or manganese-preferential subgroups. In general, cambialistic SODs have an outer sphere histidine at residue 146, whereas manganese-only SODs (including enzymes from *Bacillus stearothermophilus, Thermus thermophilus,* and *E. coli*) appear to strictly require glutamine in the second position. Thus the presence of glutamine at position 146 implies manganese specificity, glutamine at position 77 suggests iron specificity, and histidine at position 146 may be associated with either iron, manganese, or cambialistic behavior, and isolation and further characterization of the enzyme are then required.

## Cloning MnSODs for Recombinant Expression

Some of the organisms in which SOD genes have been identified (anaerobes, pathogens, hyperthermophiles) may be difficult to cultivate on a preparative scale, complicating the biochemical characterization of the endogenous SOD. Fortunately, isolation of the complete coding sequence for an SOD allows recombinant expression in a convenient host (e.g., *E. coli*) for detailed functional or structural analysis. We have found that an oxygen-inducible expression vector (pGB1)

[19] M. E. Martin, B. R. Byers, M. O. Olson, M. L. Salin, J. E. Arceneaux, and C. Tolbert, *J. Biol. Chem.* **261,** 9361 (1986).
[20] R. Gabbianelli, A. Battistoni, F. Polizio, M. T. Carri, A. De Martino, B. Meier, A. Desideri, and G. Rotilio, *Biochem. Biophys. Res. Commun.* **216,** 841 (1995).
[21] S. Yamano, Y. Sako, N. Nomura, and T. Maruyama, *J. Biochem.* **126,** 218 (1999).
[22] F. Yamakura, K. Kobayashi, S. Tagawa, A. Morita, T. Imai, D. Ohmori, and T. Matsumoto, *Mol. Biol. Int.* **36,** 233 (1995).

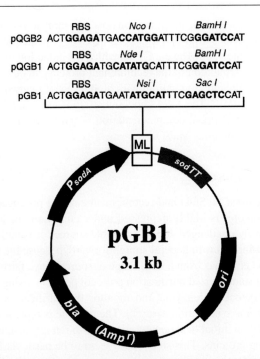

FIG. 3. Physical map of pGB1 oxygen-inducible expression vector. *bla*, β-Lactamase; ori, pUC19 ori; P*sodA*, E. coli *sod A* promoter; sodTT, E. coli *sodA* transcriptional terminator; ML, synthetic multilinker. [Based on Gao et al., Gene (Amst.) **176**, 269 (1996).]

constructed by Bao et al.[23] (Fig. 3) is particularly well suited to expression of recombinant SODs in E. coli, permitting isolation of gram quantities of pure enzyme from a single 10-liter fermentation run.

The vector backbone of pGB1, derived from pUC19, confers ampicillin resistance to transformants and gives rise to a high vector copy number, contributing to a high expression level for the recombinant protein through a gene dosage effect. The heterologous gene can be inserted in the vector between a copy of the oxygen-inducible E. coli *sodA* promoter (P$_{sodA}$) for the endogenous MnSOD gene and the rho-independent *sodA* transcriptional terminator (sodTT), making use of a variety of restriction sites (*Nsi*I and *Sac*I, *Nde*I and *Bam*HI, or *Nco*I and *Bam*HI) available in several variants[23-25] (Fig. 3). Transformants grown in Luria–Bertani (LB) medium with ampicillin selection and oxygen purging (for a dissolved oxygen level approaching 100%) efficiently express the heterologous protein. The highest levels of

[23] B. Gao, S. C. Flores, S. K. Bose, and J. M. McCord, Gene (Amst.) **176**, 269 (1996).
[24] M. M. Whittaker and J. W. Whittaker, J. Biol. Inorg. Chem. **5**, 402 (2000).
[25] M. M. Whittaker and J. W. Whittaker, J. Biol. Chem. **274**, 34751 (1999).

expression are achieved with an expression host providing a *sodA* background[26,27] (lacking the endogenous MnSOD, e.g., *E. coli* QC781,[26] available from the *E. coli* Genetics Stock Center, New Haven, CT) in which the induction by oxidative stress is most effective. Metallation of the recombinant protein can be controlled *in vivo* by supplementing the culture medium with high levels of manganese or iron salts (metal ion at 1–5 m$M$)[28] The *sodA* knockout strain is also useful for isolation of MnSOD genes by functional complementation,[27] and for the physiological characterization of heterologous MnSODs.[29]

Purification of MnSOD

Both endogenous MnSOD and recombinant enzyme produced under control of the $P_{sodA}$ promoter is highly expressed under conditions of oxidative stress, making oxidatively challenged cells a favorable source for isolating the enzyme. Purification of MnSOD is in general straightforward, because the enzyme is relatively robust and enzymes from a wide range of sources have fairly uniform properties, making a standardized purification protocol possible, using any convenient SOD assay to detect the enzyme and monitor the purification (e.g., xanthine oxidase/cytochrome *c* inhibition assay[30]). The enzyme is typically thermostable, permitting use of an initial heat denaturation step to remove a significant fraction of contaminating proteins. This step would seem to be particularly attractive for preparation of recombinant thermophilic or hyperthermophilic SODs, but may actually be a disadvantage as those enzymes are normally expressed as metal-free apoenzymes in the mesophilic expression host.[24,25] In that case, heat treatment tends to complicate the isolation of pure metalloforms. We have found that the apoenzymes undergo a thermally triggered metallation process at elevated temperatures, resulting in relatively indiscriminate incorporation of any available metal ion.[24,25] In a crude cell extract this will lead to uptake primarily of iron and to a lesser extent of manganese and other metal ions. Careful purification of the intact metal-free apoenzyme (restricting the temperature to a moderate range during all purification steps) permits relatively controlled incorporation of specific metals in a subsequent step in which the apoenzyme (0.5 m$M$ active sites) is incubated with metal salts [e.g., 10 m$M$ MnCl$_2$, or 10 m$M$ Fe(NH$_4$)$_2$(SO$_4$)$_2$ plus 5 m$M$ sodium ascorbate] in 20 m$M$ morpholineethanesulfonic acid (MOPS) buffer, pH 7 at 65–120° (e.g., in a thermal cycler or autoclave, depending on protein stability).[24,25]

---

[26] A. Carlioz and D. Touati, *EMBO J.* **5**, 623 (1986).
[27] H. M. Steinman, *Mol. Gen. Genet.* **232**, 427 (1992).
[28] M. M. Whittaker and J. W. Whittaker, *Biochemistry* **36**, 8923 (1997).
[29] T. Inaoka, Y. Matsumura, and T. Tsuchido, *J. Bacteriol.* **180**, 3697 (1999).
[30] J. M. McCord and I. Fridovich, *J. Biol. Chem.* **244**, 6049 (1969).

High-performance ion-exchange chromatography has proved particularly useful for purification of MnSOD from cell extracts.[31] Under suitable conditions, anion-exchange chromatography is capable even of resolving proteins in distinct metallation states, presumably because of slight differences in protein charge. Alternatively, chromatofocusing chromatography may be used to resolve different metalloforms[24] ($Mn_2$-, $Fe_2$-, Mn,Fe-, Mn-half apo-, Fe-half apo-, and metal-free apo-MnSOD), permitting each of these species to be isolated and characterized.

## Spectroscopic Characterization

Optical absorption spectroscopy permits simple, routine characterization of the metal center in MnSOD.[32,33] However, optical absorption is associated only with the red, oxidized Mn(III) form of the enzyme, the Mn(II) form being essentially colorless and lacking any significant visible absorption. To ensure that spectroscopic results are quantitatively significant, the purified enzyme may be converted to the oxidized Mn(III) form by treatment with strong inorganic oxidants. In practice, we have found that Mo(V) [molybdicyanide,[33] $K_3Mo(CN)_8$] and I(VII) (periodate,[34] $NaIO_4$) are particularly effective oxidants, producing only minimal side reactions. The reaction with molybdicyanide requires addition of an excess (>2 equivalents) of oxidant, and is relatively slow, but the extent of reaction may be monitored by the increase in Mn(III) absorption at 475 nm.[33] The reaction with periodate is rapid and slightly more than 1 equivalent of oxidant is generally sufficient.[34] Optical spectra recorded for the oxidized enzyme after desalting (by dialysis or gel filtration) yield the most reliable Mn(III) extinction coefficient. A typical Mn(III)SOD absorption spectrum is shown (Fig. 4, ABS). The spectrum is a broad absorption envelope, spanning the entire visible spectrum, and including all four spin-allowed ligand field ($d \rightarrow d$) electronic transitions arising from the high-spin ground state for the Mn(III) $d^4$ metal center in the protein.[33] The individual transitions are partly resolved in the first derivative of the absorption envelope (Fig. 4, DER), which contains turning points near 400, 450, 540, and 650 nm, together with multiplet fine structure features between 450 and 600 nm that rise from relatively weak spin-forbidden electronic transitions to spin-triplet excited states arising from the $^3G$ and $^3H$ free ion multiplets. The underlying structure of the absorption spectrum is more fully dissected in the circular dichroism spectrum (Fig. 4, CD), which resolves four components (dichroic extrema at 400, 435, 550, and 625 nm). Magnetic circular dichroism (MCD) spectroscopy (not shown) can provide further

---

[31] W. F. Beyer, Jr., and I. Fridovich, *J. Biol. Chem.* **266**, 303 (1991).
[32] J. A. Fee, E. R. Shapiro, and T. H. Moss, *J. Biol. Chem.* **251**, 6157 (1976).
[33] M. M. Whittaker and J. W. Whittaker, *J. Am. Chem. Soc.* **113**, 5528 (1991).
[34] R. A. Edwards, M. M. Whittaker, J. W. Whittaker, E. N. Baker, and G. B. Jameson, *Biochemistry* **40**, 15 (2001).

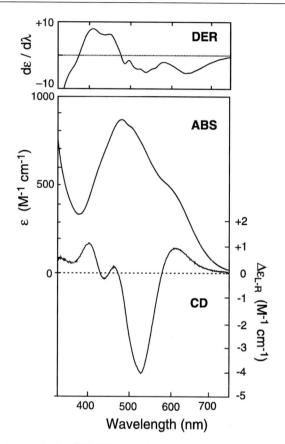

FIG. 4. Optical spectra for *E. coli* Mn(III) superoxide dismutase. ABS, Absorption; DER, absorption derivative; CD, circular dichroism. (Active sites are 1 m$M$ in 50 m$M$ potassium phosphate buffer, pH 7.)

information about assignment of the electronic transitions in the native and ligand complexes of the Mn(III) enzyme.[33] For manganese-only MnSODs, the absorption maximum typically occurs near 478 nm, associated with a molar extinction $\varepsilon_{478}$ of ~850 $M^{-1}$ cm$^{-1}$ (per manganese atom). The manganese-preferential cambialistic MnSODs are spectroscopically distinct, characteristically exhibiting a relatively blue-shifted absorption maximum ($\lambda_{max} = 450$ nm) and slightly lower extinction $\varepsilon_{450} = 660$ $M^{-1}$ cm$^{-1}$ (per manganese atom).[24] The CD spectrum also appears to be distinct for the manganese-preferential cambialistic enzyme, with a lower rotational strength for the strongest CD band ($\Delta\varepsilon_{L-R} = -2.3$ $M^{-1}$ cm$^{-1}$ for the manganese-preferred cambialistic SOD from the archaeon *Pyrobaculum aerophilum*[24] compared with $-4$ $M^{-1}$ cm$^{-1}$ for the manganese-only SOD from *E. coli*).

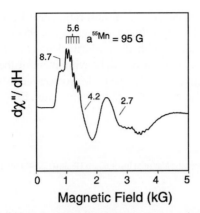

FIG. 5. EPR spectrum for *E. coli* Mn(II) superoxide dismutase. (Active sites are 1 m$M$ in 50 m$M$ potassium phosphate buffer, pH 7.) Instrumental parameters: temperature, 5 K; microwave frequency, 9.46 GHz; microwave power, 1 mW; modulation amplitude, 10 G.

Substituting iron for manganese in MnSODs[24,25,35] results in protein exhibiting optical absorption spectra nearly indistinguishable from those of authentic FeSODs,[36] with a near-UV absorption band near 350 nm at low pH (pH < 6). The enzyme also typically exhibits significant SOD activity over this lower pH range. However, the iron-substituted enzyme is relatively sensitive to pH, the UV absorption bleaches, and SOD activity is rapidly lost as the pH is increased above pH 7, as a result of inhibitory binding by hydroxide ion to Fe(III) in the enzyme.[37] This "rusting" of iron in iron-substituted MnSOD appears to be an important aspect of metal specificity in this class of enzymes.

Electron paramagnetic resonance (EPR) spectroscopy provides a complementary probe of the reduced Mn(II) mental center in MnSOD.[32,33] Although both oxidized [Mn(III), $d^4$, $S = 2$] and reduced [Mn(II), $d^5$, $S = 5/2$] forms have paramagnetic ground states and are in principle EPR active, only the reduced enzyme has an odd-electron (Kramers) ground state permitting routine EPR measurements. MnSOD is reduced by its product, hydrogen peroxide,[38] and native MnSOD (which is generally a mixture of oxidation states) may be quantitatively converted to the homogeneous Mn(II) form by treatment with a small stoichiometric excess of hydrogen peroxide to ensure the maximum sensitivity of EPR measurements.[24] A typical X-band EPR spectrum for Mn(II)SOD is shown in Fig. 5. The characteristic

---

[35] F. Yamakura, K. Kobayashi, H. Ue, and M. Konno, *Eur. J. Biochem.* **227**, 700 (1995).
[36] T. O. Slykehouse and J. A. Fee, *J. Biol. Chem.* **251**, 5472 (1976).
[37] R. A. Edwards, M. M. Whittaker, J. W. Whittaker, G. B. Jameson, and E. N. Baker, *J. Am. Chem. Soc.* **120**, 9684 (1998).
[38] C. Bull, E. C. Niederhoffer, T. Yoshida, and J. A. Fee, *J. Am. Chem. Soc.* **113**, 4069 (1991).

features of the spectrum include strong absorption near 1,500 G ($g = \sim 6$), exhibiting well-resolved nuclear hyperfine splittings ($a_{Mn} = 95$ G) arising from electron–nuclear hyperfine coupling with the $^{55}$Mn nucleus ($I = 5/2$, 100% n.a.). These spectral features reflect the nearly axial environment of the metal ion in the trigonal bipyramidal active site. The even-electron quintet Mn(III) ground state of native MnSOD has also been detected by polarization EPR methods.[39]

Acknowledgment

Support for this work from the National Institutes of Health (GM 42680) is gratefully acknowledged.

[39] K. A. Campbell, E. Yikilmaz, C. V. Grant, G. Wolfgang, A.-F. Miller, and R. D. Britt, *J. Am. Chem. Soc.* **121,** 4714 (1999).

# [9] Nickel-Containing Superoxide Dismutase

By JIN-WON LEE, JUNG-HYE ROE, and SA-OUK KANG

Introduction

Superoxide dismutases (SOD, EC 1.15.1.1) are metalloenzymes that catalyze the disproportionation of superoxide radical anion to hydrogen peroxide and molecular oxygen. The dismutation reaction catalyzed by SODs requires a metal center that is first reduced and then reoxidized by superoxide radical anion. SODs have been classified into three groups according to their metal centers: manganese (MnSOD), iron (FeSOD), and copper and zinc (Cu,ZnSOD).[1] These SODs may be derived from two evolutionary families.[2] The active site structures of these SODs show that the redox-active metal centers are ligated by a combination of histidine imidazoles, aspartate carboxylates, and water, with five-coordinate metal centers.[3–6]

[1] I. Fridovich, *Annu. Rev. Biochem.* **64,** 97 (1995).
[2] I. Fridovich, *J. Biol. Chem.* **264,** 7761 (1989).
[3] J. A. Tainer, E. D. Getzoff, J. S. Richardson, and D. C. Richardson, *Nature (London)* **306,** 284 (1983).
[4] W. C. Stallings, T. B. Powers, K. A. Pattridge, J. A. Fee, and M. L. Ludwig, *Proc. Natl. Acad. Sci. U.S.A.* **80,** 3384 (1983).
[5] M. W. Parker and C. C. F. Blake, *J. Mol. Biol.* **199,** 649 (1988).
[6] W. C. Stallings, K. A. Pattridge, R. K. Strong, and M. L. Ludwig, *J. Biol. Chem.* **259,** 10695 (1984).

In 1996, a novel type of SOD containing nickel as the only metal was found in the filamentous gram-positive bacteria *Streptomyces* spp.[7] The nickel content of this enzyme was estimated to have a molar ratio of 0.74 nickel per enzyme subunit and was thus designated as nickel-containing SOD (NiSOD). The NiSOD was first isolated from *Streptomyces seoulensis* and *Streptomyces coelicolor*,[7] and to date has been reported only in *Streptomyces* species.[8–10]

The NiSOD is distinct from the Mn, Fe, or Cu, Zn enzymes on the basis of amino acid sequence and metal ligand environment.[11] The gene for NiSOD (*sodN*) shows no apparent sequence similarity to other SODs or to other known proteins. However, the two *sodN* genes cloned from *S. coelicolor* and *S. seoulensis* share 91% homology in nucleotide sequence and 92% homology in amino acid sequence. The nickel center in *S. seoulensis* NiSOD is composed largely of S-donor ligands, which is unique among known SODs.[11] However, this cysteinate ligand environment of NiSOD places NiSOD in line with other redox-active nickel enzymes that contain thiolate ligation to the nickel center[12]—carbon monoxide dehydrogenase,[13] acetyl-coenzyme A synthase,[14] methyl coenzyme M reductase,[15] and [Ni–Fe]hydrogenase.[16] Thus NiSOD can contribute to the understanding of the biological role of nickel in performing SOD function as a member of the nickel-containing enzymes.

## Microorganisms and Culture Conditions

*Streptomyces* species are cultured as described previously.[17] For liquid culture, *Streptomyces seoulensis* IMSNU 21266$^T$ (formerly *Streptomyces* sp. IMSNU-1[18]), *Streptomyces coelicolor* ATCC 10147 (Müller), *Streptomyces*

---

[7] H.-D. Youn, E.-J. Kim, J.-H. Roe, Y. C. Hah, and S.-O. Kang, *Biochem. J.* **318,** 889 (1996).
[8] H.-D. Youn, H. Youn, J.-W. Lee, Y.-I. Yim, J.-K. Lee, Y. C. Hah, and S.-O. Kang, *Arch. Biochem. Biophys.* **334,** 341 (1996).
[9] E.-J. Kim, H.-P. Kim, Y. C. Hah, and J.-H. Roe, *Eur. J. Biochem.* **241,** 178 (1996).
[10] V. Leclere, P. Boiron, and R. Blondeau, *Curr. Microbiol.* **39,** 365 (1999).
[11] S. B. Choudhury, J.-W. Lee, G. Davidson, Y.-I. Yim, K. Bose, M. L. Sharma, S.-O. Kang, D. E. Cabelli, and M. J. Maroney, *Biochemistry* **38,** 3744 (1999).
[12] M. J. Maroney, *Curr. Opin. Chem. Biol.* **3,** 188 (1999).
[13] S. W. Ragsdale and M. Kumar, *Chem. Rev.* **96,** 2525 (1996).
[14] J. Xia, Z. Hu, C. V. Popescu, P. A. Lindahl, and E. Münck, *J. Am. Chem. Soc.* **119,** 8301 (1997).
[15] U. Ermler, W. Grabarse, S. Shima, M. Goubeaud, and R. K. Thauer, *Science* **278,** 1457 (1997).
[16] A. Volbeda, M. H. Charon, C. Piras, E. C. Hatchikian, M. Frey, and J. C. Fontecilla-Camps, *Nature (London)* **373,** 580 (1995).
[17] D. A. Hopwood, M. J. Bibb, K. F. Chater, T. Kieser, C. J. Bruton, H. M. Kieser, D. J. Lydiate, C. P. Smith, J. M. Ward, and H. Schrempf, "Genetic Manipulation of *Streptomyces*—A Laboratory Manual." John Innes Foundation, Norwich, UK, 1985.
[18] J. Chun, H.-D. Youn, Y.-I. Yim, H. Lee, M. Y. Kim, Y. C. Hah, and S.-O. Kang, *Int. J. Syst. Bacteriol.* **47,** 492 (1997).

*coelicolor* A3(2), *Streptomyces lividans* TK24, and *Streptomyces griseus* KCTC 9006 are cultured with vigorous shaking in a 500-ml baffled flask containing 100 ml of YEME medium [1% (w/v) glucose, 0.5% (w/v) Bacto-peptone, 0.3% (w/v) yeast extract, 0.3% (w/v) malt extract, 10% (w/v) sucrose, 5 m$M$ MgCl$_2$] for 40–50 hr at 30°. Inoculations are made with spores harvested from cultures grown on solid medium for 1–2 weeks and filtered through glass wool. More than $1 \times 10^7$ spores for *S. coelicolor* and *S. lividans* and $1 \times 10^6$ spores for *S. seoulensis* and *S. griseus* are inoculated per flask. For cultivation on solid medium, spores or mycelia are spread on cellophane disks overlaid on agar plates and grown for 3–4 days at 30°. Addition of 10–100 $\mu M$ NiCl$_2$ to the medium increases production of NiSOD more than 10-fold, and represses the production of Fe,ZnSOD.[19]

Assay Methods

In principle, all methods that measure SOD activity can be applied to measure NiSOD activity, and readers should consult other texts for various assay methods.[20] We describe assay methods that have been used for NiSOD.

*Indirect Assay of NiSOD Activity*

NiSOD activity has been assayed by an indirect method based on the ability of SOD to inhibit the reduction of cytochrome $c$ or nitroblue tetrazolium (NBT) by scavenging superoxide radical anion produced by the xanthine/xanthine oxidase system.

*Nitroblue Tetrazolium Assay.* SOD activity is assayed by the method described previously, with slight modifications.[21] The procedure for the standard assay is as follows: 100 $\mu$l of SOD sample is added to 0.8 ml of the reagent solution [1 m$M$ diethylenetriaminepentaacetic acid (DETAPAC), 1 U of catalase, 56 $\mu M$ NBT, and 100 $\mu M$ xanthine in 50 m$M$ sodium phosphate buffer, pH 7.4]. As a blank, 100 $\mu$l of 50 m$M$ sodium phosphate buffer (pH 7.4) is added to another tube. Xanthine oxidase is diluted in 50 m$M$ sodium phosphate buffer (pH 7.4) with 1.33 m$M$ DETAPAC to the extent that the addition of 100 $\mu$l of xanthine oxidase to the blank tube without NiSOD gives an absorbance rate between 0.02 and 0.03/min at 560 nm. The diluted xanthine oxidase (100 $\mu$l) is added to the reagent solution containing NiSOD. The rate of absorbance change at 560 nm is recorded 30 sec after adding xanthine oxidase. One unit of SOD activity is defined as the amount that gives half-maximal inhibition of the rate of NBT reduction.

*Cytochrome c Assay.* In the cytochrome $c$ assay,[22] the reaction mixture contains 0.1 m$M$ EDTA, 50 $\mu M$ xanthine, and 10 $\mu M$ cytochrome $c$ in 50 m$M$ potassium

---

[19] E.-J. Kim, H.-J. Chung, B. Suh, Y. C. Hah, and J.-H. Roe, *Mol. Microbiol.* **27**, 187 (1998).
[20] R. A. Greenwald, "CRC Handbook of Methods for Oxygen Radical Research." CRC Press, Boca Raton, Florida, 1985.
[21] C. Beauchamp and I. Fridovich, *Anal. Biochem.* **44**, 276 (1971).

phosphate buffer (pH 7.8). The reduction of cytochrome $c$ is measured at 550 nm after the addition of appropriately diluted xanthine oxidase that gives an absorbance rate of 0.025/min for the blank. One Fridovich unit of SOD activity is defined as the amount that inhibits 50% of the rate of cytochrome $c$ reduction.

*Direct Assay of NiSOD Activity*

NiSOD activity is measured directly by monitoring the disappearance of the absorbance of superoxide radical anion at 260 nm, after generation of superoxide radical anion by pulse radiolysis.[11,23]

*Activity Staining of Superoxide Dismutase in Polyacrylamide Gel*

Cell-free extracts or purified enzymes are separated on a 10% (w/v) nondenaturating polyacrylamide gel, using a buffer without sodium dodecyl sulfate (SDS). Negative activity staining for SOD is done according to the modified method of Steinman.[24] The gel is soaked first in 20 m$M$ sodium phosphate buffer (pH 7.4) containing 28 $\mu M$ riboflavin and 28 m$M$ $N,N,N',N'$-tetramethylethylenediamine (TEMED) for 10 min and then in 20 m$M$ sodium phosphate buffer (pH 7.4) containing 2.5 m$M$ NBT for 10 min. The gel is placed in a dry tray and illuminated with fluorescent light for 3–10 min. Photochemically produced superoxide radical anion reduces NBT to blue formazan. The area in the gel where SOD is present is depleted of superoxide radical anion and hence is unable to reduce NBT, leaving a clear zone in the blue background.

*Assay of Superoxide Dismutase Activity for Column Fractions*

The SOD activity in column fractions during purification is easily assayed by the modified method of positive SOD staining.[25] Fifty to 200 $\mu$l of the sample solution is added to 0.2–1 ml of assay mixture containing 2 m$M$ $o$-dianisidine and 0.1 m$M$ riboflavin in 50 m$M$ sodium phosphate buffer (pH 7.4), followed by illumination with fluorescent light for 5–30 min. The photooxidation of $o$-dianisidine develops brown color in the fractions with SOD activity. SOD increases the rate of $o$-dianisidine oxidation by removing superoxide radical anion that interacts with $o$-dianisidine radical and thereby decreases the net rate of the oxidation.

Purification of NiSOD

Two different kinds of SODs have been reported in *Streptomyces* spp.[7–9]: only NiSOD in *S. seoulensis* and NiSOD and Fe,ZnSOD in *S. coelicolor* and *S. griseus*,

---

[22] J. M. McCord and I. Fridovich, *J. Biol. Chem.* **244**, 6049 (1969).
[23] H. A. Schwarz, *J. Chem. Educ.* **58**, 101 (1981).
[24] H. M. Steinman, *J. Bacteriol.* **162**, 1255 (1985).
[25] H. P. Misra and I. Fridovich, *Arch. Biochem. Biophys.* **183**, 511 (1977).

respectively, have been detected and purified. Accordingly, the purification steps for NiSOD depend on the species.

*Purification of NiSOD from Streptomyces seoulensis*

Mycelial cells of *S. seoulensis* are harvested by aspiration on filter paper with a Büchner funnel and washed with 0.85% (w/v) KCl solution. They are suspended in buffer A (20 m$M$ sodium phosphate, pH 7.4), and disrupted with an Omni-mixer (Omni International, Warrenten, VA). The lysate is clarified by centrifugation at 12,000$g$ for 20 min at 4°. Ammonium sulfate is added slowly to 35% (w/v) saturation, and the precipitate is removed by centrifugation at 12,000$g$ for 30 min at 4°. The supernatant is loaded onto a Phenyl-Sepharose CL-4B FastFlow column (5 × 30 cm) equilibrated with buffer A containing ammonium sulfate at 35% (w/v) saturation, and then eluted with a linear gradient of ammonium sulfate between 35 and 0% (w/v) saturation. The fractions containing SOD activity are pooled and desalted on a Sephadex G-25 column (8 × 60 cm) equilibrated with buffer A. The desalted fractions are purified with a Delta Prep 4000 chromatography system (Waters, Milford, MA) with a Protein-Pak DEAE 5PW column (2.15 × 15 cm) equilibrated with buffer A. After the column has been washed with 0.1 $M$ NaCl in buffer A, the sample is eluted with a linear gradient of 0.1–0.3 $M$ NaCl in buffer A. Active fractions are pooled and concentrated with an YM10 membrane (Amicon, Danvers, MA). These fractions are purified by preparative electrophoresis with a Bio-Rad (Hercules, CA) model 491 Prep Cell apparatus using a 7.5% (w/v) nondenaturating polyacrylamide gel. After concentration of the active fractions with an YM10 membrane, these fractions are loaded onto a Superdex 200 column (HiLoad 16/60) equilibrated with buffer A and eluted with the same buffer in a fast protein liquid chromatography (FPLC) system (Amersham Pharmacia, Piscataway, NJ). The active fractions are further purified on a Mono Q column (HR 5/5) with a linear gradient elution of 0–0.3 $M$ NaCl in the a same FPLC system. Because the NiSOD loses its activity slowly under lower enzyme concentration at room temperature, it should be stored at higher concentration (more than 10 mg/ml) in the presence of 30% (v/v) glycerol at $-70°$ for long storage.

*Purification of NiSOD from Streptomyces coelicolor and Streptomyces griseus*

Mycelial cells of *S. coelicolor* and *S. griseus* are harvested by centrifugation at 8000$g$ for 20 min at 4° and washed twice with 0.85% (w/v) KCl solution. They are suspended in buffer A, and disrupted with an Omni-mixer. The lysate is clarified by centrifugation at 12,000$g$ for 20 min at 4°. Ammonium sulfate is added to 40% (w/v) saturation and the precipitate is removed by centrifugation at 12,000$g$ for 30 min at 4°. Ammonium sulfate is then added to 80% (w/v) saturation and the supernatant is removed by centrifugation at 12,000$g$ for 30 min at 4°. The protein precipitate is resuspended in buffer A and dialyzed overnight against the same buffer. The dialysate is loaded onto a DEAE-Sepharose CL-6B (4 × 40 cm) column equilibrated with buffer A and eluted with a linear gradient of 0–0.5 $M$ NaCl in buffer A. Two separate peaks of SOD activity are detected by DEAE ion-exchange chromatography; NiSOD is retrieved at about 0.1 $M$ NaCl and Fe,ZnSOD is retrieved at about 0.3 $M$ NaCl. The lower salt eluate (NiSOD) is pooled and concentrated by ultrafiltration and then further purified by chromatography on a Superdex 200 column (HiLoad 16/60) and Mono Q column (HR 5/5) as described above.

## Physicochemical Properties of NiSOD

### Molecular Properties

The gene for NiSOD (*sodN*) encodes a polypeptide of 131 amino acids (14,717 Da for *S. seoulensis*, 14,703 Da for *S. coelicolor*). However, the amino terminus of the purified NiSOD is located 14 amino acids downstream from the initiation codon of the deduced open reading frame (ORF). The molecular mass of the purified NiSOD subunit estimated by SDS–polyacrylamide gel electrophoresis (PAGE) is 13.4 kDa, in good agreement with the deduced mass of the processed polypeptide (13,188 Da for *S. seoulensis*, 13,201 Da for *S. coelicolor*). The size of the native enzyme has been estimated to be about 60 kDa by gel filtration, suggesting that the enzyme consists of four identical subunits.[7]

### Activity

The activity of NiSOD determined by pulse radiolysis indicates that the catalytic rate constant is $5.3 \times 10^9 M^{-1} \sec^{-1}$ for the tetrameric enzyme or $1.3 \times 10^9 M^{-1} \sec^{-1}$ per nickel. This catalytic rate constant is similar to that of Cu,ZnSOD ($1.2 \times 10^9 M^{-1} \sec^{-1}$ per copper or $2.4 \times 10^9 M^{-1} \sec^{-1}$ for the dimeric enzyme) under the same ionic strength conditions.[11] The pH dependence of the catalytic rate constant of NiSOD in comparison with those of bovine Cu,ZnSOD and *Escherichia coli* MnSOD turned out to be similar to those for other SODs, particularly to that of MnSOD.[11]

### Stability

NiSOD is stable up to 70°, but about 50 and 80% of the activity is lost at 75 and 80°, respectively, after a 10-min incubation. More than 80% of the enzyme activity is maintained over a broad pH range between pH 4.0 and 8.0. At pH >9.0 the enzyme activity starts to decrease rapidly,[7,11] NiSOD loses its activity slowly under dilute conditions (about 50% loss at micromolar concentrations after 3 hr at room temperature)[11]; however, the activity remains stable for several days at concentrations exceeding 10 mg/ml at room temperature.

## Inhibition Pattern

NiSOD is sensitive to $H_2O_2$, with a half-life of $8.7 \pm 1.3$ min when NiSOD at 0.5 mg/ml is incubated with 0.5 m$M$ $H_2O_2$ in 20 m$M$ potassium phosphate buffer (pH 7.4) with 0.1 m$M$ EDTA at 26°.[9] The presence of 2 m$M$ KCN, which is known to eliminate Cu,ZnSOD activity, does not eliminate NiSOD activity as assayed by activity staining in the gel. However, cyanide at 10 m$M$ in the assay mixture inhibits NiSOD activity more than 85%.[7,9] The NiSOD is only weakly inhibited by azide ion. Azide at 10 m$M$ in the assay mixture inhibits NiSOD activity by 16%, and 42 m$M$ azide is required for 50% inhibition.[9,11]

## Spectral Properties of NiSOD

### Ultraviolet–Visible Spectrum of NiSOD

NiSOD exhibits absorption bands at 278 nm ($A^{0.1\%} = 1.3$) and 378 nm ($A^{0.1\%} = 0.4$) and a broad shoulder at about 540 nm ($A^{0.1\%} = 0.06$) in 20 m$M$ sodium phosphate buffer (pH 7.4) (Fig. 1). The relatively high absorbance at 378 nm,

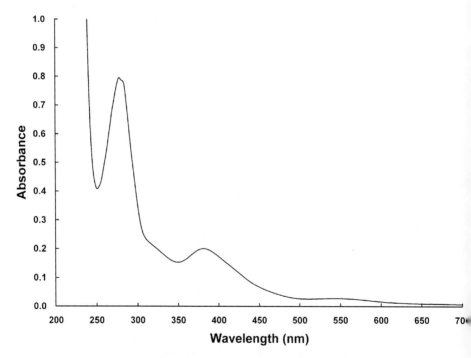

FIG. 1. UV–visible absorption spectrum of as-isolated NiSOD from *S. seoulensis*. Spectrum was measured in 20 m$M$ sodium phosphate buffer (pH 7.4) with a 620-$\mu$g/ml enzyme concentration (50 $\mu M$ NiSOD subunit).

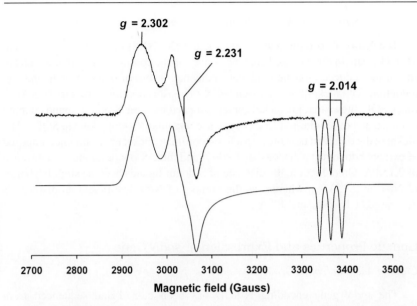

FIG. 2. *Top:* EPR spectrum of the as-isolated NiSOD from *S. seoulensis*. EPR conditions: microwave frequency, 9.482 GHz; temperature, 100 K; microwave power, 1 mW; modulation amplitude, 5 G; modulation frequency, 100 kHz. *Bottom:* Computer simulation of the NiSOD spectrum with parameters: $g_{xyz} = 2.306, 2.232$, and 2.016; $A_{xyz} = 16.2, 17.7$, and 24.6 G; and $l_{xyz} = 28, 17$, and 7.8 G.

which disappears in apoenzyme, is considered to originate from ligand–metal charge transfer (LMCT).

*Electron Paramagnetic Resonance Spectrum of NiSOD*

The electron paramagnetic resonance (EPR) spectrum of NiSOD as isolated exhibits resonances typical of a rhombically distorted $S = 1/2$ Ni(III) signal with unique g values ($g_{xyz} = 2.302, 2.231$, and 2.014) over the temperature ranges of liquid helium to liquid nitrogen (Fig. 2), which is distinct from those of other SODs: the g values of the EPR spectrum of Cu,ZnSOD[26] are $g_{\parallel} = 2.074$ and $g_{\perp} = 2.260$ and those for FeSOD[27] are $g_z = 4.90$, $g_x = 3.88$, and $g_y = 3.53$, whereas SOD containing Mn(III) is EPR silent.[28] The triplet hyperfine splitting ($A_z = 24.6$ G) originates from the nitrogen nucleus ($^{14}$N, $I = 1$), presumably from the N-terminal histidine residue. The EPR signal characteristic of the oxidized enzyme is lost on reduction with dithionite or in apoenzyme, consistent with the reduction of Ni(III) to Ni(II) or the loss of Ni(III).

[26] H. P. Misra and I. Fridovich, *J. Biol. Chem.* **247**, 3410 (1972).
[27] F. J. Yost, Jr. and I. Fridovich, *J. Biol. Chem.* **248**, 4905 (1973).
[28] J. A. Fee, E. R. Shapiro, and T. H. Moss, *J. Biol. Chem.* **251**, 6157 (1976).

## Nickel Site Structure by X-Ray Absorption Spectroscopy

The X-ray absorption near edge structure (XANES) of NiSOD[11] shows that the nickel site in the oxidized enzyme is five- or six-coordinate and that the nickel site in the reduced enzyme is planar, indicating a loss of one or two ligands on reduction. The nickel K-edge extended X-ray absorption fine structure (EXAFS) for NiSOD shows that the nickel center is ligated by three S-donor ligands in both oxidized and dithionite-reduced enzymes, being unique among known SODs. The nickel in the oxidized enzyme is in a five-coordinate nickel environment composed of two N/O donors at a distance of 1.909 Å and three S-donor ligands at a distance of 2.158 Å. On reduction, the distance of S-donor ligands become slightly shorter (3 Ni–S at 2.154 Å), whereas as the number of N/O donors and its distance is decreased (1 Ni–N/O at 1.87 Å).

## Genetic Properties and Expression of *sodN* Gene

### Cloning and Nucleotide Sequence Analysis

The *sodN* gene encoding NiSOD has been cloned and sequenced from *S. coelicolor* and *S. seoulensis* (GenBank accession numbers AF012193 and AF104994 for *S. coelicolor* and AF047461 for *S. seoulensis*).[11,19] Figure 3 shows the deduced amino acid sequences of *sodN* and the N-terminal amino acids of the

*S. seoulensis*   MLSRLFAPKVKVSA**HCDLPCGVYDPAQA**RIEAESVKAIQEKMAANDDLHF

*S. coelicolor*   MLSRLFAPKVTVSA**HCDLPCGVYDPAQA**RIEAESVKAVQEKMAGNDDPHF
                  ********** ***************************.***** *** **

*S. seoulensis*   QIRATVIKEQRAELAKHHLDVLWSDYFKPPHFESYPELHTLVNEAVKALS

*S. coelicolor*   QTRATVIKEQRAELAKHHVSVLWSDYFKPPHFEKYPELHQLVNDTLKALS
                  * ****************. ************ ***** ***. .****

*S. seoulensis*   AAKASTDPATGQKALDYIAQIDKIFWETKKA

*S. coelicolor*   AAKGSKDPATGQKALDYIAQIDKIFWETKKA
                  *** * ************************

FIG. 3. Comparison of the deduced amino acid sequences of *sodN* ORF from *S. seoulensis* and *S. coelicolor*, using single-letter notation. Identical residues are indicated by asterisks (*), and similar ones by dots (·). The amino-terminal sequences of purified NiSOD determined by Edman degradation are indicated by boldface letters.

purified enzyme determined by Edman degradation. The initiation codon of *sodN* is found 14 residues upstream of the N-terminal histidine residue of the purified NiSOD, suggesting that NiSOD is processed through proteolytic cleavage between residues 14 (alanine) and 15 (histidine) of the primary translation product. The amino acid sequence of the *sodN* product exhibits no significant similarity to other known proteins in the GenBank/EMBL/DDBJ sequence database. The sequence reveals that there are only two cysteine residues in the mature enzyme, occurring near the amino terminus as a –CXXXC– motif. Both of these residues (Cys-16 and Cys-22) are implicated as ligands to the nickel center. However, because three S-donor ligands are implicated by the EXAFS analysis of both oxidized and reduced NiSOD,[11] two active site structures are possible: a mononuclear site involving methionine ligation present in all four subunits of the enzyme, or a dinuclear, cysteinato-bridged site involving only the cysteine coordination formed at the interface of two subunits. No significant hydrophobic region is found, even in the full-length polypeptide, consistent with the distribution of NiSOD in the soluble fraction of cell-free extract.

*Effect of Nickel on sodN Transcription*

The level of *sodN* transcripts increases more than 9-fold within 1 hr of the addition of 200 $\mu M$ $NiCl_2$ as judged by S1 mapping, correlating with the increase in NiSOD activity.[19] Because the stability of *sodN* mRNA does not change on nickel addition, nickel must induce the transcription of the *sodN* gene.

*Nickel-Dependent Processing of SodN Polypeptide*

The cloned *sodN* gene in multicopy plasmids directs the overproduction of SodN polypeptides in *S. lividans* TK24 cells even in the absence of nickel, probably because of the loss of the negative regulatory site. A significant portion of the recombinant SodN polypeptide produced in the absence of nickel exists as precursor polypeptides with slower mobility on SDS–PAGE than the processed form. When $NiCl_2$ is added, all the SodN polypeptide exists as the processed form, generating active NiSOD activity.[19] Therefore, nickel can act not only as a catalytic cofactor, but also as a facilitator for proteolytic processing.

*Overproduction of Functional SodN in Escherichia coli*

When the two types of the *sodN* gene, one containing the full-length ORF and the other containing the preprocessed ORF without the N-terminal 14 amino acids, are expressed in *E. coli* using the pET vector system, the overproduced SodN proteins show no NiSOD activity on native polyacrylamide gels. When *E. coli* cells are grown in LB medium supplemented with $NiCl_2$ up to 1 m$M$, NiSOD activity is detected only in cells harboring the preprocessed gene, indicating that proteolytic cleavage of the precursor SodN polypeptide is a prerequisite for the production of active NiSOD.[19] This also indicates that *E. coli* lacks the proteolytic enzyme necessary for SodN processing. However, despite the high level of overproduced protein, only partial activity is detected in the presence of nickel, reflecting the low efficiency of NiSOD maturation in *E. coli*.

## Distribution and Expression of Superoxide Dismutases among *Streptomyces* Species

*Distribution of Superoxide Dismutases among Streptomyces Species*

The immunological cross-reactivity test using antisera against NiSOD from *S. coelicolor* and *S. seoulensis* and the SOD induction test by nickel indicate that NiSOD is present universally in *Streptomyces* species.[8,10] It is not uncommon that one organism produces more than one type of SOD. For example, aerobically grown *E. coli* contains MnSOD, FeSOD, and Cu,ZnSOD. In addition to NiSOD, some *Streptomyces* species produce Fe,ZnSOD. The Fe,ZnSOD, which contains about 0.4 mol of iron and zinc per mole of subunit, is regarded as a member of the FeSOD group on the basis of amino acid sequence and spectral properties. Until now, four Fe,ZnSOD genes (*sodF*) have been cloned from *Streptomyces* species [GenBank accession number AF141866 for *S. griseus,* AF012087 for *S. coelicolor* Müller, AF099014 and AF099015 for *S. coelicolor* A3(2)].[29–31] The deduced amino acid sequences of *sodF* genes from *S. coelicolor* Müller and *S. griseus* share 86% identity and 93% similarity with one another, and about 70% similarity with the cambialistic SOD from *Propionibacterium shermanii,* and with FeSOD or MnSOD from *Mycobacterium* spp.

*Differential Expression of NiSOD and Fe,ZnSOD in Streptomyces Species*

The relative amount of NiSOD and Fe,ZnSOD varies depending on the metal content of the growth medium. When *S. coelicolor* cells are grown in YEME medium supplemented with $NiCl_2$, 4 $\mu M$ $NiCl_2$ is sufficient to induce NiSOD expression. The higher the concentration of the added nickel, the more NiSOD is produced and the less Fe,ZnSOD is expressed. However, the presence of iron, zinc, manganese, or copper in the growth medium does not affect the expression of either NiSOD or Fe,ZnSOD.[7,32] The regulation of expression of each SOD in *S. coelicolor* by nickel is achieved by modulating the transcription of the *sodN* and *sodF* genes, most likely via negative regulation involving nickel-sensitive repressors.

[29] E.-J. Kim, H.-J. Chung, B. S. Suh, Y. C. Hah, and J.-H. Roe, *J. Bacteriol.* **180,** 2014 (1998).
[30] H.-J. Chung, E.-J. Kim, B. S. Suh, J.-H. Choi, and J.-H. Roe, *Gene* **231,** 87 (1999).
[31] J.-S. Kim, J.-H. Jang, J.-W. Lee, S.-O. Kang, K.-S. Kim, and J. K. Lee, *Biochim. Biophys. Acta* **1493,** 200 (2000).
[32] H.-J. Chung, J.-H. Choi, E.-J. Kim, Y.-H. Cho, and J.-H. Roe, *J. Bacteriol.* **181,** 7381 (1999).

## Concluding Remarks

The nickel-containing superoxide dismutases that may be common to *Streptomyces* species belong to a new class of superoxide dismutases. The NiSOD has a distinct active site metal center and exhibits no sequence homology with other types of SODs. However, its activity is comparable to those of other SODs, and its metal center configuration is similar to other nickel-containing enzymes. Thus, this new protein offers the opportunity to explore the specific role of nickel in redox chemistry in association with other nickel-containing enzymes and provides new insights in understanding the chemistry and the biological role of superoxide dismutases.

# [10] Reversible Conversion of Nitroxyl Anion to Nitric Oxide

*By* LARS-OLIVER KLOTZ and HELMUT SIES

## Introduction

A significant part of the effects of nitric oxide (NO$^\cdot$, nitrogen monoxide) in biological systems has been attributed to oxidation or reduction products, such as the nitrosonium cation (NO$^+$, nitrosyl cation), peroxynitrite [ONOO$^-$, oxoperoxonitrate(1−)], or the nitroxyl anion [NO$^-$, oxonitrate(1−)]. Toxicity of NO$^\cdot$ was proposed to be at least partly mediated by peroxynitrite[1] or nitroxyl anion,[2] both of which may be formed from NO$^\cdot$ in a sequence of seemingly simple reactions; whereas peroxynitrite is formed in the diffusion-controlled reaction with superoxide [reaction (1)], nitroxyl anion (p$K$ = 4.7) may be generated by reduction of NO$^\cdot$ [reaction (2)].

$$NO^\cdot + O_2^- \longrightarrow ONOO^- \quad k = 1.9 \times 10^{10}\, M^{-1}\, sec^{-1} \quad (1)$$

$$NO^\cdot + e_{aq}^- \longrightarrow NO^- \quad (2)$$

[For more details on reaction (1), see Refs. 3 and 4]. Direct reduction of NO$^\cdot$ can occur by the hydrated electron to form nitroxyl anion *in vitro* in pulse radiolysis studies.[3] In biological systems, more elaborate mechanisms of nitroxyl anion

---

[1] J. S. Beckman, T. W. Beckman, J. Chen, P. A. Marshall, and B. A. Freeman, *Proc. Natl. Acad. Sci. U.S.A.* **87**, 1620 (1990).
[2] D. A. Wink, M. Feelisch, J. Fukuto, D. Chistodoulou, D. Jourd'heuil, M. B. Grisham, Y. Vodovotz, J. A. Cook, M. Krishna, W. G. DeGraff, S. Kim, J. Gamson, and J. B. Mitchell, *Arch. Biochem. Biophys.* **351**, 66 (1998).
[3] R. E. Huie and S. Padmaja, *Free Radic. Res. Commun.* **18**, 195 (1993).

formation have been suggested, such as in the NO synthase reaction, which has been proposed to be primarily an $NO^-$- rather than $NO^{\cdot}$-generating system[5]; further, nitroxyl anion may be generated from S-nitrosothiols in the presence of thiols [reaction (3)][6] or result from reduction of $NO^{\cdot}$ by iron(II) in heme centers, such as those of cytochrome $c$[7] or hemoglobin[8] [reaction (4)].

$$RS-NO + R'SH \longrightarrow NO^- + RSSR' + H^+ \quad (3)$$

$$NO^{\cdot} + Fe(II)\text{-heme} \longrightarrow NO^- + Fe(III)\text{-heme} \quad (4)$$

Finally, it was shown that the reduction of $NO^{\cdot}$ may be catalyzed by Cu,Zn-superoxide dismutase (SOD)[9] [reaction (5), see Fig. 1[9]].

$$NO^{\cdot} + Cu^+SOD \rightleftharpoons NO^- + Cu^{2+}SOD \quad (5)$$

The decomposition of peroxynitrite to form singlet oxygen and nitroxyl anion has been proposed,[10] but these findings have been challenged and shown to be probably due to contamination of the peroxynitrite preparation with hydrogen peroxide,[11,12] so that this pathway can be considered ruled out. Investigation of nitroxyl anion biochemistry and of the effects of $NO^-$ in cellular systems is facilitated by the availability of substances that release $NO^-$, the most prominent being Angeli's salt ($Na_2N_2O_3$). The synthesis of a variety of nitroxyl donors has been described.[13]

## Reversible Conversion of $NO^{\cdot}$ to $NO^-$

The first evidence of a role for Cu,ZnSOD in the catalysis of interconversion between $NO^{\cdot}$ and nitroxyl anion came from experiments employing the formation of methemoglobin (MetHb) from oxyhemoglobin ($HbO_2$) as a marker for $NO^{\cdot}$ generation in $NO^{\cdot}$- and nitroxyl anion-generating systems.[9] In line with this, the

---

[4] R. Kissner, T. Nauser, P. Bugnon, P. G. Lye, and W. H. Koppenol, *Chem. Res. Toxicol.* **10,** 1285 (1997).
[5] H. H. Schmidt, H. Hofmann, U. Schindler, Z. S. Shutenko, D. D. Cunningham, and M. Feelisch, *Proc. Natl. Acad. Sci. U.S.A.* **93,** 14492 (1996).
[6] D. R. Arnelle and J. S. Stamler, *Arch Biochem. Biophys.* **318,** 279 (1995).
[7] M. A. Sharpe and C. E. Cooper, *Biochem. J.* **332,** 9 (1998).
[8] A. J. Gow and J. S. Stamler, *Nature* (London) **391,** 169 (1998).
[9] M. E. Murphy and H. Sies, *Proc. Natl. Acad. Sci. U.S.A.* **88,** 10860 (1991).
[10] A. U. Khan, D. Kovacic, A. Kolbanovskiy, M. Desai, K. Frenkel, and N. E. Geacintov, *Proc. Natl. Acad. Sci. U.S.A.* **97,** 2984 (2000).
[11] G. R. Martinez, P. Di Mascio, M. G. Bonini, O. Augusto, K. Briviba, H. Sies, P. Maurer, U. Röthlisberger, S. Herold, and W. H. Koppenol, *Proc. Natl. Acad. Sci. U.S.A.* **97,** 10307 (2000).
[12] K. M. Miranda, M. G. Espey, K. Yamada, M. Krishna, N. Ludwick, S. Kim, D. Jourd'heuil, M. B. Grisham, M. Feelisch, J. M. Fukuto, and D. A. Wink, *J. Biol. Chem.* **276,** 1720 (2001).
[13] S. B. King and H. T. Nagasawa, *Methods Enzymol.* **301,** 211 (1999).

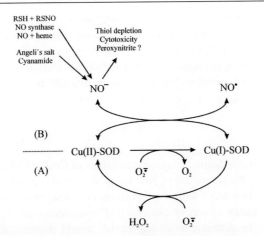

FIG. 1. Reversible interconversion of nitroxyl anion (NO⁻) and nitrogen monoxide (NO·) as catalyzed by Cu,Zn-superoxide dismutase (SOD). In the classic dismutation pathway (A), two superoxide anion molecules ($O_2^-$) disproportionate (plus $2H^+$) to one oxygen and one hydrogen peroxide. (B) Nitroxyl anion may be oxidized by SOD, leading to the formation of NO·, which in turn was shown to be reducible, also by SOD, to form nitroxyl [Murphy et al., Proc. Natl. Acad. Sci. U.S.A. **88**, 10860 (1991)]. This nonclassic SOD activity implies modulation of metabolic effects of NO· and nitroxyl: cytotoxicity, thiol oxidation, and possible formation of peroxynitrite in the presence of oxygen are examples. Nitroxyl may be formed in vitro by thermodecomposition of Angeli's salt ($Na_2N_2O_3$), or by cyanamide in the presence of $H_2O_2$ and catalase, and in vivo in the reaction of thiols with S-nitrosothiols, the action of NO synthase, and reduction of NO·, for example, by heme proteins (see text).

formation of NO· in systems containing Angeli's salt was subsequently shown to rely on the presence of Cu,ZnSOD.[5]

NO⁻, which is isoelectronic to oxygen, is similar to superoxide anion in both charge and size, and it will be attracted and channeled toward the active site of Cu,ZnSOD by the electrostatic field that has been demonstrated to be present in close proximity to the metal.[14,15] Interestingly, substrates much more distinct from superoxide than nitroxyl have been found to be accepted by Cu,ZnSOD, implying that the enzyme can react not only as a dismutase but also as a superoxide reductase (taking ferrocyanide as superoxide reductant) and superoxide oxidase (with ferricyanide as superoxide-oxidizing agent).[16]

---

[14] E. D. Getzoff, J. A. Tainer, P. K. Weiner, P. A. Kollman, J. S. Richardson, and D. C. Richardson, Nature (London) **306**, 287 (1983).
[15] J. S. Beckman, in "Nitric Oxide: Principles and Actions" (J. Lancaster, ed.), p. 1. Academic Press, San Diego, California, 1996.
[16] S. I. Liochev and I. Fridovich, J. Biol. Chem. **275**, 38482 (2000).

Some low molecular weight substances, such as $Cu^{2+} \cdot aq$ or the MnSOD mimic Mn(III)-tetrakis-(1-methyl-4-pyridyl)porphyrin, are also capable of catalyzing the formation of NO· from nitroxyl derived from Angeli's salt.[17,18] Most notably, a compound mimicking the active site of Cu,ZnSOD, a copper complex derived from putrescine and pyridine-2-aldehyde with diSchiff base-coordinated $Cu^{2+}$ (Cu-PuPy), reacted up to three times faster with NO· than Cu,ZnSOD.[19]

## Implications

Regarding the two distinct enzymatic activities of Cu,ZnSOD, the dismutation of superoxide and the formation of NO· from NO$^-$, it is evident that the enzyme may interfere in two ways with the formation of peroxynitrite, which has been implied in detrimental effects of elevated NO· concentrations[1,20]: (1) According to reaction (1), the dismutation of superoxide would impair formation of peroxynitrite from NO·. Although the reaction as catalyzed by Cu,ZnSOD occurs with a rate constant approximately three times lower than that for the formation of peroxynitrite according to reaction (1), there is some competition for superoxide anion, especially at low steady-state concentrations of NO·.[21] Intracellular levels of Cu,ZnSOD are in the range of 10 $\mu M$[21]; and (2) nitroxyl anion in its triplet state is known to react with triplet (ground state) molecular oxygen to form peroxynitrite [reaction (6)], a reaction that may be commensurate with the formation of peroxynitrite from superoxide and NO· because oxygen steady-state levels are much higher than those of superoxide.

$$NO^- + O_2 \longrightarrow ONOO^- \qquad k \approx 3\text{--}6 \times 10^7 \, M^{-1} sec^{-1} \qquad (6)$$

(see Refs. 3 and 22 for details). An assessment of the fraction of total peroxynitrite formed that is generated according to reaction (6) *in vivo*, however, is complicated by two facts, as discussed by Hughes[22]: the concentrations of nitroxyl anion available for this reaction are not known, and the spin state of nitroxyl anion is difficult to predict (only triplet- NO· efficiently reacts with $^3O_2$). Arguments is favor of the formation of peroxynitrite in cellular systems as in reaction (6) are that (1) oxygen is required for NO$^-$ toxicity[2] and (2) the cytotoxicity of Angeli's salt is higher than that of NO· derived from a variety of donors and thus reminiscent of that of peroxynitrite.[2] On the other hand, reactivities of nitroxyl derived from Angeli's

---

[17] S. Nelli, M. Hillen, K. Buyukafsar, and W. Martin, *Br. J. Pharmacol.* **131**, 356 (2000).
[18] S. Nelli, L. McIntosh, and W. Martin, *Eur. J. Pharmacol.* **412**, 281 (2001).
[19] D. Deters and U. Weser, *Biometals* **8**, 25 (1995).
[20] J. S. Beckman and W. H. Koppenol, *Am. J. Physiol.* **271**, C1424 (1996).
[21] W. H. Koppenol, *Free Radic. Biol. Med.* **25**, 385 (1998).
[22] M. N. Hughes, *Biochim. Biophys. Acta* **1411**, 263 (1999).

salt in an oxygenated environment are distinct from those of peroxynitrite[12]: although both peroxynitrite and Angeli's salt are equally effective at mediating the two-electron oxidation of dihydrorhodamine, and although the oxidation of dihydrorhodamine by Angeli's salt is oxygen dependent, the one-electron oxidation and nitration of hydroxyphenylacetate is far more efficient with peroxynitrite. With Angeli's salt, in turn, hydroxylation of benzoate is more efficient than with peroxynitrite. Angeli's salt, when thermodecomposed, releases nitroxyl in its singlet state. For a fast and efficient reaction with triplet molecular oxygen, intersystem crossing is required, turning $^1NO^-$ into $^3NO^-$. If this is the rate-limiting step, other reactions of $^1NO^-$ might prevail, as discussed by Miranda et al.[12] Irradiation of Angeli's salt with ultraviolet light of 253.7 nm, however, induces formation of triplet nitroxyl, and in the presence of oxygen, peroxynitrite has indeed been claimed to be formed.[23] As mentioned above, the spin state of nitroxyl in cellular systems is difficult to predict, and thus peroxynitrite formation from cellularly generated nitroxyl is difficult to assess.

*Biological Consequences*

Conversion of nitroxyl to nitric oxide by Cu,ZnSOD will modulate biological effects of nitroxyl if generated in excess. In Chinese hamster lung fibroblasts, cytotoxicity of $NO^-$ is preceded by strong depletion of glutathione (GSH), suggesting that glutathione may be a first line of defense against nitroxyl stress.[2] Other thiols are also targeted by nitroxyl, as has been shown for the *N*-methyl-D-aspartate receptor, which is inactivated by $NO^-$ by oxidation of two critical thiols (for review, see Lipton [24]). Wink et al.[2] found that $NO^-$ reacts some 100 times more rapidly with thiols than do other reactive nitrogen species. NADPH is another target of $NO^-$, the oxidation of which is prevented by SOD.[25] Cu,ZnSOD may therefore be regarded as a modulator of $NO^-$-mediated oxidative stress, as it oxidizes $NO^-$ to form $NO^·$. In line with this, the intracellular formation of $NO^·$ from nitroxyl has been made responsible for stimulation of migration of human neutrophils by Angeli's salt.[26]

Another significance of the conversion of nitroxyl anion to nitric oxide as catalyzed by Cu,ZnSOD has been suggested: If NO synthase indeed produces $NO^-$ rather than $NO^·$,[5] how is $NO^·$ formed? It was thus proposed that Cu,ZnSOD might act in combination with NO synthase to form $NO^·$ from intermediately generated

---

[23] C. E. Donald, M. N. Hughes, J. M. Thompson, and F. T. Bonner, *Inorg. Chem.* **25**, 2676 (1986).
[24] S. A. Lipton, *Cell Death Differ.* **6**, 943 (1999).
[25] A. Reif, L. Zecca, P. Riederer, M. Feelisch, and H. H. Schmidt, *Free Radic. Biol. Med.* **30**, 803 (2001).
[26] B. E. Vanuffelen, Z. J. Van Der, B. M. De Koster, J. Vansteveninck, and J. G. Elferink, *Biochem. J.* **330**, 719 (1998).

nitroxyl according to reaction (5) (see Fig. 1B).[5] However, this hypothesis was challenged by Xia and Zweier,[27] who unequivocally demonstrated NO· formation by purified NO synthase, employing electron spin resonance and spin traps specific for NO·.

In summary, the combination of nitroxyl anion and Cu,ZnSOD may be regarded as a relay system for the generation of NO· from NO⁻ when formed, with protective effects with regard to nitroxyl stress, and with modulating activity with regard to NO· concentrations.

Acknowledgments

Supported by Deutsche Forschungsgemeinschaft, SFB 503/B1. H.S. is a Fellow of the National Foundation for Cancer Research (NFCR, Bethesda, MD).

[27] Y. Xia and J. L. Zweier, *Proc. Natl. Acad. Sci. U.S.A.* **94,** 12705 (1997).

# [11] Purification and Determination of Activity of Mitochondrial Cyanide-Sensitive Superoxide Dismutase in Rat Tissue Extract

*By* PEDRO IÑARREA

Introduction

Mitochondria contain a tetrameric manganese-containing superoxide dismutase (MnSOD) distinct from the Cu,ZnSOD found in the cytosol of eukaryotic cells.[1] Fractionation of chicken liver mitochondria, by means of digitonin, revealed that the MnSOD was a matrix enzyme, whereas a Cu,ZnSOD resided in the intermembrane space.[2] Similar results with mitochondria of other species were subsequently reported.[3–8] The presence of a Cu,ZnSOD in mitochondria was then

[1] R. A. Weisiger and I. Fridovich, *J. Biol. Chem.* **248,** 3582 (1973).
[2] R. A. Weisiger and I. Fridovich, *J. Biol. Chem.* **248,** 4793 (1973).
[3] L. F. Panchenko, O. S. Brusov, A. M. Gerasimov, and T. D. Loktaeva, *FEBS. Lett.* **55,** 84 (1975).
[4] C. Peeters-Joris, A. M. Vandervoorde, and J. Baudhuin, *Biochem. J.* **150,** 31 (1975).
[5] L. E. Henry, R. Commack, J. P. Schwitzquebel, J. M. Palmer, and D. O. Hall, *Biochem. J.* **187,** 321 (1980).
[6] S. D. Ravindranath and I. Fridovich, *J. Biol. Chem.* **250,** 6107 (1975).
[7] D. Tyler, *Biochem. J.* **147,** 493 (1975).
[8] S. G. Ljutakova, E. M. Russanov, and S. I. Liochev, *Arch. Biochem. Biophys.* **235,** 636 (1984).

questioned by Geller and Winge,[9] whose results suggested that previous workers had been misled by mitochondrial fractions contaminated with lysosomes, which did contain Cu,ZnSOD.

Those who found Cu,ZnSOD in mitochondria had used cyanide ($CN^-$) to selectively supress the activity of Cu,ZnSOD, thus allowing the assay of both activities in extracts.[1-8] Geller and Winge used incubation at 37° with 2% (w/v) sodium dodecyl sulfate (SDS) to selectively inactivate the MnSOD and thus to achieve the same end. The activity of cytosolic Cu,ZnSOD is stable to 2% (w/v) SDS treatment in the cytosolic extract and after its purification, and they assumed that this would be the case with all Cu,ZnSODs. They did not consider the possibility of an SDS-sensitive Cu,ZnSOD activity in the crude intermembrane space extract. We now report a method for the purification of mitochondrial (mt) Cu,ZnSOD, and an assay method for determination of its activity in extracts of rat tissues.

## Materials and Methods

### Isolation of Mitochondria and Digitonin Treatment

Mitochondria are isolated from the livers of 250- to 300-g, male Sprague-Dawley rats fed *ad libitum*. The livers are homogeneized in cold 50 m$M$ Tris-HCl, 1 m$M$ EDTA, 0.25 $M$ sucrose, pH 7.4 and the organelles or crude granule fractions are isolated and repeatedly washed by differential centrifugation.[10] The crude granule fraction is further fractionated by centrifugation in a Percoll gradient,[10] which is harvested from the bottom of the tube with a peristaltic pump. The heavy mitochondrial fraction is then washed free of Percoll by 6-fold dilution and centrifugated at 10,000$g$ for 15 min at 4°. The resultant pellet is suspended and assayed for total protein[11] and diluted to 40 mg of protein per milliliter, and then treated with digitonin[12] to selectively disrupt the outer membrane. The intermembrane content is freed of mitoplasts by centrifugation at 10,000$g$ for 15 min at 4°.

### Marker Enzyme Assays

The results of the Percoll gradient fractionation are monitored by assaying for enzymes known to reside in particular subcellular fractions. Acid phosphatase activity[13] is used as the marker for lysosomes. We use $p$-nitrophenyl phosphate as the substrate and monitor the formation of $p$-nitrophenolate at 410 nm. Uricase,

---

[9] B. L. Geller and D. R. Winge, *J. Biol. Chem.* **257**, 8945 (1982).
[10] F. Gasnier, R. Rousson, F. Lerme, E. Vaganay, P. Louisot, and O. Gateau-Roesch, *Anal. Biochem.* **212**, 173 (1993).
[11] E. F. Hartree, *Anal. Biochem.* **48**, 422 (1972).
[12] T. Saidha, A. I. Stern, D. Lee, and J. A. Schiff, *Biochem. J.* **232**, 357 (1985).
[13] E. Luchter-Wasylewska, *Anal. Biochem.* **241**, 167 (1996).

monitored in terms of the consumption of urate at 295 nm, is the marker for peroxisomes.[14] Sulfite oxidase, monitored in terms of the reduction of cytochrome $c$ at 550 nm,[15] is the marker for the intermembrane space of mitochondria. MnSOD is used as the marker for mitochondrial matrix.

*Assays of Superoxide Dismutase Activity*

Total SOD activity is assayed by using the xanthine oxidase reaction to produce a flux of $O_2^-$ with either cytochrome $c$[16] or the sulfonated tetrazolium salt XTT[17] as indicating scavengers of $O_2^-$. It should be noted that the XTT assay is more sensitive than the cytochrome $c$ assay and that 1.0 cytochrome $c$ unit of SOD activity equals 2.6 XTT units. $CN^-$ at 5.0 m$M$ is used to selectively suppress the activity of Cu,ZnSOD. After incubations with SDS, the activity in samples is measured according to the procedure of Geller and Winge.[18]

*Isolation of Mitochondrial and Cytosolic Cu,ZnSODs*

The digitonin extract of the heavy mitochondria is dialyzed at 4° against three 4.0-liter changes of 50 m$M$ potassium phosphate buffer at pH 8.4, over 24 hr. It is then applied to a 1.5 × 23 cm column of Whatman (Clifton, NJ) DE-52, equilibrated and eluted with 15 m$M$ phosphate at pH 8.4. Fractions are collected at 2-min intervals; the flow rate is 100 ml/hr. Active fractions, which occur at about 50 ml of effluent, are pooled and dialyzed for 48 hr against three 4.0-liter changes of cold 25 m$M$ histidine-HCl at pH 6.0, and applied to a column of Whatman DE-52 equilibrated with 25 m$M$ histidine-HCl, pH 6.0. This column is then eluted with a descending pH gradient from pH 6.0 to 4.5 with a 13-fold dilution of Polybuffer 74 (Amersham Pharmacia Biotech, Piscataway, NJ), which is adjusted to pH 4.5 as the limiting buffer. Elution is done with 35 ml of the limiting buffer at a flow rate of 0.5 ml/min. The isoelectric point is measured, determining the pH of active fractions.

The rat liver cytosolic Cu,ZnSOD is isolated by the procedure of McCord and Fridovich.[16]

Results

*Percoll-Purified Mitochondria*

Table I presents the percentage of the total activity for each marker enzyme found in Percoll-purified mitochondria. The identity of this purified fraction as

[14] A. Hemsley, M. Pegg, D. Crane, and C. Masters, *Mol. Cell. Biochem.* **83**, 187 (1998).
[15] R. M. Garret and K. V. Rajagopalan, *J. Biol. Chem.* **269**, 272 (1994).
[16] J. M. McCord and I. Fridovich, *J. Biol. Chem.* **244**, 6049 (1969).
[17] H. Ukeda, S. Maeda, T. Ishii, and M. Sawamura, *Anal. Biochem.* **251**, 206 (1997).
[18] B. L. Geller and D. R. Winge, *Anal. Biochem.* **128**, 86 (1983).

TABLE I
PERCENTAGE OF TOTAL ACTIVITY FOR MARKER
ENZYMES AND FOR MITOCHONDRIAL Cu,ZnSOD IN
PERCOLL-PURIFIED MITOCHONDRIAL FRACTION

| Marker enzyme | % |
|---|---|
| MnSOD | 95 |
| Sulfite oxidase | 80 |
| Urate oxidase | 1 |
| Acid phosphatase | 0.5 |
| mtCu,ZnSOD | 83 |

heavy mitochondria is confirmed by the presence of 95 and 80% activity of matrix and intermembrane space mitochondrial markers, respectively. The remainder of the activity is present in the light mitochondria fraction. Likewise, the high purity of heavy mitochondria is also confirmed by the fact that lysosomal and peroxisomal marker enzymes are almost undetectable. It is noteworthy that 83% of Cu,ZnSOD activity is found in heavy mitochondria. This heavy mitochondrial fraction is used as starting material in the isolation of the mtCu,ZnSOD.

*Purification of Mitochondrial Cu,ZnSOD*

A purification of 1200-fold, with a recovery of 28%, is achieved (Table II). That the final product is homogeneous is shown by native gel electrophoresis (data not shown). The specific activity of the purified enzyme is 4200 cytochrome $c$ units/mg of protein. Purified enzyme results with a p$I$ of 5.4.

*Cyanide and Azide Sensitivity*

The determination of sensitivity to inhibition by CN$^-$ and azide (N$_3^-$) is done according to the procedures described by Borders and Fridovich[19] and Misra and Fridovich,[20] respectively. Cu,ZnSOD is sensitive to CN$^-$, and, as shown in Table III, the rat mtCu,ZnSOD is only marginally less inhibitable by CN$^-$, and N$_3^-$ than are the rat and human cytoslic Cu,ZnSODs. This supports the

TABLE II
PURIFICATION OF MITOCHONDRIAL Cu,ZnSOD

| Fraction | Volume (ml) | Total protein ($\mu$g) | Total activity (units) | Specific activity (U/mg) | Fold purification | Yield (%) |
|---|---|---|---|---|---|---|
| Digitonin treatment | 10 | 200,000 | 700 | 3.5 | 1 | 100 |
| DE-52 eluate | 55 | 1,000 | 420 | 420 | 120 | 60 |
| Chrom-focusing | 4 | 48 | 200 | 4,200 | 1,200 | 28 |

TABLE III
INHIBITION OF RAT MITOCHONDRIAL AND RAT AND HUMAN
CYTOSOLIC Cu,ZnSOD ACTIVITY BY CYANIDE AND AZIDE

| Cu,ZnSOD | IC$_{50}$, NaCN ($\mu M$) | IC$_{50}$, NaN$_3$ (m$M$) |
|---|---|---|
| Rat mitochondrial | 260 | 34 |
| Rat cytosolic | 160 | 21 |
| Human cytosolic | 200 | 22 |

conclusion that the SOD isolated from the intermembrane space of mitochondria is indeed a Cu,ZnSOD.

*Effect of Sodium Dodecyl Sulfate Treatment on Cu,ZnSODs in Crude Extract*

The bovine erythrocyte Cu,ZnSOD has been seen to be stable to SDS.[21] Thus Fig. 1 (column CytCE) shows that the SOD activity in the cytosol of rat liver is fully stable for more than 16 hr to 2% (w/v) SDS at 37°. In contrast, the Cu,ZnSOD activity present in the intermembrane space extract of heavy mitochondria is 90% inactivated by this treatment (Fig. 1, column MitCE). The activity of cytosolic Cu,ZnSOD in crude cytosolic extract retains its resistance to SDS treatment when exposed to the presence of intermembrane space extract (Fig. 1, column Mix). In this case, 140 units of the mtCu,ZnSOD in 400 $\mu$l of crude intermembrane space extract (protein at 24 mg/ml) is mixed with 320 units of cytosolic Cu,ZnSOD in 600 $\mu$l of crude cytosolic extract (protein at 40 mg/ml). Thus, it is remarkable that the behavior of mtCu,ZnSOD differs from that of cytosolic Cu,ZnSOD to 2% (w/v) SDS treatment, even when both are present in a mix of crude cytosolic and mitochondrial intermembrane space extracts. Otherwise, after purification, the mtCu,ZnSOD, like the cytosolic Cu,ZnSOD, is stable to 2% (w/v) SDS treatment (Fig. 1, columns CytSOD and MitSOD). It thus appears that other proteins, in the crude intermembrane space fractions, imposed the sensitivity to 2% (w/v) SDS.

Geller and Winge[18] have reported that MnSOD is inactivated 100% by incubation of the crude granule fraction in 2% (w/v) SDS at 37° for 30 min, when the protein concentration is less than 9.5 mg/ml. Likewise, we have studied the effect of protein concentration in the inactivation of Cu,ZnSOD present in the intermembrane space extract by SDS, under the same conditions. The intermembrane space extract is diluted to protein concentrations of 3 to 24 mg/ml. After incubation in 2% (w/v) SDS at 37° for 30 min, samples are assayed for Cu,ZnSOD activity. The results shown in Fig. 2 indicate that the 2% (w/v) SDS at 37° for 30 min is

---

[19] C. L. Borders and I. Fridovich, *Arch. Biochem. Biophys.* **241,** 472 (1985).
[20] H. P. Misra and I. Fridovich, *Arch. Biochem. Biophys.* **189,** 317 (1978).
[21] H. J. Forman and I. Fridovich, *J. Biol. Chem.* **248,** 2645 (1973).

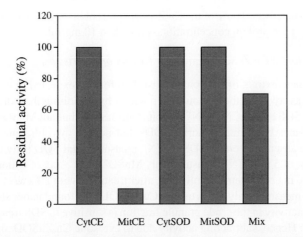

FIG. 1. Effect of SDS treatment on Cu,ZnSOD activity in extracts. SDS was added to 2% to samples of extracts, which were then incubated at 37° for 16 hr. Percent retention of Cu,ZnSOD activity was measured. Columns: CytCE, crude cytosol fraction (protein at 40 mg/ml); MitCE, crude intermembrane space extract of heavy mitochondria (protein at 24 mg/ml); CytSOD, pure cytosolic Cu,ZnSOD; MitSOD, pure mtCu,ZnSOD; Mix, 140 units of the mtCu,ZnSOD in 400 μl of crude intermembrane space extract was mixed with 320 units of cytosolic Cu,ZnSOD in 600 μl of crude cytosol fraction. This is a representative experiment of four.

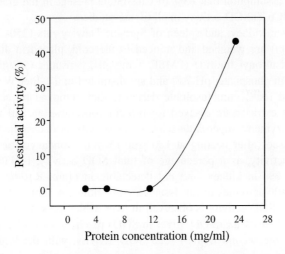

FIG. 2. Effect of protein concentration on the inactivation of Cu,ZnSOD present in mitochondrial intermembrane space extract by SDS treatment. The crude intermembrane space extract was diluted to protein concentrations of 3 to 24 mg/ml. SDS was added to 2% (w/v) to samples of extracts, which were then incubated at 37° for 30 min. Percent retention of Cu,ZnSOD activity was measured. This is a representative experiment of four.

sufficient to inactivate 100% of the Cu,ZnSOD present in the intermembrane space extract, when the protein concentration is less than 12 mg/ml.

*Determination of Cu,ZnSOD Activity in Extract of Rat Tissues*

A complete extract of rat tissues must therefore be expected to contain three SODs, and assay of all three should be possible by exploiting their different sensitivities to SDS and to $CN^-$. $CN^-$ inhibits Cu,ZnSODs but not MnSOD; whereas SDS inactivates both mitochondrial SODs, but not the cytosolic enzyme. Thus, assay in the absence of either SDS or $CN^-$ measures total SOD activity; assay in the presence of 5 m$M$ $CN^-$ measures only MnSOD; and assay of extract samples, diluted to 10 mg of protein per milliliter, after treatment with 2% (w/v) SDS at 37° for 30 min, measures only the cytosolic Cu,ZnSOD. We found in our study that all CuZnSOD activity in the crude granule fraction is sensitive to SDS treatment (data not shown). Hence, subtracting activity due to cytosolic Cu,ZnSOD and MnSOD from total SOD activity yields the activity due to Cu,ZnSOD in the crude granule fraction. Percoll gradient fractionation of the rat liver crude granule fraction reveals that at least 83% of this value represents Cu,ZnSOD activity present in heavy mitochondria. It is well known that rat liver, is compared with rat kidney and spleen, presents higher and similar specific activities of catalase and of lysosomal enzymes. Rat brain presents lower specific activities of catalase and lysosomal enzymes than the other tissues[22,23]; this could mean a lower proportional antioxidant role of peroxisome and lysosome in this tissue. In any case, calculations are made on the assumption that 83% of CuZnSOD present in the granule fraction equals the mtCu,ZnSOD activity in all tissues studied.

Brain, liver, kidney, and spleen of Sprague-Dawley rats (250–300 g, male, fed *ad libitum*) are weighed and minced in the cold, placed in 40 ml of cold phenylmethylsulfonyl fluoride (PMSF, 1 mg/ml), pepstatin (5 $\mu$g/ml), and 10 m$M$ potassium phosphate, pH 7.8, and are disrupted in a 1-liter Waring blender at high speed for 2 min. Insoluble debris is then removed by centrifugation. The resultant extracts are assayed for protein concentration, and for SOD activity. Nonenzymatic superoxide-scavenging activity is determined in samples of extract tissues after heating at 100° for 15 min. Nonenzymatic superoxide-scavenging activity, as a percentage of total SOD activity, is 43% in spleen, 8% in liver, 3% in kidney, and not detectable in brain. Results of the three SOD enzymatic activities in rat tissue extract, shown in Table IV, indicate that maximal total SOD activity is observed in liver, which presents the highest specific activity for cytCu,ZnSOD. Liver mtSODs reach 37% of total activity. Kidney presents the second highest total SOD activity, with the highest specific

[22] E. Xia, G. Rao, H. Van Remmen, A. R. Heydari, and A. Richardson, *J. Nutr.* **125**, 195 (1995).
[23] R. H. Glew, W. F. Diven, J. L. Zidian, B. B. Rankin, M. Czuczman, and A. E. Axelrod, *Am. J. Clin. Nutr.* **35**, 236 (1982).

TABLE IV
ACTIVITY OF THREE DISTINCT SODs IN EXTRACT OF RAT TISSUES

|  | %$^a$ | U/g$^{ab}$ | U/mg$^{a-c}$ |
|---|---|---|---|
| Liver |  |  |  |
| MnSOD | 20 ± 2 | 350 ± 40 | 3 ± 1 |
| cytCu,ZnSOD | 60 ± 3 | 1060 ± 8 | 10 ± 1 |
| mtCu,ZnSOD | 17 ± 1 | 295 ± 30 | 3 ± 1 |
| Kidney |  |  |  |
| MnSOD | 24 ± 2 | 380 ± 40 | 4 ± 1 |
| cytCu,ZnSOD | 45 ± 3 | 755 ± 66 | 8 ± 1 |
| mtCu,ZnSOD | 26 ± 1 | 415 ± 42 | 4 ± 1 |
| Spleen |  |  |  |
| MnSOD | 7 ± 1 | 16 ± 5 | 0.2 ± 0.1 |
| cytCu,ZnSOD | 86 ± 2 | 190 ± 25 | 2 ± 0.3 |
| mtCu,ZnSOD | 6 ± 1 | 13 ± 2 | 0.2 ± 0.1 |
| Brain |  |  |  |
| MnSOD | 30 ± 2 | 155 ± 16 | 3 ± 0.3 |
| cytCu,ZnSOD | 37 ± 2 | 195 ± 18 | 5 ± 0.3 |
| mtCu,ZnSOD | 28 ± 1 | 140 ± 11 | 3 ± 0.3 |

$^a$ Values shown represent means ± SE of five groups of four rats.
$^b$ Units (xanthine–xanthine oxidase–cytochrome $c$ reduction inhibition assay) by gram of wet tissue.
$^c$ Units (xanthine–xanthine oxidase–cytochrome $c$ reduction inhibition assay) by milligram of protein.

activity for mtCu,ZnSOD. In this tissue mtSODs reach 50% of total activity. Brain presents the highest proportional activity for mtSODs, with 58% of total activity, 28% for mtCu,ZnSOD, and with specific activity similar to that of liver. Spleen presents the lowest total SOD activity and lowest specific activity, but the highest proportional presence of cytCu,ZnSOD activity.

## Discussion

The results reported herein demonstrate that a purified heavy mitochondrial fraction of rat liver homogenate does contain a $CN^-$-sensitive SOD, presumably a Cu,ZnSOD with an intermembrane space location. This Cu,ZnSOD present in the crude intermembrane space extract was labile to 2% (w/v) SDS at 37°, whereas the cytosolic Cu,ZnSOD was not. Previous reports[4,7,8] indicated the presence in similar amounts of Cu,Zn- and Mn-containing SOD activity in the granule fraction of rat liver. However, Geller and Winge[9] reported that only a low amount of Cu,ZnSOD activity was present in this fraction. But they did not consider the possibility of an SDS-sensitive Cu,ZnSOD activity in the particulate fraction and they did not control the effect of SDS incubation on the total amount Cu,ZnSOD activity present in this fraction. In this article we report on the damaging effect of 2% (w/v) SDS incubation on the Cu,ZnSOD activity present in heavy mitochondria.

The characteristic damaging effect of 2% (w/v) SDS incubation on mtCu,ZnSOD activity permits the design of a protocol for the measurement of its activity in tissue extracts and biological samples. We now report data on mtCu,ZnSOD activity in four different rat tissues: brain, liver, kidney, and spleen. Measurements were carried out according to the XTT assay and data are expressed as units of cytochrome $c$, taking into account that a 1.0 cytochrome $c$ unit of SOD activity equals 2.6 XTT units. Protein concentration was determined in the supernatant of tissue extracts. Numerous studies of gene expression, quantitative immunodetection, and activity of SODs in homogenates and extracts of rat tissues[4,24–27] have been reported. Differences among these reports could be due to possible effects of animal age, type of diet, etc. In the particular case of SOD activity data, it is evident that differences could also be due to the nature of the SOD activity determination procedure.

Our results reveal the presence of mtCu,ZnSOD activity in brain and kidney, with 28 and 26% of total activity, respectively. Thus, mitochondrial SOD activity reaches 58 and 50% of total SOD activity in these tissues, respectively. However, its presence could be even higher, proportionally, in neurons.[22,23,28] This means that, in both tissues, the potential production of superoxide anion is located chiefly in mitochondria, and it seems possible that enough of a decrease in mtCu,ZnSOD presence or/and activity, along with enough of an increase in superoxide anion production into the intermembrane space, could produce cellular apoptosis, mitochondrial oxidative damage, and cellular malfunction and necrosis.

## Acknowledgments

This work was supported by a grant from the University of Zaragoza. We are grateful to Dr. Rosa Morales, who provided rat tissue.

---

[24] S. L. Maklund, *Biochem. J.* **222**, 649 (1984).
[25] K. Asayama and I. M. Burr, *J. Biol. Chem.* **260**, 2212 (1985).
[26] B. O. Izokun-Etiobhio, A. C. Oraedu, and E. N. A. Ugochukwo, *Comp. Biochem. Physiol. B* **95**, 521 (1990).
[27] B. Ghosh, K. Hanevold, J. K. Dobashi, J. K. Orak, and I. Singh, *Free Radic. Biol. Med.* **21**, 533 (1996).
[28] J. Lindenau, H. Noack, H. Possel, K. Asayama, and G. Wolf, *Glia* **29**, 25 (2000).

## [12] Studies of Metal-Binding Properties of Cu,Zn Superoxide Dismutase by Isothermal Titration Calorimetry

By PATRICIA L. BOUNDS, BARBARA SUTTER, and WILLEM H. KOPPENOL

## Introduction

There is renewed interest in the metal-binding properties of Cu,Zn superoxide dismutase (Cu,ZnSOD), which has been implicated as a causative agent in familial amyotrophic lateral sclerosis. Many different point mutations scattered throughout the protein have been identified in affected families,[1] and it has been suggested that the mutations might affect the zinc- and copper-binding sites of the protein,[2,3] and altered affinity for zinc ions has been postulated to play a role in the disease.[4]

Many early investigations of Cu,Zn superoxide dismutase were of the metal-binding properties of the protein. Stability constants for copper and zinc binding to bovine aposuperoxide dismutase have been determined by equilibrium dialysis,[5] and by complexometric titrations for copper binding.[6] For human wild-type superoxide dismutase, only the constants for dissociation of copper from the copper site ($6.0 \times 10^{-18}$) and of zinc from the zinc-binding site ($4.2 \times 10^{-14}$), determined at pH 7.4 by equilibrium dialysis,[4] have been reported.

The methods generally employed to determine the stability constants for binding of metal ions to Cu,ZnSOD have suffered from a number of disadvantages, including the relatively large amount of sample required and the relatively lengthy experiments. Isothermal titration calorimetry (ITC) offers a modern, automated alternative for measuring binding interactions of biomolecules, and, with the introduction of high-sensitivity, computerized instruments, it has become possible to measure binding constants with relatively low sample sizes and concentrations.[7] A further advantage of the technique is that the heat of binding and the binding constant may be determined simultaneously in a single experiment.

The ITC method is especially well suited to measure the binding of drugs or inhibitors to the active sites of small proteins. More challenging ITC experiments

---

[1] T. Siddique and H. X. Deng, *Hum. Mol. Genet.* **5**, 1465 (1996).
[2] T. J. Lyons, H. B. Liu, J. J. Goto, A. Nersissian, J. A. Roe, J. A. Graden, C. Café, L. M. Ellerby, D. E. Bredesen, E. B. Gralla, and J. S. Valentine, *Proc. Natl. Acad. Sci. U.S.A.* **93**, 12240 (1996).
[3] M. T. Carri, A. Battistoni, F. Polizio, A. Desideri, and G. Rotilio, *FEBS Lett.* **356**, 314 (1994).
[4] J. P. Crow, J. B. Sampson, Y. X. Zhuang, J. A. Thompson, and J. S. Beckman, *J. Neurochem.* **69**, 1936 (1997).
[5] J. Hirose, M. Yamada, H. Hayakawa, H. Nagao, M. Noij, and Y. Kidani, *Biochem. Int.* **8**, 401 (1984).
[6] J. A. Roe and J. S. Valentine, *Anal. Biochem.* **186**, 31 (1990).
[7] T. Wiseman, S. Williston, J. F. Brandts, and L.-N. Lin, *Anal. Biochem.* **179**, 131 (1989).

to measure protein–protein or protein–polynucleotide interactions have also been performed. The binding of metal ions to a protein, specifically of $Fe^{3+}$ to ovotransferrin,[8,9] has also been investigated. We have used ITC to determine enthalpies and stability constants for binding of copper and zinc to bovine and human Cu,Zn superoxide dismutases.

Measurements of metal-binding interactions in Cu,Zn superoxide dismutase by ITC require the preparation of metal-free protein. The affinity of metal ions for both binding sites is highly pH dependent,[10] and it was found that metal ions are most easily removed at low pH.[11] The method most commonly used to obtain apo-Cu,Zn superoxide dismutase involves lengthy dialysis against a buffer containing $N,N,N',N'$-ethylenediaminetetraacetate (EDTA) at low pH.[11,12] In an essentially similar method, holoprotein is passed slowly over a gel-filtration column (Sephadex G-25) equilibrated with a buffer containing EDTA at low pH.[12] Alternatively, the protein may be dialyzed against cyanide at pH 8, which removes metal ions but may also damage the protein.[13]

We developed a novel method to prepare apo-Cu,Zn superoxide dismutase by ion-exchange chromatography on iminodiacetic acid (IDA)-Sepharose. This method allows faster and more convenient preparation of the apoprotein in a single chromatography step followed by buffer exchange. The method may be scaled for use with variable amounts of superoxide dismutase, and would be suitable for automation. We have used the ion-exchange approach to prepare aposuperoxide dismutase from human and bovine Cu,Zn-superoxide dismutases, and the monomeric superoxide dismutase from *Escherichia coli*. We believe that a method for preparation of small quantities of apoprotein from small, precious samples of mutant superoxide dismutases could prove valuable.

Preparation of Aposuperoxide Dismutase

IDA-Sepharose, diethylaminoethyl Sephacel, Sephadex G-25, isopropyl $\beta$-D-thiogalactopyranoside, polyethylenimine, and antibiotics are purchased from Sigma, (Buchs, Switzerland). Aqueous solutions are prepared with water purified in a Millipore (Volketswil, Switzerland) Milli-Q system. All other chemicals are

---

[8] L.-N. Lin, A. B. Mason, R. C. Woodworth, and J. F. Brandts, *Biochemistry* **30**, 11660 (1991).
[9] L.-N. Lin, A. B. Mason, R. C. Woodworth, and J. F. Brandts, *Biochemistry* **33**, 1881 (1994).
[10] J. S. Valentine, M. W. Pantoliano, P. J. McDonnell, A. R. Burger, and S. J. Lippard, *Proc. Natl. Acad. Sci. U.S.A.* **76**, 4245 (1979).
[11] J. M. McCord and I. Fridovich, *J. Biol. Chem.* **244**, 6049 (1969).
[12] U. Weser, G. Barth, C. Djerassi, H.-J. Hartmann, P. Krauss, G. Voelcker, W. Voelter, and W. Voetsch, *Biochim. Biophys. Acta* **278**, 28 (1972).
[13] G. Rotilio, L. Calabrese, F. Bossa, D. Barra, A. Finazzi-Agro, and B. Mondovi, *Biochemistry* **11**, 2182 (1972).

reagent grade or better. Bovine Cu,Zn superoxide dismutase is purchased from Oxis International (Mountain View, CA). Human superoxide dismutase is expressed from a pET 3d-plasmid generously provided by the Webb-Waring Lung Institute (Denver, CO) and isolated as described.[14] *Escherichia coli* Cu,Zn superoxide dismutase, kindly provided by A. Battistoni (University of Rome "Tor Vergata," Rome, Italy), is passed over a column of IDA-Sepharose equilibrated with 10 m$M$ Tris buffer at pH 8.0 to remove excess metal ions. Protein concentrations are determined by amino acid analysis, by biuret assay,[15] or by isothermal titration calorimetry (ITC) for the human apoprotein. The bovine enzyme is determined by spectrophotometric measurement (absorptivity of the holoprotein at 258 nm, 10,300 $M^{-1}$ cm$^{-1}$;[11] absorptivity of the apoprotein at 258 nm, 2920 $M^{-1}$ cm$^{-1}$).[16] The Bradford assay[17] is used for qualitative protein determinations. Copper and zinc concentrations are determined by inductively coupled plasma mass spectrometry on a VG PlasmaQuant 2 Plus (VG Elemental, Windsfort, Cheshire, UK) or an ELAN 6100 (PerkinElmer, Norfolk, VA) equipped with an MCN 6000 (CETAC Technologies, Omaha, NE) nebulizer; $^{66}$Zn and $^{65}$Cu are used for the determinations. Copper concentrations are also measured on an ETV-atomic absorption spectrometer with Zeeman background compensation (SpectrAA-400; Varian, Mulgrave, Victoria, Australia). Samples for metal analyses are prepared with 0.1 $M$ nitric acid; indium at 20 ppb is added as internal standard for inductively coupled plasma mass spectrometry. Native polyacrylamide gel electrophoresis is performed on a PhastSystem with 8–25% acrylamide gradient gels (Amersham Pharmacia Biotech, Dübendorf, Switzerland). Protein bands on gels are visualized by treating the gel with Coomassie Blue R-250.

*Procedure*

Native superoxide dismutase (up to 1 $\mu$mol) is applied to a column (1.5 × 6.5 cm) of IDA-Sepharose preequilibrated with 17 m$M$ acetate buffer (pH 3.5) containing 10 m$M$ EDTA. At low pH, IDA-Sepharose behaves as an anion-exchange medium, and metal ions are selectively eluted with the same buffer while the apoenzyme remains bound. The EDTA is eliminated by washing with several column volumes of 50 m$M$ acetate buffer (pH 4). Aposuperoxide dismutase is subsequently obtained by gradient elution with 0 to 1 $M$ NaCl in 50 m$M$ acetate buffer at pH 4 (see Fig. 1). The protein is concentrated in a Centricon (Amicon, Wallisellen, Switzerland) concentrator device with a molecular mass cutoff of 10,000 g/mol before desalting on Sephadex G-25.

[14] B. Sutter, P. L. Bounds, and W. H. Koppenol, *Protein Expr. Purif.* **19**, 53 (2000).
[15] R. F. Itzhaki and D. M. Gill, *Anal. Biochem.* **9**, 401 (1964).
[16] J. A. Fee, *Biochim. Biophys. Acta* **295**, 87 (1973).
[17] M. M. Bradford, *Anal. Biochem.* **72**, 248 (1976).

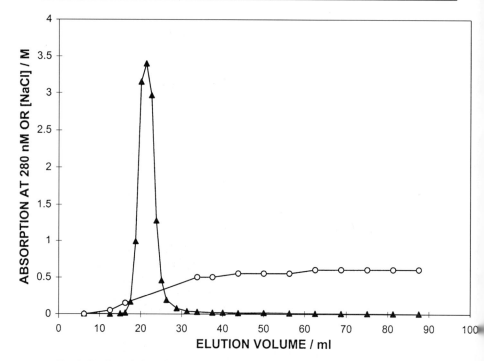

FIG. 1. Gradient elution of human aposuperoxide dismutase from IDA-Sepharose with 50 m$M$ acetic acid at pH 4.0, 0–1 $M$ NaCl. (▲) $A$ 280; (○) [NaCl]. Human SOD (1 $\mu M$) was applied to a column (1.5 × 6.5 cm) and preequilibrated with 17 m$M$ acetate buffer (pH 3.5) containing 10 m$M$ EDTA; the column was washed with 2–3 column volumes of the same buffer, followed by elimination of EDTA by washing with several column volumes of 50 m$M$ acetic acid at pH 4.0 before beginning the gradient elution.

*Results*

The results of apo-SOD preparations, summarized in Table I, show efficient removal of metal ions from SOD1. Native polyacrylamide gel electrophoresis (Fig. 2) shows that migration of apo-SOD1 is altered relative to the native protein.

Isothermal Titration Calorimetry Experiments

*Procedure*

Calorimetric measurements are performed with an MCS isothermal titration calorimeter (MicroCal, Northampton, MA) with data analysis by Origin (MicroCal).

TABLE I
TYPICAL PREPARATIONS OF APOSUPEROXIDE DISMUTASE FROM VARIOUS
SOURCES BY IDA-SEPHAROSE

|  | Cu,ZnSOD | | |
| --- | --- | --- | --- |
|  | Bovine | Human | E. coli (monomeric) |
| Amount of holo-SOD applied to column ($\mu$mol) | 0.4 | 1 | 0.6 |
| Amount of apo-SOD recovered [$\mu$mol (%)] | 0.35 (87%) | 0.70 (70%) | 0.55 (92%) |
| Metal content of holoenzyme (mol M$^{2+}$/mol SOD) | | | |
| Cu$^{2+}$ | 1.98 ± 0.02 | 9.9$^a$ | 0.80 ± 0.01 |
| Zn$^{2+}$ | 1.81 ± 0.02 | 30$^a$ | 1.20 ± 0.04 |
| Metal content of apoenzyme (mol M$^{2+}$/mol SOD) | | | |
| Cu$^{2+}$ | 0.056 ± 0.001 | 0.017 ± 0.002 | 0.0000 ± 0.0003 |
| Zn$^{2+}$ | 0.045 ± 0.002 | 0.035 ± 0.005 | 0.111 ± 0.003 |

$^a$ Reflects metal ion concentration in solution; not all metal ions are bound to the native metal-binding site, as extra Cu$^{2+}$ and Zn$^{2+}$ were added at the end of the isolation to offset depletion of metal ions.

Samples of aposuperoxide dismutase in either 50 m$M$ Tris-(hydroxymethyl)-aminomethane (Tris) or 1-methylimidazole buffer at pH 7.3 or 10 m$M$ pivalic acid buffer at pH 5.0 are diluted with the same buffer to a final concentration of approximately 25 $\mu M$ superoxide dismutase. The solutions containing copper or zinc ions are prepared by dissolving either copper sulfate or zinc acetate in the buffer to a concentration of 2 m$M$. Care is taken to use the identical buffer for both the protein and the metal ions, with readjustment of the pH of the metal ion solutions when necessary. Solutions are stirred vigorously under vacuum to remove air bubbles before being loaded into the sample cell or the syringe. In experiments in which partly metallated protein is titrated, the appropriate amount of metal ion solution is added to the apoprotein and the solution is allowed to equilibrate for several hours. In control experiments to judge the heat of mixing, protein solutions are pretreated with 0.25 m$M$ metal ion before titrating with additional metal ion. In additional control experiments, the buffer is titrated with metal ion solution. ITC experiments are performed at pH 7.3 with Tris and methylimidazole buffers, and at pH 5.0 in pivalic acid buffer. Experiments are performed at 15, 25, and 35° to probe for heat capacity effects.

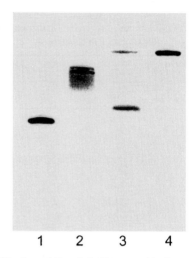

FIG. 2. Native PAGE of bovine and *E. coli* Cu,Zn superoxide dismutase, stained with Coomassie blue. Lane 1, bovine holosuperoxide dismutase; lane 2, bovine aposuperoxide dismutase; lane 3, *E. coli* holosuperoxide dismutase, a fraction of which was already present as apoprotein; lane 4, *E. coli* aposuperoxide dismutase.

The fitting routine in Origin supplied with the ITC instrument reiteratively calculates integrated curves until a best fit of the raw data from a series of injections is found. At the outset of the calculation, an initial estimate of the number of binding sites, the binding constant, and heat of binding is supplied, as well as the concentrations of SOD and metal ion. Different binding models may be considered, for example, independent or interacting binding sites.

Because the stoichiometry of metal ion interaction with the superoxide dismutase is well established, it is possible to use the ITC technique as a sensitive tool to measure the concentration of apoprotein samples. The apoprotein is loaded in the sample cell and titrated to completion with copper ions. From the titration plateaus, the concentration of the apoprotein is calculated.

## Results

*Isothermal Titration Calorimetry Experiments at pH 7.3.* In the experiments carried out at pH 7.3 in Tris and methylimidazole buffers, the metal ions are bound initially to the buffer as 1 : 4 complexes, preventing precipitation of metal–hydroxy complexes. The site-binding constants obtained from deconvolution of the isotherms reflect the difference between the constants for binding of the metal

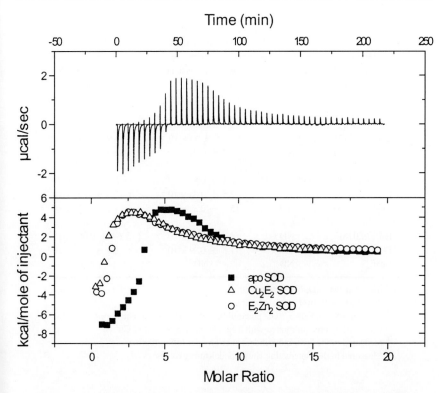

FIG. 3. ITC experiments with human SOD in 10 m$M$ pivalic acid buffer, pH 5.0, at 25°; 5-$\mu$l injections with 2 m$M$ Cu$^{2+}$. Top: Raw data for 50 injections into apo-SOD. Bottom: Integrated data after subtraction of control values: titration of apo-SOD (■), SOD preequilibrated with Zn$^{2+}$ (○), and SOD preequilibrated with Cu$^{2+}$ (△). Experiments performed at 15 and 35° are not shown.

ions to superoxide dismutase and literature stability constants for the buffer–metal complexes at 25°.[18,19,20]

*Isothermal Titration Calorimetry Experiments at pH 5.0.* In pivalate buffer at pH 5.0, formation of insoluble hydroxo complexes is negligible, and the buffer binds only weakly to the metal ions. Further, steric hindrance prevents formation of other than 1 : 1 complexes. Titrations of metal ions into apoprotein solutions are

---

[18] R. M. Smith and A. E. Martell, "Critical Stability Constants," Vol. 2: "Amines." Plenum Press, New York, 1975.
[19] A. E. Martell and R. M. Smith, "Critical Stability Constants," Vol. 3: "Other Organic Ligands." Plenum Press, New York, 1977.
[20] B. Sutter, P. L. Bounds, T. Nauser, and W. H. Koppenol, in preparation (2001).

TABLE II
BINDING CONSTANTS AND HEATS OF BINDING FOR BINDING OF $Cu^{2+}$ TO METAL-FREE SOD, MEASURED BY ITC[a]

| Metallation state[b] ([SOD], mM) | Binding sites $(n)$[c] | Binding constant $(K)$ | Heat of binding, $\Delta H$ (kcal/mol) |
|---|---|---|---|
| 15° | | | |
| $E_2E_2$ (0.020) | 4.03 ± 0.08 | $(2.1 \pm 0.7) \times 10^{6c}$ | $(-4.5 \pm 0.2) \times 10^{3c}$ |
| $E_2Zn_2$ (0.025) | 2.0 ± 0.2 | $(5 \pm 1) \times 10^{5c}$ | $(-4.9 \pm 0.1) \times 10^{3c}$ |
| $Cu_2E_2$ (0.015) | ND | $(3.2 \pm 0.2) \times 10^{5d}$ | $(-2.7 \pm 0.2) \times 10^{3c}$ |
| 25° | | | |
| $E_2E_2$ (0.018) | 4.00 ± 0.09 | $(2.1 \pm 0.8) \times 10^{6c}$ | $(-7.5 \pm 0.2) \times 10^{3c}$ |
| $E_2Zn_2$ (0.015) | 1.9 ± 0.2 | $(2.5 \pm 0.3) \times 10^{5d}$ | $(-5.0 \pm 0.2) \times 10^{3d}$ |
| $Cu_2E_2$ (0.015) | ND | $(5.1 \pm 0.3) \times 10^{4d}$ | $(-4.7 \pm 0.2) \times 10^{3d}$ |
| 35° | | | |
| $E_2E_2$ (0.028) | ND | $(4.7 \pm 0.4) \times 10^{6d}$ | $(-7.64 \pm 0.06) \times 10^{3d}$ |
| $E_2Zn_2$ (0.029) | 2.0 ± 0.1 | $(1.5 \pm 0.2) \times 10^{5d}$ | $(-6.4 \pm 0.2) \times 10^{3d}$ |
| $Cu_2E_2$ | ND | ND | ND |

*Abbreviation:* ND, Not determined.
[a] At pH 5.0 in 10 mM pivalate buffer.
[b] $E_2E_2$, apo-SOD; $E_2Zn_2$, SOD preequilibrated with a stoichiometric amount of $Zn^{2+}$; $Cu_2E_2$, SOD preequilibrated with a stoichiometric amount of $Cu^{2+}$.
[c] Measured in the "two independent sites" model-fitting routine.
[d] Measured in the "interacting sites" model-fitting routine.

carried out in 10 mM pivalic acid buffer at 25°. The results for binding of $Cu^{2+}$ are shown in Fig. 3 and Table II.

Two binding modes, "interacting sites" and "two independent sites," are considered, and the values for $K$ and $\Delta H$ calculated are similar in both models. The independent site model is used to verify the number of binding sites, which is well established for SOD. Copper ions are found to bind to all four sites: the shape of the binding curve and the integrated area are the same whether $Cu^{2+}$ is titrated into apo-SOD or SOD containing two equivalents of $Cu^{2+}$ or $Zn^{2+}$. The values of $K$ vary from $10^6$ for binding to apoprotein, to $10^5$ for binding to SOD preequilibrated with $Zn^{2+}$, to $10^4$ for SOD preequilibrated with $Cu^{2+}$. The $\Delta H$ values vary only slightly as a function of the state of metallation of SOD and as a function of temperature.

*Reconstitution of Superoxide Dismutase Activity.* The activity of reconstituted superoxide dismutase, prepared under conditions essentially identical to those of the calorimetric titrations, is determined by pulse radiolysis. The rate constants determined[20] indicate that activity is restored to SOD during the calorimetric titration.

## Discussion

IDA immobilized on Sepharose was used as an ion-exchange medium for the facile preparation of aposuperoxide dismutase from bovine, human, or *E. coli* Cu,Zn superoxide dismutase. The commercially available chromatography medium, which is marketed as affinity chromatography material for use at a higher pH to bind metal ions, provides a convenient, scalable, and reproducible preparation of apo-Cu,Zn superoxide dismutase.

Isothermal titration calorimetry determination of binding constants and heats of binding for $Cu^{2+}$ to apo-SOD at pH 5 shows that copper binds apparently to all four metal-binding sites with nearly indentical $\Delta H$ values; the curves for titration of $Cu^{2+}$ into SOD partially metallated with either $Cu^{2+}$ or $Zn^{2+}$ are perfectly overlaid and integrate to two binding sites, and the shapes of these curves fit perfectly the shape of the curve for apo-SOD that integrates to four binding sites. There is, however, considerable variation in the $K$ values as a function of the state of metallation of the protein. The ITC method provides visualization of the integrated areas and, thus, a richer appreciation of the similarity of these binding interactions in spite of the dissimilarity in the binding constants.

# [13] Superoxide Reductase from *Desulfoarculus baarsii*

*By* VINCENT NIVIÈRE and MURIELLE LOMBARD

## Introduction: Superoxide Reductase as New Enzymatic System Involved in Superoxide Detoxication

Superoxide radical ($O_2{^{\cdot-}}$) is the univalent reduction product of molecular oxygen and belongs to the group of so-called toxic oxygen derivatives. For years the only enzymatic system known to catalyze the elimination of superoxide was the superoxide dismutase (SOD),[1] which catalyzes dismutation of superoxide radical anions to hydrogen peroxide and molecular oxygen:

$$O_2{^{\cdot-}} + O_2{^{\cdot-}} + 2H^+ \longrightarrow H_2O_2 + O_2$$

A new concept concerning the mechanisms of cellular defense against superoxide has emerged.[2,3] It was discovered that elimination of $O_2{^{\cdot-}}$ could also

---

[1] I. Fridovich, *Annu. Rev. Biochem.* **64**, 97 (1995).
[2] M. Lombard, D. Touati, M. Fontecave, and V. Nivière, *J. Biol. Chem.* **275**, 27021 (2000).
[3] F. E. Jenney, Jr., M. J. M. Verhagen, X. Cui, and M. W. W. Adams, *Science* **286**, 306 (1999).

occur by reduction, a reaction catalyzed by an enzyme thus named superoxide reductase (SOR):

$$O_2^{\cdot -} + 1e^- + 2H^+ \longrightarrow H_2O_2$$

Up to now, SOR has been characterized mainly from anaerobic microorganisms, sulfate-reducing bacterial,[2] an archaeon,[3] and in a microaerophilic bacterium.[4,5] *In vivo*, SOR was shown to be an efficient antioxidant protein from the observation that a *Desulfovibrio vulgaris* (Hildenborough) mutant strain lacking the *sor* gene became more oxygen sensitive during transient exposure to microaerophilic conditions.[6] In addition, although SOR is not naturally present in *Escherichia coli*, it was demonstrated that expression of SORs from *Desulfoarculus baarsii* and *Treponema pallidum* in a *sodA sodB E. coli* mutant strain could totally replace the SOD enzymes to overcome a superoxide stress.[4,5,7]

Two classes of SOR have been described. Class I SORs are small metalloproteins found in anaerobic sulfate-reducing and microaerophilic bacteria, initially called desulfoferrodoxins (Dfx). There are homodimers of $2 \times 14$ kDa, which have been extensively studied for their structural properties.[8,9] The monomer is organized in two protein domains.[9] The N-terminal domain contains a mononuclear ferric iron, center I, coordinated by four cysteines in a distorted rubredoxin-type center. In the SOR from *T. pallidum*, three of the four N-terminal cysteine residues involved in iron chelation are lacking and center I is missing.[4,5] The C-terminal domain, which carries the active site of SOR,[2] contains a different mononuclear iron center, center II, consisting of an oxygen-stable ferrous iron with square-pyramidal coordination to four nitrogens from histidines as equatorial ligands and one sulfur from a cysteine as the axial ligand.[9] Its midpoint redox potential has been reported to be about $+ 250$ mV.[5,8] The iron center II reduces superoxide efficiently, with a second-order rate constant of about $10^9 \, M^{-1} \, \sec^{-1}$.[2] It does not exhibit significant SOD activity.[2] In addition, the active site of the SOR is specific for $O_2^{\cdot -}$, because the reduced iron center II is not oxidized by $O_2$ and only slowly by $H_2O_2$.[2]

A class II SOR has been characterized from the anaerobic archaeon, *Pyrococcus furiosus*.[3,10] The homotetrameric protein presents strong homologies to

---

[4] M. Lombard, M. Fontecave, D. Touati, and V. Nivière, *J. Biol. Chem.* **275**, 115 (2000).
[5] T. Jovanovic, C. Ascenso, K. R. O. Hazlett, R. Sikkink, C. Krebs, R. Litwiller, L. M. Benson, I. Moura, J. J. G. Moura, J. D. Radolf, B. H. Huynh, S. Naylor, and F. Rusnak, *J. Biol. Chem.* **275**, 28439 (2000).
[6] J. K. Voordouw and G. Voordouw, *Appl. Environ. Microb.* **64**, 2882 (1998).
[7] M. J. Pianzzola, M. Soubes, and D. Touati, *J. Bacteriol.* **178**, 6736 (1996).
[8] P. Tavares, N. Ravi, J. J. G. Moura, J. LeGall, Y. H. Huang, B. R. Crouse, M. K. Johnson, B. H. Huynh, and I. Moura, *J. Biol. Chem.* **269**, 10504 (1994).
[9] A. V. Coelho, P. Matias, V. Fülöp, A. Thompson, A. Gonzalez, and M. A. Coronado, *J. Biol. Inorg. Chem.* **2**, 680 (1997).
[10] P. Y. Andrew, Y. Hu, F. E. Jenney, M. W. W. Adams, and D. C. Rees, *Biochemistry* **39**, 2499 (2000).

neelaredoxin (Nlr), a small protein containing a single mononuclear center, earlier characterized from sulfate-reducing bacteria. The amino acid sequence and the overall protein fold are similar to that of the iron center II domain of class I SORs and the structure of the unique mononuclear iron center is similar to that of the iron center II.[10] The main difference is that class II SORs do not contain the N-terminal domain that chelates the iron center I in class I SORs.

## Superoxide Reductase from *Desulfoarculus baarsii*: Class I Superoxide Reductase

*Overexpression and Purification of Enzyme*

SOR from *Desulfoarculus baarsii* has been overexpressed in *Escherichia coli*, by infecting *E. coli* with a vector carrying the *sor* structural gene under the control of a *tac* promoter.[2] Induction is achieved by adding 1 m$M$ isopropyl-$\beta$-D-thiogalactopyranoside (IPTG) at the beginning of the exponential phase, and best expression is obtained when cells reach the stationary phase. SOR is overexpressed to about 10–15% of the total soluble proteins from *E. coli*. No significant formation of inclusion bodies occurs. Metallation of SOR is best achieved when the cells are grown on minimal medium M9, complemented with 0.4% (w/v) glucose and 100 $\mu M$ FeCl$_3$. Under these conditions, SOR is fully metallated. Overexpression of SOR in Luria–Bertani medium results in a protein that is partially demetallated.

Purification of SOR is monitored by UV–visible spectrophotometry and sodium dodecyl sulfate–polyacrylamide gel electrophoresis (SDS–PAGE). The $A_{280 \text{ nm}}/A_{503 \text{ nm}}$ ratio, characteristic of the oxidized iron center I of SOR, increases during purification until it reaches a factor of 7 for the purified protein. During purification, the iron center II remains in a reduced state and does not contribute to the visible spectrum. It can be oxidized with a slight excess of Fe(CN)$_6$ and then exhibits an absorbance band centered at 644 nm (Fig. 1). Purified SOR exhibits a value of $\epsilon_{503 \text{ nm}}$ of 4400 $M^{-1}$ cm$^{-1}$. The difference spectrum of the Fe(CN)$_6$-oxidized minus the as-isolated SOR provides the absorption bands associated with the iron center II with a $\epsilon_{644 \text{ nm}}$ of 1900 $M^{-1}$ cm$^{-1}$ (Fig. 1). It should be noticed that SDS–PAGE analysis of the fully purified SOR exhibits several polypeptide bands. These artifacts could arise from partial oxidation of cysteine residues during electrophoresis migration. A single polypeptide band is obtained when the reduced thiols are omitted from the SDS–PAGE loading buffer.

Purification of the overexpressed SOR can usually be achieved with two chromatography steps.[2] In a typical experiment, 20 g of IPTG-induced *E. coli* cells are sonicated in 70 ml of 0.1 $M$ Tris-HCl, pH 7.6, and centrifuged at 180,000$g$ for 90 min at 4°. Soluble extracts (400 mg) are first treated with 3% (w/v) streptomycin sulfate, centrifuged at 40,000$g$ for 15 min at 4°, and then precipitated with 75% (w/v) ammonium sulfate. The pellet is dissolved in 0.1 $M$ Tris-HCl, pH 7.6, and

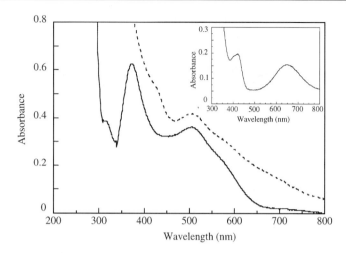

FIG. 1. Absorption spectra of the recombinant *D. baarsii* SOR (82 $\mu M$). Spectrum of the as-isolated SOR (—), or treated with 100 $\mu M$ potassium ferricyanide (– – –). *Inset:* Difference spectrum of (– – –) minus (—).

the solution is loaded onto a gel-filtration ACA 54 column (360 ml) equilibrated with 10 m$M$ Tris-HCl, pH 7.6 (buffer A). The low molecular weight fractions with an $A_{280\,nm}/A_{503\,nm}$ ratio of about 20 are pooled (115 mg) and loaded into an anion-exchange UNO-Q column (6 ml; Bio-Rad, Hercules, CA), equilibrated with buffer A. A linear gradient is applied (0–0.2 $M$ NaCl in buffer A) for 60 ml. Thirty milligrams of pure, fully metallated SOR is eluted with about 50 m$M$ NaCl.

*Assay for Superoxide Reductase Activity*

The reaction catalyzed by SOR can be described by the sum of two half-reactions at its active site:

$$SOR_{red} + O_2^{\cdot-} + 2H^+ \longrightarrow H_2O_2 + SOR_{ox} \quad (1)$$

$$SOR_{ox} + 1e^- \longrightarrow SOR_{red} \quad (2)$$

$$O_2^{\cdot-} + 1e^- + 2H^+ \longrightarrow H_2O_2 \quad (3)$$

In the first half-reaction [Eq. (1)], the reduced iron center II of SOR transfers one electron to $O_2^{\cdot-}$ to produce $H_2O_2$ and the oxidized form of the iron center II. This reaction is rapid and occurs near the limit of diffusion of molecules in solution. In the second half-reaction [Eq. (2)], the oxidized iron center II is reduced by an electron donor to regenerate the active form of SOR. The physiological electron donors for SOR from *D. baarsii* have not yet been identified, although in *E. coli* it was found that the NADPH flavodoxin reductase (Fpr) can provide electrons efficiently to SOR.[11]

Direct measurement of global SOR activity [Eq. (3)] may be difficult for several reasons. The absolute requirement for $O_2$ in the assay to generate $O_2{\cdot}^-$ (usually generated by the xanthine–xanthine oxidase system) may strongly interfere with the electron donor. In addition, a possible use of nonautoxidable electron donors, like reduced cytochrome $c$, may not give an accurate measurement of full SOR activity. As a matter of fact, the overall rate constant for the global reaction is limited by the electron transfer step to the oxidized form of SOR, which would generally occur 100 to 1000 times more slowly than the reduction of $O_2{\cdot}^-$ by SOR. Measurement of global SOR activity will then provide only kinetic information about the reduction of SOR by the electron donor rather than the reaction with superoxide.

All these remarks make difficult or almost impossible the measurement of SOR activity in crude extracts. However, for a purified SOR preparation, it is possible to measure its ability to reduce superoxide by the following assay.

We have developed an SOR assay that allows determination of the rate constant of the first half-reaction of SOR [Eq. (1)], corresponding to a stoichiometric reduction of superoxide by SOR.[2] We have used methodology developed in the case of several dehydratases, such as aconitase and fumarase, which are known to react with $O_2{\cdot}^-$ rapidly.[12] The general principle of this SOR assay is to monitor spectrophotometrically the kinetics of oxidation of SOR by superoxide, generated by the xanthine–xanthine oxidase system. This is possible at 644 nm, which is a characteristic absorption band of the oxidation of the active site of SOR (Fig. 1). As shown in Fig. 2, the kinetics are monitored for several minutes, in the absence or in the presence of different amounts of SOD, which compete with SOR for superoxide. A sufficient amount of SOD induces inhibition of the kinetics of oxidation of SOR. As explained in detail elsewhere,[2] under these experimental conditions the velocity of oxidation of the active site of SOR by $O_2{\cdot}^-$ ($v_{ox}$) can be expressed as shown in Eq. (4):

$$\frac{1}{v_{ox}} = \frac{1}{cte} + \frac{k_{SOD}}{ctek_{SOR}[SOR]}[SOD] \qquad (4)$$

where $k_{SOR}$ and $k_{SOD}$ are the second-order rate constants of the reaction of SOR and SOD with superoxide, respectively. The cte term represents the rate of synthesis of $O_2{\cdot}^-$ by the xanthine oxidase system. Under these conditions, when the initial rate of oxidation of center II is decreased by 50% because of the competition with SOD for $O_2{\cdot}^-$, it can be written as[2]

$$k_{SOD}[SOD] = k_{SOR}[SOR] \qquad (5)$$

---

[11] V. Nivière and M. Lombard, in preparation (2001).
[12] D. H. Flint, J. F. Tuminello, and M. H. Emptage, *J. Biol. Chem.* **268**, 22369 (1993).

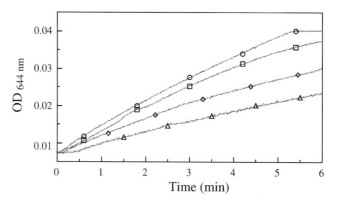

FIG. 2. Kinetics of oxidation of center II from the *D. baarsii* SOR by $O_2^{\cdot-}$. Oxidation of center II was monitored spectroscopically, at 25°, by the increase in absorbance at 644 nm, in the presence of various [Cu,ZnSOD]: (○) 0 μM; (□) 1.1 μM; (◇) 5.7 μM; and (△) 8.6 μM. The cuvette contains (1-ml final volume) 19 μM SOR, 50 mM Tris-HCl (pH 7.6), 400 μM xanthine, catalase (500 U/ml), and different amounts of Cu,ZnSOD. The oxidation was initiated by adding 0.013 U of xanthine oxidase.

The concentration of SOD that decreases by 50% the rate of oxidation of center II is graphically determined from Eq. (4), as illustrated in Fig. 3. By taking into account the known second-order rate constant of the reaction of $O_2^{\cdot-}$ with SOD ($2 \times 10^9$ $M^{-1}$ sec$^{-1}$ for the Cu,ZnSOD from bovine erythrocytes or $3.2 \times 10^8$ $M^{-1}$ sec$^{-1}$ for FeSOD from *E. coli*), the second-order rate constant of

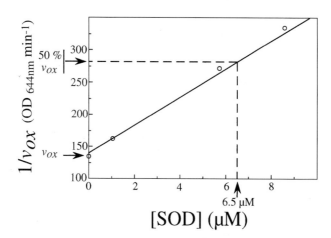

FIG. 3. Determination of the SOD concentration that inhibits by 50% the rate of oxidation of iron center II by superoxide. Shown is the reciprocal of the initial velocity of the oxidation of center II ($v_{ox}$) calculated from Fig. 2 as a function of [Cu,ZnSOD], according to Eq. (4). A value of 6.5 μM Cu,ZnSOD is found.

the oxidation of the SOR center II by $O_2^{\cdot-}$ can be now calculated, using Eq. (5). A value of $7 \times 10^8 \, M^{-1} \, \text{sec}^{-1}$ was obtained for the SOR from *D. baarsii*.
With the same assay, the SOR from *T. pallidum* exhibits a slightly higher value of the rate constant for its reaction with $O_2^{\cdot-}$, $1 \times 10^9 \, M^{-1} \, \text{sec}^{-1}$.[4]

Conclusions

So far there is no suitable enzymatic assay for monitoring the purification of an SOR enzyme from cell extracts. However, the protein can be purified in all cases according to its specific absorbance band at about 650 nm when oxidized with a strong oxidant (e.g., ferricyanide).

The overall SOR activity [Eq. (3)] is probably rate limited by the efficiency of the electron donors [Eq. (2)]. However, within the cell, taking into account the great instability of $O_2^{\cdot-}$ in solution (spontaneous dismutation rate constant of $5 \times 10^5 \, M^{-1} \, \text{sec}^{-1}$) and that dehydratase enzymes such as aconitase and fumarase react rapidly with superoxide (rate constant of about $10^7 \, M^{-1} \, \text{sec}^{-1}$),[12] the efficiency of SOR as an antioxidant resides in the large rate constant for its reaction with $O_2^{\cdot-}$ [Eq. (1)]. A slower reaction rate for Eq. (1) would probably make a useless SOR enzyme, which would have no chance to react with $O_2^{\cdot-}$ *in vivo*.

# [14] Enzyme-Linked Immunosorbent Assay for Human and Rat Manganese Superoxide Dismutases

*By* KEIICHIRO SUZUKI, TOMOMI OOKAWARA, YASUHIDE MIYAMOTO, and NAOYUKI TANIGUCHI

Introduction

There are in mammals three types of superoxide dismutases, designated Cu,ZnSOD, extracellular SOD, and MnSOD. Among them MnSOD is localized in mitochondria and is distinct from the other two isozymes in terms of localization, chromosomal localization, and antigenicity. The significance of the location of MnSOD in the mitochondria is that the enzyme acts as the first line of defense against superoxide generated as a by-product of oxidative phosphorylation.

To analyze the roles of SODs, several mouse lines with targeted inactivation of SODs have been generated. To analyze the role of Cu,ZnSOD, two groups generated mouse lines with targeted inactivation of Cu,ZnSOD.[1,2] Contrary to

---

[1] A. G. Reaume, J. L. Elliott, E. K. Hoffman, N. W. Kowall, R. J. Ferrante, D. F. Siwek, H. M. Wilcox, D. G. Flood, M. F. Beal, R. H. Brown, Jr., R. W. Scott, and W. D. Snider, *Nat. Genet.* **13**, 43 (1996).

their expectations, these mice developed normally and showed no phenotypic abnormality under normal conditions. On the other hand, MnSOD-deficient mice showed dilated cardiomyopathy, accumulation of lipid in liver and skeletal muscle, and metabolic acidosis.[3] These mice were exceedingly hypotonic, hypothermic, and paler compared with wild mice. In addition, they died within 10 days of birth. A severe reduction in succinate dehydrogenase (complex II) and aconitase activities was also found. Another MnSOD-deficient mouse line (deletion of exons 1 and 2) exhibited reduced growth rate and survival (up to 3 weeks).[4] Likewise, they exhibited severe anemia, dilated hearts, degeneration of the neurons in the basal ganglia and brainstem, and progressive motor disturbance (weakness, rapid fatigue, and circling behavior). These findings indicate that MnSOD is important for maintaining the integrity of mitochondrial enzymes and that its deficiency leads to an increased susceptibility to oxidative mitochondrial injury in many tissues, such as cardiac myocytes, skeletal muscles, central nervous system, and hepatocytes. Extracellular SOD (EC-SOD)-deficient mice, also generated, developed normally and remained healthy up to the age of 14 months or more.[5]

MnSOD is an inducible enzyme, and various cytokines such as interleukin 1 (IL-1), IL-6, and tumor necrosis factor $\alpha$ (TNF-$\alpha$) induce the MnSOD gene in various types of cells *in vitro* as well as *in vivo*.[6–9] Human MnSOD contains a GC-rich and TATA/CAAT-less promoter, which is characteristic of a housekeeping gene. MnSOD is also induced by 12-*O*-tetradecanoylphorbol-13-acetate (TPA).[10] Its induction is due to the phosphorylation of a CREB/ATF-1-like factor via the protein kinase C pathway.[11] AP-1 is also an important regulator of transcriptional activities of the MnSOD gene.[12] Extensive studies of the gene expression of MnSOD *in vitro* or *in vivo* have been carried out by Northern blotting analysis, but

---

[2] T. Kondo, A. G. Reaume, T. T. Huang, E. Carlson, K. Murakami, S. F. Chen, E. K. Hoffman, R. W. Scott, C. J. Epstein, and P. H. Chan, *J. Neurosci.* **17**, 4180 (1997).
[3] Y. Li, T. T. Huang, E. J. Carlson, S. Melov, P. C. Urcell, J. L. Olson, L. J. Noble, M. P. Yoshimura, C. Berger, P. K. Chan, D. C. Wallace, and C. J. Epstein, *Nat. Genet.* **11**, 376 (1995).
[4] R. M. Lebovitz, H. Zhang, H. Vogel, J. Cartwright, Jr., L. Dionne, N. Lu, S. Huang, and M. M. Matzuk, *Proc. Natl. Acad. Sci. U.S.A.* **93**, 9782 (1996).
[5] L. M. Carlsson, J. Jonsson, T. Edlund, and S. L. Marklund, *Proc. Natl. Acad. Sci. U.S.A.* **92**, 6264 (1995).
[6] G. H. W. Wong and D. V. Goeddel, *Science* **242**, 941 (1988).
[7] A. Masuda, D. L. Longo, Y. Kobayashi, E. Appella, J. J. Oppenheim, and K. Matsushima, *FASEB J.* **2**, 3087 (1988).
[8] G. A. Visner, W. C. Dougall, J. M. Willson, I. A. Burr, and H. S. Nick, *J. Biol. Chem.* **265**, 2856 (1990).
[9] M. Ono, H. Kohda, T. Kawaguchi, M. Ohhira, C. Sekiya, M. Namiki, A. Takeyasu, and N. Taniguchi, *Biochem. Biophys. Res. Commun.* **182**, 1100 (1992).
[10] J. Fujii and N. Taniguchi, *J. Biol. Chem.* **266**, 23142 (1991).
[11] H. P. Kim, J. H. Roe, P. B. Chock, and M. B. Yim, *J. Biol. Chem.* **274**, 37455 (1999).
[12] Y. S. Ho, A. J. Howard, and J. D. Crapo, *Am. J. Respir. Cell Mol. Biol.* **4**, 278 (1991).

studies of the expression of MnSOD protein yielded more interesting observations in terms of pathophysiology. MnSOD is composed of four identical subunits, each containing one molecule of manganese. MnSOD is a mitochondrial enzyme with a signal sequence, but is released into the cytoplasm or plasma under pathological conditions.[13]

In this article we describe procedures for a quantitative immunoassay of MnSOD proteins by the sandwich enzyme-linked immunosorbent assay (ELISA) technique, using polyclonal and monoclonal antibodies against human and rat MnSOD proteins. The use of this technique may provide new insight into the role of MnSOD protein in humans, mice, and rats. The enzyme assay of MnSOD is usually carried out via an inhibition assay of superoxide generation, using a xanthine–xanthine oxidase system, a nitroblue tetrazolium (NBT) method, or a pulse radiolysis assay.[14] MnSOD is not sensitive to cyanide and therefore its activity can be distinguished from that of Cu,ZnSOD, which is sensitive to cyanide. However, a problem arises because nonspecific reactions are possible in crude extracts or tissues, which may lead to controversial results in terms of pathophysiology. For example, it is well known that in cancer cells MnSOD levels are low,[15] but quantitative analyses of MnSOD protein by ELISA indicate that this is not the case and, in fact, most tumor tissues express MnSOD proteins.[16,17] These conflicting results are probably due to erroneous calculations of the experimental data because of nonspecific enzyme activity or the presence of endogenous inhibitors. The method of analysis described here is useful and reliable, and can be used in various types of experimental and clinical procedures that involve the use of human and rat material.

## General Procedures

*Purification and Production of Polyclonal Antibodies against MnSOD*

Rat MnSOD is purified from 800 g of rat liver by the same procedure used for human MnSOD.[18]

Immunization for the preparation of polyclonal antibodies against rat MnSOD is performed in rabbits and the resulting antibodies are purified by precipitation

---

[13] T. Nakata, K. Suzuki, J. Fujii, M. Ishikawa, and N. Taniguchi, *Int. J. Cancer* **55,** 646 (1993).
[14] N. Taniguchi and J. M. C. Gutteridge, "Experimental Protocols for Reactive Oxygen and Nitrogen Species." Oxford University Press, London, 2000.
[15] L. W. Oberley and G. R. Buettner, *Cancer Res.* **39,** 1141 (1979).
[16] S. Iizuka, N. Taniguchi, and A. Makita, *J. Natl. Cancer Inst.* **72,** 1043 (1984).
[17] N. Taniguchi, in "Advances in Clinical Chemistry," Vol. 29, p. 1. Academic Press, San Diego, California, 1992.
[18] Y. Matsuda, S. Higashiyama, Y. Kijima, K. Suzuki, K. Kawano, M. Akiyama, S. Kawata, S. Tarui, H. F. Deutsch, and N. Taniguchi, *Eur. J. Biochem.* **194,** 713 (1990).

with 50% (w/v) saturated ammonium sulfate, DEAE-cellulose chromatography, and immunoaffinity column chromatography with purified MnSOD as an adsorbent.

*Enzyme-Linked Immunosorbent Assay Using a Polyclonal Antibody against Rat MnSOD*

Here we demonstrate an ELISA system using polyclonal antibodies.[19]

*Reagents*

Dilution buffer: 20 m$M$ phosphate-buffered saline (PBS, pH 7.4) containing 0.1% (w/v) bovine serum albumin (BSA); this buffer is used for dilution of standard SOD, samples, and related materials
Washing buffer: 20 m$M$ PBS, pH 7.4, containing 0.05% (v/v) Tween 20
Blocking buffer: 20 m$M$ PBS, pH 7.4, containing 1% BSA
Peroxidase-labeled avidin solution: Horseradish peroxidase–avidin D solution (commercially available; dilute 1 : 5000 with dilution buffer)
First antibody solution: Dilute antibody with 50 m$M$ NaHCO$_3$, pH 9.6 (typical concentration, 1 to 10 μg/ml)
Second antibody solution: Dilute a biotinated antibody with dilution buffer (typical concentration: 1 to 5 μg/ml)

Biotination of antibody is as follows.

1. An antibody solution is dialyzed with 0.1 $M$ NaHCO$_3$, pH 8.0, at 4°.
2. Dissolve biotin-$N$-hydroxysuccinimide in dimethyl sulfoxide (1 mg/ml).
3. Mix antibody solution with biotin solution (antibody : biotin, 1 : 1).
4. Allow the mixture to stand for 4 hr at room temperature.
5. Dialyze against 20 m$M$ PBS, pH 7.4, at 4°.

OPD solution: Combine 0.1 $M$ citrate buffer (pH 5.0), 15 ml; $o$-phenylenediamine (OPD), 10 mg; and 30% (v/v) H$_2$O$_2$, 6 μl

*Protocol*

1. Add 100 μl of the first antibody solution to each well of a microtiter plate (96 wells).
2. Incubate at 4°, overnight.
3. Discard the first antibody solution and wash three times with washing buffer.
4. Add 100 μl of blocking buffer.
5. Incubate at 4°, overnight (or at 37° for 1 hr).

[19] K. Suzuki, N. Miyazawa, T. Nakata, H. G. Seo, T. Sugiyama, and N. Taniguchi, *Carcinogenesis* **14**, 1881 (1993).

6. Discard blocking buffer and wash three times with washing buffer.
7. Add 100 μl of standard or sample.
8. Incubate at room temperature for 2 hr or at 37° for 1 hr.
9. Discard and wash three times with washing buffer.
10. Add 100 μl of the second antibody solution.
11. Incubate at room temperature for 2 hr or at 37° for 1 hr.
12. Discard the second antibody solution and wash three times with washing buffer.
13. Add 100 μl of peroxidase-labeled avidin solution.
14. Incubate at room temperature for 10 to 15 min.
15. Discard and wash four or five times
16. Add 100 μl of OPD solution.
17. Add 50 μl of 1 $N$ $H_2SO_4$ after 10 to 15 min.
18. Measure the absorbance of each well, using a microplate reader (490 nm).

*Preparation of Monoclonal Antibody against Human MnSOD and Its Epitope*

A monoclonal antibody against human MnSOD has been raised in mice.[20] The purified MnSOD (20 μg) is dissolved in 100 μl of saline, emulsified with an equal volume of Freund's complete adjuvant, and then injected subcutaneously into BALB/c mice followed by another injection in Freund's incomplete adjuvant after 2 weeks. The final injection without the adjuvant is given intraperitoneally 2 weeks after the second injection.

The specificity of the monoclonal antibody is examined with various synthetic peptides. To localize the antibody-binding epitope, synthetic peptides of the $NH_2$-terminal (residues 1–16) and COOH-terminal (residues 182–189, 190–196, and 182–196) regions of the enzyme are synthesized, and their effects on the binding are then studied by an ELISA method. Synthetic peptides of human MnSOD from residues 188–193 did inhibit the binding of MnSOD to the monoclonal antibody, indicating that these residues are involved in an epitope to this monoclonal antibody. In conclusion, a monoclonal antibody against MnSOD recognizes several peptides of the COOH termini of MnSOD.

*Cell Culture Conditions, Cell Fusion, Hybridoma Production, and Screening of Hybridoma Cells*

Mouse plasmacytoma cell lines NS-1 and PSU1 are grown in RPMI 1640 medium supplemented with 20% fetal calf serum and glutamine. Myeloma cells and hybridoma cells are routinely maintained at 37° under a humidified atmosphere mixture of 5% $CO_2$ and 95% air.

---

[20] T. Kawaguchi, S. Noji, T. Uda, Y. Nakashima, A. Takeyasu, Y. Kawai, H. Takagi, M. Tohyama, and N. Taniguchi, *J. Biol. Chem.* **264**, 5762 (1989).

## Enzyme-Linked Immunsorbent Assay Using a Monoclonal Antibody Against Human MnSOD

The ELISA is carried out essentially as described previously with PG-11, the monoclonal antibody raised against human liver MnSOD.[21] In summary, an IgG fraction is obtained by precipitation with 50% (w/v) saturated ammonium sulfate and DEAE-Sepharose chromatography of ascites fluid. Aliquots (100 $\mu$l) of the IgG fraction (10 $\mu$g/ml; redissolved in PBS) are added to the wells of flat-bottomed polystyrene microtiter plates and coating is allowed to proceed at 4° overnight. The wells are then washed twice with PBS, filled with 0.1% (w/v) bovine serum albumin in PBS, and then maintained for 30 min at room temperature. Unbound protein is removed by washing with PBS. Human sera are diluted 10-fold with PBS containing 0.1% (w/v) bovine serum albumin and 100-$\mu$l aliquots are then added to the antibody-coated wells. After incubation for 1 hr at room temperature, unbound antigen is removed by washing three times with PBS containing 0.05% (v/v) Tween 20, and 100 $\mu$l of PG-11 conjugated with horseradish peroxidase (IgG at approximately 500 ng/ml) is then added to each well. The enzyme is conjugated by using sodium $m$-periodate.[22]

After 1 hr at room temperature, the wells are washed four times with PBS containing 0.05% (v/v) Tween 20. The substrate for the horseradish peroxidase is then added to the wells [100 $\mu$l of 0.003% (v/v) $H_2O_2$ in 0.1 $M$ sodium citrate buffer, pH 5.0, containing 0.6 mg of $o$-phenylenediamine per milliliter]. The enzymatic reaction is stopped after 15 min at room temperature by the addition of 50 $\mu$l of 2 $N$ sulfuric acid. Absorbance is measured at 492 nm with an Immuno-Reader MTP 32 (Corona Electric, Ibaragi, Japan).

## Serum and Tissue Samples for Human and Rat MnSODs

Pooled serum from normal healthy individuals (Nescol-X) is purchased from Nihon Shoji (Osaka, Japan). Normal serum samples are also obtained from 194 healthy male and 207 healthy female volunteers (age range, 19–79 years) whose liver function test results are normal.

Serum samples from patients with various diseases are randomly selected for this study, without taking age, sex, therapy regimen, or stage of the disease into account. They are stored at $-20°$ for several weeks or at $-80°$ for up to 2 years. The diagnosis of malignant disease is confirmed by X-ray analysis findings, angiography, endoscopy, and computed tomography, as well as by biopsy or surgery.

For rat serum MnSOD, the collected sera are directly used without dilution because of the low amount of MnSOD in rat serum. Tissues are dissected out

---

[21] T. Kawaguchi, K. Suzuki, Y. Matsuda, T. Nishiura, T. Uda, M. Ono, C. Sekiya, M. Ishikawa, S. Lino, Y. Endo, and N. Taniguchi, *J. Immunol. Methods* **127,** 249 (1990).
[22] M. J. O'Sullivan and V. Marks, *Methods Enzymol.* **73,** 147 (1981).

and frozen at $-70°$ until use. For assaying tissue MnSOD levels, the tissues are thawed and homogenized in 4 volumes of PBS, pH 7.4, containing 1 m $M$ EDTA and centrifuged at $900g$ for 10 min at 4°. The resulting supernatant is sonicated and again centrifuged at $900g$ for 10 min at 4°. An aliquot of the centrifugate is used for the determination of MnSOD levels by ELISA and protein concentrations.

*Enzyme-Linked Immunosorbent Assay for Human and Rat MnSODs*

*Standard Curve for Human and Rat MnSODs.* MnSOD purified from human liver, obtained as described above, is serially diluted and then introduced into the wells of microtiter plates precoated with the MnSOD antibody. The assay is then completed as described above. A typical standard curve for the ELISA is shown in Fig. 1. The lower limit of detection is 2 ng/ml and the working range is 2–200 ng/ml.

The standard curve for the rat MnSOD ELISA indicates that the working range is from 1 to 50 ng/ml in samples. The following data have been obtained mainly from the human MnSOD ELISA.

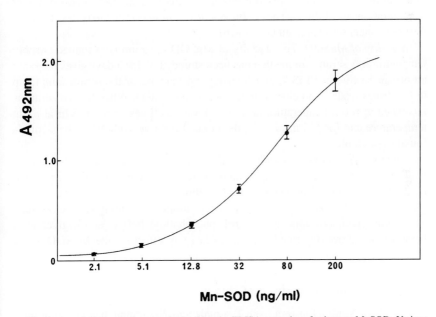

FIG. 1. A typical standard curve obtained by the ELISA procedure for human MnSOD. Various concentrations of pure MnSOD from human liver were added to the wells of a polystyrene microtiter plate that had been previously coated with PG-11 antibody. The ELISA procedures was completed as described in General Procedures. Each point represents the mean ± SD.

*Specificity of Enzyme-Linked Immunosorbent Assay.* The specificity of the ELISA method for human MnSOD has been examined. None of the major proteins found in human serum or erythrocytes, such as human serum albumin, human $\alpha$-, $\beta$-, and $\gamma$-globulins, Cu,ZnSOD, and hemoglobin, show any evidence of binding to the coated plate (data not shown).

*Effect of Serum Dilution on Enzyme-Linked Immunosorbent Assay.* To test the possibility of assaying the enzyme in human sera, three different concentrations of MnSOD are added to normal human sera and the effect of dilution on the ELISA is examined. A linear relationship is observed between the immunoreactive MnSOD level and the serum dilution over the range of 8- to 128-fold.

*Additive Experiments.* A constant amount of purified MnSOD is added to pooled normal sera (Nescol-X) and immunoreactive MnSOD is then quantitated by ELISA. The recovery is observed to be excellent. These results suggest that monoclonal antibody PG-11 is able to detect MnSOD in serum accurately and specifically.

*Reproducibility and Within-Run Reproducibility of Enzyme-Linked Immunosorbent Assay.* Repeat assays of a high control sample on six different days showed the following reproducibility. The mean ± SD is 152 ± 4.6 ng/ml. The coefficient of variation is 3%. The within-run reproducibility is examined at two different concentrations of MnSOD. The coefficient of variation is found to be less than 5% for both concentrations (data not shown).

*Stability of MnSOD.* The stability of MnSOD in serum is examined at room temperature. A serum sample that has been stored at 4° for 5 days gives the same absorbance value in ELISA, and freezing and thawing of the serum sample up to five times leads to no change in the absorbance value. When human serum is incubated at room temperature or at 4° for various times or for 1–5 hr at room temperature and for 1–5 days at 4°, the immunoreactive MnSOD in the serum is found to be stable.

*Effects of Hemolysis, Jaundice, and Lipids on Enzyme-Linked Immunosorbent Assay.* The effect of hemolysis on the determination of MnSOD is examined by means of ELISA. The hemolysis of erythrocytes is found to have no effect. The effects of jaundice and lipids are also examined by adding substances such as ascorbic acid, bilirubin, cholesterol, phosphatidylcholine, and triglycerides. None of the above-described substances has any effect on the MnSOD levels in sera.

## MnSOD Levels in Normal Healthy Controls

Levels of MnSOD in sera from 194 male and 207 female healthy adult individuals are examined. Serum MnSOD levels in normal adult males, as determined by the ELISA method, are found to be normally distributed (Fig. 2, top). On the other hand, the data for normal adult females are found to be slightly skewed

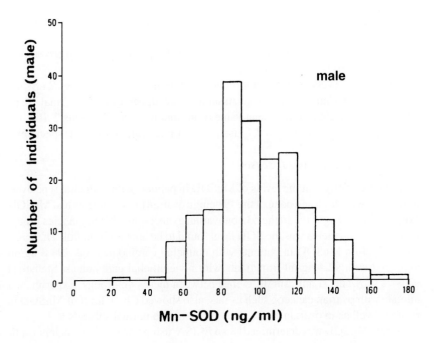
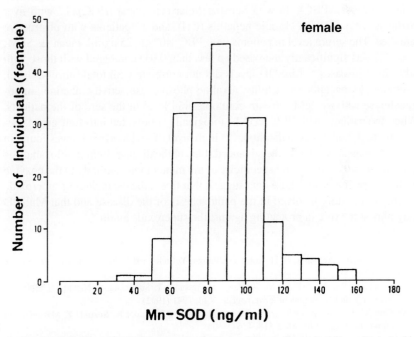

FIG. 2. Distribution of MnSOD in normal male (*top*) and female (*bottom*) adults, as determined by the ELISA procedure.

(Fig. 2, bottom), although plotting the cumulative frequency on normal probability paper gives a nearly straight line.
The mean levels and SD for males and females are 99.8 ± 24.8 and 88.8 ± 20.8 ng/ml, respectively. Assuming the upper limit of normal to be 150 ng/ml (equivalent to the mean value of normal male subjects plus 2 SD), the percentage of false positives is 2.1 and 1.0% for males and females, respectively.

*MnSOD Levels in Various Diseases*

It is noteworthy that the levels of MnSOD in patients with epithelial type ovarian cancers were high and none of the 19 benign ovarian tumor patients had MnSOD levels above 150 ng/ml.[23] In the 63 nonovarian gynecological malignancies group, in only 1 of 33 patients was the SOD level found to be above 150 ng/ml. The mean value of serum MnSOD in patients with epithelial carcinomas was 194.8 ng/ml compared with 92.4 ng/ml in patients with nonepithelial carcinomas. Statistical analysis showed a significant difference between these two groups. Experimental animals with ovarian cancers, such as rats, also showed a high level of MnSOD in serum as well as in ovarian tissues as compared with normal subjects.[24]

Serum MnSOD was determined by an ELISA, using a specific monoclonal antibody, in patients with various liver diseases, including 31 patients with primary biliary cirrhosis (PBC), 46 with hepatocellular carcinoma (HCC), 17 with liver cirrhosis (LC), 23 with chronic hepatitis (CH), and 12 patients with obstructive jaundice. The serum level in patients with PBC (407 ± 35 ng/ml, mean ± SEM; $n = 31$) was significantly increased ($p$ less than 0.01) compared with those with other liver diseases.[25] MnSOD levels did not correlate with total bilirubin level, $\gamma$-glutamyltranspeptidase activity, alkaline phosphatase activity, alanine aminotransferase activity, IgM, or with ceruloplasmin level in the sera of the patients. When the patients with PBC were histologically subdivided into four groups according to Scheuer's classification,[26] a high level of serum MnSOD was found in the early stage as well as in the advanced stage of the disease. Immunoblot analysis confirmed the reactivity and specificity of the monoclonal antibody to the enzyme protein in patient sera. These data suggest that free radicals, including superoxide anion, are possibly involved in the pathogenesis of the disease and that MnSOD may play some role in protecting against the superoxide anion.

---

[23] M. Ishikawa, Y. Yaginuma, H. Hayashi, T. Shimizu, Y. Endo, and N. Taniguchi, *Cancer Res.* **50**, 2538 (1990).
[24] T. Nakata, K. Suzuki, J. Fujii, M. Ishikawa, H. Tatsumi, T. Sugiyama, T. Nishida, T. Shimizu, M. Yakushiji, and N. Taniguchi, *Carcinogenesis* **13**, 1941 (1992).
[25] M. Ono, C. Sekiya, M. Ohhira, M. Ohhira, M. Namiki, Y. Endo, K. Suzuki, Y. Matsuda, and N. Taniguchi, *J. Lab. Clin. Med.* **118**, 476 (1991).
[26] P. J. Scheuer, in "Liver Biopsy Interpretation" (P. J. Scheuer, ed.), 3rd Ed., p. 47. Baillière Tindall, London, 1980.

Moreover, patients with acute myocardial infarction showed high MnSOD levels, suggesting that MnSOD is released from heart tissue.[27] In contrast, no elevated values were found in patients with chronic heart failure.

Adult respiratory distress syndrome (ARDS) can develop as a complication of various disorders, including sepsis, but it has not been possible to identify which of the patients at risk are likely to develop this serious disorder. In 26 patients with sepsis, at the initial diagnosis of sepsis (6–24 hr before the development of ARDS), serum MnSOD concentration was higher in the 6 patients who subsequently developed ARDS than in the 20 patients who did not. These changes in antioxidant enzymes, therefore, can serve as predictors of the development of ARDS in septic patients. Measurement of MnSOD should facilitate the identification of patients at highest risk for ARDS and allow appropriate treatment.[28]

Discussion

In the present study immunoreactive MnSOD in tissues and sera of humans and rats has been assayed by a sandwich ELISA technique. Unexpectedly, it was found that a single monoclonal antibody, PG-11, could be used both as the primary antibody and the secondary antibody. This is because the MnSOD is composed of four identical subunits of molecular weight 85,300[29]; that is, PG-11, as the primary antibody, recognized one of the subunits and was also able, as the secondary antibody, to interact with the other subunit. Therefore, with this technique it is possible to determine the immunoreactive MnSOD level in human serum.

Serum MnSOD levels in various diseases (Fig. 3) were examined and the findings show that patients with primary hepatoma, gastric cancer, and acute myocardial infarction had relatively high levels of the enzyme. However, in some cases of primary liver cirrhosis, relatively higher values were also found. Therefore, the issue of whether immunoreactive MnSOD can be used as a marker for the diagnosis and monitoring of primary hepatoma remains unclear. We also found that in patients with acute myeloid leukemia, MnSOD levels were increased, whereas in patients with acute lymphocytic leukemia the levels were normal.[30] This suggests that myeloid cells have the capacity to synthesize MnSOD. Wong and Goeddel[6] and Masuda et al.[7] independently reported that MnSOD synthesis is induced by

---

[27] K. Suzuki, N. Kinoshita, Y. Matsuda, S. Higashiyama, T. Kuzuya, T. Minamino, M. Tada, and N. Taniguchi, *Free Radic. Res. Commun.* **15,** 325 (1992).
[28] J. A. Leff, P. E. Parsons, C. E. Day, N. Taniguchi, M. Jochum, H. Fritz, F. A. Moore, E. E. Moore, J. M. McCord, and J. E. Repine, *Lancet* **341,** 777 (1993).
[29] J. M. McCord, J. A. Boyle, E. D. Day, Jr., L. J. Rizzolo, and M. L. Salin, in "Superoxide and Superoxide Dismutase" (A. M. Michelson, J. M. McCord, and I. Fridovich, eds.), p. 129. Academic Press, London, 1977.
[30] T. Nishiura, K. Suzuki, T. Kawaguchi, H. Nakao, N. Kawamura, M. Taniguchi, Y. Kanayama, T. Yonezawa, S. Iizuka, and N. Taniguchi, *Cancer Lett.* **62,** 211 (1992).

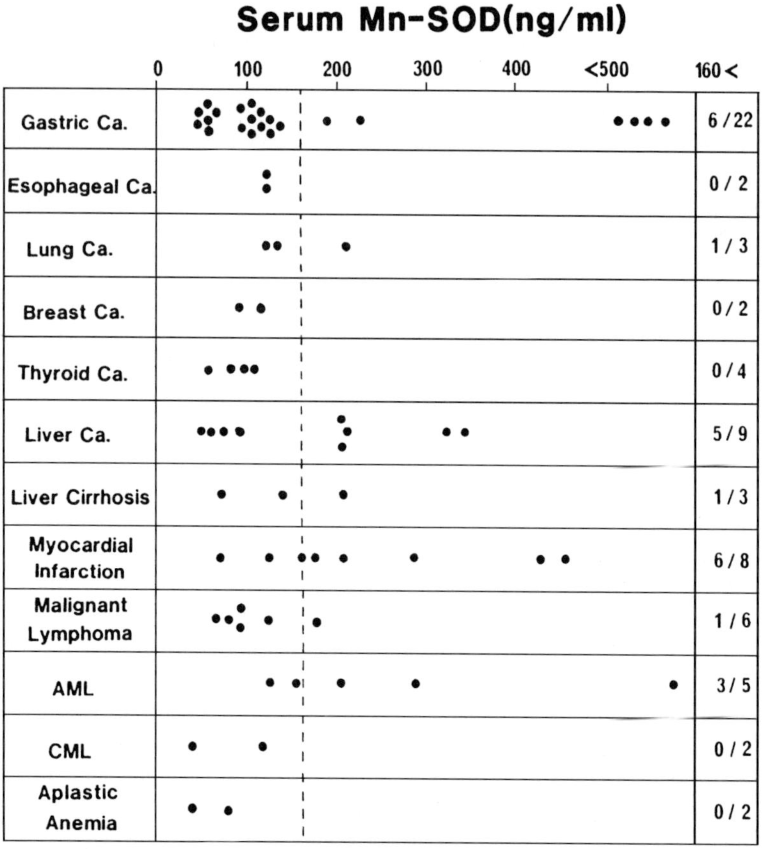

## Serum Mn-SOD(ng/ml)

| Disease | 0 | 100 | 200 | 300 | 400 | <500 | 160< |
|---|---|---|---|---|---|---|---|
| Gastric Ca. | ••••• ••••••• | •• | | | | •••• | 6/22 |
| Esophageal Ca. | •• | | | | | | 0/2 |
| Lung Ca. | •• | • | | | | | 1/3 |
| Breast Ca. | •• | | | | | | 0/2 |
| Thyroid Ca. | • ••• | | | | | | 0/4 |
| Liver Ca. | •••• | ••• | | •• | | | 5/9 |
| Liver Cirrhosis | • • | • | | | | | 1/3 |
| Myocardial Infarction | • • •• • | • | | •• | | | 6/8 |
| Malignant Lymphoma | ••• • • | | | | | | 1/6 |
| AML | • •• • | • | | | | • | 3/5 |
| CML | • • | | | | | | 0/2 |
| Aplastic Anemia | • • | | | | | | 0/2 |

FIG. 3. Immunoreactive MnSOD levels, determined by the ELISA procedure, in sera from patients with various diseases. The vertical dashed line denotes the upper normal limit for immunoreactive serum MnSOD, taken as 150 ng/ml (mean + 2 SD for 194 normal male subjects). *Abbreviations:* Ca., cancer; AML, acute myeloid leukemia; CML, chronic myeloid leukemia. Each group included both male and female adult subjects.

interleukin 1 and tumor necrosis factor. We also reported that IL-6 induces MnSOD levels in various cells.[9] Moreover, it has been reported that interleukin 1 acts as an autocrine factor in the case of acute myeloid cells.[31] This suggests that the synthesis of interleukin 1 in acute myeloid cells induces MnSOD production in the cells. This is reflected in the high levels of serum MnSOD in patients with acute myeloid leukemia.

[31] F. T. Cordingley, A. Bianchi, A. V. Hoffbrand, J. E. Reittie, H. E. Heslop, A. Vyakarnam, M. Turner, A. Meager, and M. K. Brenner, *Lancet* **1,** 969 (1988).

We also found that serum levels of MnSOD are increased in patients with acute myocardial infarction, suggesting that the enzyme is released from the mitochondrial matrix of the heart tissue.

Using the rat MnSOD ELISA, it would be possible to measure MnSOD from rat and even mouse tissues, which contain MnSOD, in various experimental protocols, which would be useful for understanding the role of MnSOD *in vivo* as well as *in vitro*.

The measurement of serum-immunoreactive MnSOD levels would be helpful in diagnosis, and a comprehensive assessment of changes in tissue levels of MnSOD will likely provide new information concerning biochemical processes involved in various diseases. An ELISA kit for human MnSOD is commercially available.[32]

---

[32] Sceti Company, Tokyo, Japan (fax, +81-3-3404-4472).

# Section II
# Mutants, Knockouts, Transgenics

# [15] Investigating Phenotypes Resulting from a Lack of Superoxide Dismutase in Bacterial Null Mutants

By DANIÈLE TOUATI

## Introduction

Superoxide dismutases (SODs) are present in virtually all aerobes and in several anaerobes,[1] with both evolutionarily distinct SOD families found. Mn- and FeSODs (*sodA* and *sodB* genes), and Cfe,MnSOD (cambialistic), are generally located in the cytoplasm, but a few are secreted. Cu,ZnSODs (*sodC*) have so far been found only in gram-negative bacteria and are located in the periplasm. As superoxide ($O_2^{·-}$) cannot cross the inner bacterial membrane at physiologically neutral pH, the protective effects of SODs should be related to SOD localization and are directed against endogenous or exogenous $O_2^{·-}$-mediated oxidative stress. Studies of SOD null mutant phenotypes revealed that, despite a common enzymatic function, bacterial SODs display a broad spectrum of biological roles ranging from minor to essential, depending on specific bacterial metabolism, environmental conditions, and stage in the life cycle.

## Obtaining Bacterial Null Mutants

There is no positive selection system for SOD-deficient mutants,[2] and it is difficult to screen for negative phenotypes. This makes it virtually impossible to identify null mutants obtained by random insertional mutagenesis. However, improvements in genetic engineering techniques and the development of sequencing programs have provided powerful tools for constructing mutants. Strategies for the construction of *sod* mutants are not specific and depend on the available genetic tools in the bacterium of interest. Generally, the cloned gene is mutated (internal deletion and/or insertion of an antibiotic resistance cassette) and the mutated allele is exchanged with the wild-type chromosomal allele to produce stable mutants.

It is important to check that exchange has actually occurred, as there is always a risk that the mutated allele will integrate into another part of the chromosome. The mutant should display a lack of the corresponding SOD activity and the expected changes in DNA digestion pattern as shown by Southern blotting or, if

---

[1] D. Touati, in "Oxidative Stress and the Molecular Biology of Antioxidant Defenses" (J. Scandalios, ed.), p. 447. Cold Spring Harbor Laboratory Press, Cold Spring Harbor, New York, 1997.

[2] A positive effect of SOD deficiency has just been described in *Corynebacterium melassecola* [M. Merkamm and A. Guyonvarch, *J. Bacteriol.* **183**, 1284 (2001)], but the underlying metabolic mechanism is still unclear. Its understanding might provide a new useful tool to select *sod* mutants in some bacteria.

sequences flanking the cloned region are available, by polymerase chain reaction (PCR) amplification of the region.

Although *sod* genes have now been cloned and sequenced for a large number of bacteria, the number of *sod* mutants available is still small. This is partly because the genetic tools required to obtain mutants are not available for several bacteria, including pathogens such as mycobacteria. Other strategies, currently used predominantly in eukaryotes, such as the inhibition of gene expression by antisense RNA, may be necessary to determine the role of SOD in these difficult-to-handle microorganisms.

POTENTIAL PROBLEMS

1. The *sod* mutation may be lethal. However, the failure to obtain a *sod* mutant is never sufficient in itself to conclude that the lack of SOD is lethal. In such a case, the chromosomal mutant should be constructed in the presence of expression of a wild-type gene (e.g., carried on a plasmid). To conclude that the mutation is lethal, it is necessary to demonstrate that it is maintained only when a wild-type copy is expressed. An example is the failure to obtain a mutant of the single cytoplasmic FeSOD of *Legionella pneumophila*, a strict aerobe living in a highly oxygenated environment.[3]

2. There are several known examples of two SODs (Fe and Mn) being present in the same compartment of a bacterium. But one can fail to detect it on an activity gel. Indeed, some SODs are expressed only under particular conditions, such as the MnSOD from *Bordetella pertussis*, which is expressed only under strict iron starvation.[4,5] Despite the high level of conservation between Fe- and MnSOD amino acid sequences, differences in codon usage between bacterial species may result in failure to detect a new *sod* gene on Southern blot analysis. Only BLAST sequence analysis of a complete genome sequence, when available, can ensure that no cryptic SOD has been missed.

Cytoplasmic *sod* Mutants

Cytoplasmic SODs protect against the toxicity caused by excess $O_2^{·-}$ generation in the cell. This may result from natural endogenous $O_2^{·-}$ production, principally via the respiratory chain, or by metabolic transformation of a xenobiotic.

Extensive studies have been carried out in *Escherichia coli*, from which the first cytoplasmic SOD mutants were isolated and characterized.[6] In those mutants,

---

[3] A. B. Sadosky, J. W. Wilson, H. M. Steinman, and H. A. Shuman, *J. Bacteriol.* **176**, 3790 (1994).
[4] N. Khelef, D. DeShazer, R. L. Friedman, and N. Guiso, *FEMS Microbiol. Lett.* **142**, 231 (1996).
[5] H. Graeff-Wohlleben, S. Killat, A. Banemann, N. Guiso, and R. Gross, *J. Bacteriol.* **179**, 2194 (1997).
[6] A. Carlioz and D. Touati, *EMBO J.* **5**, 623 (1986).

lacking both cytoplasmic Fe- and MnSOD, steady-state $O_2^{\cdot-}$ levels were found to be 100 to 1000 times higher than in their $SOD^+$ counterpart. Some of the observed defects were clearly shown to be due to direct damage to specific targets, but most of the damage, including that to DNA and presumably also to membrane, was caused by hydroxyl radicals. These were produced by the Fenton reaction, which was triggered by the excess of $O_2^{\cdot-}$, via an increase in the intracellular free iron pool.[7]

Only a limited number of studies have been carried out in other bacteria and have tended to focus exclusively on the effect of SOD deficiency on a particular function of interest. As endogenous $O_2^{\cdot-}$ production and the targets sensitive to $O_2^{\cdot-}$ mediated stress depend on the specific metabolism and mode of life of the bacterium, there is no general method for assaying the physiological consequences of an SOD defect in bacterial species.

*General Remarks*

1. Each of the experiments described below should be repeated at least three times to confirm the overall trend.

2. Any phenotype linked to an *sod* mutation should be oxygen dependent. This can easily be (and, indeed, has been) demonstrated with anaerobes and facultative anaerobes, in which the phenotype is abolished in anaerobiosis.

*Phenotypes Related to General $O_2^{\cdot-}$ Stress Conditions*

*Impairment of Aerobic Growth and Aerotolerance*

$O_2^{\cdot-}$ is produced only in the presence of oxygen. Thus oxygen becomes toxic if $O_2^{\cdot-}$ is not eliminated at a high rate by SOD. Comparison of the growth rate and yield of *sod* mutants and wild type in rich medium in air supports this view. However, as illustrated in Table I, the sensitivity of *sod* mutants to oxygen varies greatly from one bacterial species to another.

*Sensitivity to Compounds Generating Endogenous $O_2$*

*sod* mutants generally show greater sensitivity to compounds producing $O_2^{\cdot-}$ by redox cycling (Table I). Paraquat is the most commonly used, because it does not generate other toxic by-products. A crude assay (disk assay) involves spotting 10 $\mu$l of various concentrations of paraquat on small paper disks placed over a lawn of bacteria (0.1 $\mu$l of saturated culture) spread on a Luria–Bertani (LB) agar plate, and measuring the diameter of the inhibition zone after a period of growth.

A more precise measurement involves evaluating the effect on the exponentially growing cells. So far as possible, a protocol adapted from that described below (for *E. coli*) is recommended. Precultures in LB (or minimal) medium, inoculated with

---

[7] K. Keyer and J. A. Imlay, *Proc. Natl. Acad. Sci. U.S.A.* **93**, 13635 (1996).

TABLE I
SENSITIVITY TO OXYGEN AND PARAQUAT OF CYTOPLASMIC sod MUTANTS

| Species | sod gene | Impairment of aerobic growth or loss of aerotolerance | Sensitivity to paraquat | Refs. |
|---|---|---|---|---|
| Escherichia coli | sodA | No | Yes | 6 |
| | sodB | No | Yes, low | 6 |
| | sodA sodB | Yes, slight | Yes, high | 6 |
| Pseudomonas aeruginosa | sodA | No | Yes, low | a, b |
| and P. putida | sodB | Yes | Yes, high | a, b |
| | sodA sodB | Yes | Yes, high | a, b |
| Bordetella pertussis | sodB | Yes | Yes | 4 |
| | sodA | No | ND | 5 |
| Streptococcus pneumoniae | sodA | Yes, slight | Yes, high | c |
| Streptococcus pyogenes | sod | Yes, strong | Yes, high | d |
| Legionella pneumophila | sodB$^e$ | Lethal | — | 3 |
| Sinorhizobium meliloti | sodA$^e$ | Yes, slight | Yes, low | 17 |
| Synechococcus species | sodB$^e$ | Yes, slight | Yes, high | f |
| Haemophilus influenzae | sodA$^e$ | Yes, strong | Yes, high | g |
| Lactococcus lactis | sodA$^e$ | Yes, strong | ND | h |
| Campylobacter jejuni, C. coli | sodB$^e$ | No | ND | i |
| Porphoryonas gingivalis | sodA$^e$ | Lysis | — | j |

$^a$ D. J. Hasset, H. P. Schweizer, and D. E. Ohman, *J. Bacteriol.* **177,** 6330 (1995).
$^b$ Y. C. Kim, C. D. Miller, and A. J. Anderson, *Appl. Environ. Microbiol.* **66,** 1460 (2000).
$^c$ H. Yesilkaya, A. Kadioglu, N. Gingles, J. E. Alexander, T. J. Mitchell, and P. W. Andrew, *Infect. Immun.* **68,** 2819 (2000).
$^d$ C. M. Gibson and M. G. Caparon, *J. Bacteriol.* **178,** 4688 (1996).
$^e$ Reported as single cytosolic SOD.
$^f$ G. Samson, S. K. Herbert, D. C. Fork, and D. E. Laudenbach, *Plant Physiol.* **105,** 287 (1994).
$^g$ R. A. D'Mello, P. R. Langford, and J. S. Kroll, *Infect. Immun.* **65,** 2700 (1997).
$^h$ J. W. Sanders, K. J. Leenhouts, A. J. Haandrikman, G. Venema, and J. Kok, *J. Bacteriol.* **177,** 5254 (1995).
$^i$ E. C. Pesci, D. L. Cottle, and C. L. Pickett, *Infect. Immun.* **62,** 2687 (1994).

single colonies, are grown to an optical density (OD) of 1 (end of exponential growth phase) and then rapidly chilled in a water–ice mixture and kept cold for up to 40 hr. Cultures are prepared by inoculating the same medium, prewarmed, with precultures. The dilution is chosen to allow the growth of three generations before challenge with paraquat (at an OD of about 0.2). No growth lag is observed under these conditions. Growth is recorded by OD measurement at various times. The range of paraquat concentrations used results in only mild growth inhibition for the wild-type strain (1 to 100 $\mu M$ for *E. coli*). A control, to ensure that the correspondence between OD and colony-forming units per milliliter (CFU/ml) is maintained, should be done at several points by plating bacteria and counting colonies after growth.

POTENTIAL PROBLEMS

1. The initial volume of cultures should be sufficient compared with the total volume of samples withdrawn to maintain approximately constant aeration of the culture.

2. Paraquat penetrates cells by means of active transport (demonstrated in *E. coli* and presumably also true in other bacteria). Thus (a) the intracellular concentration depends on both the concentration of paraquat in the medium and the number of bacteria. Only experiments done at the same bacterial density can be compared; and (b) bacteria vary greatly in their wild-type sensitivity to paraquat because of differences in penetration.

3. Starting with a diluted overnight culture results in a growth lag that is not uniform from one cell to another. This may result in paraquat being added to a heterogeneous population giving less reproducible results unless the overnight culture is highly diluted and allowed to go through a large number of generations before challenge.

*Specific Characterized Damage and Associated Phenotypes*

*[4Fe–4S] Cluster Sensitivity*

The inactivation by $O_2^{·-}$ of a group of dehydratases containing [4Fe–4S] clusters, via the oxidative excision of iron, was first demonstrated *in vitro*.[8] Enzymes containing these clusters (dihydroxy acid dehydratase, aconitase, and gluconate dehydratase),[6,9,10] are inactivated in *E. coli* mutants deficient in both cytoplasmic SODs, as shown by direct enzyme activity measurements and defective associated phenotypes (branched amino acid auxotrophy, growth impairment on succinate, gluconate). The presence and role of [4Fe–4S]-containing enzymes varies depending on the specific metabolism of each bacterium, and putative targets must be identified in each case.[11]

POTENTIAL PROBLEM. Impairment of aconitase activity can be measured in many bacterial *sod* mutants. However, some bacteria (such as *E. coli*) can partially compensate for the oxidative inactivation of aconitase by an increase in aconitase gene expression under oxidative stress conditions.

*DNA Damage*

$O_2^{·-}$ cannot directly react with DNA, but there is now considerable evidence that an increase in $O_2^{·-}$ levels leads to DNA damage. Several studies provide strong support for the theory that $O_2^{·-}$ triggers Fenton reaction ($H_2O_2 +$

---

[8] D. H. Flint, J. F. Tuminello, and M. H. Emptage, *J. Biol. Chem.* **268,** 22369 (1993).
[9] P. R. Gardner and I. Fridovich, *J. Biol. Chem.* **266,** 1478 (1991).
[10] P. R. Gardner and I. Fridovich, *J. Biol. Chem.* **267,** 8757 (1992).
[11] S. I. Liochev and I. Fridovich, *Free Radic. Biol. Med.* **16,** 29 (1994).

$Fe^{2+} \rightarrow OH^{\cdot} + Fe^{3+} + OH^{-}$) by the oxidative release of iron from iron-containing molecules, such as those containing [4Fe–4S] clusters, and thereby increasing the pool of free iron available to catalyze hydroxyl radical production.[10] Studies of mutants lacking cytosolic SOD provided the first *in vivo* evidence that excess $O_2^{\cdot-}$ causes DNA damage.[12,13]

The hallmarks of DNA damage in SOD-deficient mutants are as follows.

1. An oxygen-dependent increase in spontaneous mutagenesis (methods for measuring mutagenesis in *sod* mutants are described in a previous volume of this series).[14]

2. A lethal aerobic phenotype of strains lacking cytosolic SOD and unable to repair DNA double-strand breaks by homologous recombination.[15]

3. An increase in $H_2O_2$-mediated killing via the Fenton reaction.[6,13]

MEASUREMENT OF KILLING BY $H_2O_2$. A midexponential phase culture (OD 0.3 to 0.5) in LB, prepared as for paraquat challenge, is dispensed into Erlenmeyer flasks and $H_2O_2$ is added to various concentrations (0 to 20 m$M$). After incubation with shaking for 20 min at growth temperature, the treatment is stopped by adding catalase (1000 U/ml; Boehringer Mannheim, Indianapolis, IN) and chilling. Surviving cells are counted by plating on LB plates, after dilutions into cold $10^{-2}$ $M$ MgSO$_4$ [or phosphate-buffered saline (PBS)] containing catalase. To check that the increase in killing is mediated by Fenton chemistry, the experiment can be repeated in the presence of a cell-permeable iron chelator (1 m$M$ 2,2-dipyridyl or diethylenetriaminepentaacetic acid (DTPA) added to the cell culture a few minutes before challenge and in the dilution buffer), which should abolish the increase.

POTENTIAL PROBLEMS

1. Killing should be determined as a function of $H_2O_2$ concentration. $H_2O_2$ kills cells in two modes (demonstrated in *E. coli*,[16] but also observed in other bacteria). Only mode 1 killing, which occurs at lower $H_2O_2$ concentration, is well characterized and has been shown to be due to DNA damage mediated by the Fenton reaction. The sensitivity of *sod* mutants to $H_2O_2$ should therefore be measured in the range of concentrations corresponding to mode 1 killing.

2. The sensitivity of wild-type strains differs greatly between species (e.g., similar levels of killing are observed with 250 $\mu M$ in *Salmonella typhimurium* and 2 m$M$ in *E. coli*).

3. The reliability of measurements of killing by $H_2O_2$ due to excess $O_2^{\cdot-}$ implies that $O_2^{\cdot-}$ does not interfere with catalase and peroxidase activities. Although

---

[12] S. B. Farr, R. D'Ari, and D. Touati, *Proc. Natl. Acad. Sci. U.S.A.* **83**, 8268 (1986).
[13] K. Keyer, A. Strohmeier Gort, and I. A. Imlay, *J. Bacteriol.* **177**, 6782 (1995).
[14] D. Touati and S. B. Farr, *Methods Enzymol.* **186**, 646 (1990).
[15] D. Touati, M. Jacques, B. Tardat, L. Bouchard, and S. Despied, *J. Bacteriol.* **177**, 2305 (1995).

this is generally true, exceptions exist. In *Sinorhizobium meliloti* catalase and presumably a global defense against $H_2O_2$ is strongly induced by $O_2^{\cdot-}$, leading to higher resistance to $H_2O_2$ in the *sod*A mutant than in the wild type.[17]

*Metabolic Defects, Impairment of Growth in Minimal Medium*

Growth in minimal medium requires numerous synthethic pathways and presents many targets susceptible to oxidative damage. Thus, *sod* mutants display growth impairment. However, targets are not easy to identify and differ depending on the metabolic pathway affected. *Escherichia coli sod* mutants were found to be unable to grow on minimal medium unless provided with all amino acids.[6] Studies in several laboratories demonstrated that this general auxotrophy was due to several types of damage: a defect in branched amino acid synthesis due to dehydratase inactivation,[18] interference with aromatic biosynthesis by transketolase inactivation,[19] which acts at an early step of biosynthesis, and cell envelope defects leading to sulfite leakage and cysteine auxotrophy.[20] Similar general amino acid auxotrophy has been observed in *sod* mutants in other bacteria, suggesting similar defects. However, in some cases the growth impairment on minimal medium was not reversed by providing amino acids, indicating other (or additional) metabolic defects.[16]

*Membrane Damage*

Although there is evidence of membrane damage in *sod* mutants,[21] the nature of the lesions has not been identified and is difficult to investigate. Lipid peroxidation is unlikely to be solely responsible for the damage, because there are almost no unsaturated lipids in most bacterial membranes.

## Mutants in Noncytoplasmic Superoxide Dismutases

No sources of $O_2^{\cdot-}$ have yet been found in the periplasm and outer bacterial membranes. Thus, periplasmic, secreted, and outer membrane-associated SODs are thought to protect against external $O_2^{\cdot-}$. In particular, SODs are expected to protect against the oxidative burst produced by the host during infection, thereby contributing to pathogenicity and virulence. No mutants have yet been obtained for secreted Fe- and MnSODs. Mutants have been obtained only for the periplasmic Cu,ZnSODs (*sodC*). The expected phenotypes were, as follows: resistance to endogenous $O_2^{\cdot-}$ (as generated by paraquat), sensitivity to exogenous $O_2^{\cdot-}$ (as generated by hypoxanthine–xanthine oxidase, or pyrogallol), and attenuation

---

[16] J. A. Imlay and S. Linn, *J. Bacteriol.* **166,** 519 (1986).
[17] R. Santos, D. Hérouart, A. Puppo, and D. Touati, *Mol. Microbiol.* **38,** 750 (2000).
[18] C. F. Kuo, T. Mashino, and I. Fridovich, *J. Biol. Chem.* **262,** 4724 (1987).
[19] L. Benov and I. Fridovich, *J. Biol. Chem.* **274,** 4202 (1999).
[20] L. Benov, N. M. Kredich, and I. Fridovich, *J. Biol. Chem.* **271,** 21037 (1996).
[21] J. A. Imlay and I. Fridovich, *J. Bacteriol.* **174,** 953 (1992).

of virulence and pathogenicity (sensitivity to killing by macrophages, attenuated virulence in animal models).

All $sodC$ mutants display levels of resistance to paraquat similar to those of the wild type, consistent with periplasmic SOD affording no protection against cytoplasmic $O_2^{\cdot-}$.

In pathogens, $sodC$ mutants generally display attenuation of virulence (reviewed in Lynch and Kuramitsu),[22] consistent with a lower level of protection against oxidative burst ($O_2^{\cdot-}$, $H_2O_2$, $NO^{\cdot}$).

$sodC$ mutants have been reported to be sensitive to exogenous $O_2^{\cdot-}$. However, the sources of $O_2^{\cdot-}$ used produced both $O_2^{\cdot-}$ and $H_2O_2$ and most experiments were done in the absence of addition of catalase. The addition of this enzyme is necessary to ensure that the observed effects are attributable to exogenous $O_2^{\cdot-}$. Reports that the sensitivity of $E.$ $coli$ $sodC$ mutant to hypoxanthine–xanthine oxidase was abolished by catalase raises further doubt about the mechanism of $sodC$ protection.[23]

The phenotypes of $sodC$ mutants observed to date suggest that periplasmic SOD may protect against both endogenous and exogenous $O_2^{\cdot-}$. The source of endogenous $O_2^{\cdot-}$ remains unclear, but this aspect should not be neglected in future studies of $sodC$ mutants.

## Bacterial $sod$ Null Mutants as Tools

### Cloning $sod$ Genes from Other Organisms

It has been found that the production of SOD from any organism in the cytoplasm of the $sodA$ $sodB$ mutant of $E.$ $coli$ fully complements the deficiencies of this mutant. Thus, complementation assays have been widely used to clone $sod$ genes from other organisms. This strategy is still useful in the absence of a genomic sequence. Such assays also led to the isolation of another $O_2^{\cdot-}$-scavenging enzyme, $O_2^{\cdot-}$ reductase (see Ref. 24 and [13] in this volume[25]). Two tests are available.

*Complementation of Lack of Growth on Minimal Medium.* A $sodA$ $sodB$ strain is transformed with a genomic library. Transformants are scraped from the selection plates and pooled, suspended in $10^{-2}$ $M$ cold $MgSO_4$ buffer, washed twice, and plated on minimal glucose medium (M63 or M9) to which $10^{-6}$ $M$ paraquat and appropriate antibiotic (used for transformant selection) have been added. Putative $sod^+$ colonies appear after 2 to 3 days of incubation at 37°. *Note:* External suppressors of the amino acid auxotrophy phenotype occur with high frequency, but these colonies are nonetheless usually sensitive to paraquat.[21,26] A preliminary assay should be done to evaluate the maximum number (usually at least $10^5$) of

---

[22] M. Lynch and H. Kuramitsu, *Microbes Infect.* **2**, 1245 (2000).
[23] A. Strohmeier Gort, D. M. Ferber, and J. A. Imlay, *Mol. Microbiol.* **32**, 179 (1999).
[24] M. J. Pianzzola, M. Soubes, and D. Touati, *J. Bacteriol.* **178**, 6736 (1996).
[25] V. Nivière and M. Lombard, *Methods Enzymol.* **349**, [13], 2002 (this volume).
[26] S. Maringanti and J. A. Imlay, *J. Bacteriol.* **181**, 3792 (1999).

TABLE II
OTHER SPECIFIC PHENOTYPES ASSOCIATED WITH sod MUTANTS

| Defect condition | Mutant | Phenotype | Species | Refs. |
|---|---|---|---|---|
| Normal growth | sodB | Sensitivity of photosystem to moderate light | Cyanobacterium | a |
| | sodA sodB | Pyocyanin production abolished | Pseudomonas aeruginosa | b |
| | sodB | Reduced production of toxin and adhesin (adenylate cyclase, hemolysin, pertactin) | Bordetella pertussis | 4 |
| Stationary phase | sodC | Rapid decrease in survival | Legionella pneumophila | c |
| | sodC | Sensitivity to peroxynitrite | Salmonella typhimurium | d |
| | sodA | Does not survive long-term starvation | Staphylococcus aureus | e |
| | sodB | Loss of viability | Campylobacter coli | f |
| | sodA | Less resistant to oxidative stress; growth resumes slowly from stationary phase | Pseudomonas aeruginosa | g |
| | sodA | Decrease in sporulation frequency; defect in spore coat assembly | Bacillus subtilis | h |
| Infection | Cytosolic or periplasmic | Attenuated virulence depending on mode and location of infection, and/or resistance to killing by macrophages | Numerous pathogens | 22 |
| Symbiosis | sodA | Reduced nodulation; defect in bacteroid differentiation; drastic decrease in nitrogen fixation | Sinorhizobium meliloti | 17 |
| Chilling | sodB | Sensitivity to moderate chilling in light | Cyanobacterium | i |
| Acid stress | sodA | Increase in sensitivity to acid stress | Staphylococcus aureus | e |

[a] G. Samson, S. K. Herbert, D. C. Fork, and D. E. Laudenbach, *Plant Physiol.* **105,** 287 (1994).
[b] D. J. Hasset, H. P. Schweizer, and D. E. Ohman, *J. Bacteriol.* **177,** 6330 (1995).
[c] G. St. John and H. M. Steinman, *J. Bacteriol.* **178,** 1578 (1996).
[d] M. A. De Groote, U. A. Ochsner, M. U. Shiloh, C. Nathan, J. M. McCord, M. C. Dinauer, S. J. Libby, A. Vasquez-Torres, Y. Xu, and F. C. Fang, *Proc. Natl. Acad. Sci. U.S.A.* **94,** 13997 (1997).
[e] M. O. Clements, S. P. Watson, and S. J. Foster, *J. Bacteriol.* **181,** 3898 (1999).
[f] D. Purdy, S. Cathraw, J. H. Dickinson, D. G. Newell, and S. F. Park, *Appl. Environ. Microbiol.* **65,** 2540 (1999).
[g] B. Polack, D. Dacheux, I. Delic-Attree, B. Toussaint, and P. M. Vignais, *Infect. Immun.* **64,** 2216 (1996).
[h] A. O. Hentiques, L. R. Melsen, and C. P. Moran, Jr., *J. Bacteriol.* **180,** 2285 (1998).
[i] D. J. Thomas, J. B. Thomas, S. D. Prier, N. E. Nasso, and S. K. Herbert, *Plant Physiol.* **120,** 275 (1999).

*sodA sodB* bacteria that can be plated without residual growth. A slightly higher concentration of paraquat can be used if there is residual growth.

*Restoration of Aerobic Growth of sodA sodB recA Strain.* The test requires access to an anaerobic chamber. A *sodA sodB recA* strain (air sensitive) is transformed with the genomic library under anaerobiosis. Pooled transformants are plated on LB plates under aerobiosis and incubated overnight at 37° under aerobiosis. The plating of $10^4$ to $10^5$ SOD⁻ bacteria should result in no survival under these conditions.

POTENTIAL PROBLEM. The aerobic growth of a *sodA sodB recA* strain also can be restored by recombination (cloning of a $rec^+$ gene). Thus, the putative $sod^+$ clones should be further tested for complementation in minimal medium.

*Identifying New $O_2^{\cdot-}$ Sensitive Targets and Situations of Endogenous Oxidative Stress*

Other defects observed in *sod* mutants of various bacteria under various conditions are listed in Table II. In all cases, the specific damage has not been clearly characterized. Defects may result from the vulnerability to $O_2^{\cdot-}$-mediated stress of a particular function or from an increase in oxidative stress in particular conditions. Thus, the low level of nitrogen fixation in the *sodA S. meliloti* mutant is probably due to direct inactivation of the nitrogenase complex, but the reasons for the defect in nodulation remain unclear.[16] The attenuation of virulence in various cytoplasmic *sod* mutants may result from damage due to the high metabolic rate during multiplication of bacteria in their host and associated endogenous $O_2^{\cdot-}$ production, whereas attenuation in periplasmic *sod* mutants seems to be related to protection against oxidative burst.[22]

## Concluding Remarks

Studies of bacterial *sod* null mutants have already identified several $O_2^{\cdot-}$ targets and shed light on the mechanism of $O_2^{\cdot-}$ toxicity. The effects of SOD deficiency in bacteria differ greatly according to SOD location and also because of the diversity of bacterial modes of life. For a particular bacterium, the lack of SOD may have minor effects in one phase of the life cycle, but drastic effects in another (e.g., free-living and symbiotic *S. meliloti*).[17] Many of the effects of SOD deficiency are still unclear. Studies of additional *sod* null mutants in various bacteria will help uncover new protective effects, to identify new $O_2^{\cdot-}$-sensitive targets and possibly new sources of $O_2^{\cdot-}$, and to obtain further insight into the mechanisms, both direct and indirect, of $O_2^{\cdot-}$ toxicity.

## Acknowledgments

We thank the Association pour la Recherche sur le Cancer for supporting our work on bacterial superoxide dismutase mutants (Grant 5581).

# [16] Bacterial Superoxide Dismutase and Virulence

By PAUL R. LANGFORD, ASSUNTA SANSONE, PIERA VALENTI,
ANDREA BATTISTONI, and J. SIMON KROLL

## Introduction

Three major forms of superoxide dismutase (SOD), encoded by *sod* genes, are found in bacterial pathogens. Although there are rare examples of pathogenic bacteria lacking any functional SOD, for example, some strains of *Neisseria gonorrhoeae*,[1] others may encode as many as five (some serovars of *Salmonella enterica*).[2] These are classified according to the metal cofactor at their active site: iron (FeSOD), manganese (MnSOD), and copper and zinc (Cu,ZnSOD). In the latter the zinc is structural rather than catalytic. Mn-, Fe-, and Cu,ZnSODs are encoded by *sodA*, *sodB*, and *sodC* genes, respectively. With rare exceptions, Fe- and/or MnSOD is to be found in all aerobic organisms. FeSODs and MnSODs are located in the cytosol and have a primary role in minimizing the deleterious effects of oxygen-free radicals leaking from the confines of the respiratory chain during aerobic metabolism. Once considered a curiosity in bacteria, Cu,ZnSODs have been found increasingly evident through improvements in methods for screening for enzyme activity[3] and the development of genetic methods for finding *sodC* genes.[4] With the availability of whole genome sequences, it is clear that they are widespread in gram-negative and gram-positive organisms. In contrast to Fe- and MnSODs, bacterial Cu,ZnSODs are exported from the cytosol to the periplasm or beyond,[3,5–11] free or perhaps membrane associated, and, in at least one case, apparently exposed on the bacterial surface.[12] Because superoxide generated within the cytosol cannot cross the cytoplasmic membrane, and no endoperiplasmic superoxide-generating

---

[1] F. S. Archibald and M. N. Duong, *Infect. Immun.* **51**, 631 (1986).
[2] N. Figueroa-Bossi, S. Uzzau, D. Maloriol, and L. Bossi, *Mol. Microbiol.* **39**, 260 (2001).
[3] L. T. Benov and I. Fridovich, *J. Biol. Chem.* **269**, 25310 (1994).
[4] J. S. Kroll, P. R. Langford, K. E. Wilks, and A. D. Keil, *Microbiology* **141**, 2271 (1995).
[5] F. C. Fang, M. A. DeGroote, J. W. Foster, A. J. Baumler, U. Ochsner, T. Testerman, S. Bearson, J. C. Giard, Y. Xu, G. Campbell, and T. Laessig, *Proc. Natl. Acad. Sci. U.S.A.* **96**, 7502 (1999).
[6] J. S. Kroll, P. R. Langford, and B. M. Loynds, *J. Bacteriol.* **173**, 7449 (1991).
[7] P. R. Langford, B. M. Loynds, and J. S. Kroll, *Infect. Immun.* **64**, 5035 (1996).
[8] G. St. John and H. M. Steinman, *J. Bacteriol.* **178**, 1578 (1996).
[9] T. J. Stabel, Z. Sha, and J. E. Mayfield, *Vet. Microbiol.* **38**, 307 (1994).
[10] H. M. Steinman, *J. Biol. Chem.* **257**, 10283 (1982).
[11] K. E. Wilks, K. L. Dunn, J. L. Farrant, K. M. Reddin, A. R. Gorringe, P. R. Langford, and J. S. Kroll, *Infect. Immun.* **66**, 213 (1998).
[12] C. H. Wu, J. J. Tsai-Wu, Y. T. Huang, C. Y. Lin, G. G. Lioua, and F. J. Lee, *FEBS Lett.* **439**, 192 (1998).

system has been discovered, this suggests that Cu,ZnSODs provide protection from exogeneously derived superoxide. In the context of bacteria pathogenic for mammalian hosts, one obvious source might be the respiratory burst of phagocytic cells. This defines a plausible role for the enzyme in promoting bacterial survival in the face of oxygen free radical-based host defense, contributing to virulence.

To determine whether a particular SOD contributes to the virulence of a pathogen, investigators have constructed defined mutants and compared their phenotype with that of the isogenic wild-type strain under a variety of *in vitro* and *in vivo* conditions. *In vitro* studies have included assessment of comparative sensitivity to reactive oxygen and nitrogen intermediates (ROIs, RNIs) and to killing by polymorphonuclear granulocytes and monocyte-macrophages, freshly derived from various hosts or from established cell lines (reviewed in Lynch and Kuramitsu[13]). Comparative virulence studies *in vivo* have been carried out with a variety of animal models of infection, and, in at least two cases, in the natural host.[14,15] In a review, Lynch and Kuramitsu[13] provide a comprehensive résumé of the results of comparative virulence experiments performed with bacterial strains mutated with respect to a variety of *sod* genes, encoding cytosolic and periplasmic enzymes. Although many of these studies have shown differences in the pathogenic behavior of wild-type and *sod* mutants, interpretation is not always straightforward. We have argued that the central metabolic function of the cytosolic enzymes makes it, at the least, questionable whether the attenuation of *sodA/sodB* mutants can be regarded as establishing such SODs as virulence factors.[16] Rather, attenuation of such mutants suggests that at some point in the pathogenic sequence of host–microbe interaction, organisms are subject to significant aerobic metabolic stress, requiring intact cytosolic detoxifying mechanisms for normal survival.

In this article we present a number of generic methods that have been developed to study bacterial SODs. In the context of pathogenicity studies, we have confined our attention to Cu,ZnSODs and, for specific examples, the enzyme from *S. enterica*. Although the experimental protocols provided have been tailored to *Salmonella*, with minor adaptations they should be applicable to a wide range of pathogens.

## Analysis of Bacterial Superoxide Dismutases

As an initial step in the exploration of the role of SOD in the biology of a pathogen, the number of enzymes expressed, their type, and their subcellular localization need to be determined. Methods for cell fractionation and identification of SOD are provided below.

[13] M. Lynch and H. Kuramitsu, *Microbes Infect.* **2,** 1245 (2000).
[14] J. L. Farrant, A. Sansone, J. R. Canvin, M. J. Pallen, P. R. Langford, T. S. Wallis, G. Dougan, and J. S. Kroll, *Mol. Microbiol.* **25,** 785 (1997).
[15] B. J. Sheehan, P. R. Langford, A. N. Rycroft, and J. S. Kroll, *Infect. Immun.* **68,** 4778 (2000).
[16] R. A. D'Mello, P. R. Langford, and J. S. Kroll, *Infect. Immun.* **65,** 2700 (1997).

## Preparation of Whole Cell Extracts

The bacterium of interest is grown in 25 ml of broth medium to $\sim 10^9$ CFU/ml under appropriate conditions (generally overnight with aeration). The growth phase may be critical for yield: for example, *Escherichia coli* Cu,ZnSOD is not expressed during logarithmic phase[17] and *Salmonella sodC* genes are maximally expressed in the stationary phase.[4] Cells are harvested by centrifugation (3000$g$, 10 min, 4°) and washed once with 10 m$M$ Tris (pH 7.8)–1 m$M$ benzamidine (a protease inhibitor) because some bacterial SODs are protease sensitive.[18] After recentrifugation, the supernatant is discarded and the pellet is resuspended in 1 ml 10 m$M$ Tris (pH 7.8)–1 m$M$ benzamidine. Cellular constituents may be released by one of a variety of methods, for example, multiple freeze–thaw cycles, French press, MiniBead beater, Braun homogenizer, sonication, or lysozyme treatment. Where applicable, optimal machine settings may have to be determined empirically. Excessive heating must be avoided, as it might result in loss of enzyme activity. The conditions required to lyse particular bacteria vary depending on the organism and the conditions under which they were grown. For example, Cu,ZnSOD activity was found only in cellular lysates of *E. coli* prepared by osmotic shock or by using the Braun homogenizer and not when sonication or freeze–thawing were used.[3] After cellular disruption, the sample is centrifuged at 13,500$g$ for 60 min at 4° to remove cellular debris and, in the case of gram-negative bacteria, outer membrane material. This aids band resolution on visualization of enzyme activity in SOD gels. Another sample preparative technique that can result in improved band resolution (by reducing streaking effects) and be of help in the identification of expressed SODs is the use of a chloroform–ethanol extraction step,[19] which removes lipid material and denatures a large number of proteins present in the extract. Cu,ZnSODs and some FeSODs (but not MnSODs) are resistant to this treatment. The procedure is as follows. Sonicates are mixed with 0.5 volume of cold chloroform–ethanol (3 : 5, v/v) at 4° and after 1 hr on ice are centrifuged at 6500$g$ for 25 min at 4°. The upper layer is removed for further analysis, care being taken to avoid taking precipitated protein from the solvent interface and the lower organic phase. Material can be stored at $-20°$ for future use.

## Preparation of Periplasmic Extracts from Salmonella enterica for Superoxide Dismutase Analysis

The method is based on that described by Higgins and Hardie,[20] with a subsequent ammonium sulfate precipitation step.[10] To a 5-ml overnight culture of *S. enterica*, add 0.5 ml of 0.5 $M$ Tris-HCl (pH 7.8), incubate for 10 min, and

---

[17] A. S. Gort, D. M. Ferber, and J. A. Imlay, *Mol. Microbiol.* **32**, 179 (1999).
[18] A. Battistoni and G. Rotilio, *FEBS Lett.* **374**, 199 (1995).
[19] P. V. Dunlap and H. M. Steinman, *J. Bacteriol.* **165**, 393 (1986).
[20] C. F. Higgins and M. M. Hardie, *J. Bacteriol.* **155**, 1434 (1983).

harvest the bacteria by centrifugation (3000$g$, 10 min, 21°). Resuspend the pellet in 0.8 ml of 40% (w/v) sucrose, 2 m$M$ ethylenediaminetetraacetic acid (EDTA), and 30 m$M$ Tris-HCl (pH 7.8) and incubate for 10 min at 30°. Recentrifuge and resuspend the bacterial pellet in 2 ml of ice-cold water and leave on ice for 10 min. Recover the supernatant fluid by further centrifugation. We have found the following step helpful to concentrate the periplasmic proteins. Add ammonium sulfate (650 g/liter), incubate for 18 hr at 4°, centrifuge (12,000$g$, 30 min, 4°) and resuspend the protein pellet in 1 ml of water. Periplasmic proteins may be further concentrated by using microcentrifuge filters [nominal molecular weight cutoff, 5000; Millipore (Bedford, MA); Costar (Cambridge, MA); Vivascience (Stonehouse, UK)] according to the manufacturer instructions. Periplasmic extracts may be stored for future analysis at −20°.

*Superoxide Dismutase Separation*

For the separation of SODs, sodium dodecyl sulfate–polyacrylamide gel electrophoresis (SDS–PAGE), nondenaturing PAGE, or isoelectric focusing (IEF) methods can be used. SDS–PAGE has the advantage of allowing an approximate molecular mass to be determined, whereas the other methods are commonly used to identify, by an SOD-specific colorimetric assay, the different enzymes present in the extracts. SDS–PAGE is typically carried out with 10% (w/v) polyacrylamide separating gels [10% (w/v) acrylamide with a ratio of acrylamide to $N,N'$-methylene-bisacrylamide of 30 : 0.8 (w/w), 375 m$M$ Tris-HCl (pH 8.9), 0.1% (w/v) ammonium persulfate (APS), 0.04% (v/v) $N,N,N',N'$-tetramethylethylenediamine (TEMED), 0.05% (w/v) SDS] with a stacking gel containing 4.5% (w/v) acrylamide, 125 m$M$ Tris-HCl (pH 6.8), 0.1% (w/v) APS, 0.1% (v/v) TEMED, and 0.1% (w/v) SDS. For nondenaturing PAGE, SDS is omitted from all solutions, the pH of the stacking gel is increased to pH 8.3, and a separating gel of 6–10% (w/v) is used. However, not all SODs are stable at the higher pH. For example, the *E. coli* Cu,ZnSOD cannot be visualized on pH 8.8 gels but is clearly identified when using a continuous system in which the gels and buffers contain 50 m$M$ Tris–acetate, pH 7.8.[18] The loading buffer for nondenaturing gels should contain 5% (w/v) sucrose and 0.1% (w/v) bromphenol blue, with the addition of 1% (w/v) SDS and 10% (v/v) 2-mercaptoethanol for SDS–polyacrylamide gels. In the latter case samples are also boiled for 5 min. SDS–PAGE has been used successfully to determine an approximate mass of bacterial Cu,Zn-, Fe-, and MnSODs,[21] but its applicability to the enzyme of interest needs to be determined empirically and it is not advised for initial screening purposes. The running buffer used should contain 40 m$M$ glycine and 50 m$M$ Tris base (pH 8.9) (for nondenaturing PAGE); 0.05% (w/v) SDS is included for SDS–PAGE. Gels should be run at a constant current of

---

[21] J. Chen, C. Liao, S. J. Mao, T. Chen, and C. Weng, *J. Biochem. Biophys. Methods* **47**, 233 (2001).

15 mA until the dye front reaches 1 cm from the bottom of the gel. Higher currents can be used, but with caution, as this may result in excessive heating with protein streaking and denaturation. It is advisable to perform nondenaturing PAGE at low temperatures (in a cold chamber or with a circulating coolant around the gel) to avoid heat-induced denaturation of specific enzymes.

In nondenaturing PAGE, the following method allows determination of an approximate molecular mass.[22] The SODs of interest are separated on nondenaturing polyacrylamide gels of differing percentages, visualized by activity staining (see below), and their relative electrophoretic mobilities calculated and converted to molecular masses by comparison with standard SODs, using the method of Hedrick and Smith.[23] Alternatively, gel-filtration chromatography may be used.[18]

IEF separates proteins on the basis of their net charge and is useful when SODs cannot be separated either by nondenaturing gel electrophoresis or SDS–PAGE. Good separation can be obtained with IEF Ready Gels pH 3–10 (Bio-Rad, Hercules, CA), using 7 m$M$ phosphoric acid (anodic) and 20 m$M$ lysine plus 20 m$M$ arginine (cathodic) as the buffers. The running conditions are (in turn) 100 V (constant) for 1 hr, 250 V (constant) for 1 hr, and 500 V (constant) for 30 min. With SDS–PAGE, nondenaturing PAGE, and IEF we have obtained good separation of SODs with the Bio-Rad MiniProtean II system with, typically, 10–20 $\mu$g of a cellular protein lysate protein giving satisfactory results for an abundant SOD. Larger gel formats may be necessary for optimum separation.

*Superoxide Dismutase Activity Staining*

SOD activity can be visualized in polyacrylamide gels, using the method of Beauchamp and Fridovich[24] as modified by Steinman.[25] After SDS–PAGE it is advisable first to soak the gel in 50 m$M$ phosphate buffer, pH 7.5, or water[21,26] twice (10 min each) to remove the SDS. For SDS–PAGE, nondenaturing PAGE, and IEF, incubate the gel in 50 m$M$ phosphate buffer, pH 7.5, containing in addition 28 m$M$ TEMED and 0.028 m$M$ riboflavin, and incubate in the dark for 10 min with gentle shaking. Next, after pouring away the TEMED–riboflavin solution, soak the gel in the same buffer, now containing 2.5 m$M$ nitroblue tetrazolium (NBT) and incubate in the dark for a further 10 min. Pour away this solution and place the gel under a bench lamp for 5–20 min. Under these conditions, photochemically reduced flavins generate superoxide that reduces NBT, and insoluble dark blue formazan is deposited in the gel. SODs, by removing superoxide, prevent the formation of formazan and their activity is visualized as an achromatic band (Fig. 1).[15] Once

---

[22] P. R. Langford, B. M. Loynds, and J. S. Kroll, *J. Gen. Microbiol.* **138**, 517 (1992).
[23] J. L. Hedrick and A. J. Smith, *Arch. Biochem. Biophys.* **126**, 155 (1968).
[24] C. Beauchamp and I. Fridovich, *Anal. Biochem.* **44**, 276 (1971).
[25] H. M. Steinman, *J. Bacteriol.* **162**, 1255 (1985).
[26] J. R. Chen, C. N. Weng, T. Y. Ho, I. C. Cheng, and S. S. Lai, *Vet. Microbiol.* **73**, 301 (2000).

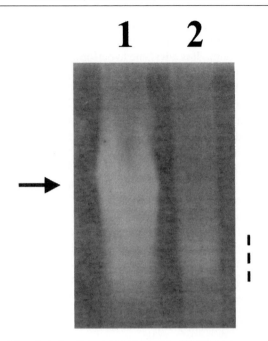

FIG. 1. SOD activity of whole cell sonicates of *Actinobacillus pleuropneumoniae* 4074 wild type (lane 1) and its isogenic *sodC* mutant (lane 2) [B. J. Sheehan, P. R. Langford, A. N. Rycroft, and J. S. Kroll, *Infect. Immun.* **68,** 4778 (2000)]. Cu,ZnSOD activity in the wild type is indicated by the arrow and the diffuse MnSOD, present in both lanes, is indicated by the dashed line.

development is complete the reaction is stopped by addition of 7% (v/v) acetic acid. Gels can be dried under vacuum onto filter paper or between cellulose film (Promega, Madison, WI).

Staining of gels for SOD activity provides a powerful method for establishing the identity of the type(s) of enzymes expressed by the pathogen of interest, as well as for verifying the successful inactivation of a specific *sod* gene. A first indication of SOD type is provided by detection of activity in specific subcellular fractions. Periplasmic extracts are highly enriched in Cu,ZnSODs (where it is present) and contain only small amounts of contaminating cytoplasmic Fe- and/or MnSODs. Differential inhibitors of SOD activity can be added to the riboflavin–TEMED solution to allow identification of specific forms of the enzyme. Potassium cyanide (2 m$M$) can be used to inhibit Cu,ZnSODs, although extreme care is needed with its use because of its highly toxic nature. A safer, although sometimes less effective, alternative is the copper chelator diethyldithiocarbamate (DDC).[3] With this agent, best results are obtained by presoaking the gel in 50 m$M$ phosphate buffer, pH 7.5, plus DDC for 20–30 min and then proceeding to the riboflavin–TEMED/NBT steps without added SOD inhibitor. Hydrogen peroxide at 5 m$M$ concentration may be used to inhibit FeSODs. Failure of inhibition should be interpreted cautiously, as occasional examples of the enzyme are partially resistants to this agent.[27] Cu,ZnSODs may also be (partially) inactivated by this agent.[28] As a further potential complication, the failure of an SOD to be inhibited by hydrogen peroxide may be due to a comigrating catalase. This can be determined by staining duplicate nondenaturing or IEF gels for both catalase[29] and SOD activities. MnSODs are resistant to all these inhibitors. To sort out identification of SODs by the use of a set of inhibitors, controls are useful. Commercially available bovine erythrocyte Cu,ZnSOD and *E. coli* Fe- and MnSODs (Sigma, St. Louis, MO) are satisfactory for this.

The interpretation of results is usually straightforward, particularly if the sample preparation has been thorough. On occasion a number of bands of similar electrophoretic mobility are observed that all disappear with the same inhibitor. These are most likely isoforms of the same SOD. With *Pseudomonas maltophila* two separate bands of Cu,ZnSOD activity were observed on a gel.[25] It was thought that this was due to instability of the dimer Cu,ZnSOD at pH 8.9 such that a dimer and monomer fraction were separated on gel electrophoresis. The monomer redimerized at pH 7.8 during the SOD visualization, resulting, with the dimer, in the appearance of two bands, both of which were potassium cyanide sensitive. With a continuous system using 50 m$M$ Tris–phosphate (pH 8.3) only one zone of activity was seen.

*Assay of Superoxide Dismutase Activity in Solution*

It is also possible to screen for SOD activity in bacterial extracts by a variety of solution assays.[3,18,30] These methods are rapid and sensitive, but not so convenient in many circumstances as the gel assay.[25] However, because some Cu,ZnSODs are inactivated during electrophoresis carried out under nondenaturing conditions,[3,18] the use of a solution assay in combination with activity staining in gels is strongly recommended when hunting for Cu,ZnSOD activity in pathogenic bacteria. Solution assays have a further advantage in allowing quantitation of active enzyme expressed under specific conditions. A simple method to detect SOD activity in cellular extracts is that based on the autoxidation of pyrogallol.[31] The color reaction commences on the addition of 10 $\mu$l of 20 m$M$ pyrogallol (dissolved in 10 m$M$ HCl) to 1 ml of buffer containing 20 m$M$ Tris-HCl–1 m$M$ diethylenetriaminepentaacetic

---

[27] R. Gabbianelli, A. Battistoni, C. Capo, F. Polticelli, G. Rotilio, B. Meier, and A. Desideri, *Arch. Biochem. Biophys.* **345**, 156 (1997).
[28] A. Battistoni, G. Donnarumma, R. Greco, P. Valenti, and G. Rotilio, *Biochem. Biophys. Res. Commun.* **243**, 804 (1998).
[29] E. M. Gregory and I. Fridovich, *Anal. Biochem.* **58**, 57 (1974).
[30] P. Langford, A. E. Williams, and J. S. Kroll, *FEMS Microbiol. Lett.* **61**, 347 (1991).
[31] S. Marklund and G. Marklund, *Eur. J. Biochem.* **47**, 469 (1974).

acid, pH 8.2. Advancing superoxide-induced autoxidation of pyrogallol is monitored at 420 nm at 30° for 2 min. The reaction is inhibited by SOD, 1 unit of activity being defined as the amount of enzyme necessary to achieve 50% inhibition. For sample assay, up to 50 $\mu$l of cellular extracts may be added to the buffer before the addition of pyrogallol. In the case of periplasmic extracts, preincubation with 2m$M$ DDC abolishes (or at the least strongly decreases) activity due to Cu,ZnSOD.

### Functional Assays: Chemical Susceptibility Assays with *Salmonella enterica* Strains

The cytoprotective properties of SODs in different bacterial compartments can be assessed by different superoxide-generating systems as follows.

*Susceptibility to Cytoplasmically Generated Superoxide*

The susceptibility of wild-type and *sod* mutant strains to ROIs and NOIs can be assessed by the disk diffusion method developed by Bauer et al.[32] and De Groote et al.[33] Mutants in cytosolic SOD are usually more sensitive to the redox cycling superoxide-generating drug paraquat (methyl viologen) when grown aerobically. Paraquat is reduced univalently by cytoplasmic NADPH diaphorases with the production of superoxide and also regeneration of the paraquat dication.[34]

Stationary-phase *S. enterica* wild-type and *sod* mutant strains ($10^6$ CFU in 100-$\mu$l volumes) are spread onto separate L-agar or M9 minimal medium plates [$Na_2HPO_4$ (7 mg/ml), $KH_2PO_4$ (3 mg/ml), NaCl (0.5 mg/ml), $NH_4Cl$ (1 mg/ml), thiamine (5 $\mu$g/ml), $MgSO_4$ (0.12 mg/ml), $CaCl_2$ (0.015 mg/ml), 0.2% (w/v) glucose, and 1.5% (w/v) agar]. The volume of each plate must be the same. Fifteen microliters of 1.9% (v/v) paraquat is spotted onto 0.6-cm round filter paper disks, which are then placed on the plates. After incubation overnight at 37°, the sizes of the zones of inhibition are a direct measure of paraquat sensitivity. Hydrogen peroxide (3%, v/v), S-nitrosoglutathione (500 m$M$), and 3-morpholinosydnonimine (500 m$M$) have been tested in a similar manner.[33] Some investigators have chosen to test the susceptibility of wild-type and *sod* mutant strains to paraquat by monitoring viable counts of bacteria after exposure in solution[35] or by plating out the same number of bacteria onto agar plates containing paraquat.[16]

---

[32] A. W. Bauer, W. M. Kirby, J. C. Sherris, and M. Turck, *Am. J. Clin. Pathol.* **45,** 493 (1996).
[33] M. A. De Groote, U. A. Ochsner, M. U. Shiloh, C. Nathan, J. M. McCord, M. C. Dinauer, S. J. Libby, A. Vazquez-Torres Y. Xu, and F. C. Fang, *Proc. Natl. Acad. Sci. U.S.A.* **94,** 13997 (1997).
[34] S. Liochev, A. Hausladen, W. Beyer, Jr., and I. Fridovich, *Proc. Natl. Acad. Sci. U.S.A.* **91,** 1328 (1994).
[35] A. Carlioz and D. Touati, *EMBO J.* **5,** 623 (1986).

## Susceptibility to Exogenously Generated Superoxide and Nitric Oxide

Periplasmic *sodC* mutants have typically been found to be more sensitive than wild type to exogenously derived superoxide, and in the case of *S. enterica*, in particular to the combination of superoxide and nitric oxide. Such susceptibility can be assessed with xanthine–xanthine oxidase (X–XO)[36] and 2,2′-(hydroxynitrosohydrazono) bisethanamine (SPER–NO)[33] as superoxide- and nitric oxide-generating systems, respectively. XO is a metalloflavoprotein, producing urate and reduced oxygen metabolites by oxidation of X. SPER–NO is a diazeniumdiolate compound, producing nitric oxide by dissociation.[37] Washed stationary-phase *S. enterica* strains are resuspended to a concentration of $10^5$ CFU/ml in phosphate-buffered saline (PBS, pH 7.4). X is added to 1 m$M$ and superoxide formation is initiated by the addition of XO (0.1 unit/ml). Duplicate experiments in which catalase is added to the assay mixture may be carried out to determine whether bacterial death is directly due to superoxide or to the hydrogen peroxide formed by the spontaneous dismutation of the superoxide anion under the assay conditions.[17] With SPER–NO, a 100 m$M$ stock in 0.1 $N$ sodium hydroxide on ice is prepared and added to a final concentration of 1 m$M$. The change in pH initiates nitric oxide release. Bacterial suspensions are incubated aerobically with shaking (150 rpm) at 37° and aliquots are removed at timed intervals and assayed for viable counts.[14,33]

## Macrophage Infection Assays

### Isolation and Preparation of Murine Peritoneal Macrophages

Peritoneal macrophages may be obtained from genetically *Salmonella*-susceptible (Ity$^s$) or *Salmonella*-resistant (Ity$^r$), 6- to 8-week old mice (e.g., BALB/c and C3H/HeN strains, respectively) by eliciting a sterile chemical peritonitis with 1 ml of 5 m$M$ sodium periodate injected intraperitoneally. After 3 days animals are humanely killed by $CO_2$ asphyxiation. Three milliliters of sterile PBS containing heparin (5 U/ml) is injected into the peritoneal cavity of each mouse, and after gentle massage the peritoneal exudate cells are collected by aspiration.

After centrifugation (10 min, 600$g$, 4°) cells are resuspended in 1 ml of ice-cold Iscove's modified Dulbecco's medium (IMDM) containing 10% (v/v) fetal calf serum (FCS). Ten microliters of the cell suspension is diluted in 90 $\mu$l of IMDM and 100 $\mu$l of trypan blue reagent [0.25% (w/v) trypan blue, 0.85% (w/v) NaCl] and placed in a hemocytometer, and cell viability is determined by light microscopy. Macrophages, which take up trypan blue, are nonviable. Viability is routinely >95%. For cell–bacterial interaction studies, macrophages are resuspended in IMDM containing 10% (v/v) FCS and placed as 100-$\mu$l aliquots

---

[36] I. Fridovich, *J. Biol. Chem.* **245**, 4053 (1970).
[37] L. K. Keefer, R. W. Nims, K. M. Davies, and D. A. Wink, *Methods Enzymol.* **268**, 281 (1996).

containing $\sim 5 \times 10^4$ cells in wells of 96-well tissue culture plates. Macrophages must be left for 2 hr at 37° in an atmosphere of 5% $CO_2$, to allow cells to adhere and form monolayers.

A modification of this protocol has been used by De Groote et al.[33] for experiments in which cells are additionally to be stimulated with interferon $\gamma$ (IFN-$\gamma$). Instead of IMDM, the medium used is based on RPMI 1640, containing in addition to 10% (v/v) FCS, IFN-$\gamma$ at 20 U/ml. Monolayers are incubated for 2–24 hr in the presence of IFN-$\gamma$.

*Preparation of Normal Murine Serum*

Blood is collected from BALB/c or C3H/HeN mice by cardiac puncture under terminal anesthesia. Blood from mice within each breed is pooled and incubated for 1 hr at 37° followed by 2 hr at 4°. Normal murine serum (NMS) is separated from cells, clotted by centrifugation (10 min, 600$g$, 4°), and stored in aliquots at $-20°$.

*Infection of Macrophage Monolayers*

*Salmonella enterica* harvested from exponential- or stationary-phase cultures is washed in IMDM plus 10% (v/v) FCS and resuspended to a concentration of $5 \times 10^6$ CFU/ml. Where appropriate, bacteria can be incubated with 10% (v/v) NMS at 37° for 30 min on a rolling platform set at 120 rpm. Bacteria are added to the monolayers to give a multiplicity of infection of 5–10 bacteria per macrophage (verified by viable count of the inoculum). Infected monolayers are incubated for 1 hr at 37° in an atmosphere of 5% $CO_2$, and then the bacteria are centrifuged down onto the monolayer (10 min, 150$g$, 4°) and incubated for 15 min (or the desired length of the experiment) at 37° in an atmosphere of 5% $CO_2$.

*Bacterial Uptake and Killing*

To measure *S. enterica* uptake, wash the infected monolayers twice with PBS to remove superficially adherent bacteria and lyse with 0.1% (w/v) sodium deoxycholate in PBS. The number of viable organisms in the lysates can be determined by plating serial dilutions on MacConkey agar containing suitable selective agents. Bacterial killing can be assessed by a modification of the method described by Buchmeier and Heffron.[38] For experiments with unstimulated macrophages, wash multiple monolayers with PBS, cover with IMDM containing 10% (v/v) FCS and gentamicin (100 mg/liter), and incubate at 37° in an atmosphere of 5% $CO_2$. At successive time points, replace the medium in subsets of wells

[38] N. A. Buchmeier and F. Heffron, *Infect. Immun.* **57**, 1 (1989).

with IMDM containing 10% (v/v) FCS and gentamicin (10 mg/liter) (the antibiotic is included to inhibit extracellular bacterial growth) and incubate as described above. After 1 hr (or other suitable time period) wash the monolayers twice with sterile PBS and then lyse with 0.1 ml of 0.1% (w/v) sodium deoxycholate. Determine the number of viable organisms by plating serial dilutions on MacConkey agar containing selective agents. The protocol is the same as in the study of IFN-$\gamma$-stimulated murine macrophages, except that RPMI 1640 containing 10% (v/v) FCS is used instead of IMDM containing 10% (v/v) FCS.

## Epithelial Infection Assays

Facultative intracellular pathogens such as *S. enterica* are able to penetrate host cells and survive within nonprofessional phagocytes. It is well known that the oxidative burst contributes to bacterial killing and, more recently, it has been shown that Cu,ZnSOD can enhance bacterial survival also within nonphagocytic cells.[34] Invasion and intracellular survival assays as described in the following sections are suitable for investigation of this aspect of pathobiology.

*Invasion Assay*

Epithelial cells derived from a human colonic carcinoma (Caco-2) are grown as monolayers at 37° in minimum essential medium (MEM) supplemented with bicarbonate (1.2 g/liter), 2 m$M$ glutamine, penicillin (100 U/ml), streptomycin (0.1 mg/ml), and 10% (v/v) heat-inactivated (56° for 30 min) FCS in 5% $CO_2$.

For infection experiments semiconfluent monolayers are washed three times with MEM (to remove antibiotics) and then FCS is added at a concentration of 2% (v/v).[39] Cells are infected with bacteria (prewashed in PBS) grown to exponential or stationary phase, at a multiplicity of infection (MOI) of 100 bacteria per cell. Depending on the organisms under consideration, this infection phase of the experiment—incubating at 37°—lasts for 1 or 2 hr. After extensive washing with PBS without $Ca^{2+}$ and $Mg^{2+}$, cells are incubated in fresh medium containing gentamicin (100–200 $\mu$g/ml, depending on the minimal inhibiting concentration for the bacteria under study) for 2 hr in order to kill extracellular bacteria. After this period, infected cells are washed three times with MEM and lysed by addition of 1.0 ml of ice-cold 0.1% (v/v) Triton X-100 in water. Cell lysates are diluted in PBS and plated on selective media to quantify viable intracellular bacteria.

[39] A. Battistoni, F. Pacello, S. Folcarelli, M. Ajello, G. Donnarumma, R. Greco, M. G. Ammendolia, D. Touati, G. Rotilio, and P. Valenti, *Infect. Immun.* **68,** 30 (2000).

*Intracellular Survival Assay*

After bacterial infection, cell monolayers are washed and fresh medium containing gentamicin (20–50 µg/ml), 2% (v/v) FCS, bicarbonate (1.2 g/liter), and 2 m$M$ glutamine is added. The cells are then incubated for 4, 6, 24, and 48 hr at 37°. At each time point cells are washed and lysed as described above, and the number of viable intracellular bacteria is evaluated as colony-forming units.

## Experimental Infection of Mice

*Oral and Intraperitoneal Inoculation of Mice, and Assessment of Burden of Experimental Infection*

Infect mice (8–10 weeks old) by oral administration, typically with $10^6$ CFU, or intraperitoneally, typically with 100 CFU of *S. enterica* suspended in 0.1 ml of 0.9% (w/v) saline. Sham-inoculated mice receive saline only. At suitable time intervals postinfection, or as mice began to exhibit symptoms of systemic disease, they are humanely killed by $CO_2$ asphyxiation.

To assess the burden of infection, target organs (liver and spleens) are removed, weighed, and homogenized (UltraTurrax homogenizer) in 0.9% (w/v) saline containing 1% (v/v) Triton X-100. Determine the number of viable organisms in the organs by plating serial dilutions on MacConkey agar containing selective agents. Replica plate on appropriate selective media to exclude the theoretical possibility that mutant strains might have reverted to wild type.

## Acknowledgments

This work was supported by grants from the Wellcome Trust, BBSRC (P.R.L. and J.S.K.), and the Meningitis Research Foundation (J.S.K.), and by a MURST-PRIN grant and the CNR target project on Biotechnology (A.B.).

# [17] Superoxide Dismutase Null Mutants of Baker's Yeast, *Saccharomyces cerevisiae*

By LORI A. STURTZ and VALERIA CIZEWSKI CULOTTA

## General Handling and Maintenance of *Saccharomyces cerevisiae* sod Mutants

The baker's yeast *Saccharomyces cerevisiae* provides an ideal model system in which to study the oxidative stress consequences associated with loss of copper- and zinc-containing superoxide dismutase 1 (Cu,ZnSOD1) and/or MnSOD2. The ease with which knockout mutations can be introduced in this organism, together with the capacity for yeast growth under both anoxic and aerobic conditions, allows rapid scoring for SOD-related phenotypes that are specific to oxygen toxicity. In general, a greater number of metabolic defects have been noted with strains lacking the more abundant and largely cytosolic Cu,ZnSOD1 (so-called *sod1*Δ strains) than those devoid of the manganese mitochondrial form of the enzyme. Furthermore, virtually all the phenotypes associated with *sod1*Δ cells are shared with *lys7*Δ yeast lacking the copper chaperone for SOD1 (CCS).[1] Because yeast Cu,ZnSOD1 is absolutely dependent on CCS for acquiring its copper ion *in vivo*, strains lacking CCS accumulate inactive yeast SOD1 polypeptide and are virtually identical in phenotype to *sod1*Δ mutants.

As a caution, it is important to note that when handling *sod1*Δ (and *lys7*Δ) strains, fresh cultures should be repeatedly obtained from single colonies and frozen stocks. Do not utilize *sod1*Δ cells that have been cultured at 30° for >2 days or stored at 4° for >2 weeks. Yeast *sod1*Δ cells tend to accumulate suppressor mutations at the *PMR1* locus at an extremely high rate ($10^{-6}$), and these suppressor mutants confer a substantial growth advantage, allowing them to rapidly overpopulate *sod1*Δ cell cultures.[2] Because *pmr1* suppressors rescue virtually all the oxygen toxicity of *sod1*Δ mutants, it is easy to test for their presence by scoring for symptoms of *sod1* deficiency, for example, aerobic lysine and methionine auxotrophy (see below). To minimize accumulation of *sod1*Δ suppressors, cells should be cultured under anaerobic conditions, except as needed for oxygen-dependent experiments. For these purposes, anaerobic culture jars (BBL GasPak; Becton Dickinson Labware, Lincoln Park, NJ) work well. These chambers can hold up to 12 (small chambers) or 40 (large chambers) $100 \times 15$ mm plates and generate

---

[1] V. C. Culotta, L. W. Klomp, J. Strain, R. L. Casareno, B. Krems, and J. D. Gitlin, *J. Biol. Chem.* **272**, 23469 (1997).

[2] P. J. Lapinskas, K. W. Cunningham, X. F. Liu, G. R. Fink, and V. C. Culotta, *Mol. Cell. Biol.* **15**, 1382 (1995).

a $CO_2$-enriched, oxygen-depleted atmosphere. For best results, the GasPak *Plus* envelopes with built-in palladium catalyst should be avoided as this system does not sufficiently deplete oxygen to suppress oxidative damage in yeast cells lacking SOD1. The palladium catalyst should be added separately and may be regenerated for repeated use by heating to 160–170° for 2 hr.

### Sensitivity to Oxidants

Redox cycling agents such as paraquat and menedione effect increases in intracellular levels of superoxide, and, as such, both *sod1*Δ and *sod2*Δ strains are exquisitely sensitive to growth in the presence of these agents. The precise cytotoxic concentrations must be determined empirically for each genetic background and growth medium. Typically, strains that are wild type for SOD are tolerant to paraquat concentrations up to 50 m$M$. By comparison, single *sod1*Δ and double *sod1*Δ *sod2*Δ strains fail to thrive on medium containing paraquat in excess of ~10 $\mu M$. The *sod2*Δ strains are somewhat more resistant (tolerance up to ~100 $\mu M$) when grown on glucose, but exhibit a high sensitivity (~10 $\mu M$) in medium containing nonfermentable carbon sources such as ethanol and glycerol.[3] Both liquid and plate tests are appropriate for monitoring sensitivity toward redox cycling agents. Paraquat should be added from freshly made stocks after autoclaving growth media.

### Amino Acid Auxotrophies

An easily scored marker for oxidative damage in *sod1*Δ (but not *sod2*Δ) strains is a dual auxotrophy for the amino acids methionine and lysine. Because of oxidative inactivation of methionine and lysine metabolic pathways, *sod1*Δ cells cannot biosynthesize these amino acids when grown in air.[4,5] Similar blocks in amino acid biosynthesis are not shared with mammalian null models for SOD1 because lysine and methionine are essential amino acids (not biosynthesized) for mammals. The methionine and lysine defects of yeast *sod1*Δ strains are thought to arise from independent oxygen-related defects. The lysine auxotrophy may well result from superoxide inactivation of the 4Fe–4S cluster of homoaconitase, an aconitase-like enzyme in the lysine biosynthesis pathway.[6] By comparison, the methionine defect of *sod1*Δ cells has been proposed to result from depletion of available NADPH, an essential cofactor for methionine biosynthesis enzymes.[7]

---

[3] E. B. Gralla and D. J. Kosman, *Adv. Genet.* **30**, 251 (1992).
[4] T. Bilinski, Z. Krawiec, L. Liczmanski, and J. Litwinska, *Biochem. Biophys. Res. Commun.* **130**, 533 (1985).
[5] E. Chang and D. Kosman, *J. Bacteriol.* **172**, 1840 (1990).
[6] S. D. Irvin and J. K. Bhattacharjee, *J. Mol. Evol.* **46**, 401 (1998).

To assay for amino acid deficiencies, cells are grown in minimal medium devoid of methionine or lysine (liquid or plates), and growth is monitored in air and also under anaerobic conditions as a control. It is important in these assays to seed the cells at a low density (e.g., $OD_{600}$ of $<0.2$ in liquid cultures), as the amino acid auxotrophies are less pronounced under high density conditions. Furthermore, growth should be scored within 24 hr of incubation at 30° because beyond this period, *pmr1* suppressors of SOD1 deficiency will begin to populate cultures.

## Stationary-Phase Survival

The growth cycle of *S. cerevisiae* begins with log phase and progresses through the diauxic shift to stationary phase. On entry into stationary phase, reactive oxygen species (ROS) production by the electron transport chain is enhanced.[8] The ability of a yeast cell to survive during stationary phase is dependent on its ability to detoxify superoxide radicals produced by the mitochondrial respiratory chain. Yeast lacking Cu,ZnSOD1 and MnSOD2 exhibit reduced stationary phase survival when compared with wild-type yeast.[9]

For experiments employing the stationary-phase viability assay, cells are inoculated at an $OD_{600}$ of 0.1 in minimal medium. Cultures are grown with shaking (220 rpm) at 30° for extended periods (e.g., 5 days to 1 month). At each 24-hr interval, the $OD_{600}$ is monitored and viability is tested by serially diluting cells and spot planting 5 $\mu$l of $10^5$, $10^4$, $10^3$, and $10^2$ cells on enriched YPD medium. The plates are then incubated in an anaerobic chamber at 30°, and survival is measured by counting the number of growing colonies. There are two methods for stationary-phase viability tests. For short-term survival assays (less than 1 week), cells are maintained in spent minimal medium throughout the study. As an alternative approach, after 2 days in culture, the cells may be spun down, washed twice in sterile distilled water, and then resuspended in sterile distilled water for the remainder of the assay. This latter procedure enables the study of long-term stationary-phase survival (1 month or longer). In the short-term survival assays, after 2 days in stationary phase, *sod1* mutants exhibit less than 1% of the viability of wild-type cells. With long-term stationary-phase tests in water, wild-type cells retain greater than 40% viability for more than 30 days, whereas *sod1*$\Delta$ cells show $<0.1\%$ viability after 10 days; *sod2*$\Delta$ mutants appear somewhat more tolerant and do not show viability defects until 15 days.[9]

---

[7] K. H. Slekar, D. Kosman, and V. C. Culotta, *J. Biol. Chem.* **271**, 28831 (1996).
[8] V. D. Longo, L. L. Liou, J. S. Valentine, and E. B. Gralla, *Arch. Biochem. Biophys.* **365**, 131 (1999).
[9] V. D. Longo, E. B. Gralla, and J. S. Valentine, *J. Biol. Chem.* **271**, 12275 (1996).

## Heat Sensitivity

Both *sod1*Δ and *sod2*Δ mutants of *S. cerevisiae* are hypersensitive to killing by heat treatment at 50°.[10] The precise duration of heat treatment needed to discern SOD-related effects varies with strain background; however, treatment for 15–45 min will generally result in 1–10% viability of wild-type cells, and >2 orders of magnitude less viability of *sod* mutants. This assay is typically conducted in a thermal cycler, where rapid changes in temperature can be achieved.

To conduct the heat sensitivity test, cultures are pregrown to midlog phase in enriched YPD medium, and then diluted in YPD to a final $OD_{600}$ of 0.3. A 100-$\mu$l volume of this culture is then aliquoted to a 500-$\mu$l thin-walled polymerase chain reaction (PCR) tube and placed in a thermal cycler. The temperature is quickly raised to 50° for the desired time (e.g., 30 min), followed by rapid cooling to room temperature. Cell viability is then determined as described above for stationary-phase survival tests.

## Hyperoxia

Yeast lacking either Cu,ZnSOD1 or MnSOD2 exhibit exquisite sensitivity to oxygen. Mutants of SOD1, but not SOD2, grow slowly under atmospheric oxygen and both mutants display low viability when grown under hyperoxic conditions.[3] Presumably, high levels of oxygen result in an increased propensity for superoxide formation in the cell, hence the SOD-associated sensitivity toward high oxygen tensions. Sensitivity of *sod1*Δ or *sod2*Δ cells to oxygen can be determined by growing cells under hyperoxic conditions and by comparing the survival of these mutants with that of wild-type cells. Hyperoxia is defined herein as exposure to 100% oxygen.

To test for hyperoxic sensitivity, the desired test strains are precultured in enriched medium in an anaerobic chamber at 30°. Cells are then serially diluted in sterile water and 5-$\mu$l aliquots containing $1 \times 10^5$, $1 \times 10^4$, $1 \times 10^3$, and $1 \times 10^2$ cells are spotted on enriched medium (YPD). The plates are then placed in an airtight chamber (Modular incubator chamber; Billups-Rothenberg, Del Mar, CA) and flushed with 100% $O_2$ at 4 lb/in$^2$ for 5 min. The $O_2$ source is turned off and the outlet of the chamber is quickly clamped. The inlet is then clamped and the chamber is incubated at 30° for 1–2 days. Duplicate plates are also incubated in an anaerobic chamber and in air at 30°. The viability of cells grown under hyperoxic conditions is determined relative to the survival of cells grown anaerobically or in air. In general, strains wild type for SOD are somewhat sensitive to hyperoxia and exhibit 1–10% viability under these conditions. By comparison, viability of a

---

[10] J. F. Davidson, B. Whyte, P. H. Bissinger, and R. H. Schiestl, *Proc. Natl. Acad. Sci. U.S.A.* **93**, 5116 (1996).

*sod2* mutant is an order of magnitude lower than that of an isogeneic wild type, whereas no viable colonies of *sod1* mutants are formed during hyperoxia (C. Giap and V. C. Culotta, unpublished, 2000).

## Oxidative Damage to Proteins

In addition to the oxygen-related growth defects described above, biochemical assays for the products of oxidative damage may be employed as markers for SOD deficiency. Protein carbonyl formation is one easily monitored product of damage from reactive oxygen. Metal-catalyzed protein oxidation introduces carbonyl groups to the side chains of lysine, arginine, proline, and threonine residues. Carbonyl side chains can be detected by derivatization with DNPH (dinitrophenylhydrazine) followed by immunoblot analysis with an anti-DNP antibody. The Oxy-Blot protein oxidation detection kit manufactured by Intergen (Purchase, NY) has been developed to detect such protein oxidation. As expected, yeast lacking SOD1 have been found to have enhanced protein oxidation as compared with wild-type yeast in both the cytosolic and mitochondrial fractions.[11]

For monitoring production of protein carbonyls, cell extracts are obtained from yeast cells that have been grown with shaking (220 rpm) at 30° to midlog phase in either enriched or minimal medium. Lysates are prepared from yeast spheroplasts (yeast lacking a cell wall) as follows: Approximately $10^9$ cells are harvested (~0.15 g) and washed in sterile water, and the cell pellet is resuspended in 1.2 $M$ sorbitol–20 m$M$ HEPES buffer, pH 7.4, to a final concentration of 0.15 g of cells per milliliter. The cell wall of the yeast is then digested by the addition of 3 mg of Zymolyase per gram of cells and subsequent incubation at 30° for 30–60 min. The spheroplasts are then harvested by centrifugation at 3000 rpm in a microcentrifuge, gently washed twice in the sorbitol–HEPES buffer, and resuspended in ice-cold SEM lysis buffer [250 m$M$ sucrose, 1 m$M$ EDTA, 10 m$M$ morpholinepropanesulfonic acid (MOPS, pH 7.2), 0.5% (w/v) bovine serum albumin (BSA), 0.5 m$M$ phenylmethylsulfonyl fluoride (PMSF), and 1.0 m$M$ dithiothreitol (DTT) or 0.1% (v/v) 2-mercaptoethanol]. Two milliliters of SEM buffer is used per original gram of cells. The cells are then lysed by applying about 15 strokes with a microcentrifuge pestle (Bel-Art, Pequannock, NJ). All remaining steps should be carried out at 4°. The cell lysates are subject to two consecutive centrifugations at 3000 rpm for 5 min and the soluble material, representing the total cell extract, is collected. For experiments in which mitochondrial versus postmitochondrial supernatant fractions (largely cytsol) are desired, the total cell extract is subjected to centrifugation at 13,000 rpm for 10 min. The resultant supernatant represents the postmitochondrial supernatant (largely cytosol) whereas the pellet contains a crude

---

[11] L. A. Sturtz, K. Diekert, L. T. Jensen, R. Lill, and V. C. Culotta, *J. Biol. Chem.* **276**, 38084 (2001).

preparation of mitochondria. The mitochondria are then washed and resuspended in SEM buffer to a final volume of 20–30 $\mu$l/0.15 g of original yeast cell weight. The supernatant is also subjected to a second round of centrifugation to help remove any contaminating mitochondria. Protein concentration measurements are obtained with all samples.

To derivatize the proteins with DNP, a 5- to 10-$\mu$l sample containing 10–20 $\mu$g of protein is first denatured by addition of 5 $\mu$l of 12% (w/v) sodium dodecyl sulfate (SDS); derivatization of the protein samples is then performed by adding 10 $\mu$l of 1× DNPH (2,4-dinitrophenylhydrazine; as provided by manufacturer), and incubation at room temperature for 15 min. The reaction is then halted by addition of 7.5 $\mu$l of neutralization solution (provided by manufacturer) followed by vigorous shaking. At this point, the samples can be stored at 4° for up to 1 week.

After neutralization, 25 $\mu$l of the derivatized samples can be directly loaded onto a 12% (w/v) SDS–polyacrylamide gel. Addition of standard protein gel loading buffer is not necessary, and the proteins should not be boiled. After SDS–PAGE, the proteins are transferred onto a nitrocellulose membrane, which is subsequently probed with a primary rabbit anti-DNP antibody (diluted 1 : 150) and horseradish peroxidase (HRP)-conjugated anti-rabbit IgG (as a secondary antibody). Immunodetection can employ the ECL detection kit (Amersham Pharmacia Biotech, Piscataway, NJ). Wild-type yeast, as well as *sod1*Δ cells, exhibit protein oxidation as revealed by specific banding patterns on the Western blot. However, the degree of oxidation is greatly enhanced in those cells lacking SOD1. The number and molecular weight of the bands visualized by this procedure vary according to the strain and growth conditions used. Nevertheless, the majority of the proteins detected by this method fall between 30,000 and 120,000 Da.[11]

Acknowledgments

We thank Laran Jensen for critical reading of this review. This work was supported by the JHU NIEHS center and by NIH Grant GM 50016 (to V.C.C.) L.S. is supported by NIEHS Training Grant ES 07141.

# [18] Measurement of "Free" or Electron Paramagnetic Resonance-Detectable Iron in Whole Yeast Cells as Indicator of Superoxide Stress

By CHANDRA SRINIVASAN and EDITH BUTLER GRALLA

## Introduction

Yeast, like most eukaryotes, contain manganese superoxide dismutase (MnSOD; product of the *SOD2* gene) in the mitochondria and copper, zinc superoxide dismutase (Cu,ZnSOD; product of the *SOD1* gene) in the cytoplasm, mitochondrial intermembrane space, and nucleus. There is no evidence of a separate extracellular SOD in yeast, as there is in mammals. Strains of *Saccharomyces cerevisiae* that lack either or both superoxide dismutases have been available for many years and have provided much useful information. Experiments in our laboratory have highlighted the important roles of these two enzymes in aging (stationary-phase survival) and have helped define the cellular sources of superoxide and the differing roles of the two enzymes.[1-3] Selecting genetic suppressors of the *sod*Δ mutant phenotypes has led to important discoveries relating to metabolism of copper and manganese,[4-8] and, more recently, iron.[9,10] We discovered that excess superoxide in the cell (due either to the absence of SOD in the *sod*Δ mutants or in wild-type cells treated with the redox cycling drug paraquat) leads to accumulation of a form of iron that is detectable by electron paramagnetic resonance (EPR) spectroscopy at $g = 4.3$ (see below).[10] In this article we summarize the methods for growing the *sod*Δ yeast strains and for analyzing their "free" iron content by Fe(III) EPR. It should be noted that *in vivo* there in no such thing as free iron—it would certainly be bound to something—so we use the term "free" in quotes to denote iron that is loosely bound and accessible to chelator, or,

---

[1] V. D. Longo, E. B. Gralla, and J. S. Valentine, *J. Biol. Chem.* **271**, 12275 (1996).
[2] V. D. Longo, L. M. Ellerby, D. E. Bredesen, J. S. Valentine, and E. B. Gralla, *J. Cell Biol.* **137**, 1581 (1997).
[3] V. D. Longo, L. L. Liou, J. S. Valentine, and E. B. Gralla, *Arch. Biochem. Biophys.* **365**, 131 (1999).
[4] X. F. Liu and V. C. Culotta, *Mol. Cell. Biol.* **14**, 7037 (1994).
[5] X. F. Liu and V. C. Culotta, *J. Biol. Chem.* **274**, 4863 (1999).
[6] S. J. Lin and V. C. Culotta, *Mol. Cell. Biol.* **16**, 6303 (1996).
[7] S. J. Lin and V. C. Culotta, *Proc. Natl. Acad. Sci. U.S.A.* **92**, 3784 (1995).
[8] V. C. Culotta, *Metal Ions Biol. Syst.* **37**, 35 (2000).
[9] J. Strain, C. R. Lorenz, J. Bode, S. Garland, G. A. Smolen, D. T. Ta, L. E. Vickery, and V. C. Culotta, *J. Biol. Chem.* **273**, 31138 (1998).
[10] C. Srinivasan, A. Liba, J. A. Imlay, J. S. Valentine, and E. B. Gralla, *J. Biol. Chem.* **275**, 29187 (2000).

in the case of yeast, already existing as Fe(III) in a high-spin rhombic form, but not bound to enzyme active sites or to heme.

Fe(III) EPR spectra that are useful in biological systems fall into three $g$ value classes: those with features in the $g = 6$ region for heme iron, those with three $g$ features ($x$, $y$, $z$) clustered around $g = 2$ for low-spin iron species, and those at $g = 4.3$ for nonheme high-spin iron.[11] ($g$ defines the position of the signal and the $g$ value $= h\nu/\beta H$, where $h$ is Planck's constant, $\nu$ is the frequency, $\beta$ is the Bohr magneton, and H is the external magnetic field at resonance.) The signal at $g = 4.3$ is characteristic of ferric iron in a high-spin rhombic complex, and is the one we are concerned with here. Desferrioxamine-bound iron gives a signal at this $g$ value, as do ferric citrate, other simple chelates, and oxidized [Fe(III)] transferrin. H is about 1550 G for the Fe(III) EPR signal under our conditions. Liquid nitrogen temperatures are needed for detecting the Fe(III) EPR signal at $g = 4.3$ and liquid helium temperatures are required for the other two $g$ regions. The ferrous form of iron cannot be detected in biological systems, as the signal is too broad.

The Fe(III) EPR signal at $g = 4.3$ in the presence and absence of desferrioxamine has been monitored as an indication of the "free," or loosely bound iron in several organisms. In most organisms, the addition of desferrioxamine is necessary to observe the Fe(III) EPR signal at $g = 4.3$, indicating that the "free" iron present is in the Fe(II) state and can be detected only after chelation and conversion to Fe(III) by this avid ligand. This is true in *Escherichia coli*[12] as well as in some mammalian systems, where a pool of chelatable or "labile" iron has been demonstrated that can be detected by EPR after treatment with desferrioxamine[13] or by fluorescence methods with other chelators such as calcein.[14–16] The mammalian pool is thought to be the iron that is sensed by the iron regulatory protein (IRP), which binds a specific motif in mRNA (IRE) to regulate the synthesis of ferritin and transferrin receptor in response to changes in iron availability.[17] In *E. coli*, the chelatable iron was shown to be greatly increased by superoxide stress, that is, in mutants lacking superoxide dismutase (*sodA sodB*).[12,18] This increase may be attributable to release of iron from superoxide-sensitive iron–sulfur cluster proteins such as aconitase, as was originally proposed by Liochev and Fridovich.[19]

---

[11] M. C. R. Symons and J. M. C. Gutteridge, "Free Radicals and Iron: Chemistry, Biology, and Medicine." Oxford University Press, Oxford, 1998.
[12] K. Keyer and J. A. Imlay, *Proc. Natl. Acad. Sci. U.S.A.* **93**, 13635 (1996).
[13] A. V. Kozlov, A. Bini, D. Gallesi, F. Giovannini, A. Iannone, A. Masini, E. Meletti, and A. Tomasi, *Biometals* **9**, 98 (1996).
[14] Z. I. Cabantchik, H. Glickstein, P. Milgram, and W. Breuer, *Anal. Biochem.* **233**, 221 (1996).
[15] W. Breuer, S. Epsztejn, and Z. I. Cabantchik, *FEBS Lett.* **382**, 304 (1996).
[16] S. Epsztejn, O. Kakhlon, H. Glickstein, W. Breuer, and I. Cabantchik, *Anal. Biochem.* **248**, 31 (1997).
[17] W. Breuer, S. Epsztejn, and I. Cabantchik, *J. Biol. Chem.* **270**, 24209 (1995).
[18] K. Keyer and J. A. Imlay, *J. Biol. Chem.* **272**, 27652 (1997).

In collaboration with the laboratory of J. A. Imlay, we developed a similar EPR method for yeast and found that superoxide stress causes accumulation of EPR-detectable iron in yeast as well, but, surprisingly, the iron is already in the Fe(III) state, and the bulk of it can be detected without the addition of desferrioxamine.[10]

The elevation of EPR-detectable or "free" iron in yeast correlates with conditions of elevated superoxide, and thus is a symptom of superoxide stress. In addition, it may cause some of the detrimental effects of the lack of SOD in *sod* mutant strains, because "free" iron could catalyze many destructive oxidation reactions via Fenton chemistry.[19] The Fe(III) EPR method we describe can provide an *in vivo* assessment of oxidative stress status in yeast and *E. coli*, and we are currently testing whether this is true for other, more complex, organisms as well. One important advantage of this method is that the analysis is performed on whole cells, eliminating uncertainties about the effects of cell lysis on the results. Processing of samples is minimal, consisting only of a short incubation in desferrioxamine (which often can be omitted) and some quick washing steps. Prepared samples are stored in EPR tubes at $-70°$ until they can be analyzed. This article outlines culture methods for these strains and describes how the Fe(III) EPR samples are prepared and analyzed.

## Methods

*Growth of sod$\Delta$ Yeast Strains*

*Saccharomyces cerevisiae* lacking Cu,ZnSOD (*sod1*$\Delta$) or MnSOD (*sod2*$\Delta$) and *sod1*$\Delta$/*sod2*$\Delta$ double knockouts have been constructed in our laboratory in the DBY746 (EG103) background and have been widely used.[20,21] The individual *sod* knockout strains in the BY4741 genetic background can be purchased from Research Genetics (Huntsville, AL). Phenotypes of the mutants in either backgrounds are similar.

*Media.* These strains are grown in normal yeast media—YP-based rich media and SC defined medium[22]—with a few modifications[1] as described below. Because we have noticed some effects of the medium on the growth of the *sod* mutants, we are including some details about our medium preparation. YPD medium is the standard yeast culture medium and consists of 1% (w/v) yeast extract, 2% (w/v) peptone, and 2% (w/v) dextrose (glucose). For plates, 2% (w/v) Difco (Detroit, MI) Bacto-agar is included. Other carbon sources can be substituted for the dextrose,

---

[19] S. I. Liochev and I. Fridovich, *Free Radic. Biol. Med.* **16,** 29 (1994).
[20] E. B. Gralla and D. J. Kosman, *Adv. Genet.* **30,** 251 (1992).
[21] E. B. Gralla, in "Oxidative Stress and the Molecular Biology of Antioxidant Defenses" (J. Scandalios, ed.), p. 495. Cold Spring Harbor Laboratory Press, Cold Spring Harbor, New York, 1997.
[22] C. Kaiser, S. Michaelis, and A. Mitchell, "Methods in Yeast Genetics." Cold Spring Harbor Laboratory Press, Cold Spring Harbor, New York, 1994.

for example, 3% (v/v) glycerol (making YPG), 2% (v/v) ethanol (YPE), 2% (w/v) galactose (YPGal), or a combination of glycerol and ethanol (YPGE). To avoid darkening (burning) of the carbon source when it is autoclaved in the medium, it is best to sterilize the carbon source separately, as a 10× solution, either by autoclaving or filtering through a 0.22-$\mu$m pore size filter, and then add it to the YP mix (made at 0.9 volume) when it has cooled somewhat. SC medium (synthetic complete, or defined medium) is based on Difco yeast nitrogen base (YNB) [per liter: 1.8 g of YNB without amino acids or ammonium sulfate, 5 g of ammonium sulfate, 1.4 g of $NaH_2PO_4$, dry amino acid mix (described below)]. Ammonium sulfate is the nitrogen source; a carbon source [usually 2% (w/v) dextrose, but others are fine, too] is sterilized separately and added as a 10× solution. Amino acids, adenine, and uracil are added to the medium as a dry mix to improve growth and compensate for auxotrophic markers the strain may carry. The $sod1$ mutants grow better with increased amounts of certain amino acids, so we use the following amounts for all yeast culture (in mg/liter): Ade, 80; Ura, 80; Trp, 80; His-HCl, 80; Arg-HCl, 40; Met, 80; Tyr, 40; Leu, 120; Ile, 60; Lys-HCl, 60; Phe, 60; Glu, 100; Asp, 100; Val, 150; Thr 200; Ser, 400. It is convenient to make a dry mix of amino acids with enough for 10 liters of medium. The amino acids are crushed and mixed with a mortar and pestle, weighed out as needed (1.73 g/liter of the complete mix), and added to the bulk SD medium before autoclaving. Dropout mixes can be made by leaving out the desired component and adjusting the amount to be added. Additional information on various medium preparations and basic techniques used in yeast culture are described in several places.[22–24]

Because the $sod1\Delta$ strains in particular are weakened and oxygen sensitive, it is easy inadvertently to select second-site suppressors during growth and/or during storage on plates. In addition, both $sod1\Delta$ and $sod2\Delta$ mutant strains die quickly on plates. To avoid problems and increase reproducibility each experiment is started with freshly streaked frozen stocks. Strains carrying the $sod1\Delta$ mutation are revived from frozen cultures on YPD plates in a microaerophilic atmosphere, using BBL Campy pouches (Becton Dickinson Microbiology Systems, Sparks, MD). These bags reduce oxygen to 5 to 10% and increase the $CO_2$ levels, and they improve growth of $sod1\Delta$ strains without the added complications and metabolic alterations that true anaerobiosis induces. The culture process for a typical EPR experiment is outlined below.

1. *Day 1.* Streak cells from frozen stocks onto YPD agar plates to obtain single colonies. Place the $sod\Delta$ yeast plates in a BBL Campy pouch (Becton Dickinson)

---

[23] C. Guthrie and G. R. Fink, (eds.), "Guide to Yeast Genetics and Molecular Biology," Vol. 194. Academic Press, San Diego, California, 1991.
[24] F. M. Ausubel, R. Brent, R. E. Kingston, D. D. Moore, J. G. Seidman, J. A. Smith, and K. Struhl, in "Current Protocols in Molecular Biology," Chap. 13. John Wiley & Sons, New York, 1998.

after adding the liquid activating reagent. Incubate the plates at 30° for 2 to 3 days, until colonies are visible on all the plates.

2. *Day 3.* Inoculate starter cultures. Starter cultures are grown in 16 × 150 mm tubes in 4 to 5 ml of the medium in which the experiment will be conducted (usually SDC, but SGC, YPD, or various SD dropout media also can be used). Remove the plates from the Campy pouch. Using a sterile wire loop, toothpick, or applicator stick, transfer a single colony into the liquid medium and grow overnight (12–18 hr) at 30° with shaking at 220 rpm. It is best to pick an average-sized colony from the *sod* mutant plates—large ones are likely to be suppressors. Make duplicate cultures for each strain, using a different colony for each overnight culture. These growth conditions are not fully aerated, so the mutant strains are able to grow well and are not overly stressed. It is important not to overgrow the overnights as $sod\Delta$ strains begin to lose viability on entry into stationary phase.

3. *Day 4.* Start experimental cultures. The density of each overnight culture is measured by diluting 50 $\mu$l of (fully resuspended) culture to 1 ml in distilled water and reading the optical density at 600 nm ($OD_{600}$) in an ultraviolet–visible wavelength (UV–Vis) spectrophotometer. Spectrophotometer readings should be below 0.5, as the correlation between cell density and "absorbance" is not linear above this limit, so cells should be diluted until a low enough reading is obtained. In our hands, and $OD_{600}$ of 1 is equivalent to $1 \times 10^7$ cells/ml. For SDC growth the cells usually reach a maximum density of between $5 \times 10^7$ and $10 \times 10^7$ cells/ml, or an $OD_{600}$ of 5 to 10; for YPD growth it can be much higher ($20 \times 10^7$ to $30 \times 10^7$ cells/ml). The $sod1\Delta$ mutants generally grow to a lower final density than the wild type. The amount of overnight culture needed to obtain a starting $OD_{600}$ of 0.1 ($10^6$ cells/ml) in 50 ml of medium is calculated and 250-ml culture flasks containing 50 ml of medium are inoculated with the calculated amount of each overnight. Flasks are placed in an incubator at 30°, with shaking at 220 rpm. Other flask sizes can be used, but it is important that the ratio of flask volume to medium volume be at least 5 : 1 to maintain good aeration. Good aeration for the experimental cultures is essential as the phenotype of the mutants is expressed only under aerobic conditions. Typically, cells are harvested after 24 or 72 hr of growth and prepared for EPR as described below. If log-phase cells are desired a similar protocol is used, but much larger volumes will be needed as cells should be harvested at an $OD_{600}$ of 1 or below.

*Electron Paramagnetic Resonance Sample Preparation*

The Fe(III) EPR method we are using currently was developed for *E. coli* by Keyer and Imlay[12,18] and modified for our yeast system. The original procedure relied on the ability of desferrioxamine to enter the cell and chelate loosely bound iron or iron not specifically bound to proteins or other high-affinity or biologically protected sites, converting it to Fe(III) and rendering it detectable by EPR at

$g = 4.3$. In *E. coli* the desferrioxamine was essential to detect a signal, indicating the most loosely bound iron was present as Fe(II) in that organism. In yeast, on the other hand, the iron is detectable whether or not the desferrioxamine was added, indicating that the iron is already in the Fe(III) state.[10]

*Day 5 or 7.* Typically, cultures are grown for 24 or 72 hr because at these times we see the largest differences in EPR-detectable iron between wild type and *sod* mutants. At 24 hr cultures in SDC are saturated, but are still transitioning out of active growth; by 72 hr they are entering stationary phase and are set up for long-term survival without growth. The most consistent results are obtained with 72-hr cultures.

After the required growth period, the $OD_{600}$ is recorded for each culture as described above and the cell harvest procedure is begun. From this point on it is important to work as quickly as possible. Each culture is processed in duplicate. Transfer 10-ml aliquots of culture to each of two centrifuge tubes. (Disposable 50-ml conical screw-cap centrifuge tubes work well.) Spin at 4° for 5 min at 4000 rpm. Discard the supernatant, resuspend the pellets in 9 ml of YP or SC medium (no sugar added), and transfer the contents into 50-ml sterile Erlenmeyer flasks. To one flask add 1 ml of 0.2 $M$ desferrioxamine and to the other add 1 ml of buffer. [Make a fresh stock of 0.2 $M$ stock solution of desferrioxamine mesylate salt (Sigma/Aldrich, St. Louis, MO) in water and adjust to pH 8 with 1 $M$ KOH.] Incubate both samples at 30° with shaking for 12–14 min. Transfer the samples to centrifuge tubes and spin at 4° for 5 min at 4000 rpm. Discard the supernatant and suspend the cell pellets in 10 ml of cold 20 m$M$ Tris-HCl, pH 7.4, by pipetting up and down a few times. Centrifuge the suspension for 5 min at 4000 rpm and 4°. Carefully remove the supernatant without disturbing the cell pellet. Add 200 $\mu$l of 20 m$M$ Tris-HCl, pH 7.4, containing 10% (v/v) glycerol to the pellet and resuspend well. Measure and record the total volume of the suspension. Transfer exactly 200 $\mu$l of this sample to a 4-mm diameter EPR tube (Wilmad Glass, Buena, NJ), making sure that the cell suspension reaches the bottom of the tube. Immediately freeze samples on dry ice, and store at $-70°$ until EPR measurements can be performed. The volume of culture used, the final $OD_{600}$ of the culture, and the total volume of EPR sample before transferring of the 200 $\mu$l to the EPR tube must be recorded for calculation purposes.

*Whole-Cell Low-Temperature Fe(III) Electron Paramagnetic Resonance*

EPR spectra of the "free" and desferrioxamine-chelated iron(III) are measured by Fe(III) EPR. We use a Bruker (Billerica, MA) X-band spectrometer (model ESP300E), although other instruments can be used as well. Samples are maintained at $-125°$ during the recording of the spectra, using a finger Dewar (Wilmad Glass) filled with liquid nitrogen. Before running samples each day, the spectrum of an iron standard of known concentration (31 $\mu M$ in our case) is recorded and the EPR instrument is tuned and the position of the finger Dewar adjusted to maximize the signal intensity. The iron standard is prepared by taking a solution of ferric sulfate of known concentration and diluting it in 20 m$M$ Tris-HCl buffer, pH 7.4, containing 1 m$M$ desferrioxamine to make a solution approximately 0.1 m$M$ in Fe(III). The exact concentration of this solution is then determined by measuring the absorbance of an aliquot of this solution at 420 nm (extinction coefficient = 2865 $M^{-1}$ cm$^{-1}$). This concentrated solution is further diluted in 20 m$M$ Tris-HCl–10% (v/v) glycerol, to make a standard of the desired concentration. The same iron standard can be used over and over again by storing at $-70°$. It is a good idea to acquire spectra for this standard several times over the course of the day to test the performance of the instrument. To compare the values obtained for different samples with the standard it is important to keep all EPR parameters unchanged, including factors such as number of scans and instrument gain. After inserting each sample, the instrument is autotuned and acquisition is started.

Parameters used for whole cell low-temperature Fe(III) EPR are as follows: frequency, 9.27 GHz; center field, 1500 G; sweep width, 500 G; microwave power, 20 mW; attenuation, 10 dB; modulation amplitude, 20.0 G; modulation frequency, 100 kHz; receiver gain, $1 \times 10^5$; sweep time, 41.9 sec; time constant, 20.48 msec; conversion time, 10.24 msec; resolution, 4096 points; number of scans, 16.

*Electron Paramagnetic Resonance Data Processing and Calculations*

EPR data processing is done with the Bruker WinEPR program. Spectra of samples and the iron standard are filtered to decrease noise and the baseline is corrected. The signal at $g = 4.3$ is quantitated by double integration, using the same software, and the double integral (DI) value is noted.

The total free iron (in moles) in the resuspended sample is calculated as follows:

$$\text{"Free" Fe} = \left(\frac{DI_x [Fe]_{std}}{DI_{std}}\right) V_{resusp}$$

where $DI_x$ is the double integral of the sample, $DI_{std}$ is the double integral of the standard, and $V_{resusp}$ is the volume of the whole sample (of which 200 $\mu$l is transferred to the EPR tube).

The number of cells in the resuspended sample is calculated from the $OD_{600}$ of the culture when it is harvested, the volume in ml of culture used ($V_{cult}$), and the conversion factor $1 \times 10^7$ cells per OD of 1.

$$\text{No. of cells} = OD_{600} \times 10^7 \times V_{cult}$$

Substituting these numbers in the following equation gives the intracellular concentration of EPR-detectable iron ([free Fe]). We assume that there is little or no iron in the suspension buffer, because of the washing steps, so any iron in the sample is located inside the cells. The volume of a single haploid yeast cell is 70 fl; that of a single diploid cell is about 150 l.[23]

$$[\text{Free Fe}]_{\text{intracellular}} = \frac{\text{free Fe/no. of cells}}{\text{single-cell volume}}$$

We hope that this method proves useful for studies of superoxide stress, other kinds of oxidative stress, and iron metabolism.

Acknowledgments

We express our sincere gratitude to Dr. James A. Imlay for helping us initiate these studies of the yeast system, and to Drs. Barney Bales and Miroslav Peric for their help with EPR. This research was supported by NIH Grant DK46828 to Dr. Joan S. Valentine.

# [19] Transgenic Superoxide Dismutase Overproducer: Murine

*By* SERGE PRZEDBORSKI, VERNICE JACKSON-LEWIS, DAVID SULZER, ALI NAINI, NORMA ROMERO, CAIPING CHEN, and JULIA ARIAS

Introduction

Reactive oxygen species (ROS), such as superoxide and hydroxyl radicals, are produced constantly during normal cellular metabolism, and presumably even more so in a number of pathological conditions. Defense mechanisms, however, exist to limit the levels of ROS and the damage they inflict on cellular components such as lipids, proteins, and DNA. It has been hypothesized that the finely tuned balance between the production of ROS and their destruction is skewed in a number of diseases and in normal aging, resulting in oxidative damage that leads to severe cellular dysfunction and, ultimately, to cell death.[1]

Among the many ROS-scavenging enzymes, superoxide dismutase (SOD) has received the lion's share of attention not only because of its key role in ROS metabolism, but also because mutations in a member of the SOD family can cause an inherited form of the paralytic disorder amyotrophic lateral sclerosis (ALS). SOD comes in three isoforms: two copper, zinc SODs (cytosolic SOD1 and extracellular SOD3) and a manganese SOD (mitochondrial SOD2).[2] All three

[1] S. Przedborski and V. Jackson-Lewis, in "Free Radicals in Brain Pathophysiology" (G. Poli, E. Cadenas, and L. Packer, eds.), p. 273. Marcel Dekker, New York, 2000.
[2] I. Fridovich, *Annu. Rev. Biochem.* **64**, 97 (1995).

isoforms catalyze the same reaction, namely the dismutation of superoxide radicals, with comparable potency but in different cellular compartments, thus providing a tight and comprehensive protection against ROS attack.[2] Nullifying SOD activity in knockout mice has been illuminating in that the ablation of SOD1, SOD2, and SOD3 has revealed the relative importance of these three isoenzymes with respect to cell viability. The reverse experiments, that is, the overexpression of SOD isoenzymes in transgenic mice, have also generated invaluable information about the role of ROS by demonstrating that "extra" SOD activity confers greater resistance to oxidative-mediated insults. In this article we summarize information about transgenic mice expressing wild-type SOD and review several practical methods used in studies involving transgenic SOD mice.

## Transgenic Superoxide Dismutase Mice

To our knowledge, the first transgenic SOD mice produced were those expressing human SOD1. These mice were developed not to explore the effects of ROS but rather to unravel the role of increased SOD1 activity in the neurobiology of Down's syndrome.[3] In this disease, affected children are dysmorphic and mentally retarded, and have higher SOD1 activity due to an additional copy of chromosome 21 (or long arm of chromosome 21), which carries the SOD1 gene. Transgenic SOD1 mice have also been extensively used as controls for transgenic mice expressing mutant SOD1 (mSOD1).[4] In this mouse model of ALS, mSOD1 protein retains significant catalytic activity, which raises the following dilemma: Is neurodegeneration in transgenic mSOD1 mice due to the mutant protein or to the increased SOD1 activity? The transgenic wild-type SOD1 mouse is an ideal tool with which to explore this problem. In our opinion, however, the most significant contributions of transgenic SOD1 mice have been and still are toward elucidating the role of ROS in a variety of models of pathological conditions such as stroke, 1-methyl-4-phenyl-1,2,3,6-tetrahydropyridine (MPTP) intoxication, cold-induced brain edema, drug-induced diabetes, and myocardic ischemia–reperfusion injury. Similarly, transgenic SOD2 and SOD3 mice have proved to be of the utmost importance in exploring the importance of ROS in different experimental models of diseases. Unquestionably, these different lines of transgenic mice provide precious tools with which to test the effects of ROS under a given experimental condition and to allow dissection of the actual compartment (i.e., intracellular vs.

---

[3] C. J. Epstein, K. B. Avraham, M. Lovett, S. Smith, O. Elroy-Stein, G. Rotman, C. Bry, and Y. Groner, *Proc. Natl. Acad. Sci. U.S.A.* **84,** 8044 (1987).

[4] M. E. Gurney, H. Pu, A. Y. Chiu, M. C. Dal Canto, C. Y. Polchow, D. D. Alexander, J. Caliendo, A. Hentati, Y. W. Kwon, H.-X. Deng, W. Chen, P. Zhai, R. L. Sufit, and T. Siddique, *Science* **264,** 1772 (1994).

extracellular and cytosolic vs. mitochondrial) within which ROS-mediated injury originates or culminates.

## Characteristics of Superoxide Dismutase Transgenic Mice

Currently, well-characterized and viable lines of transgenic mice expressing each of the three eukaryotic SOD isoenzymes are available.[3-6] To date, only a few lines of transgenic SOD1 mice are obtainable from commercial sources (e.g., Jackson Laboratories, Bar Harbor, ME), whereas all others can be acquired from their respective founding laboratories. Most have been produced by using purified human genomic DNA fragments containing the *SOD1*, *SOD2*, or *SOD3* gene or human *SOD1*, *SOD2*, or *SOD3* cDNA, most usually driven by promoters such as normal human SOD or $\beta$-actin promoter, which assure robust and generalized expression of SOD. In the transgenic SOD1 mice developed by C. Epstein and collaborators,[3] we have previously found that all organs, with the exception of the liver, contain significantly higher SOD1 activity compared with their nontransgenic littermates. In the brains of these transgenic mice,[7] whereas overall SOD1 activity is markedly increased, substantial regional variations exist: the highest activity is found in the hippocampus and striatum whereas the lowest is found in the cerebellum. It is also useful to know that in these mice, expression of the human SOD1 transgene is already detectable by embryonic day 15 and remains stable throughout aging, at least until 1.5 years of age. More rare are lines of transgenic SOD mice in which the transgene is driven by a tissue-specific promoter, which provides human SOD expression restricted to specific organs or cell types.[8] Although most researchers will welcome generalized expression of the SOD transgene, in certain situations expression restricted to a specific type of cells may be more suitable. There are at least two other factors that must be taken into account when selecting a line of transgenic SOD mice. First, most transgenic SOD mice have different genetic backgrounds. Because the genetic background may affect the outcome of an experiment, it is essential that, before selecting a line of transgenic SOD mice, its genetic background be established as suitable for the planned investigation. Second, the level of transgenic expression among lines can vary tremendously. In most of the available lines of transgenic SOD mice, there is a good correlation between the number of copies of the transgene and the magnitude of increased

---

[5] T. D. Oury, Y. S. Ho, C. A. Piantadosi, and J. D. Crapo, *Proc. Natl. Acad. Sci. U.S.A.* **89**, 9715 (1992).

[6] H. C. Yen, T. D. Oberley, S. Vichitbandha, Y. S. Ho, and D. K. St. Clair, *J. Clin. Invest.* **98**, 1253 (1996).

[7] S. Przedborski, V. Jackson-Lewis, V. Kostic, E. Carlson, C. J. Epstein, and J. L. Cadet, *J. Neurochem.* **58**, 1760 (1992).

[8] R. J. Folz, A. M. Abushamaa, and H. B. J. Suliman, *J. Clin Invest.* **103**, 1055 (1999).

SOD activity. Yet, there is a wide range of transgene copy number and increased SOD activity among transgenic SOD mice. For example, among the transgenic SOD1 mouse lines, the increase in enzymatic activity in the brain varies from 1.5- to 10-fold. The problem here is that while a 2- to 3-fold increase in SOD1 may be beneficial, greater increases may, at least in certain systems such as the neuromuscular junction, have deleterious effects. It is therefore essential to keep in mind that the success of an experiment may rely on the judicious selection of the line that provides the appropriate magnitude of increased SOD expression in the necessary organ and/or cell.

## Genotyping

Usually, experiments involving transgenic mice require that all mice be genotyped at some point. It is our experience that genotyping should occur when the mice are between 2 days and 2 weeks of age. Several laboratories genotype transgenic SOD mice by Southern blot analysis; in our laboratory, however, we obtain reliable and reproducible genotyping results by polymerase chain reaction (PCR), using a pair of primers that specifically amplifies a fragment of the human SOD DNA.[9] We have had excellent and reliable results by amplifying a fragment of exon 4 of the human SOD1 gene, the sequence of which is given in the next section. Conceivably, primers for human-specific sequences from any of the other exons should provide the same good results. Genotyping for human SOD1 can also be successfully achieved by activity gel electrophoresis.[10] As illustrated below, the procedure is straightforward and can provide reliable genotyping information in a few hours. The basic principle of this assay is that, in a nondenaturing gel, SOD1 retains its catalytic activity, and that human SOD1 migrates in acrylamide gel differently than mouse SOD1.

*Polymerase Chain Reaction Method*

The extraction of genomic DNA is performed by a standard method, using about 1 cm of mouse tail. To date, several available commercial kits (e.g., QIAamp DNA Mini kit; Qiagen, Valencia, CA) provide a simple and reliable procedure for this extraction. Fifty microliters of PCR reaction mixture contains 1 $\mu$l of DNA (usually at 100 ng/$\mu$l), 10 pmol of each primer (forward primer, 5'-CAT-CAG-CCC-TAA-TCC-ATC-TGA-3'; reverse primer, 5'-CGC-GAC-TAA-CAA-TCA-AAG-TGA-3'), 200 $\mu M$ dNTP (Pharmacia Biotech, Piscataway, NJ), 1 unit of *Taq*

---
[9] V. Kostic, V. Jackson-Lewis, F. De Bilbao, M. Dubois-Dauphin, and S. Przedborski, *Science* **277**, 559 (1997).
[10] S. Przedborski, V. Kostic, V. Jackson-Lewis, A. B. Naini, S. Simonetti, S. Fahn, E. Carlson, C. J. Epstein, and J. L. Cadet, *J. Neurosci.* **12**, 1658 (1992).

polymerase (GIBCO-BRL, Rockville, MD), and 1× PCR buffer (10 m$M$ Tris-HCl, 1.5 m$M$ MgCl$_2$, 50 m$M$ KCl, pH 8.3). Thermal cycling parameters are: 1 cycle of 3 min at 95°; then 25 cycles consisting of 30 sec at 95°, 1.5 min at 55°, and 45 sec at 72°; followed by 1 cycle of 7-min final extension at 72°. Then, 1/10 of the reaction product is loaded onto a 0.8% (w/v) agarose gel containing ethidium bromide (10 µg/ml) and run at 100 V for 30 min. For each run, normal human and mouse genomic DNA are included as positive and negative controls. Only transgenic mice produce a PCR fragment of 213 bp in length.[9]

*Activity Gel Procedure*

On the day of the assay, 20 µl of blood from the tail of each mouse is collected, using 75-mm heparinized microhematocrit tubes (Clay Adams, Parsippany, NJ) and diluted into 1.5-ml plastic microcentrifuge tubes containing 500 µl of 36 m$M$ phosphate-buffered saline (PBS, pH 7.4). Blood cells are then pelleted by centrifugation at 10,000$g$ for 5 min at 4°. The supernatant is discarded and 100 µl of cold distilled water is added and the mixture is vortexed to lyse the blood cells. Lysate (30 µl) is then mixed with 30 µl of stacking gel buffer (see below for composition) containing 20% (v/v) glycerol and bromphenol blue (200 µg/ml) and loaded onto an acrylamide gel that is prepared as outlined below. Gels and running buffers are without sodium dedocyl sulfate and samples are not heated before electrophoresis. Electrophoresis is performed at 250 V for 3–4 hr or until the loading buffer has migrated 8–10 cm from the stacking gel (i.e., bromphenol blue marker dye has swept through most of the gel). Gently remove the gel and soak it in 0.2% (w/v) nitroblue tetrazolium (NBT) in distilled water for 20 min, followed by immersion for 15 min in a solution containing 28 m$M$ tetramethylethylene diamine (TEMED), 28 µ$M$ riboflavin, and 36 m$M$ potassium phosphate (pH 7.8) for 50 min at 25°. The gel is then placed on a light source (e.g., a regular light box) and illuminated for 5–15 min. During the illumination, the gel becomes uniformly blue except at the position containing the SOD, which remains yellowish (i.e., achromatic zone). As a positive control, we routinely include in each gel a lane with human blood lysate. Pups with an extra SOD enzymatic activity band on the nondenaturing gel that comigrated with the human SOD1 enzyme are designated transgenic; nontransgenic mice show the mouse SOD1 enzymatic activity band only.

Below are the reagents and solutions needed for the activity gel. Unless stated otherwise, all the reagents are obtenable from any standard commercial source and should have the highest purity. Although any similar gel electrophoresis apparatus is suitable, we use the Bio-Rad (Hercules, CA) Protean II xi Cell with a long glass plate of 18 × 20 cm and a short glass plate of 16 × 20 cm. Glass plates should be sealed with 0.5% (w/v) agarose. Test the sealing by pouring water through the plates to check for leaks.

Separation gel buffer: Prepare by dissolving 18.15 g of Tris base in distilled water, adjusting to pH 8.8 with 12 $M$ HCl, and then making up the volume to 100 ml with water

Stacking gel buffer: Prepare by dissolving 6.05 g of Tris base in distilled water, adjusting to pH 6.8 with 12 $M$ HCl, and then making up the volume to 100 ml with water

Separation gel [10% (w/v) acrylamide solution]: To prepare 40 ml, mix thoroughly 10 ml of separation gel buffer, 16.50 ml of distilled water, 13.25 ml of 30% (w/v) acrylamide in water, 0.15 ml of 10% (w/v) ammonium persulfate (APS) in water, and 0.10 ml of concentrated TEMED, and then cast the gel by pouring the solution between the two glass plates. A 20% space at the top must be left for the stacking gel. Let the gel polymerize for 15 min

Stacking gel [4.5% (w/v) acrylamide solution]: To prepare 15 ml, mix thoroughly 3.80 ml of stacking gel buffer, 8.80 ml of distilled water, 2.25 ml of 30% (w/v) acrylamide in water, 0.05 ml of 10% (w/v) APS, and 0.10 ml of TEMED.

*Note.* SOD1 is a homodimer; it is thus usual to see on the activity gel from transgenic SOD1 mouse tissue three bands: mouse homodimer, human–mouse heterodimer, and human homodimer.[10] When blood is used, the presence of hemoglobin masks the mouse–human heterodimer and only human homodimer (below hemoglobin) and mouse homodimer (above hemoglobin) are seen; if desired, hemoglobin can be eliminated from samples by ethanol–chloroform extraction.[11]

Superoxide Dismutase Activity

Often, the next steps in a study using transgenic SOD mice are to confirm that SOD activity is indeed increased in the regions of interest and to demonstrate whether it is modified by the experimental factor under investigation. Some researchers have addressed these questions by using activity gels. Although feasible, great care must be taken when using SOD activity gels for quantitative purposes. In our hands, more reliable and reproducible quantitative data have been generated by spectrophotometric assays such as that developed specifically for biological samples by Spitz and Oberley.[12] Using this method, we reproducibly showed that SOD1 activity is 3-fold higher in midbrain homogenates from SF-218 transgenic SOD1 mice (267.6 ± 17.5 U/mg protein, $n = 3$) than in their nontransgenic littermates (84.7 ± 10.7 U/mg protein; $n = 3$; $p = 0.0009$).

[11] S. Przedborski, D. Donaldson, P. L. Murphy, O. Hirsch, D. J. Lange, A. B. Naini, D. McKenna-Yasek, and R. H. Brown, Jr., *Neurodegeneration* **5**, 57 (1996).
[12] D. R. Spitz and L. W. Oberley, *Anal. Biochem.* **179**, 8 (1989).

Outlined below is the procedure for quantifying SOD1 and SOD2 activity in biological fluids and tissue samples.

*Procedures*

For the SOD isoenzyme activity assay, frozen samples are homogenized in 5 volumes of buffer [10 m$M$ Tris-HCl (pH 7.4), 250 m$M$ sucrose]. Homogenates are centrifuged at 2000$g$ for 5 min at 4°. Pellets are discarded and supernatants are centrifuged at 15,000$g$ for 10 min at 4°C. Supernatants of the second centrifugation are stored at $-80°$ until assayed for cytosolic SOD. Pellets are gently resuspended and used for mitochondrial purification, using a Ficoll gradient as previously described.[13] At the end of the purification procedure, mitochondria are resuspended and subjected to three cycles of freeze–thawing before centrifugation at 100,000$g$ for 1 hr at 4°. Supernatants are stored at $-80°$ until they are assayed for mitochondrial SOD.

Total and SOD2 activities are determined by the spectrophotometric method described by Spitz and Oberley.[12] This method is based on the reduction of NBT by superoxide radicals monitored at 560 nm at 25°; xanthine–xanthine oxidase is utilized to generate a superoxide radical flux. The reaction mixture (total volume, 1 ml) contains 0.15 m$M$ xanthine, 0.6 m$M$ NBT, 1 m$M$ diethylenetriaminepentaacetic acid, 1 unit of catalase, 0.13 mg of defatted bovine serum albumin (BSA), 0.25 m$M$ bathocuproinedisulfonic acid, and 100 $\mu$l of sample. After a 3-min preincubation, the reaction is initiated by the addition of xanthine oxidase [amount necessary to achieve a reference rate (i.e., the rate of NBT reduction in the absence of tissue) of 0.02; i.e., $\Delta A_{560}$ of 0.2/min]. One unit of SOD activity is defined as the amount of enzyme that inhibits the reaction rate by 50%. SOD2 activity is distinguished from that of cyanide-sensitive SOD1 by the inclusion of 5 m$M$ NaCN in the reaction mixture; the sample is incubated in the reaction mixture at 25° for 30 to 60 min before the reaction is started, to ensure complete inhibition of the SOD1. SOD1 activity is calculated by subtracting cyanide-resistant SOD activity from total SOD activity.

*Note.* In most tissues, the contribution of SOD3 to cyanide-sensitive SOD activity is less than 10% and thus does not represent a significant problem in measuring SOD1. However, if SOD3 activity is to be assessed or eliminated, it may be separated from SOD1 and SOD2 by passing the homogenate or fluid through a concanavalin A-Sepharose column and, if necessary, eluted from the column by $\alpha$-methylmannoside as described by Marklund.[14]

---

[13] S. Przedborski, V. Jackson Lewis, U. Muthane, H. Jiang, M. Ferreira, A. B. Naini, and S. Fahn, *Ann. Neurol.* **34,** 715 (1993).
[14] S. L. Marklund, *Methods Enzymol.* **186,** 260 (1990).

## Whole Mouse versus Cell Culture

For certain experiments, it is more suitable to use a complex system as provided by a living mouse. In this context, it is worth mentioning that several findings with far-reaching implications have been generated in transgenic SOD mice subjected to various insults produced either by toxins or by mechanical injuries. In these specific cases, the impact of these findings would not have been as significant had they been performed in *in vitro* systems. Conversely, certain investigations require a simplified framework in order to eliminate confounding factors. For instance, we have previously demonstrated in primary neuronal cultures grown from transgenic SOD1 mice that increased SOD1 activity promotes dopaminergic neuronal survival and nerve fiber growth[15] and confers greater resistance to ROS-mediated injury.[16] It is unlikely that such striking and unequivocal results could have been obtained by using a system other than this type of primary cell culture.

In these studies, we first used immunohistochemistry for tyrosine hydroxylase (TH), the rate-limiting enzyme in dopamine synthesis, in order to define a set of anatomical dopaminergic landmarks to assure reproducible ventral mesencephalic dissection in mouse pups. We also found that optimal growth of mouse postnatal dopaminergic neurons is obtained with tissue originating from mouse pups no older than 2 days of age. Under these conditions, in contrast to embryonically derived midbrain cultures, which usually contain $\sim$1% TH-positive neurons, our culture system contains $\sim$40% dopaminergic neurons.[15]

Transgenic and nontransgenic cultures were plated onto cortically derived astrocyte monolayers at the same density (80,000 cells per culture). At 24 hr after plating, both genotypes had the same number of TH-positive neurons. Thereafter, the number of both transgenic and nontransgenic midbrain neurons significantly decreased. However, the number of transgenic TH-positive neurons decreased more slowly than the number of nontransgenic neurons, and by 4 weeks, there were two to three times more TH-positive neurons in transgenic as compared to nontransgenic cultures.[15]

Modulation of SOD activity in cell cultures modifies the magnitude of death by apoptosis. In examining our cultures, we found that a small fraction of the dying cells underwent apoptosis in both groups. However, throughout the entire experimental period, the percentage of apoptotic cells was significantly smaller in transgenic compared to nontransgenic cultures.[15] Below are outlined the different key methods necessary to prepare our mouse postnatal midbrain cultures and to assess neuronal viability and death.

---

[15] S. Przedborski, U. Khan, V. Kostic, E. Carlson, C. J. Epstein, and D. Sulzer, *J. Neurochem.* **67,** 1383 (1996).

[16] M. A. Mena, U. Khan, D. M. Togasaki, D. Sulzer, C. J. Epstein, and S. Przedborski, *J. Neurochem.* **69,** 21 (1997).

## Animals

For these studies, we have used hemizygote male (aged 2 to 8 months) SF-218 mice (C57/BL × SJL/J from the laboratory of C. J. Epstein) as breeding pairs to produce transgenic and nontransgenic offspring. The day of birth is counted as postnatal day 0. On postnatal day 1, 20 μl of blood from the tail is collected from each mouse pup and used for SOD activity gel electrophoresis as described above.

## Cell Culture Preparation

Neonatal mouse culture procedures have been adapted from techniques developed for neonatal rats.[17] Polystyrene petri dishes (50 mm; Falcon, Becton Dickinson Labware, Lincoln Park, NJ) are punched with a 1.0-cm-diameter stamp, and glass coverslips (Carolina Biological Supply, Burlington, NC) are glued to the bottom surface to provide a well. Coverslips are coated with poly-D-lysine (100 μg/ml) and, on the day of plating, recoated with laminin (10 μg/ml) in minimal essential medium [chemical components from Sigma (St. Louis, MO) except where noted].

Astrocyte monolayers are prepared from the rostral half of the cerebral cortex in ice-cold calcium- and magnesium-free phosphate-buffered saline and cultured for 2–3 weeks before plating neurons. In our standard laboratory protocol, we generally use mouse or rat cortex for astrocytes; we have found that the mouse neurons survive well on rat astrocytes.

For production of glial monolayers, the dissociated cortex is plated at a density of 150,000 cells/well and fed with medium containing 90% (v/v) minimal essential medium, 10% (v/v) calf serum, 0.33% (w/v) glucose, bovine pancreatic insulin (5 μg/ml), 500 μ$M$ glutamine, penicillin (6 U/ml), and streptomycin (60 μg/ml). The medium is replaced 24 hr postplating with ice-cold fresh medium. At 4–5 days postplating, 5-fluorodeoxyuridine (11 μ$M$ with 55 μ$M$ uridine) is added to prevent the proliferation of oligodendrocytes.

On postnatal day 2, mouse pups are anesthetized by intraperitoneal injection of ketamine (10 μl of a 100-mg/ml solution; Fort Dodge Laboratories, Fort Dodge, IA), followed by cooling on ice. Chunks of tissue from the designated ventral mesencephalon are dissociated with papain (10 U/ml; Worthington, Freehold, NJ) in 0.88 m$M$ cysteine, 0.01% (w/v) phenol red, 500 μ$M$ kynurenate, 116 m$M$ NaCl, 5.4 m$M$ KCl, 26 m$M$ NaHCO$_3$, 2 m$M$ NaH$_2$PO$_2$, 1 m$M$ MgPO$_4$, 500 μ$M$ EDTA, 25 m$M$ glucose, pH 7.3. Dissociation is performed under 5% CO$_2$/95% O$_2$ at 36° on a magnetic stirrer. By following anatomical landmarks verified by TH immunostaining, a block of tissue containing the ventral tegmental area and

---

[17] S. Rayport, D. Sulzer, W.-X. Shi, S. Sawasdikosol, J. Monaco, D. Batson, and G. Rajendran, *J. Neurosci.* **12**, 4264 (1992).

the substantia nigra is dissected. The borders of the block are defined dorsally by a slice halfway between the midbrain flexure and the cerebral aqueduct, rostrally by the caudal end of the mammillary bodies, and caudally by the juncture of the pons and midbrain flexure. The tissue is dissociated as described above and 80,000 cells are plated per well. The growth medium contains 47% (v/v) minimal essential medium, 40% (v/v) Dulbecco's modified Eagle's medium, 10% (v/v) Ham's F12 nutrient medium, glucose (3.4 mg/ml), 0.25% (w/v) albumin, 500 $\mu M$ glutamine, transferrin (100 $\mu$g/ml), 15 $\mu M$ putrescine, 30 n$M$ Na$_2$SeO$_3$, 30 n$M$ triiodothyronine, insulin (25 $\mu$g/ml), 200 n$M$ progesterone, 125 n$M$ cortisol, SOD1 (5 $\mu$g/ml), catalase (10 $\mu$g/ml), 500 $\mu M$ kynurenate, and 1% (v/v) supplemented heat-inactivated calf serum (HyClone, Logan, UT). In our current laboratory protocols, we add, once, glial-derived neurotrophic factor (Intergen, Purchase, NY) at 10 ng/ml, 30 min after plating; this promotes survival, neurite outgrowth, and dopamine release from TH-positive neurons.[18] At 24 hr, cultures are exposed to 5-fluorodeoxyuridine as described above. Cultures are maintained at 36° in 5% CO$_2$.

*Immunostaining*

SOD-transgenic and nontransgenic cultures are fixed with 4% (w/v) paraformaldehyde.[15] After several washes in 0.1 $M$ Tris-HCl (pH 7.4) containing NaCl at 9 g/liter (TBS), cultures are incubated for 48 hr in a humid chamber at 4° with one of the following mixtures of antibodies in TBS containing 0.1% (v/v) Triton X-100 (TBS-Tx), 1% (v/v) normal goat serum, and 1% (v/v) normal horse serum: (1) mouse monoclonal anti-TH antibody (1 : 1000; Chemicon, Temecula, CA) and rabbit polyclonal anti-$\gamma$-aminobutyric acid (GABA) antibody (1 : 1000, Sigma), (2) mouse monoclonal anti-human SOD1 antibody (1 : 1000, Sigma) and rabbit polyclonal anti-TH antibody (1 : 1000, Eugene Tech, Ridgefield Park, NJ), or (3) mouse monoclonal anti-human SOD1 antibody and rabbit polyclonal anti-GABA antibody. After several washes in TBS-Tx, dishes are incubated for 1 hr at 25° with a mixture of goat anti-rabbit antibody conjugated to fluorescein (1 : 100; Vector Laboratories, Burlingame, CA) and horse anti-mouse antibody conjugated to Texas Red (sulforhodamine 101, 1 : 100; Vector) in TBS. To further study human SOD1 distribution in the cell cultures, instead of Texas Red-conjugated antibody, in some experiments, we use biotinylated–conjugated polyclonal horse anti-mouse antibody (1 : 200; Vector), and a horseradish-conjugated avidin–biotin complex (Vector); immunostaining is then revealed with diaminobenzidine–H$_2$O$_2$. To assess the specificity of TH, GABA, and Human SOD1 immunostaining, the primary or the secondary antibody is omitted. None of these conditions are associated with identifiable specific immunofluorescence.

[18] R. E. Burke, M. Antonelli, and D. Sulzer, *J. Neurochem.* **71,** 517 (1998).

## Cell Death Quantification and Morphological Characterization

To quantify the number of living and dead cells, we use the membrane-impermeant DNA dye ethidium homodimer to identify cells in which plasma membrane integrity is disrupted. The membrane-permeant dye calcein-AM is used to label live cells; it is metabolized by cytoplasmic esterases and yields the membrane-impermeant dye calcein, which is thus retained only in cells with an intact plasma membrane. The cultures are briefly washed twice with physiologic saline at 30° and then incubated with 4 $\mu M$ ethidium homodimer and 2.5 $\mu M$ calcein-AM for 10 min at room temperature according to the supplier's recommendations (Molecular Probes, Eugene, OR) before examination with an inverted fluorescence microscope ($\times 200$).

To study the morphology of dying cells *in situ*, we use the membrane-permeant bisbenzimide dye Hoechst 33342 (Molecular Probes), which stains nuclei in both live and dead cells. A concentrated stock (2 mg/ml) is made in water and sterilized by filtration. The stock is diluted 1 : 50 in sterile PBS and 200 $\mu$l is added directly to 4% (w/v) paraformaldehyde-fixed cultures maintained in 2 ml of PBS. Cultures are then incubated with the Hoechst dye for 10 min at room temperature and examined with an inverted fluorescence microscope ($\times 200$–400).

In preliminary studies, we have determined the optimal concentrations of ethidium homodimer, calcein-AM, and Hoechst 33342 to avoid spurious fluorescence. Using the procedure described above, we have found that only negligible autofluorescence is observed and excellent morphological definition of cell bodies and neuronal processes is achieved.

## Acknowledgments

This work was supported by the NIH/NINDS Grants R29 NS37345, RO1 NS38586, NS42269, DA07418 and DA10154, and P50 NS38370, the U.S. Department of Defence (DAMD 17-99-1-9474), the Parkinson's Disease Foundation, the Lowenstein Foundation, the Soldman Foundation, the Muscular Dystrophy Association, the ALS Association, and Project ALS.

## [20] Transgenic and Mutant Mice for Oxygen Free Radical Studies

By TING-TING HUANG, INES RAINERI, FAYE EGGERDING, and CHARLES J. EPSTEIN

### Introduction

Animal models with altered levels of antioxidant enzymes have been widely used to study the roles of oxygen free radicals in tissue injury, degenerative diseases, and the aging process. The most frequently used models include transgenic and mutant animals, with increased and reduced levels, respectively, of copper–zinc superoxide dismutase (Cu,ZnSOD), manganese superoxide dismutase (MnSOD), catalase (CAT), and glutathione peroxidase (GPx-1). The model organisms include *Drosophila*, mouse, and rat, with the mouse being the most widely used. In this article we shall limit our discussion to the methodology used in the creation and characterization of Cu,ZnSOD and MnSOD transgenic and mutant mice and of Cu,ZnSOD transgenic rats.

*Transgenic Mice and Rats*

In designing the constructs to be used to create transgenic animals, the transgenes can be derived from either genomic or cDNA sequences. Either native or heterologous promoters can be used, the latter to achieve tissue-specific or developmental stage-specific expression. Although they can be taken from a wide range of organisms, transgenes of human, mouse, and rat origin are most frequently used. Genomic DNA usually produces more stable messages than does cDNA and hence allows higher level transgene expression. However, some genomic DNAs are large, which may make cloning and production of transgenic animals difficult. By contrast, cDNAs are smaller in size and are easier to handle, but the messages are generally less stable. The stability problem can be circumvented in many instances by the insertion of generic intron sequences into the cDNA.[1-4]

Transgenic mice are usually generated by pronuclear injection (with the designation of TgN for the transgenic strains that are obtained), and the level of transgene

---

[1] T. Choi, M. Huang, C. Gorman, and R. Jaenisch, *Mol. Cell. Biol.* **11**, 3070 (1991).
[2] C. B. Whitelaw, A. L. Archibald, S. Harris, M. McClenaghan, J. P. Simons, and A. J. Clark, *Transgenic Res.* **1**, 3 (1991).
[3] A. J. Clark, A. L. Archibald, M. McClenaghan, J. P. Simons, R. Wallace, and C. B. Whitelaw, *Philos. Trans. R. Soc. Lond. B Biol. Sci.* **339**, 225 (1993).
[4] R. L. Brinster, J. M. Allen, R. R. Behringer, R. E. Gelinas, and R. D. Palmiter, *Proc. Natl. Acad. Sci. U.S.A.* **85**, 836 (1988).

expression is assessed after germ line transmission of the transgene is achieved. Alternatively, it is possible to assess expression levels before the production of transgenic mice by means of an embryonic stem (ES) cell approach (with the designation of TgE). This is done by electroporation of the transgene, together with an antibiotic resistance gene as a selection marker, into mouse ES cells.[5] Provided that the promoter of the transgene is functional in ES cells, clones can be screened for different levels of transgene expression and the desired clones used for transgenic animal production. The *Sod2* transgenic strain TgE(*Sod2*)11Cje was generated by this approach after identification of an ES clone with high-level MnSOD activity. As a result, the resulting transgenic mice have expression levels much higher than found in mice generated by direct pronuclear injection.

The same piece of human genomic (*SOD1*) DNA for Cu,ZnSOD with a 1.5-kb native promoter has been used by several laboratories to achieve ubiquitous transgene expression.[6,7] However, the relative increase in Cu,ZnSOD from the transgene is not necessarily proportional to the endogenous levels. For example, in TgN(*SOD1*)3Cje (previously known as Tg218-3) mice, the increases in Cu,ZnSOD actually are only 60 to 80% in tissues such as liver, kidney, and lung with high endogenous levels. By contrast, activities in tissues with lower endogenous levels, such as brain, heart, skeletal muscle, spleen, and red blood cells, can increase as much as 250%.[8] The discrepancy may be due to the differences in gene regulation between the endogenous gene and the transgenes. Transgenes usually have a relatively short promoter and may not have all the control elements required for regulated gene expression. Therefore, it may be harder for transgenes to compete with endogenous gene for the necessary transcription factors. In addition, transgene insertion into the genome is random, and the copy number varies from one transgenic line to another. Therefore, transgene expression can be affected by the nature of the insertion site, the copy number, and other unknown factors.

Neuron-specific expression of Cu,ZnSOD has also been achieved by replacing the native promoter of *SOD1* with the neuron-specific enolase (NSE) promoter.[9] In addition, after discovery of *SOD1* mutations in a subset of familial amyotrophic lateral sclerosis (ALS) patients, transgenic mice with specific ALS-linked mutations in human *SOD1* transgenes were generated for the study of disease mechanisms.[10–12]

---

[5] A. Gossler, T. Doetschman, R. Korn, E. Serfling, and R. Kemler, *Proc. Natl. Acad. Sci. U.S.A.* **83**, 9065 (1986).
[6] P. H. Chan, M. Kawase, K. Murakami, S. F. Chen, Y. Li, B. Calagui, L. Reola, E. Carlson, and C. J. Epstein, *J. Neurosci.* **18**, 8292 (1998).
[7] C. J. Epstein, K. B. Avraham, M. Lovett, S. Smith, O. Elroy-Stein, G. Rotman, C. Bry, and Y. Groner, *Proc. Natl. Acad. Sci. U.S.A.* **84**, 8044 (1987).
[8] T. T. Huang, E. J. Carlson, A. M. Gillespie, Y. Shi, and C. J. Epstein, *J. Gerontol. A Biol. Sci. Med. Sci.* **55**, B5 (2000).
[9] Y. Li, E. Carlson, K. Murakami, J. C. Copin, R. Luche, S. F. Chen, C. J. Epstein, and P. H. Chan, *J. Neurosci. Methods* **89**, 49 (1999).

MnSOD transgenic mice have been generated by using either mouse genomic (*Sod2*) DNA with its native promoter or human *SOD2* cDNA with a human actin promoter.[13,14] The native promoter leads to ubiquitous expression of the transgene with variable expression levels in different tissues. On the other hand, the actin promoter leads to expression of the *SOD2* cDNA predominantly in the heart and skeletal muscle.[13]

*Mutant Mice*

In addition to the *Sod1* and *Sod2* mutant mice generated in our laboratory, several different lines of *Sod1* and *Sod2* mutant mice have been independently generated by other laboratories.[15–19] These mice differ from one another in several respects, including the region of the gene that is deleted, the nature of the selection markers employed, and the genetic background on which the phenotypes of the mutant mice are ascertained. These factors, especially genetic background effects, probably account for most of the phenotypic discrepancy reported in the literature.

Strain Maintenance

Once transgenic and mutant mice are successfully generated, it is often desirable and standard practice to backcross the mice into different inbred strains to generate congenic mice. By convention, 10 generations of backcrossing are necessary to achieve congenic status. However, by the fifth generation, the mice should have obtained more than 95% (96.875%) of their genetic material from

---

[10] P. C. Wong, C. A. Pardo, D. R. Borchelt, M. K. Lee, N. G. Copeland, N. A. Jenkins, S. S. Sisodia, D. W. Cleveland, and D. L. Price, *Neuron* **14,** 1105 (1995).
[11] M. E. Ripps, G. W. Huntley, P. R. Hof, J. H. Morrison, and J. W. Gordon, *Proc. Natl. Acad. Sci. U.S.A.* **92,** 689 (1995).
[12] M. E. Gurney, H. Pu, A. Y. Chiu, M. C. Dal Canto, C. Y. Polchow, D. D. Alexander, J. Caliendo, A. Hentati, Y. W. Kwon, H. X. Deng, W. Chen, P. Zhai, R. L. Sufit, and T. Siddique, *Science* **264,** 1772 (1994).
[13] H. C. Yen, T. D. Oberley, S. Vichitbandha, Y. S. Ho, and D. K. St. Clair, *J. Clin. Invest.* **98,** 1253 (1996).
[14] Y. S. Ho, R. Vincent, M. S. Dey, J. W. Slot, and J. D. Crapo, *Am. J. Respir. Cell Mol. Biol.* **18,** 538 (1998).
[15] M. M. Matzuk, L. Dionne, Q. Guo, T. R. Kumar, and R. M. Lebovitz, *Endocrinology* **139,** 4008 (1998).
[16] T. Kondo, A. G. Reaume, T. T. Huang, E. Carlson, K. Murakami, S. F. Chen, E. K. Hoffman, R. W. Scott, C. J. Epstein, and P. H. Chan, *J. Neurosci.* **17,** 4180 (1997).
[17] T. T. Huang, M. Yasunami, E. J. Carlson, A. M. Gillespie, A. G. Reaume, E. K. Hoffman, P. H. Chan, R. W. Scott, and C. J. Epstein, *Arch. Biochem. Biophys.* **344,** 424 (1997).
[18] R. M. Lebovitz, H. Zhang, H. Vogel, J. Cartwright, Jr., L. Dionne, N. Lu, S. Huang, and M. M. Matzuk, *Proc. Natl. Acad. Sci. U.S.A.* **93,** 9782 (1996).
[19] Y. S. Ho, M. Gargano, J. Cao, R. T. Bronson, I. Heimler, and R. J. Hutz, *J. Biol. Chem.* **273,** 7765 (1998).

the inbred parent and are generally suitable for phenotype assessment at that point. Expression of a particular mutation or transgene on different genetic backgrounds may sometimes result in markedly different phenotypes, as has been observed, for example, with $Sod2$ mutant mice.[20] This opens the possibility for identifying genetic modifiers[21–25] and helps to further the understanding of the underlying mechanisms of the effects of the genetic alterations.

Frequently used inbred strains include C57BL/6, DBA/2, C3H, and BALB/c. In addition, $F_1$ hybrids from two different inbred strains are frequently useful, as the mice are still genetically uniform while lacking some of the deleterious effects often associated with inbreeding. Sometimes it is also practical to have transgenic and mutant mice on outbred backgrounds, such as CD1. Outbred mice are generally larger in size and more robust than inbred mice. This can facilitate experiments that may require surgical intervention, such as microsurgery, to produce brain ischemia in the study of ischemia–reperfusion injury.[26]

For colony maintenance, hemizygous transgenic and heterozygous mutant mice are bred with either inbred or outbred strains purchased from commercial suppliers. The resulting offspring are genotyped by polymerase chain reaction (PCR) or by enzymatic assays. Inbreeding of transgenic or mutant strains should be avoided for several reasons: unique inbred strains that are no longer genetically comparable with existing inbred stains may be produced; the segregation of recessive traits from inbreeding can be mistaken as a specific phenotype associated with the genetic alteration; and infertility and subsequent loss of the entire stock can occur. Therefore, although it is tempting to generate homozygous transgenic and mutant mice and use them as breeders (provided that the homozygous mice can develop normally into adulthood) to bypass the need for genotyping, this practice is not advisable. In addition to leading to inbreeding, it makes it impossible to have genetically comparable nontransgenic or nonmutant control animals. To avoid these difficulties, homozygous transgenic or mutant mice for experimental purposes, along with the appropriate heterozygous or hemizygous and wild-type controls, should be generated from crosses between hemizygous transgenics or heterozygous mutants.

Multiple strains of transgenic mice differing in copy number, expression levels, and sites of transgene insertion are frequently generated and maintained, and it is often of advantage to use multiple strains for the same study. By using different

---

[20] T. T. Huang, E. J. Carlson, I. Raineri, A. M. Gillespie, H. Kozy, and C. J. Epstein, *Ann. N.Y. Acad. Sci.* **893,** 95 (1999).
[21] E. S. Lander and D. Botstein, *Genetics* **121,** 185 (1989).
[22] E. S. Lander and N. J. Schork, *Science* **265,** 2037 (1994).
[23] A. Darvasi, *Mamm. Genome* **8,** 163 (1997).
[24] W. G. Hill, *Genetics* **148,** 1341 (1998).
[25] A. M. Beebe, S. Mauze, N. J. Schork, and R. L. Coffman, *Immunity* **6,** 551 (1997).
[26] H. Kinouchi, C. J. Epstein, T. Mizui, E. Carlson, S. F. Chen, and P. H. Chan, *Proc. Natl. Acad. Sci. U.S.A.* **88,** 11158 (1991).

transgenic strains with similar levels of transgene expression, the possibility that the observed phenotypic effects are related to the site of transgene insertion rather than to transgene expression can be ruled out. Furthermore, by using several transgenic strains spanning a wide range of expression levels, information regarding threshold and dosage effects of transgene expression can be obtained.

In contrast to multiple transgenic strains, a single mutant strain is usually propagated for analysis of phenotype. Nevertheless, it should be a general practice to inject multiple independent mutant ES clones for the generation of mutant mice. Chimeric mice derived from independent ES clones should be bred for germ line transmission and initial phenotype assessment. This practice will help to rule out epigenetic changes or *de novo* mutations inherited from the culturing of ES cells *in vitro*.

## Genotyping of Transgenic and Mutant Mice

Genotyping of transgenic and mutant mice is crucial for colony maintenance and generation of animals for study. Important information regarding the mode of genetic segregation and the male-to-female ratio can be obtained from routine genotyping. Furthermore, litter size should be monitored. Abnormality in one of these parameters may signal pre- or perinatal lethality or some other interesting phenotype associated with developing fetuses. Decreasing litter size with breeding to a inbred strain may also signify the development of parental infertility.

PCR and Southern blot DNA hybridization are the techniques most commonly used for genotyping transgenic and mutant animals. However, activity gel assays relying on the differing electrophoretic mobilities of Cu,ZnSOD from human, mouse, and rat are effective in identifying Cu,ZnSOD transgenic animals and are useful for quantitating the relative expression of the endogenous genes and transgenes. The assay methods commonly used in our laboratory for genotyping are listed in Table I, and the detailed procedures for nondenaturing polyacrylamide gel electrophoresis (nondenaturing PAGE) and nondenaturing isoelectric focusing (nondenaturing IEF) are described.

TABLE I
GENOTYPING METHODS FOR SUPEROXIDE DISMUTASE
TRANSGENIC AND MUTANT ANIMALS

| Mouse strain | Method |
| --- | --- |
| Cu,ZnSOD transgenic mice | Nondenaturing PAGE |
| Cu,ZnSOD transgenic rats | Nondenaturing IEF, pH 4.0–6.5 |
| Cu,ZnSOD mutant mice | PCR and nondenaturing PAGE |
| MnSOD transgenic mice | PCR |
| MnSOD mutant mice | PCR |

Although all the methods listed below are quick and relatively simple, none is effective for identifying homozygous transgenic mice. The genomic sequences encompassing the transgenes are often unknown, making it impossible to design a PCR or Southern blot protocol to distinguish between hemizygous and homozygous transgenic mice. As a consequence, either chromosome fluorescence *in situ* hybridization (FISH), using the transgene as the DNA probe, or test breeding is usually the only definitive method for identifying homozygous transgenic mice.[27] Measurement of enzyme activities in the blood or other tissues can often be helpful but is not always reliable.

*Nondenaturing Polyacrylamide Gel Electrophoresis*

Nondenaturing PAGE is used to identify Cu,ZnSOD transgenic mice by taking advantage of the different migration rates of human and mouse Cu,ZnSOD on a nondenaturing gel. Because Cu,ZnSOD is a dimeric protein, the homodimers of human and mouse Cu,ZnSOD and the heterodimer of human–mouse Cu,ZnSOD can be separated from each other and visualized as three distinctive bands in the gel. The gel system is also used for identification and confirmation of $Sod1^{-/-}$ mice. To separate human and rat Cu,ZnSOD, nondenaturing IEF is used because nondenaturing PAGE cannot resolve these two species.

*Sample Preparation.* Collect tail blood into 20–40 $\mu$l of a solution consisting of 1 m$M$ EDTA–0.5% (v/v) Nonidet P-40 (NP-40). Although the absolute amount of blood is not critical, a bright red color needs to be achieved to have sufficient Cu,ZnSOD for the gel. This amounts to roughly 5 to 10 $\mu$l of whole blood. Infant mice can be easily bled from their tails between 7 and 10 days of age, and it is often helpful to keep the pups warm during the process.

*Acrylamide Gels.* Samples can be run in 4.5% (w/v) stacking and 10% (w/v) separation gels measuring 15 × 15 × 0.075 cm. In addition, precast gels are readily available from several companies and are convenient alternatives. The buffers and the gel composition are as follows.

Separation gel buffer: 375 m$M$ Tris base, pH 8.8; prepare a stock as 4×
Stacking gel buffer: 125 m$M$ Tris base, pH 6.8; prepare a stock as 4×
Acrylamide (30%, w/v): 0.8 g of bisacrylamide and 29.2 g of acrylamide in 100 ml of $H_2O$
Electrophoresis buffer: 25 m$M$ Tris base and 192 m$M$ glycine
Sample loading buffer (2×): 20% (v/v) glycerol and 0.01% (w/v) bromphenol blue in 2× stacking gel buffer
Separation gel: 4 ml of 4× separation gel buffer, 5.3 ml of 30% (w/v) acrylamide, 6.7 ml of $H_2O$, 60 $\mu$l of 10% (w/v) ammonium persulfate (APS), and 8 $\mu$l of tetramethylethylenediamine (TEMED)

[27] Y. P. Shi, T. T. Huang, E. J. Carlson, and C. J. Epstein, *Mamm. Genome* **5**, 337 (1994).

Stacking gel: 3.75 ml of 4× stacking gel buffer, 2.25 ml of 30% (w/v) acrylamide, 9 ml of $H_2O$, 45 $\mu$l of APS, and 15 $\mu$l of TEMED

*Sample Loading and Gel Electrophoresis.* Mix equal amounts of sample and loading buffer, and load between 10 and 20 $\mu$l per sample depending on the size of the sample wells. Perform electrophoresis at 200–250 V for 1.5 to 2 hr or until the blue dye is near the bottom of the gel.

*Gel Staining Solution.* For every 100 ml of 0.036 $M$ phosphate buffer, pH 7.8, add 1.13 mg of riboflavin (equivalent to 34.5 $\mu M$), 16.4 mg of nitroblue tetrazolium (equivalent to 0.19 m$M$), 20 $\mu$l of 0.5 $M$ EDTA, pH 8.0 (equivalent to 0.1 m$M$), and 175 $\mu$l of TEMED. Prepare the solution immediately before use and protect from light.

*Staining the Gel.* The staining procedure is a minor modification from that described by White et al.[28] Soak the gel in the staining solution with gentle shaking for 10 min in the dark, and then remove it and place it on a clean glass plate. Expose the gel to light on a light box for 5 to 10 min to develop the dye color. Transfer the gel to water to stop the reaction once good contrast is achieved. For archiving purposes, gels can be dried on a gel dryer or the image can be photographed or scanned into a computer.

*Interpreting the Gel.* The excitation of riboflavin by light (catalyzed by TEMED) generates superoxide radicals, which convert the yellow, water-soluble NBT (nitroblue tetrazolium) into the blue, water-insoluble formazan. However, in the regions in which SOD is present, the superoxide radicals are rapidly removed and formazan formation is prevented. SOD is thereby revealed in clear bands on a blue background. Under the conditions specified above, the human homodimer migrates the fastest, followed by human–mouse heterodimer and then by mouse homodimer.

*Nondenaturing Isoelectric Focusing Gel Electrophoresis*

Nondenaturing IEF gel electrophoresis is used for identification of *SOD1* transgenic rats, as well as for estimation of Cu,ZnSOD and MnSOD activities in tissues. Therefore, both procedures are discussed here, and the differences for the two applications are highlighted.

*Sample Preparation.* For identification of transgenic rats, collect tail blood directly into lysis buffer containing 0.1 m$M$ EDTA and 0.5% (v/v) NP-40 as mentioned above. For quantitation of Cu,ZnSOD and MnSOD, homogenize tissues in 4 volumes (1:4, w/v) of 0.05 $M$ phosphate buffer, pH 7.8, using a Tekmar (Cincinnati, OH) tissue homogenizer, followed by three short pulses of sonication

---

[28] C. W. White, D. H. Nguyen, K. Suzuki, N. Taniguchi, L. S. Rusakow, K. B. Avraham, and Y. Groner, *Free Radic. Biol. Med.* **15,** 629 (1993).

to disrupt mitochondria. Spin the samples for 5 min in a microcentrifuge at 4°, 14,000 rpm, and store the supernatants at −80°. To adjust for protein loading on the gels, thaw the tissue samples, dilute an aliquot 1 : 10 in 1% (w/v) glycine, and measure protein concentrations in triplicate 15-$\mu$l volumes, using the bicinchorinic acid (BCA) protein assay reagent (Pierce, Rockford, IL).

*Nondenaturing Isoelectric Focusing Gels.* Precast polyacrylamide gels (Ampholine PAGplate) for analytical isoelectric focusing are purchased from Amersham Pharmacia Biotech (Piscataway, NJ). The pH 4.0–6.5 gels are used for identification of *SOD1* transgenic rats, and pH 3.5–9.5 gels are used for analysis of Cu,ZnSOD and MnSOD activities.

*Gel Electrophoresis.* A Multiphor II electrophoresis unit (Amersham Pharmacia Biotech) connected to a refrigerated circulating water bath with temperature adjusted to 10° is used for electrophoresis. For each transgenic rat sample, mix 10 $\mu$l of blood lysate with 10 $\mu$l of 1% (w/v) glycine and load the entire 20 $\mu$l into a pH 4.0–6.5 gel. Place the sample application strip 2.5 cm from the cathode. Cathode and anode solutions are 0.1 $M$ glutamic acid in 0.5 $M$ $H_3PO_4$ and 0.1 $M$ $\beta$-alanine, respectively. Set the power supply at 2000 V, 25 mA, and run the gel for 2.5 hr. For tissue lysates, dilute each sample to a total volume of 20 $\mu$l at the desired concentration (see Table II) in 1% (w/v) glycine and use pH 3.5–9.5 gels. For optimal resolution, place the application strip 4 cm from the cathode. The cathode and anode solutions are 1 $M$ $H_3PO_4$ and 1 $M$ NaOH, respectively. Set the power supply at 1500 V, 50 mA, and run the gel for 1.5 hr. We routinely evaluate three or four different protein concentrations (at 2-fold dilutions of the highest concentration) for each sample. Because of the wide variation of baseline

TABLE II
SUGGESTED PROTEIN LOADING FOR QUANTITATION BY GEL ELECTROPHORESIS OF Cu,ZnSOD, MnSOD, AND ACONITASE IN TISSUES

| Tissue | Cu,ZnSOD[a] ($\mu$g) | MnSOD[a] ($\mu$g) | Aconitase[a] ($\mu$g) |
|---|---|---|---|
| Brain | 4 | 60 | 20 |
| Heart | 4 | 12 | 10 |
| Liver | 5 | 40 | 40 |
| Skeletal muscle | 8 | 30 | 10 |
| Lung | 2 | 40 | 40 |
| Kidney | 2.5 | 20 | 20 |

[a] The highest concentrations are listed. For example, a serial dilution from the suggested level would result in the loading of 60, 30, and 15 $\mu$g of total protein for quantitation of MnSOD in the brain. Mice younger than 2 weeks may require higher protein loading.

Cu,ZnSOD and MnSOD activities in individual tissues, the optimal amount of protein loading varies significantly among different tissues (Table II). Also, it must be taken into consideration whether young (<3 weeks) or older animals are investigated, as we and others have observed a significant increase in MnSOD activity in mature animals.[29–31]

*Staining and Preservation of Gels.* Gels are placed in staining solution as mentioned previously for 15 min in the dark, and the color reaction is then developed on a light box. To archive, fix gels in 10% (v/v) acetic acid–20% (v/v) methanol for at least 30 min. Next, place the gels in 10% (v/v) methanol–3% (v/v) glycerin for at least 1 hr, cover them with cellophane sheets, and air dry. We capture the gel image with a CCD camera and analyze the intensity of MnSOD and Cu,ZnSOD signals with GelExpert software (Nucleotech, San Mateo, CA). The intensities are then normalized to the amount of protein loaded. Mouse MnSOD runs toward the cathode and Cu,ZnSOD toward the anode. In contrast to MnSOD, Cu,ZnSOD is sensitive to inactivation by KCN. Therefore, 5 m$M$ KCN can be added to the staining solution to differentiate between the two types of SOD. For the interpretation of IEF gels for *SOD1* transgenic rats, nontransgenic rat and human blood samples should be loaded as controls onto the same gel.

## Characterization of Transgenic and Mutant Mice

Cu,ZnSOD and MnSOD transgenic and mutant mice, as well as Cu,ZnSOD transgenic rats, have been used in different experimental systems to study the relationship between superoxide radicals and oxidative tissue injury or cell death. In particular, *SOD1* transgenic mice have been used in a wide range of experimental paradigms. The most important contribution of these mice has been to demonstrate the role of superoxide radicals in tissue injury and degenerative processes. A summary of studies carried out with *SOD1* transgenic mice to show the effects of increased Cu,ZnSOD activity on the response of the CNS to acute oxidative stress can be found in Huang *et al.*[20] A wide variety of related research articles using *SOD* transgenic or mutant mice as the animal model can also be found in the published literature. Here we present some protocols for characterization of *SOD* transgenic and mutant animals that provide a basic understanding of the oxidant/antioxidant state of the animals or of cells derived from the animals.

*Cu,ZnSOD and MnSOD Activities in Different Tissues*

In transgenic animals, it is important to know how much of an increase is achieved in different tissues of any given transgenic line. In our experience, the level

---

[29] G. M. Carbone, D. K. St. Clair, Y. A. Xu, and J. C. Rose, *Pediatr. Res.* **35**, 41 (1994).
[30] Y. Chen and L. Frank, *Pediatr. Res.* **34**, 27 (1993).
[31] I. Raineri, unpublished data (2001).

of transgene expression may vary with age and genetic background. Therefore, it becomes important to determine the activity levels as transgenic mice are gradually backcrossed to different inbred strains. In the mutant animals, it is important to know whether the activity of the mutant gene is completely knocked out and whether the remaining copy is upregulated in heterozygous mutant animals under different experimental paradigms.

In addition to the established kinetic assays,[32,33] a rapid way to assess Cu,ZnSOD and MnSOD activities in different tissues is by activity gel electrophoresis. Using nondenaturing IEF, only small amounts of tissue lysates are required and multiple samples can be accommodated in the same gel. The procedure is described above.

*Aconitase Activity*

Because of its sensitivity to direct inactivation by superoxide radicals, aconitase activity is used as an indicator of increased intracellular oxidative stress.[34] There are two forms of aconitase: the cytoplasmic form, which is also an iron response protein (IRP-1), and the mitochondrial form, which is one of the enzymes in the tricarboxylic acid (TCA) cycle. Because the two aconitases are compartmentalized in different locations, changes in the activities can also indicate where the superoxide radicals are being generated. The two forms of aconitase can be differentiated by fractionation of tissue or cell lysates or by electrophoresis. The method presented below is a modification of a previously described procedure[35,36] with multiple concentrations of protein loading of each sample for assessing relative aconitase activities. The procedure requires relatively small amounts of tissues and bypasses laborious tissue fractionation process. This is especially important when the amounts of tissues or cells are limited in quantity.

*Sample Preparation.* Homogenize tissues in 3 volumes (1 : 3, w/v) of lysis buffer containing 50 m$M$ Tris-HCl (pH 8.0), 0.6 m$M$ MnCl$_2$, and 2 m$M$ citric acid. We use a hand-held tissue grinder [Kontes (Vineland, NJ) microscale tissue grinder] to gently homogenize the tissues. Transfer tissue lysates into 1.5-ml microcentrifuge tubes and freeze–thaw three times between liquid nitrogen and water at room temperature to break open the mitochondrial membranes. Centrifuge the lysates at 10,000$g$ at 4° for 10 min, transfer the supernatants to new tubes, measure protein concentrations, and store at −80°. The tissue lysates can be thawed and used for aconitase gel or enzyme assays for specific activities. For quantitation, serial 1 : 2 dilutions of tissue lysates with lysis buffer are carried out to yield three

---

[32] J. D. Crapo, J. M. McCord, and I. Fridovich, *Methods Enzymol.* **53**, 382 (1978).
[33] L. W. Oberley and D. R. Spitz, *Methods Enzymol.* **105**, 457 (1984).
[34] P. R. Gardner and I. Fridovich, *J. Biol. Chem.* **266**, 19328 (1991).
[35] A. L. Koen and M. Goodman, *Biochim. Biophys. Acta* **191**, 698 (1969).
[36] R. K. Selander, D. A. Caugant, H. Ochman, J. M. Musser, M. N. Gilmour, and T. S. Whittam, *Appl. Environ. Microbiol.* **51**, 873 (1986).

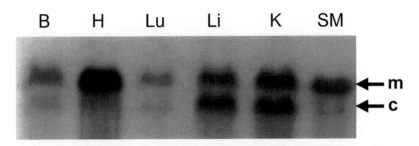

FIG. 1. Electrophoretic pattern of aconitases in mouse tissues. B, brain; H, heart; Lu, lung; Li, liver; K, kidney; SM, skeletal muscle; c, cytoplasmic aconitase; m, mitochondrial aconitase.

different concentrations for gel loading. Because of the variation of aconitase levels in different tissues, the recommended highest protein concentration for each tissue is listed in Table II.

*Cellulose Gels.* Cellogel (Accurate Chemical & Scientific, Westbury, NY), 300 μm thick (16.5 × 14 cm), provides the best resolution. Each gel is sufficient for 15 or 16 samples. The gels are stored in 30% (v/v) methanol and must be equilibrated in electrophoresis buffer before use.

*Electrophoresis.* Electrophoresis is carried out in 20 m$M$ potassium phosphate and 3.6 m$M$ citric acid, pH 6.5, in a horizontal gel chamber with the two ends of the cellulose gel draped into electrophoresis buffer. Load samples (1 μl) at the cathode end. Because there are no sample wells on the gel, a pipette tip is used to draw short, straight lines of sample across the gel. Run the gel at 130 V at room temperature for 1 hr. At pH 6.5, both forms of aconitase run from cathode (−) to anode (+), with the cytoplasmic form running faster than the mitochondrial form (Fig. 1).

*Staining Solution*

| | |
|---|---|
| Potassium phosphate buffer (1 $M$), pH 8.0 | 1 ml |
| NADP (10 m$M$, 7.65 mg/ml $H_2O$; Sigma, St. Louis, MO): Prepare fresh every time. | 1 ml |
| $MgCl_2$, 1 $M$ | 50 μl |
| *cis*-Aconic acid (25 m$M$, 4.35 mg/ml $H_2O$; Sigma): Store at 4°. | 1 ml |
| MTT (5 mg/ml $H_2O$; Sigma): Keep in the dark and store at 4°. | 1 ml |
| Phenazine methosulfate (1 mg/ml $H_2O$; Sigma): Keep in the dark and store at 4°. | 1 ml |
| Isocitrate dehydrogenase (Sigma): Aliquot and store at −20°. | 100 U |
| $H_2O$ | to 10 ml |

*Staining the Gel.* Cut off the sides of the cellulose gel that drape into the buffer, so that the gel will lay flat. Pour staining solution onto the middle of a clean glass plate (18 × 16 cm) and lay the gel (permeable side down) onto the staining solution.

Make sure no air bubbles become trapped underneath the gel. Cover the whole assembly so that the gel remains in the dark and stain at room temperature for 6 to 10 min. Submerge the gel in $H_2O$ to stop the process and allow it to shake in the dark for 10 to 30 min to remove stain from the background. Photograph or scan the gel into the computer immediately. The gel can be kept for several days by wrapping it in plastic wrap and storing it in the dark at 4°.

*Quantitation.* Capture the gel image with a CCD camera and analyze the intensity of cytoplasmic and mitochondrial aconitase with the image analysis program as described for IEF gels. With the exception of lung, liver, and kidney, most tissues have a relatively small amount of cytoplasmic aconitase (Fig. 1). Therefore, the gel method is most useful for assessing the relationship between the cytoplasmic and mitochondrial forms of aconitase in these three tissues.

## Detection of Human Cu,ZnSOD Gene Expression by mRNA in Situ Hybridization

In situ hybridization (ISH) is a powerful method for examining the spatial expression of mRNAs within tissues and their differential expression in transgenic animals. The usefulness of any ISH method depends on both its sensitivity and the accuracy with which ISH signals reflect actual local concentrations of the target mRNAs. We have developed a high-stringency *in situ* hybridization protocol, using both radiolabeled and nonisotopically labeled riboprobes to examine the pattern of human Cu,ZnSOD transcription in transgenic mouse tissues. Antisense and control sense RNAs transcribed *in vitro* using either T3 or T7 RNA polymerase in the presence of digoxigenin–UTP (DIG–11-UTP) or [$^{35}$S]UTP are used as hybridization probes. Although both isotopic and nonisotopic ISH are of comparable sensitivity, the nonradioactive method is simpler, much faster, and gives precise cellular localization.

### Fixation and Preparation of Tissue Sections

SOLUTIONS

Paraformaldehyde (PFA; 4%, w/v), freshly prepared
Phosphate-buffered saline (PBS, pH 7.2): 3× and 1×
Ethanol (30, 60, 80, 95, and 100%, v/v)

The goal of fixation is to retain maximal levels of cellular target mRNAs while maintaining adequate morphologic detail and accessibility of probe to target mRNA. Tissues prepared for mRNA *in situ* hybridization should be frozen or fixed immediately after surgical removal. Remove organs and place them on a block of dry ice, cover tissue with finely powdered dry ice to quick freeze, and store tissues frozen at $-80°$. Before sectioning tissue blocks, equilibrate them in the cryostat chamber at $-20°$ for at least 1 hr; cut 4- to 8-$\mu$m sections on the cryostat microtome. Place sections on acid-cleaned slides (1 $M$ HCl) coated with poly-L-lysine, 500 $\mu$g/ml in water (Sigma). Fix cryostat sections in 4% (w/v) paraformaldehyde (PFA) in PBS in Coplin jars for 20 min at room temperature, rinse slides once in 3× PBS, pH 7.2, and twice in 1× PBS, pH 7.2. Sections are dehydrated through graded alcohols [30, 60, 80, 95, and 100% (v/v) ethanol], air dried, and either used immediately or stored with desiccant in a sealed container at −80°. Although it is better to use fresh frozen tissues for RNA *in situ* studies, protocols are available for using fixed paraffin-embedded sections.[37] The main advantage of paraffin sections is that archival materials can be used. In addition, paraffin embedding is the way tissues are routinely prepared for pathology, and the blocks can be stored for long periods.

*Synthesis of Hybridization Riboprobes*

MATERIALS

pBluescript II KS (+) phagemid (Stratagene, La Jolla, CA)
Plasmid Midi kit (Qiagen, Valencia, CA)
Uridine 5′-[α-$^{35}$S]thiotriphosphate, triethylammonium salt (specific activity, >1000 Ci/mmol; Amersham Pharmacia Biotech)
RNA transcription kit (Stratagene)
DIG–11-UTP (Roche Molecular Biochemicals, Indianapolis, IN)
DIG RNA-labeling kit (Roche Molecular Biochemicals)
DNase I, RNase free (Roche Molecular Biochemicals)

Hybridization riboprobes are synthesized from a pBluescript phagemid containing a 620-bp human Cu,ZnSOD cDNA insert representing the complete coding and 3′-untranslated sequences.[38] Linearize an amount of phagemid vector equivalent to 1 μg of insert DNA, using appropriate restriction enzymes (*Hin*dIII and *Bam*HI). The phagemid DNA used for riboprobe synthesis should be of high quality and free of RNA or RNase contamination. Antisense and sense RNA probes are generated as runoff transcripts with T7 or T3 polymerase, respectively. Generation of both antisense (positive probe, complementary to Cu,ZnSOD mRNA) and sense (negative probe) strand probes provides a built-in control for each probe. *In vitro* transcription is performed in the presence of either [$^{35}$S]UTP or DIG–11-UTP.

$^{35}$S-Labeled cRNAs are synthesized with the Stratagene RNA transcription kit according to the manufacturer instructions. After *in vitro* transcription is complete, the DNA template may be removed by addition of 10 U of RNase-free DNase to the reaction mix and incubation for an additional 15 min at 37°. After DNase digestion, extract the sample with phenol–chloroform (1 : 1, v/v) and precipitate the cRNAs by addition of a 1/10 volume of 3 *M* sodium acetate, pH 5.2, and 2.5 volumes of 100% ethanol. Resuspend the cRNA pellet in 50 μl of 20 m*M*

---

[37] J. G. Seidman, in "Current Protocols in Molecular Biology" (F. M. Ausubel, ed.), Chap. 14. John Wiley & Sons, New York, 2000.
[38] J. Lieman-Hurwitz, N. Dafni, V. Lavie, and Y. Groner, *Proc. Natl. Acad. Sci. U.S.A.* **79,** 2808 (1982).

dithiothreitol (DTT) and determine the counts per minute per microliter and the percent incorporation of label. Under the conditions of synthesis, more than 70% of the label is incorporated.

DIG-labeled cRNAs are synthesized with the Roche Molecular Biochemicals DIG RNA-labeling kit, according to the manufacturer instructions. Phagemid DNA template can be removed by digestion with RNase-free DNase. However, this step is often not needed as the amount of cRNA synthesized far exceeds the amount of template present. cRNA probes are purified by ethanol precipitation. DIG-labeled cRNA is precipitated by addition of a 1/10 volume of 4 $M$ LiCl and a 3× volume of ethanol, pelleted by centrifugation, and resuspended in 100 $\mu$l of diethylpyrocarbonate (DEPC)-treated water. Labeling efficiency should be checked for each reaction by direct detection and comparison with the labeled control RNA supplied with the kit. Avoid phenol–chloroform extraction of DIG-labeled RNA as the digoxigenin is hydrophobic and may partition in the organic layer during phenol extraction.

Before hybridization, the cRNA probes may be subjected to mild alkali hydrolysis by heating at 60° for 20 min in 100 m$M$ carbonate buffer, pH 10.2 (40 m$M$ NaHCO$_3$–60 m$M$ Na$_2$CO$_3$) to reduce their size to about 100–300 bases (exact incubation time depends on the initial cRNA probe length).[39] As the linkage between DIG and UTP is resistant to alkali, DIG-labeled cRNA probes can be fragmented by alkali treatment. After limited alkali degradation, the length of the cRNA probes can be determined by electrophoresis in 8% (w/v) polyacrylamide denaturating gels. After alkali fragmentation, neutralize samples and ethanol precipitate as described above.

Caution should be exercised to prevent RNase contamination throughout the entire procedure and particularly during the synthesis of the riboprobe. Up to 10 $\mu$g of cRNA can be synthesized in a single reaction, providing enough probe for several hundred hybridizations. Riboprobe can be aliquoted and stored at −20° in the presence of an RNase inhibitor or stored in hybridization solution.

*In Situ Hybridization*

SOLUTIONS

Triethanolamine (TEA) buffer, freshly prepared
Acetic anhydride
Glycine, 2 mg/ml in PBS
SSC, 20× : 1× SSC is 0.15 $M$ NaCl plus 0.015 $M$ sodium citrate
Deionized formamide (Fluka, Rokonkoma, NY)
Hybridization buffer: 50% (v/v) formamide, 2× SSC, 20 m$M$ Tris-HCl (pH 8.0), 1 m$M$ EDTA, 1× Denhardt, 10% (w/v) dextran sulfate, 10 m$M$ DTT, yeast tRNA (0.5 mg/ml)

[39] K. H. Cox, D. V. DeLeon, L. M. Angerer, and R. C. Angerer, *Dev. Biol.* **101**, 485 (1984).

PREHYBRIDIZATION. The hybridization procedure consists of a prehybridization treatment step, a probe hybridization step, and a posthybridization washing procedure. Bring slides that have been frozen at $-80°$ to room temperature before opening and beginning the pretreatment procedure. Some protocols require Pronase digestion of the tissues before hybridization. However, we have found this to be unnecessary. In fact, the treatment often results in loss of all sections from the slides. Rinse the slides briefly in PBS containing glycine (2 mg/ml) and in two changes of PBS followed by a 1-min rinse in $2\times$ SSC. Acetylate the slides by preparing fresh a 0.1 $M$ solution of TEA and immediately before use add acetic anhydride to 0.25% (v/v). Immerse the slides in this solution and acetylate the slides for 10 min. Rinse the slides in two changes of $2\times$ SSC and then dehydrate them for 2 min each in a series of graded alcohols [30, 60, 80, 95, and 100% (v/v) ethanol]. Air dry the slides, cover with 300 $\mu$l of hybridization solution, and incubate them for 1 hr at room temperature in a humid chamber.

PROBE HYBRIDIZATION. Add riboprobes to the hybridization buffer to give a final probe concentration of 0.1 to 0.2 $\mu$g/ml ($2 \times 10^5$ to $1 \times 10^6$ cpm per reaction for $^{35}$S-labeled riboprobes). The optimal probe concentration depends on mRNA abundance and must be determined in each case. The same concentrations of sense and antisense probes are used for each hybridization in order to assess nonspecific hybridization to sense cRNA probes. Add $^{35}$S-labeled riboprobe to hybridization buffer lacking DTT; heat $^{35}$S- or DIG-labeled riboprobes for 1 min at 80° and immediately place on ice. After heating, adjust $^{35}$S-labeled riboprobe–hybridization solution to 10 m$M$ DTT. Remove the prehybridization solution from the sections and apply 20–30 $\mu$l of riboprobe–hybridization buffer mix to each section and cover with an acid-cleaned coverslip. Incubate for 12–16 hr at 50° in a chamber moistened with hybridization buffer.

POSTHYBRIDIZATION WASH. Wash the slides in 50% (v/v) formamide–$2\times$ SSC at 55° for 1–2 hr with several changes of wash solution. Wash in $2\times$ SSC for 30 min at 37°. Include 20 m$M$ DTT in wash solution when using $^{35}$S-labeled probes. Incubate the slides at 37° for 1 hr in RNase A (20 $\mu$g/ml; Sigma) and RNase T1 (1 U/ml; Sigma) in $2\times$ SSC. After RNase treatment, wash the slides in 50% (v/v) formamide–$2\times$ SSC for 1 hr at 55° with several changes of wash solution. Wash the slides in $0.5\times$ SSC and then in $0.1\times$ SSC with several changes at 37°.

*Detection of in Situ Hybridization*

MATERIALS

Kodak (Rochester, NY) photographic emulsion (NTB2)
Kodak D19 developer
Kodak fixer
Kodak safe light filter no. 2
Water bath, 45°

Hematoxylin and eosin stain
DIG nucleic acid detection kit (Roche Molecular Biochemicals)

RADIOACTIVE DETECTION SYSTEM. Before initiating emulsion autoradiography dehydrate the slides through graded alcohols [30, 60, 80, and 95%, (v/v)] containing 0.3 $M$ ammonium acetate, for 1 min each. Rinse the slides in two changes of 100% ethanol followed by two changes of xylene and air dry the slides. Emulsion autoradiography must be performed in complete darkness or at least 4 ft from a safe light. Melt photographic emulsion (NTB2) in a 45° water bath; add the melted emulsion to an equal volume of 0.6 $M$ ammonium acetate heated to 45° and mix carefully. Dip the slides slowly and steadily into the emulsion and place each slide vertically (frosted side down) to dry for 1–2 hr. Place dry slides in a light-tight box with desiccant and expose at 4° for the appropriate time and away from any radioactive substances. Let the slides stand at room temperature for 1 hr before developing. Develop in darkness for 2.5 min in Kodak D19, briefly dip in water, and then fix in Kodak fixer for about 3–4 min. Rinse the slides in water for 10 min in the light and counterstain the slides with hematoxylin and eosin.

DIGOXIGENIN DETECTION SYSTEM. Digoxigenin is a steroid hapten that is not present in biologic materials. Hybridized riboprobes are detected by an enzyme-linked immunoassay using highly purified sheep anti-digoxigenin Fab fragments conjugated to alkaline phosphatase. Incubation with 5-bromo-4-chloro-3-indolyl phosphate and nitroblue tetrazolium (NBT) at an alkaline pH initiates the color reaction. Formation of a dark purple–blue–brown signal in the presence of dye substrates indicates the location of the hybridized probe. Immunologic detection is done with the Roche Molecular Biochemicals nucleic detection kit according to the manufacturer instructions as given in a Technical Bulletin.[40–42]

*Cu,ZnSOD Gene Expression in Transgenic Animals.* Striking overexpression of Cu,ZnSOD is present in homozygous transgenics compared with nontransgenic mice in which Cu,ZnSOD transcripts are minimally, if at all, detectable above background. Prominent antisense hybridization signals are seen in brain, liver, and lymphoid tissues. No signal above background level is detected for sections hybridized with control sense riboprobes. The most prominent hybridization signals are seen in the brain, in structures with a high density of neuronal cell bodies, notably the hippocampal formation and the molecular layer and Purkinje cells of the cerebellum (Fig. 2).

---

[40] Roche Molecular Biochemicals, "Technical Bulletin: *In Situ* Hybridization with Tissue Sections, Using Digoxigenin-Labeled Oligonucleotide Probes." Roche Molecular Biochemicals, Indianapolis, Indiana, 2000.
[41] F. Baldino, M. F. Chesselet, and M. E. Lewis, *Methods Enzymol.* **168,** 761 (1989).
[42] H. J. Holtke and C. Kessler, *Nucleic Acids Res.* **18,** 5843 (1990).

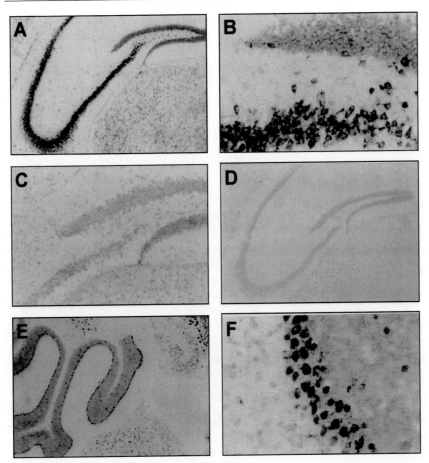

FIG. 2. RNA *in situ* hybridization of hippocampus and cerebellum. Results with DIG-labeled probes are shown. (A) Low-power view (×4) of the hippocampal formation from a transgenic mouse analyzed with antisense Cu,ZnSOD probe. (B) Higher power view (×25) of (A). (C) Hybridization with sense probe showed no hybridization signals. (D) Low-power view of the hippocampal formation from a nontransgenic mouse hybridized with antisense probe. (E) Low-power view of the cerebellum from a transgenic mouse analyzed with antisense probe. (F) Higher power view of (E). Similar results were obtained with $^{35}$S-labeled riboprobe.

## Derivation of Fetal Fibroblasts for in Vitro Studies

Primary fetal fibroblasts can be readily derived from day 13 (E13) fetuses and used for *in vitro* studies of paraquat and $H_2O_2$ toxicity, radiation sensitivity, and replicative life span. Although it is often easier to use immortalized cell lines for *in vitro* studies because the cell lines are readily available and can be passaged indefinitely, most such cells are aneuploid, with amplification or deletion in multiple chromosome regions. Such genetic perturbations lead to deviations from normal patterns of gene expression, making it difficult to generalize the study results. By contrast, primary cells have a normal genetic constitution and, therefore, are better representatives of how normal cells would respond to the experimental treatments.

*Culture Medium*

DMEM–F12 (1 : 1, v/v) with 15 m$M$ HEPES (GIBCO-BRL, Gaithersburg, MD), 10% (v/v) fetal calf serum, penicillin (100 U/ml), and streptomycin (100 $\mu$g/ml)

*Timed Mating for Embryonic Day 13 Fetuses.* Cage estrous females with males (one female per male) and check the females the next morning for vaginal plugs. Separate females from males and sacrifice pregnant females 13 days (E13) from the plug date. Remove the fetuses from the uterus and quickly rinse twice with sterile PBS. Place the fetuses in the wells of six-well plates, one fetus per well. If the fetuses are not to be processed immediately, they need to be placed on ice. Aseptically remove a piece of liver or tail for genotyping. This can be easily done with microdissecting forceps (Roboz Surgical Instrument, Gaithersburg, MD) or Dumont tweezers no. 7 (Roboz Surgical Instrument) sterilized by wiping the tips with alcohol pads.

*Derivation of Fetal Fibroblasts.* Dispense 3 ml of culture medium into each well to cover the fetuses. Use a 3- or 6-ml syringe with an 18-gauge needle, and pass the fetus through the needle several times to dissociate the tissues into small pieces. The solution should appear turbid. Label a set of 75-mm$^2$ flasks and dispense 17 ml of culture medium into each flask. Transfer the tissue suspension into the flasks, one fetus per flask. Incubate at 37° until confluent, which usually takes 4 to 5 days. The culture medium is changed on day 3 to remove floating cells. Wild-type and transgenic cells can be split 1 : 2 to 1 : 4 at each passage. However, $Sod2^{-/-}$ cells can be split only 1 : 2 because of the high sensitivity to ambient oxygen, which leads to low plating efficiency. $Sod1^{-/-}$ cells are even more sensitive to ambient oxygen and are difficult to propagate. Reduction of the oxygen concentration to 3 to 5% will greatly improve the viability of $Sod1^{-/-}$ and $Sod2^{-/-}$ cells. The cell population is almost entirely fibroblasts by the second passage (P2) and can be plated out for *in vitro* studies. If cells are not to be used immediately, they are usually stored in liquid nitrogen at P1. A common mistake is to split the cells too soon. This is especially devastating for $Sod1^{-/-}$ and $Sod2^{-/-}$ cells.

*Fetal Liver Hematopoietic Stem Cell Studies*

Hematopoietic stem cells can be obtained from day 15 (E15) fetal livers and used for *in vitro* studies to evaluate the growth properties of stem cells from

$Sod2^{-/-}$, $Sod2^{-/+}$, and $Sod2^{+/+}$ mice. In addition, the stem cells can be injected into irradiated hosts for *in vivo* evaluation of stem cell survival in a normal host. A method for isolation of fetal liver hematopoietic stem cells and data from the *in vitro* evaluation of granulocyte-macrophage and erythroid- colony formation from *Sod2* mutant mice are presented below.

*Isolation of Fetal Liver Hematopoietic Stem Cells.* Carry out timed matings of $Sod2^{-/+}$ mice and sacrifice pregnant females on day 15 to 16 of pregnancy. Aseptically remove individual fetal livers and place into 3 ml of PBS containing 1% (v/v) heat-inactivated fetal calf serum (FCS; HyClone, Logan, UT), penicillin (100 U/ml), streptomycin (100 $\mu$g/ml), and amphotericin B (Fungizone, 2.5 $\mu$g/ml). In addition, set aside a small piece of tissue from each fetus for PCR analysis of genotype. After serial passage of livers through 19-, 21-, and 23-gauge needles, layer the cells on a Ficoll gradient (Ficoll Paque; Amersham Pharmacia Biotech) and centrifuge at 400$g$ for 20 min at 4°. Remove the cells at the interface and wash three times with PBS. Pellet the cells at 200$g$ for 5 min at 4° after each wash. Resuspend the cells after the last wash in McCoy's 5A medium containing only FCS, penicillin, and streptomycin, and count viable cells by trypan blue exclusion.

*Culture Media*

Granulocyte-macrophage colony-stimulating factor (GM-CSF)-dependent colony formation (CFU-GM): McCoy's 5A medium with 20% (v/v) FCS, penicillin (100 U/ml), streptomycin (100 $\mu$g/ml), 0.3% (w/v) agar, GM-CSF (20 ng/ml; R & D Systems, Minneapolis, MN)

Colony-forming units erythroid (CFU-E) and burst-forming units erythroid (BFU-E): Iscove's modified Dulbecco's medium containing 30% (v/v) FCS, penicillin (100 U/ml), streptomycin (100 $\mu$g/ml), 100 $\mu M$ 2-mercaptoethanol, 10% (w/v) bovine serum albumin, 1% (v/v) spleen-conditioned medium, 0.85% (w/v) methyl cellulose (all from Stem Cell Technologies, Vancouver, BC, Canada), and erythropoietin (EPO, 3 U/ml; R & D Systems)

*Derivation and Culture of Fetal Liver Hematopoietic Cells.* For analysis of CFU-GM, CFU-E, and BFU-E, plate out $1.5 \times 10^5$ cells in 1 ml of their respective media in triplicate in 35-mm petri dishes (35-mm petri dishes with 2-mm grid; Nune, Roskilde, Denmark). Incubate cells in a humidified chamber at 37° with 7% $CO_2$ and 5% (using a sealed incubation chamber) or 20% $O_2$. The morphology of individual colonies is assessed for CFU-GM after 7 and 14 days, and colonies containing more than 20 cells are counted. The erythroid colonies are assessed on day 7 for CFU-Es and again on day 14 for BFU-Es. Colonies containing at least 10 mature erythroblasts are counted.

TABLE III
In Vitro GROWTH CHARACTERISTICS OF FETAL LIVER HEMATOPOIETIC
STEM CELLS FROM Sod2 MUTANT MICE[a]

| Genotype | n[b] | 20% Oxygen | 5% Oxygen | p Value[c] |
|---|---|---|---|---|
| | | CFU-GM[d] | | |
| Controls[e] | 12 | 63 ± 5 | 57 ± 5 | 0.1 |
| $Sod2^{-/-}$ | 6 | 5 ± 1 | 28 ± 1 | <0.001 |
| p Value[f] | | <0.001 | <0.001 | |
| | | CFU-E[d] | | |
| $Sod2^{+/+}$ | 6 | 51 ± 3 | 51 ± 2 | 0.94 |
| $Sod2^{-/-}$ | 9 | 30 ± 1 | 49 ± 2 | <0.001 |
| p Value[f] | | <0.001 | 0.45 | |
| | | BFU-E[d] | | |
| $Sod2^{+/+}$ | 6 | 55 ± 3 | 53 ± 3 | 0.43 |
| $Sod2^{-/-}$ | 9 | 7 ± 1 | 31 ± 3 | <0.001 |
| p Value[f] | | <0.001 | <0.001 | |

[a] Results from one representative experiment are shown.
[b] n, Number of fetuses analyzed.
[c] Within-group comparison between 20 and 5% oxygen.
[d] Number of colonies formed per $1 \times 10^5$ cells seeded.
[e] Data pooled from cells derived from $Sod2^{+/+}$ and $Sod2^{-/+}$ fetuses.
[f] Between-group comparison between $Sod2^{-/-}$ and controls.

*Study Results.* Fetal liver hematopoietic stem cells isolated from Sod2 mutant mice on a CD1 background were used for this study. We observed a marked reduction of CFU-GM, CFU-E, and BFU-E formation by stem cells derived from $Sod2^{-/-}$ fetuses when the study was carried out in 20% oxygen (Table III). Colony formation by cells derived from $Sod2^{-/+}$ did not differ from that derived from $Sod2^{+/+}$. In contrast, lowering the oxygen tension to 5% greatly enhanced the formation of CFU-GM, CFU-E, and BFU-E by stem cells derived from $Sod2^{-/-}$ (Table III). The data indicate the increased sensitivity of $Sod2^{-/-}$ hematopoietic stem cells to oxygen toxicity.

## "Mix and Match" of Transgenic and Mutant Mice for Specific Studies

Most studies with transgenic or mutant mice are done with single genetic alterations. However, there are situations in which combinations of different transgenes and/or mutations would create a unique animal model and help to answer specific questions. For example, to determine the true expression levels of Sod2 transgenes and to determine whether low-level Sod2 expression could rescue the lethal phenotype of $Sod2^{-/-}$ mice, we crossed the Sod2 transgene into the $Sod2^{-/-}$

TABLE IV
GENETIC SEGREGATION AND SURVIVAL OF Sod2 MUTANT MICE FROM RESCUE STUDY WITH Sod2 TRANSGENES[a]

| Sod2 mutant genotype | Sod2 transgenes[a] | Expected segregation (%) | Observed segregation at 5 days[b] | Survival to adulthood[c] | p Value[d] |
|---|---|---|---|---|---|
| $Sod2^{-/-}$ | Tgn(Sod2)262Cje | 12.5 | 9.7% (26/268) | Yes | 0.166 |
| | Tgn(Sod2)265Cje | | | | |
| | Tgn(Sod2)275Cje | | | | |
| | Nontransgenic | 12.5 | 6.7% (18/268) | No[e] | 0.004 |
| $Sod2^{-/+}$ | Tgn(Sod2)262Cje | 25 | 22.4% (60/268) | Yes | 0.323 |
| | Tgn(Sod2)265Cje | | | | |
| | Tgn(Sod2)275Cje | | | | |
| | Nontransgenic | 25 | 27.2% (73/268) | Yes | 0.397 |
| $Sod2^{+/+}$ | Tgn(Sod2)262Cje | 12.5 | 17.5% (47/268) | Yes | 0.013 |
| | Tgn(Sod2)265Cje | | | | |
| | Tgn(Sod2)275Cje | | | | |
| | Nontransgenic | 12.5 | 16.4% (44/268) | Yes | 0.052 |

[a] $Sod2^{-/+}$Cje mice were first crossed with each of the three independently derived Sod2 transgenic lines (262, 265, and 275) with low-level expression of the Sod2 transgene to obtain breeders that were $Sod2^{-/+}$/Sod2tg. These mice were then crossed with $Sod2^{-/+}$BCM to obtain $Sod2^{-/-}$, $Sod2^{-/+}$, and $Sod2^{+/+}$ with or without the transgenes. Because the MnSOD transgene is a genomic sequence and originated from mouse, it is not possible to differentiate the genotype between $Sod2^{-/+}$ and $Sod2^{-/-}$/Sod2tg by PCR. Therefore, we took advantage of the difference in the mutant alleles between Sod2Cje and Sod2BCM [Lebovitz et al., Proc. Natl. Acad. Sci. U.S.A. **93**, 9782 (1996)] and introduced Sod2BCM into this study to facilitate PCR genotyping of Sod2 mutant alleles and Sod2 transgenes. Data for the genetic segregation and survival from the three independent crosses were pooled.
[b] A total of 268 mice were generated for this study, and the mice were genotyped on day 5 by PCR. Because the three mouse strains used for the breeding here had a different mix of genetic backgrounds (Sod2Cje was N7 B6, Sod2BMC was a mixture of B6 and 129, and Sod2tg were a mixture of B6D2F1 and CD1), the genetic background of these pups is a mixture of B6, 129, D2, and CD1.
[c] All adult mice were killed at about 6 weeks of age for biochemical analysis. No overt phenotype was observed in $Sod2^{-/-}$/tg mice.
[d] $Sod2^{-/-}$Cje mice on a CD1 background could survive for up to 10 days [Li et al., Nat. Genet. **11**, 376 (1995)], whereas $Sod2^{-/-}$BMC mice on a mixed B6/129 background could survive for up to 21 days [Lebovitz et al., Proc. Natl. Acad. Sci. **93**, 9782 (1996)].
[e] $\chi^2$ analysis was carried out for the genetic segregation. A significantly lower segregation rate was observed in the $Sod2^{-/-}$ population, suggesting a prenatal death of $Sod2^{-/-}$ mice. On the other hand, all $Sod2^{-/-}$ mice with Sod2 transgenes were able to survive to adulthood, indicating a successful rescue of the neonatal lethality phenotype by the three Sod2 transgenic lines used in this study.

background, assessed survival, and measured the levels of MnSOD activity in different tissues of these mice. The data indicate that as little as 15, 33, and 4% of the normal level of MnSOD in the brain, heart, and kidney, respectively, are sufficient to rescue the lethal phenotype of the Sod2 mutation[18,43] (Tables IV

[43] Y. Li, T. T. Huang, E. J. Carlson, S. Melov, P. C. Ursell, J. L. Olson, J. L. Noble, M. P. Yoshimura, C. Berger, P. H. Chan, D. C. Wallace, and C. J. Epstein, Nat. Genet. **11**, 376 (1995).

TABLE V
MnSOD ACTIVITY IN TISSUES OF <$Sod2^{-/-}$>/<$Sod2$tg> MICE

| Genotype | Brain | | Heart | | Kidney | |
|---|---|---|---|---|---|---|
| | Mean ± SEM[a] | %C[b] | Mean ± SEM[a] | %C[b] | Mean ± SEM[a] | %C[b] |
| $Sod2^{+/+}$ | 0.59 ± 0.06 | | 51.4 ± 4.6 | | 0.94 ± 0.04 | |
| $Sod2^{-/-}$/TgN($Sod2$)262[c] | 0.22 ± 0.02 | 37 | 34.5 ± 3.4 | 67 | 0.13 ± 0.0 | 14 |
| $Sod2^{-/-}$/TgN($Sod2$)265[c] | 0.091 ± 0.01 | 15 | 17.0 ± 3.1 | 33 | 0.04 ± 0.001 | 4 |
| $Sod2^{-/-}$/TgN($Sod2$)275[c] | 0.25 ± 0.02 | 41 | 32.9 ± 2.8 | 64 | 0.27 ± 0.02 | 28 |

[a] Values represent units per milligram total protein. MnSOD activities were measured in 6-week-old mice as described.[33] Results represent means ± SEM of five or six mice per group.
[b] Percent $Sod2^{+/+}$ value.
[c] Homozygous MnSOD mutant mice ($Sod2^{-/-}$) with $Sod2$ transgenes from strain 262, 265, or 275.

and V). Cu,ZnSOD transgenic mice have also been crossed into the $Sod2^{-/-}$ background to determine whether increased Cu,ZnSOD can rescue the lethal phenotype of the $Sod2$ mutation. This increase has no effect on the life span of $Sod2^{-/-}$ mice, presumably because Cu,ZnSOD does not enter the mitochondria where the abnormalities resulting from the $Sod2$ mutation occur (Fig. 3).

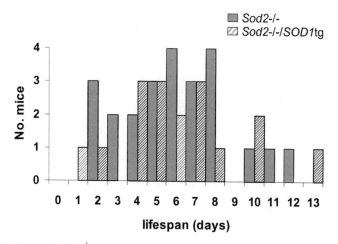

FIG. 3. Life span of $Sod2^{-/-}$ mice with or without a $SOD1$ transgene. Mice were generated from the crosses between $Sod2^{-/+}$/TgN($SOD1$)218Cje and $Sod2^{-/+}$, and were screened by PCR for the $Sod2$ alleles and by nondenaturing PAGE for the $SOD1$ transgenes. The mice were monitored daily from birth and the life span recorded. Twenty-four pups in the $Sod2^{-/-}$ group were monitored, and had a mean life span of 6.0 ± 0.6 (SD) days; 17 pups in the $Sod2^{-/-}$/$SOD1$tg group were monitored, and had a mean life span of 6.1 ± 0.7 days.

Similarly, Bruijn et al.[44] crossed *SOD1* transgenic mice with an ALS-linked mutation with *Sod1*$^{-/-}$ or *SOD1* transgenic animals to eliminate or increase, respectively, wild-type Cu,ZnSOD. These manipulations did not lead to any change in the development of ALS-like symptoms in these mice, suggesting that the disease mechanism is independent of superoxide-mediated oxidative stress.

Acknowledgments

This work was supported by NIH Grants AG14694, AG16998, and AG16633.

---

[44] L. I. Bruijn, M. K. Houseweart, S. Kato, K. L. Anderson, S. D. Anderson, E. Ohama, A. G. Reaume, R. W. Scott, and D. W. Cleveland, *Science* **281**, 1851 (1998).

# [21] Overexpression of Cu,ZnSOD and MnSOD in Transgenic *Drosophila*

By ROBIN J. MOCKETT, WILLIAM C. ORR, and RAJINDAR S. SOHAL

## Introduction

Elevation of superoxide dismutase (SOD) activity in transgenic *Drosophila* can be achieved by incorporation of an extra copy of the SOD gene into the nuclear, chromosomal DNA, using the *P* element technology initially described by Spradling and Rubin.[1,2] The autonomous 2.9-kb *P* element contains the *P* transposase gene sequence, and has 31-bp perfect inverted repeats at its termini. These terminal repeats are the target sequence for excision of the *P* element from nuclear or plasmid DNA by the transposase.[3] Transposase activity results in integration of sequences flanked by and including the repeats into the nuclear DNA at random sites.

The *P* element was initially exploited for transgenic research by constructing a plasmid containing the transposase gene and a plasmid containing a marker gene, such as *rosy* or *lacZ*, flanked by *P* element termini.[2] Simultaneous microinjection of the two plasmids into *Drosophila* embryos resulted in expression of the transposase and germ line transformation of embryonic chromosomal DNA with the reporter sequence. In subsequent studies, the transposase sequence was supplied

---

[1] A. C. Spradling and G. M. Rubin, *Science* **218**, 341 (1982).
[2] G. M. Rubin and A. C. Spradling, *Science* **218**, 348 (1982).
[3] R. E. Karess and G. M. Rubin, *Cell* **38**, 135 (1984).

by a plasmid with one partially deleted $P$ element end (the "wings-clipped" helper plasmid, p$\pi$25.7wc).[3] Consequently, the transposase sequence was not translocated and the inserted reporter gene was not subject to unwanted remobilization in subsequent generations.

In addition to the marker gene, an additional gene sequence can also be inserted. In the case of Cu,ZnSOD overexpression, the *Drosophila* genomic Cu,ZnSOD sequence, including the native promoter domain,[4,5] and the bovine Cu,ZnSOD cDNA, fused with the constitutive actin 5C promoter,[6] have been incorporated between the $P$ element ends, using either G418 antibiotic resistance or $white^+$ gene expression as a marker. More recently, the effects of targeted Cu,ZnSOD expression have been examined in transgenic *Drosophila*, using an additional transgene to control the timing[7] or spatial pattern[8] of SOD overexpression. In the former case, the yeast FLP recombinase sequence was introduced as an additional $P$ element insert. After a transient heat pulse early in adult life, the recombinase was expressed and removed a transcriptional stop sequence near the 5′ end of the Cu,ZnSOD gene insert, permitting overexpression of SOD. In the latter study,[8] spatial regulation was effected by using the GAL4/UAS transgenic system to target human Cu,ZnSOD expression specifically to the motor neurons in adult *Drosophila*.

Overexpression of the *Drosophila* MnSOD gene has also been achieved with native regulatory sequences, in a manner analogous to Cu,ZnSOD overexpression.[9] However, the effects of regulated overexpression of MnSOD have not yet been reported.

## Genetics of Superoxide Dismutase Overexpression

### DNA Preparation

The SOD transgene and regulatory sequences are assembled in a transposable plasmid, that is, one with $P$ element ends flanking the insert and marker sequences, such as p*CaSpeR* or one of its derivatives.[10] A fresh solution is prepared containing a 0.5-mg/ml concentration of transposable plasmid DNA and a 0.15-mg/ml concentration of helper plasmid DNA. The injection buffer is filter-sterilized 5 m$M$

---

[4] N. O. L. Seto, S. Hayashi, and G. M. Tener, *Proc. Natl. Acad. Sci. U.S.A.* **87,** 4270 (1990).
[5] W. C. Orr and R. S. Sohal, *Arch. Biochem. Biophys.* **301,** 34 (1993).
[6] I. Reveillaud, A. Niedzwiecki, K. G. Bensch, and J. E. Fleming, *Mol. Cell. Biol.* **11,** 632 (1991).
[7] J. Sun and J. Tower, *Mol. Cell. Biol.* **19,** 216 (1999).
[8] T. L. Parkes, A. J. Elia, D. Dickinson, A. J. Hilliker, J. P. Phillips, and G. L. Boulianne, *Nat. Genet.* **19,** 171 (1998).
[9] R. J. Mockett, W. C. Orr, J. J. Rahmandar, J. J. Benes, S. N. Radyuk, V. I. Klichko, and R. S. Sohal, *Arch. Biochem. Biophys.* **371,** 260 (1999).
[10] C. S. Thummel, A. M. Boulet, and H. D. Lipshitz, *Gene* **74,** 445 (1988).

potassium chloride, 0.1 m$M$ potassium phosphate, pH 7.8. Sufficiently clean DNA can be obtained by ultracentrifugation on a cesium chloride gradient containing ethidium bromide, or by preparation of DNA on a Qiagen (Chatsworth, CA) column, provided that the column is not overloaded. The quantity and purity of the DNA are assessed by measuring the $OD_{260\,nm}/OD_{280\,nm}$ ratio, which should be near 1.8, and by fractionating an aliquot by agarose gel electrophoresis, to check for nicking or degradation of the DNA, or contamination with excess salts, protein, or RNA.

*Microinjection*

Ordinarily, $P$ element microinjection is performed with a recipient strain that lacks existing $P$ elements, because transposase activity is suppressed in cells that already possess integrated $P$ elements.[11] The basis for this suppression is thought to involve differential splicing of the third transposase intron, which also restricts transposase activity to the germ line.[12]

The mechanics of microinjection have been described in detail elsewhere[11] and are only summarized here. Injections must be performed within about 2 hr of egg laying, before sequestration of the germ line by pole cell formation. During the first hour, embryos are collected on agar plates containing fruit juice. These plates serve as replaceable lids on a 1-ft$^3$ cage, which houses about 2000 flies, approximately 3–14 days posteclosion. During the remaining hour, the embryos are gently removed from the plates and dechorionated with 50% bleach for 2–3 min, and then rinsed thoroughly with several exchanges of deionized water, until the odor of bleach is entirely removed. The embryos are then blotted gently to remove excess water, and aligned individually on double-stick tape on standard microscope slides. The slides are next placed into a dish containing desiccant for 2–7 min, depending on the softness of the embryos, and the embryos are subsequently overlayered with halocarbon oil to prevent further desiccation.

Microinjection into the posterior pole is performed with an inverted, phase-contrast microscope with an adjustable stage, fitted with an adjustable micromanipulation system including the needle, which contains the DNA solution. The size of the needle aperture must be optimal. If the opening is too small, it will become clogged quickly. If it is too large, damage to the embryo may occur. Although motorized systems are available, mechanical injections produce a reasonable yield of transformants at a fraction of the cost.

After injection, the slides are placed on a level surface to prevent drifting of the oil and desiccation of the embryos. They are ideally kept at 18–20°, because the

---

[11] A. C. Spradling, in *"Drosophila: A Practical Approach"* (D. B. Roberts, ed.), p. 175. IRL Press, Oxford, 1986.
[12] F. A. Laski, D. C. Rio, and G. M. Rubin, *Cell* **44**, 7 (1986).

yield of larvae is slightly lower at warmer temperatures. The slides are monitored after approximately 24, 36, and 48 hr, and hatched larvae are transferred to standard food vials. Surviving $G_0$ adults are collected as virgins and back-crossed to the parental strain. The $G_1$ progeny are scored for marker gene expression.

After the creation of stable stocks, the presence of the Cu,ZnSOD or MnSOD transgene is confirmed by Southern analysis. The chromosomal location of the transgene is determined from the segregation patterns of the transgene and dominant markers after crosses of transgenic males to *yw*; *CyO* and *yw*; *TM3, Sb* females. The overexpression of SOD protein may be assessed by immunoblot analysis, using rabbit anti-*Drosophila* Cu,ZnSOD or MnSOD antibodies. However, the litmus test for SOD overexpression is the measurement of enzymatic activity. This topic is treated in [28] in this volume.[13]

*Transgene Remobilization*

The overexpression of SOD for studies of aging requires the generation of multiple independent lines. This process is laborious, especially when large constructs are used to provide native regulatory sequences and insulate the transgene from regulatory sequences near the insertion site, because the frequency of transformation is generally much lower with larger inserts.[11] To avoid the need to inject thousands of embryos to obtain a sufficient number of transgenic lines, the transgenes in the initial lines can be subjected to transient remobilization after crosses with the strain *yw*; *ry*$^{506}$ *Sb P[ry$^+$ Δ2-3](99B)/TM6, Ubx*. Remobilization is effected with the Δ2-3 element,[14] which contains an endogenous transposase gene from which the third intron has been removed, and which does not readily remobilize itself. Genetic Scheme 1 can be used to move a transgene, *P[w$^+$ SOD]*, from the X chromosome to an autosome.

In Scheme 1, the $w^+$ phenotype in $G_2$ male flies reveals transgene mobilization, because transgenes remaining on the X chromosome would be transmitted exclusively to female progeny. Consequently, the *Sb* Δ2-3 element can be selected against in $G_2$. New stocks are developed by crossing these $G_2$ males to virgins of the parental strain.

If there are no lines with inserts on the X chromosome, or the X chromosome inserts are not readily mobilized, transgenes can also be moved from chromosome 2 to chromosome 3, and vice versa. This strategy should be used as a second choice, because (1) one extra generation is required, and (2) the likelihood of moving the transgene to an autosome free of markers is decreased. Moreover, when the insert is initially located on chromosome 3, the Δ2-3 element is retained with the mobilized

[13] R. J. Mockett, A.-C. V. Bayne, B. H. Sohal, and R. S. Sohal, *Methods Enzymol.* **349**, [28], 2002 (this volume).
[14] H. M. Robertson, C. R. Preston, R. W. Phillis, D. M. Johnson-Schlitz, W. K. Benz, and W. R. Engels, *Genetics* **118**, 461 (1988).

## Scheme 1

$G_0$: $\dfrac{P[w^+\ SOD]}{P[w^+\ SOD]}; \dfrac{+}{+}; \dfrac{+}{+}$ ♀ × $\dfrac{y\,w}{Y}; \dfrac{+}{+}; \dfrac{Sb\Delta2\text{-}3}{TM6,\ Ubx}$ ♂

$G_1$: $\dfrac{y\,w}{y\,w}; \dfrac{+}{+}; \dfrac{+}{+}$ ♀ × $\dfrac{P[w^+\ SOD]}{Y}; \dfrac{+}{+}; \dfrac{Sb\,\Delta2\text{-}3}{+}$ ♂

$G_2$: $\dfrac{y\,w}{y\,w}; \dfrac{+}{+}; \dfrac{+}{+}$ ♀ × $\dfrac{y\,w}{Y}; \dfrac{P[w^+\ SOD]}{+}; \dfrac{+}{+}$ ♂

SCHEME 1

transgene during the extra generation. With inserts initially on chromosome 3, the crosses are as shown in Scheme 2.

Mobilization of the transgene from its chromosome 3 homolog is detected by scoring $G_2$ males for the presence of both $w^+$ and $Sb$. If the transgene has moved to chromosome 2, these two markers can be separated in $G_3$ males. If the transgene is initially located on chromosome 2, then $CyO$ is introduced (the $G_0$ males are $CyO/Sp$; $Sb\ \Delta2\text{-}3/TM6$, $Ubx$), and the $G_2$ males are scored for both $w^+$ and $Cy$.

Incorporation of the transposase gene into the *Drosophila* chromosomal DNA can also be used to facilitate the microinjection process. Injection of the $y\,w$; $Sb\ \Delta2\text{-}3/TM6$, $Ubx$ strain, instead of the parental $y\,w$ strain, allows the helper plasmid to be dispensed with. An advantage of this approach is that a sufficient level of

## Scheme 2

$G_0$: $\dfrac{y\,w}{y\,w}; \dfrac{+}{+}; \dfrac{P[w^+\ SOD]}{P[w^+\ SOD]}$ ♀ × $\dfrac{y\,w}{Y}; \dfrac{+}{+}; \dfrac{Sb\Delta2\text{-}3}{TM6,\ Ubx}$ ♂

$G_1$: $\dfrac{y\,w}{y\,w}; \dfrac{+}{+}; \dfrac{+}{+}$ ♀ × $\dfrac{y\,w}{Y}; \dfrac{+}{+}; \dfrac{Sb\Delta2\text{-}3}{P[w^+\ SOD]}$ ♂

$G_2$: $\dfrac{y\,w}{y\,w}; \dfrac{+}{+}; \dfrac{+}{+}$ ♀ × $\dfrac{y\,w}{Y}; \dfrac{P[w^+\ SOD]}{+}; \dfrac{Sb\Delta2\text{-}3}{+}$ ♂

$G_3$: $\dfrac{y\,w}{y\,w}; \dfrac{+}{+}; \dfrac{+}{+}$ ♀ × $\dfrac{y\,w}{Y}; \dfrac{P[w^+\ SOD]}{+}; \dfrac{+}{+}$ ♂

SCHEME 2

endogenous transposase activity can be assumed, and a slightly higher frequency of transformation may be achieved. One drawback is that transgenes inserted on chromosome 3 cannot be separated from the dominant markers located on both chromosome 3 homologs.

The loss of transgenes on chromosome 3 reduces the net yield, but in some circumstances provides an added benefit. The endogenous Cu,ZnSOD gene resides on chromosome 3, and an endogenous SOD null background is created by a series of crosses introducing two nontransformed chromosome 3 homologs bearing the $x16$ and $x39$ deletions.[15] Thus, the transgene can be introduced into an endogenous Cu,ZnSOD null background only if it is not also located on chromosome 3. Exclusion of chromosome 3 inserts immediately after injection therefore facilitates screening for transgene inserts that can be crossed into a SOD null background.

## Experimental Conditions for Life Span Determination in *Drosophila*

The main objective in most studies of SOD overexpression has been to determine its effects on longevity and, by extension, on the aging process. Accordingly, the remainder of this article is devoted to those aspects of experimental design that affect the accuracy and interpretation of life span data in *Drosophila*.

*Position Effects*

The problem that has attracted the greatest amount of attention in studies of SOD overexpression is probably insertional position effects. These are effects on transgene expression and other phenotypic traits that are attributable to the chromosomal locus at which the transgene is integrated, but not to the transgene itself. Position effects can influence not only the level and tissue specificity of transgene expression, but also the expression of genes at neighboring loci. A quantitative effect can be advantageous, because it allows any effect of SOD gene overexpression to be correlated with the level of SOD activity. Changes in tissue-specific expression, which would likely complicate analysis, have been minimized by the nature of the transgene constructs, except where tissue-specific expression was the main objective of the study. The impact on neighboring loci is usually a nuisance effect, although it can lead to fortuitous discoveries.[16]

The basic strategy to distinguish between direct phenotypic effects of transgene expression and position effects is to replicate the experiments, using multiple lines with transgenes inserted at different chromosomal loci. The numbers of lines

---

[15] J. P. Phillips, J. A. Tainer, E. D. Getzoff, G. L. Boulianne, K. Kirby, and A. J. Hilliker, *Proc. Natl. Acad. Sci. U.S.A.* **92**, 8574 (1995).

[16] B. Rogina, R. A. Reenan, S. P. Nilsen, and S. L. Helfand, *Science* **290**, 2137 (2000).

required to detect a 10% difference in life span have been estimated by Tatar.[17] This number depends largely on the variation among lines within a given set, which arises as a result of both position effects and variations among individual cohorts with the same genotype. If these effects can be minimized, then the number of lines required will be reduced and smaller differences in life span can be detected with a given number of lines. Although the exact numbers of lines cannot be determined in advance, use of about 10–15 independent lines for each treatment (SOD transgenics vs. control lines transformed with the marker but no SOD gene) is prudent.

To minimize adverse position effects, a good policy whenever practicable is to use stocks in which the transgene can be driven to fixation. The transgene is initially placed over a balancer chromosome. If the balancer chromosome disappears within several generations, and the marker phenotype is not lost, then the transgenic stock is homozygous viable. The homozygous stocks are then backcrossed with the parental strain or outcrossed into a different genetic background, yielding experimental flies heterozygous for the transgene. The impact of position effects is reportedly lowered by up to an order of magnitude with heterozygous versus homozygous inserts,[18] and complications associated with inbreeding depression are also avoided. Furthermore, where homozygous stocks are viable, the transgene in a heterozygous state almost never has a pronounced negative effect on survival times. Such negative effects are sometimes observed when the chromosome bearing the transgene does not outcompete the balancer chromosome. Although this approach does not attempt to control for random mutation and genetic drift among the lines, the differences in life span attributable to position effects are smaller in the authors' laboratories than the variations between multiple cohorts with the same genotype.

*Cohort Variations*

The key to avoiding undue cohort variations is good animal husbandry. The variation in life span among lines of a given type is significantly elevated if the condition of fly stocks is in any way suboptimal. Suboptimal conditions also decrease the absolute length of life in *Drosophila*. These effects are important, because there is a significant inverse relationship between the effects of antioxidative enzyme overexpression and the length of life of the control flies serving as a baseline.[19] The fly medium should contain an acid mixture [0.33% (v/v) propionic acid, 0.033% of 85% (v/v) phosphoric acid, final concentration] to inhibit bacterial

---

[17] M. Tatar, *Am. Nat.* **154**, S67 (1999).
[18] T. F. C. Mackay, R. F. Lyman, and M. S. Jackson, *Genetics* **130**, 315 (1992).
[19] R. S. Sohal, R. J. Mockett, and W. C. Orr, *in* "Results and Problems in Cell Differentiation" (S. Hekimi, ed.), Vol. 29, p. 45. Springer-Verlag, Berlin, 2000.

growth, and methylparaben (2 g/liter) as a mold inhibitor. If excess bacteria are observed, the health of the stock can be improved by more prompt transfers to fresh bottles, or by adding antibiotics to the medium temporarily, and then allowing at least one generation for recovery on regular medium before continuing with experimental crosses.

Death of a small fraction of pupae is a clear sign of excess bacterial growth. In severe cases, death of third instar larvae and some young adults is also observed. Such flies are useless for survival experiments. The problem is avoided by removing the adult flies 2–3 days after brooding stocks, and especially after performing crosses in which experimental flies are raised. Any dead adult flies must be removed from the bottles at this stage, because the larvae will feed on the adult carcasses and become sick. The health of the cross is further improved by transferring the parents to fresh food regularly before the cross, and by crossing young adult flies, roughly 3–14 days of age, because their fecundity is maximal and they deposit bacteria on the food surface at a slower rate than older flies.

The severe, confounding effects of bacterial contamination must also be minimized with adult experimental flies. This is accomplished in part by collecting adults promptly after eclosion, and by placing dry rayon onto the surface of the medium before adult eclosion. The amount of rayon depends on humidity and larval density, which determine the degree of liquefaction of the food surface. Another crucial parameter is the frequency of transfer of adult flies to fresh food vials. The flies should be transferred every 36–48 hr during the first 3–4 weeks posteclosion, and every 24 hr thereafter. When this schedule is not followed, the absolute length of life is curtailed drastically by bacterial contamination. However, if this schedule is followed, the life span is not extended further by adding antibiotics to the food vials provided to healthy adult flies during the aging process.

### Acknowledgment

The authors thank J. Benes and P. Benes for helping to establish the experimental conditions described herein.

# Section III
# Superoxide Dismutase Mimics

# [22] Manganese Porphyrins and Related Compounds as Mimics of Superoxide Dismutase*

*By* INES BATINIĆ-HABERLE

## Introduction

The manganese porphyrins, because of their exceptionally high metal–ligand stability, the variety of structural modifications they offer, the accessibility of several oxidation states of their metal center, and their reactivity toward different reactive oxygen and nitrogen species, have provoked extensive *in vitro* and *in vivo* studies.[1–3] Along with the manganese macrocyclic polyamines[4] and manganese salen compounds,[5] manganese porphyrins are used to treat the conditions and to test the mechanisms of oxidative stress injuries.

*Abbreviations:* $O_2^{\cdot-}$, Superoxide anion; $NO^{\cdot}$, nitric oxide; $ONOO^-$, peroxynitrite anion; $ONOOCO_2^-$, 1-carboxylato-2-nitrosodioxidane; SOD, superoxide dismutase; NHE, normal hydrogen electrode; $H_2P$, any porphyrin protonated at its basic pyrrolic nitrogens; $Mn^{II}Br_8TM$-4-$PyP^{4+}$, manganese(II) 2,3,7,8,12,13,17,18-octabromo-5,10,15,20-tetrakis(N-methylpyridinium-4-yl); $Mn^{II}Cl_5TE$-2-$PyP^{4+}$, manganese(II) β-pentachloro-5,10,15,20-tetrakis(N-ethylpyridinium2-yl); $Mn^{III}TPP^+$, manganese(III) 5,10,5,20-tetrakis(phenyl)porphyrin; $Mn^{III}T(PFP)P^+$, manganese(III) 5,10,15,20-tetrakis(pentafluorophenyl)porphyrin; $Mn^{III}T$-2(4)-$PyP^+$, manganese(III) 5,10,15,20-tetrakis(2(4)-pyridyl)porphyrin; M, methyl; E, ethyl; $Mn^{III}TM(E)$-2(3,4)-$PyP^{5+}$, manganese(III) 5,10,15,20-tetrakis(N-methyl(ethyl)pyridinium-2(3,4)-yl)porphyrin; $Mn^{III}BM$-2-$PyP^{3+}$, manganese(III) 5,10,15,20-bis(2-pyridyl)-bis(N-methylpyridinium-2-yl)porphyrin; $Mn^{III}TrM$-2-$PyP^{4+}$, manganese(III) 5-(2-pyridyl)-10,15,20-tris(N-methylpyridinium-2-yl)porphyrin; $Mn^{III}T(TMA)P^{5+}$, manganese(III) 5,10,15,20-tetrakis(N,N,N-trimethylanilinium-4-yl) porphyrin; $Mn^{III}T(TFTMA)P^{5+}$, manganese(III) 5,10,15,20-tetrakis(2,3,5,6-tetrafluoro-N,N,N-trimethylanilinium-4-yl)porphyrin; $Mn^{III}TCPP^{3-}$, manganese 5,10,15,20-tetrakis(4-carboxylatophenyl)porphyrin; $Mn^{III}T(2,6-Cl_2-3-SO_3-P)P^{3-}$, manganese(III) 5,10,15,20-tetrakis(2,6-dichloro-3-sulfonatophenyl)porphyrin; $Mn^{III}T(2,6-F_2-3-SO_3-P)P^{3-}$, manganese(III) 5,10,15,20-tetrakis(2,6-difluoro-3-sulfo-natophenyl)porphyrin; $Mn^{III}Pc^+$, manganese(III) phthalocyanine; $Mn^{III}T$-2,3-$PyPz^+$, manganese(III) tetrakis(2,3-pyridino)porphyrazine; $Mn^{III}T$-3,4-$PyPz^+$ manganese(III) tetrakis(3,4-pyridino)porphyrazine; $\{Mn^{III}BV^{2-}\}_2$, manganese(III) biliverdin IX; $\{Mn^{III}BVDME\}_2$, manganese(III) biliverdin IX dimethyl ester; $\{Mn^{III}MBVDME\}_2$, manganese(III) mesobiliverdin IX dimethyl ester; $\{Mn^{III}BVDT^{2-}\}_2$, manganese(III) biliverdin IX ditaurate.

[1] R. F. Pasternack, A. Banth, J. M. Pasternack, and C. S. Johnson, *J. Inorg. Biochem.* **15,** 261 (1981); R. F. Pasternack and B. Halliwell, *J. Am. Chem. Soc.* **101,** 1026 (1979).

[2] I. Batinić-Haberle, I. Spasojević, P. Hambright, L. Benov, A. L. Crumbliss, and I. Fridovich, *Inorg. Chem.* **38,** 4011 (1999).

[3] I. Spasojević and I. Batinić-Haberle, *Inorg. Chim. Acta* **317,** 230 (2001).

[4] D. P. Riley, *Adv. Supramol. Chem.* **6,** 217 (2000); S. Cuzzocrea, E. Mazzon, L. Dugo, A. P. Caputi, K. Aston, D. P. Riley and D. Salvemini, *Br. J. Pharmacol.* **132,** 19 (2001).

[5] S. Melov, J. Ravenscroft, S. Malik, M. S. Gill, D. W. Walker, P. E. Clayton, D. C. Wallace, B. Malfroy, S. R. Doctrow, and G. J. Lithgow, *Science* **289,** 1567 (2000).

Superoxide anion $O_2^{\cdot-}$, a common endogenous by-product of aerobic metabolism, and its progeny (OH·, ONOO⁻, and $H_2O_2$) play major roles in oxidative stress. Superoxide dismutases (SODs)[6] are the primary defense against the damage that can be caused by $O_2^{\cdot-}$. They disproportionate (dismute) $O_2^{\cdot-}$ to $O_2$ and $H_2O_2$, the latter being removed by catalases and peroxidases. Both half-reactions occur at the same rate ($2 \times 10^9 \ M^{-1} \ \text{sec}^{-1}$ for Cu,ZnSOD).[7,8] Hence, independently of the type of the metal center of the enzyme, the disproportionation occurs at ~+0.30 V versus normal hydrogen electrode (NHE) (pH 7).[9,10] This potential is about midway between the potentials for the oxidation (−0.16 V vs NHE)[11] and for the reduction of $O_2^{\cdot-}$ (+0.89 V vs NHE, pH 7.0).[11]

The metalloporphyrin-based SOD mimics, whose metal-centered redox potentials lie between −0.16 and +0.89 V (pH 7) versus NHE, reduce and oxidize $O_2^{\cdot-}$ at different rates, the dismutation rate $k_{cat}$ being determined by the rate-limiting step. The essential requirement for a metal complex to efficiently and catalytically dismute $O_2^{\cdot-}$ is to have the accessible redox couple close to the midpoint potential. The unsubstituted metalloporphyrins $\text{Mn}^{\text{III}}\text{TPP}^+$ ($E_{1/2} = -0.27$ V vs. NHE)[3] and $\text{Mn}^{\text{III}}\text{T-2-PyP}^+$ ($E_{1/2} = -0.28$ V vs. NHE)[3] have metal-centered redox potentials too negative to be effective SOD mimics. Addition of the electron-withdrawing substituents in either *meso* (N-alkylpyridyls) or β positions (halogens, formyl, cyano, and nitro) on the porphyrin ring shifts the metal-centered redox potential to the positive as compared with the unsubstituted porphyrins, thereby increasing the catalytic rate constant.[2,3]

The water-soluble manganese porphrins have been actively sought, and were designed to target $O_2^{\cdot-}$ in the aqueous medium of living cells. More lipophilic compounds that would reside in the membranes and would also be able to pass the blood–brain barrier are being investigated as well.

*Water-Soluble Manganese Porphyrins*

The structure–activity relationship was established for aquamanganese(III)[2,3] and monohydroxoiron(III)[2] porphyrins whereby log $k_{cat}$ was found to vary linearly with metal-centered redox potential ($E_{1/2}$) as shown for aquamanganese(III) porphyrins in Fig. 1A. Each 120-mV increase in $E_{1/2}$ imparted a 10-fold increase in $k_{cat}$. Such behavior is in accord with the Marcus equation[12] for outer-sphere electron transfer, suggesting the same mechanism to be operative for iron and manganese

---

[6] I. Fridovich, *Annu. Rev. Biochem.* **64**, 97 (1995); I. Fridovich, *J. Biol. Chem.* **264**, 7761 (1989).
[7] L. M. Ellerby, D. E. Cabelli, J. A. Graden, and J. S. Valentine, *J. Am. Chem. Soc.* **118**, 6556 (1996).
[8] G. D. Lawrence and D. T. Sawyer, *Biochemistry* **18**, 3045 (1979).
[9] C. K. Vance and A.-F. Miller, *J. Am. Chem. Soc.* **120**, 461 (1998).
[10] D. Klug-Roth, I. Fridovich, and J. Rabani, *J. Am. Chem. Soc.* **95**, 2786 (1973).
[11] P. M. Wood, *Biochem. J.* **253**, 287 (1988).
[12] R. A. Marcus, *Annu. Rev. Phys. Chem.* **15**, 155 (1964).

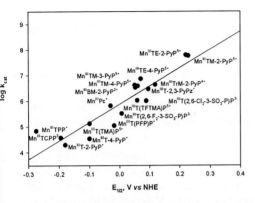
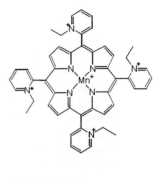

FIG. 1. (A) Reactivity of manganese(III) porphyrins (see abbreviations footnote) expressed in terms of log $k_{cat}$ for $O_2^{\cdot-}$ dismutation versus metal-centered redox potential, $E_{1/2}$. (B) The structure of $Mn^{III}TE$-2-$PyP^{5+}$. Reprinted with permission from Spasojevic et al., Inorg. Chim. Acta **317**, 230 (2001). Copyright 2001 from Elsevier Science.

porphyrins. The plot also suggests that any increase in $E_{1/2}$ might be favorable with respect to the increased SOD activity. However, along with increasing $E_{1/2}$ (stabilization of +2 oxidation state of a metal center), the electron-withdrawing substituents weaken the metal–ligand bond. Indeed, the compounds that have highly electron-withdrawing groups, for example, β-octabrominated ($Mn^{II}Br_8TM$-4-$PyP^{4+}$)[13] and β-pentachlorinated manganese meso-tetrakis(N-alkylpyridyl)-porphyrins ($Mn^{II}Cl_5TE$-2-$PyP^{4+}$),[14] despite exhibiting favorably high redox potentials and high SOD activities, proved to be unstable under physiological conditions. Moreover, a too-positive $E_{1/2}$ may promote the prooxidant properties of the compound.

Several singly charged Mn(III) porphyrins were found to obey the structure–activity relationship fairly well (Fig. 1A).[3] However, the deviations from the plot indicate that the electrostatics and the overall structure of the complexes have an impact on the catalysis.[3]

The rate constant for the reduction of $O_2^{\cdot-}$ by $Mn^{II}TM$-4-$PyP^{4+}$ ($E_{1/2}$ = +0.06 V vs NHE)[2,3] is $4 \times 10^9$ $M^{-1}$ $sec^{-1}$,[15] so that the catalytic rate constant

[13] I. Batinić-Haberle, S. Liochev, I. Spasojević, and I. Fridovich, Arch. Biochem. Biophys. **343**, 225 (1997).
[14] R. Kachadourian, I. Batinić-Haberle, and I. Fridovich, Inorg. Chem. **38**, 391 (1999); R. Kachadourian, I. Batinić-Haberle, and I. Fridovich, Free Radic. Biol. Med. **25**(Suppl. 1), S17 (1998); 5th Annual Meeting of the Oxygen Society, Washington, D.C., 1998.
[15] M. Faraggi, in "$O_2^{\cdot-}$ Dismutation Catalyzed by Water-Soluble Porphyrins: A Pulse Radiolysis Study" (W. Bors, M. Saran, and D. Taits, eds.), p. 419. Walter de Gruyer: Berlin, 1984.

($k_{cat} = 3.8 \times 10^6 \, M^{-1} \, \text{sec}^{-1}$)$^2$ is governed by the rate of the oxidation of $O_2^{\cdot-}$ by Mn(III) porphyrin. Once the $E_{1/2}$ becomes too positive, the +2 metal oxidation state is stabilized as is the case with $Mn^{II}Br_8TM\text{-}4\text{-}PyP^{4+}$ (+0.48 V)[13] and $Mn^{II}Cl_5TE\text{-}2\text{-}PyP^{4+}$ (+0.56 V),[14] and the reduction of $O_2^{\cdot-}$ by Mn(II) porphyrin becomes the rate-limiting step.

The metal-centered redox potential of the metalloporphyrins is also linearly related to the acid dissociation constant of the acidic pyrrolic nitrogen of the parent ligands, $pK_{a3}$ ($H_3P^+ \longleftrightarrow H^+ + H_2P$).[2] Consequently, either property, $E_{1/2}$ or $pK_{a3}$, may be useful in predicting the SOD activity of a particular manganese porphyrin and in justifying its evaluation in rodent and mammalian models of oxidative stress injuries.

Thus far, the most potent metalloporphyrin-based compounds, with respect to *in vitro* SOD activity, are the isomeric manganese *N*-alkylpyridylporphyrins. *In vivo*, in a simple *Escherichia coli* model of oxidative stress injury, a fairly positive metal-centered redox potential of $\geq +0.05$ V versus NHE (or $pK_{a3} \leq 2$) proved to be a necessary, yet not sufficient, requirement for a catalytic scavenger of $O_2^{\cdot-}$ to allow the SOD-deficient *E. coli* to grow aerobically.[2] The bulkier ortho ($E_{1/2} = +0.22$ V vs. NHE) and meta ($E_{1/2} = +0.05$ V vs. NHE) isomers of manganese(III) tetrakis(*N*-methylpyridyl)porphyrin appear protective whereas the rather planar para isomer ($E_{1/2} = +0.06$ V vs NHE) is lacking the protection because of its association with nucleic acids.[2] Also, iron(III) porphyrins were toxic, presumably because of their interactions with amino acid residues of the proteins and enzymes.[2]

On the basis of the structure–activity relationship manganese(III) tetrakis(*N*-methyl(ethyl)-pyridinium-2-yl)porphyrin, $Mn^{III}TM(E)\text{-}2\text{-}PyP^{5+}$ (Fig. 1B), was first postulated and then proved to be the most potent SOD mimic *in vitro* and in the *E. coli* model. The compound has further been proved effective in offering protection in a variety of oxidative stress injuries such as liver ischemia,[16] diabetes,[17a] radiation,[17b] and most notably in stroke.[18] (Replacement of an *N*-methyl by an *N*-ethyl substituent on *meso* pyridyl groups does not significantly affect the *in vitro* and *in vivo* reactivity of this manganese porphyrin.) Therefore the potency of a compound in a simple *E. coli* model of oxidative stress provides good guidance

---

[16] D. Parks, I. Fridovich, V. B. O'Donnell, Z. Wang, I. Batinić-Haberle, B. J. Day, P. H. Chumley, and B. A. Freeman, *Free Radic. Biol. Med.* **25**(Suppl. 1), S36 (1998); 5th Annual Meeting of the Oxygen Society, Washington, D.C., 1998.

[17a] J. D. Piganelli, S. C. Flores, C. Cruz, J. Koepp, I. Batinić-Haberle, J. D. Crapo, B. J. Day, R. Kachadourian, R. Young, B. Bradley, and K. Haskins, *Diabetes*, in press (2002).

[17b] Ž. Vujašković, I. Batinić-Haberle, I. Spasojević, I. Fridovich, T. V. Samulski, M. W. Dewhirst, and M. S. Anscher, Paper presented at the Annual Meeting of the Radiation Research Society, San Juan, Puerto Rico, 2001.

[18] G. B. Mackensen, M. Patel, H. Sheng, C. C. Calvi, I. Batinić-Haberle, B. J. Day, L. P. Liang, I. Fridovich, J. D. Crapo, R. D. Pearlstein, and D. S. Warner, *J. Neurosci.* **21**, 4582 (2001).

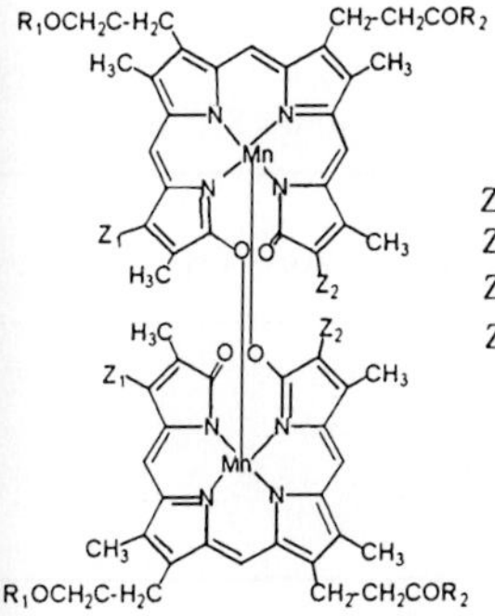

FIG. 2. Manganese(III) complexes with biliverdin IX and its derivatives.

$Z_1 = Z_2$ = vinyl, $R_1 = R_2 =$ OH, biliverdin IX, $\{Mn^{III}BV^{2-}\}_2$
$Z_1 = Z_2$ = vinyl, $R_1 = R_2 =$ OCH$_3$, biliverdin IX dimethyl ester, $\{Mn^{III}BV^{2-}\}_2$
$Z_1 = Z_2$ = ethyl, $R_1 = R_2 =$ OCH$_3$, mesobiliverdin IX dimethyl ester, $\{Mn^{III}MBVDME\}_2$
$Z_1 = Z_2$ = vinyl, $R_1 = R_2 =$ NHCH$_2$CH$_2$SO$_3^-$, biliverdin IX ditaurate, $\{Mn^{III}BVDT^{2-}\}_2$

to its efficacy in other *in vivo* models. Aside from using it as an SOD mimic, the well-characterized Mn$^{III}$TE-2-PyP$^{5+}$ is also conveniently used as a standard in *in vitro* and *in vivo* studies involving prospective manganese porphyrins and related compounds.

*Lipophilic Manganese Porphyrin-Related Compounds*

A new class of manganese(III) porphyrin-related complexes is emerging, in which manganese(III) is pentacoordinated inside a dimeric unit (Fig. 2).[3,19] Each manganese is coordinated to four pyrrolic nitrogens of one biliverdin and to an enolic oxygen of another biliverdin molecule. Such a coordination leads to unique reactivity of the metal center of Mn(III) biliverdin, in which the +4 metal oxidation state is stabilized. Consequently, the O$_2^{\cdot-}$ dismutation is facilitated through Mn(IIII)/Mn(IV) redox cycling at $E_{1/2} = +0.45$ V versus NHE.[19] Although UV–Vis, electrochemical characteristics, and catalytic rate constants for the dismutation of O$_2^{\cdot-}$ have been determined for all the compounds given in Fig. 2,[3,19] more data are available on the properties of $\{Mn^{III}BVDME\}_2$ as it was isolated in the solid state and studied and proved effective in the *E. coli* model.[3,19] The use and study of Mn(III) biliverdins are intriguing because such compounds may be involved in heme oxygenase-related antioxidant activity. Furthermore, using $\{Mn^{III}BVDME\}_2$, which is based on a naturally occuring biliverdin molecule, may impose a lower toxicity profile in *in vivo* studies.

*Future Directions*

The new thermodynamic and kinetic data that are constantly emerging make us aware of the variety of reactions in which manganese porphyrins may rapidly

[19] I. Spasojević, I. Batinić-Haberle, R. D. Stevens, P. Hambright, A. N. Thorpe, J. Grodkowski, P. Neta, and I. Fridovich, *Inorg. Chem.* **40**, 726 (2001).

TABLE I
ULTRAVIOLET–VISIBLE DATA, METAL-CENTERED REDOX POTENTIALS, AND SECOND-ORDER RATE CONSTANTS FOR REACTIONS OF MANGANESE COMPLEXES WITH REACTIVE OXYGEN AND NITROGEN SPECIES[a]

| Compound | log $\varepsilon$ ($\lambda_{max}$)[b] | $E_{1/2}$ ($Mn^{III/II}$)[c] | log $k(H_2O_2)$[d] | log $k_{cat}(O_2^{·-})$ | log $k(ONOO^-)$ | log $k(NO^·)$ |
|---|---|---|---|---|---|---|
| $Mn^{III}TE$-2-$Pyp^{5+}$ | 5.14 (453.8)[e] | +0.23[e] | 0.11 | 7.76[e] | 7.49[f] | −0.9[g] |
| $Mn^{II}TE$-2-$Pyp^{4+}$ | 5.26 (438.4)[g] | +0.23[e] | −0.63[h] | 7.76[e,i] | >7[f] | 6[g] |
| $Mn^{III}TCPP^{3-}$ | 4.96 (468.2) | −0.19[e] | −1.10[j] | 4.56[e] | 4.83[k,l] | m |
| {$Mn^{III}BVDME$}$_2$ | 4.49 (390.0)[n] | +0.45[n] | −3.21[n] | 7.70[n] | o | p |

[a] Obtained at 25 (±1)° unless otherwise specified.
[b] UV–Vis data were obtained in water ($Mn^{III}TE$-2-$Pyp^{5+}$); phosphate-buffered saline, pH 7.4 ($Mn^{II}TE$-2-$Pyp^{4+}$ 0.05 $M$ phosphate buffer, pH 7.8 ($Mn^{III}TCPP^{3-}$); and in methanol ({$Mn^{III}BVDME$}$_2$).
[c] $E_{1/2}$ values were determined in 0.05 $M$ phosphate buffer, (0.1 $M$ NaCl), pH 7.8, except in the case o {$Mn^{III}BVDME$}$_2$ [methanol–0.05 $M$ Tris, pH 7.8 (9:1, v/v) buffer]. In the latter case the given $E_{1/2}$ valu was extrapolated from methanol/aqueous to phosphate-buffered solution, pH 7.8 (0.1 $M$ NaCl). $E_{1/2}$ value are given as volts versus NHE and relate to the Mn(III)/Mn(II) couple except in the case of {$Mn^{III}BVDME$}$_2$ where the Mn(III)/Mn(IV) redox is meant.
[d] Rate constant corresponds to the oxidative degradation (bleaching) of the metalloporphyrin in the presence o $H_2O_2$, pH 7.8.
[e] Batinić-Haberle et al., Inorg. Chem. **38**, 4011 (1999).
[f] Ferrer-Sueta et al., Chem. Res. Toxicol. **12**, 442 (1999). Data were obtained at 37°, pH 7.4.
[g] Spasojević et al., Nitric Oxide Biol. Chem. **4**, 526 (2000).
[h] Data were obtained in the presence of 1 m$M$ ascorbic acid, 0.1 to 1.0 m$M$ $H_2O_2$, 10 $\mu M$ porphyrin, 0.05 $M$ Tris buffer, pH 7.8 [I. Batinić-Haberle, unpublished data (2000)].
[i] log $k_{cat}$ = 7.76 is determined by rate-limiting oxidation of $O_2^{·-}$ by Mn(III) complex. The log $k$ for the reduc tion of $O_2^{·-}$ by $Mn^{II}TE$-2-$Pyp^{4+}$ is estimated to be ∼8. Estimation is based on the reported data for the par isomer of the N-methylated compound. [Spasojević et al., Inorg. Chim. Acta **317**, 230 (2001); Spasojevi et al., Nitric Oxide Biol. Chem. **4**, 526 (2000)].
[j] Batinić-Haberle, unpublished data (2000).
[k] Data obtained at 37°, pH 7.2.
[l] Quijano et al., J. Biol. Chem. **276**, 11631 (2001).
[m] No reaction observed. When reduced with threefold excess of $Na_2S_2O_4$ (aerobically), 11 $\mu M$ $Mn^{II}TCPP^{4-}$ reacts readily with 11 $\mu M$ NO.
[n] Spasojević et al., Inorg. Chem. **40**, 726 (2001). $\varepsilon$ and rate constants are calculated per manganese.
[o] No data available.
[p] No reaction observed.

interact with reactive oxygen and nitrogen species employing +2, +3, +4, and +5 metal oxidation states.[2,3,19–21] Thus, although the $O_2^{·-}$ dismutation by $Mn^{III}TM$ (E)-2-$Pyp^{5+2}$ occurs at a high rate compared with its reaction with $ONOO^-$,[20] $ONOOCO_2^-$,[20] $NO^·$,[21] or $H_2O_2$[2,22] (Table I), the tissue localization and the

[20] G. Ferrer-Sueta, C. Quijano, I. Batinić-Haberle, and R. Radi, Paper presented at the Third Int. Conf. on Peroxynitrite and Reactive Nitrogen Species in Biology and Medicine, Pacific Grove, CA, 2001; G. Ferrer-Sueta, I. Batinić-Haberle, I. Spasojević, I. Fridovich, and R. Radi, Chem. Res. Toxicol. **12**, 442 (1999).

concentration of these species in oxidative stress injuries would determine the actual *in vivo* mode of action of this metalloporphyrin. Indeed, $Mn^{III}TM(E)$-2-PyP$^{5+}$ [20] and its para analog $Mn^{III}TM$-4-PyP$^{5+}$ [23] have been suggested as peroxynitrite-scavenging agents *in vivo*. $Mn^{III}TM(E)$-2-PyP$^{5+}$ is easily reducible by cellular antioxidants, and thus its rate-limiting reduction by $O_2^{\cdot-}$ may be avoided. Consequently, the $O_2^{\cdot-}$ scavenging would take place solely via reductive elimination of $O_2^{\cdot-}$ with a high rate constant that has been estimated to be $\sim 10^8 \, M^{-1} \, \text{sec}^{-1}$.[3] The $H_2O_2$ produced could be further eliminated through the actions of catalase or peroxidase. [The bleaching of $Mn^{II}TM(E)$-2-PyP$^{4+}$ with $H_2O_2$ under biologically relevant ascorbic acid concentrations proceeds at a negligibly low rate.[22]] Whether eliminating $O_2^{\cdot-}$ via its reduction or via its dismutation, the manganese porphyrins, which are *in vitro* dismuting $O_2^{\cdot-}$ at high rate constants, may be regarded as *in vivo* mimics of superoxide dismutase; particularly in the light of suggestions that Cu,ZnSOD may scavenge $O_2^{\cdot-}$ not only through its dismutation but also through its reduction,[24] similar to the proposed mode of action of desulfoferrodoxin.[25]

Rapidly emerging data on high *in vivo* potency of $Mn^{III}TE$-2-PyP$^{5+}$ justify employing the structure–activity relationship as a decisive factor in selecting a powerful antioxidant for *in vivo* studies. As it becomes thermodynamically and kinetically better characterized, $Mn^{III}TE$-2-PyP$^{5+}$, accompanied by equally SOD-potent {$Mn^{III}BVDME$}$_2$, is gaining an additional and important role in elucidating the mechanisms of oxidative stress injuries.

Once a compound has been selected as a prospective catalytic scavenger of $O_2^{\cdot-}$ on the basis of the structure–activity relationship, the reactivity of its metal center should be explored with regard to all reactive oxygen and nitrogen species. Extensive and complex studies are necessary to understand the impact of charge, stereochemistry, type of metal center, tissue localization, and its reactivity on the behavior of manganese complexes in each particular injury inside complex biological systems. Negatively charged porphyrins, such as $Mn^{III}TCPP^{3-}$, have been occasionally[26–29] reported as being beneficial in different injury models on an as yet poorly understood basis. Further work is needed to understand

---

[21] I. Spasojević, I. Batinić-Haberle, and I. Fridovich, *Nitric Oxide Biol. Chem.* **4**, 526 (2000).
[22] I. Batinić-Haberle, unpublished data (2000).
[23] J. P. Crow, *Arch. Biochem. Biophys.* **371**, 41 (1999); J. Lee, J. A. Hunt, and J. T. Groves, *J. Am. Chem. Soc.* **120**, 6053 (1998).
[24] S. I. Liochev and I. Fridovich, *J. Biol. Chem.* **275**, 38482 (2000).
[25] E. D. Coulter, J. P. Emerson, D. M. Kurtz, Jr., and D. E. Cabelli, *J. Am. Chem. Soc.* **122**, 11555 (2000).
[26] D. Liu, X. Ling, J. Wen, and J. Liu, *J. Neurochem.* **75**, 2144 (2000).
[27] S. Kotamraju, E. A. Konorev, J. Joseph, and B. Kalyanaraman, *J. Biol. Chem.* **275**, 33585 (2000).
[28] E. Konorev, M. Claire Kennedy, and B. Kalyanaraman, *Arch. Biochem. Biophys.* **368**, 421 (1999).
[29] M. Patel and B. J. Day, *Trends Pharmacol. Sci.* **20**, 359 (1999).

the origin of these effects because neither $E_{1/2}$ ($pK_{a3}$),[2] nor log $k_{cat}$($O_2^{\cdot-}$),[2] nor log $k(ONOO^-)$[30] (Fig. 1A and Table I), nor *E. coli*[2] data could be easily applied to rationalize its mode of action. However, the *in vivo* data thus far reported justify that such compounds be evaluated in particular models of oxidative stress injuries even though they would not be expected to be SOD-active compounds on the basis of the structure–activity relationship. So far we can only speculate that either the anionic charge and/or protonation equilibria of Mn$^{III}$TCPP$^{3-}$ at its carboxylic groups, or its redox status, may be responsible for the observed effects. Namely, because of its low $E_{1/2}$ the compound in its reduced state (Mn$^{II}$TCPP$^{4-}$) is a powerful reducing agent. Thus, in a reducing environment it may effectively catalyze the reduction of $O_2^{\cdot-}$.

Future work on {Mn$^{III}$BVDME}$_2$[3] and on those of its derivatives that would differ from Mn$^{III}$TE-2-PyP$^{5+}$ in their lipophilicity[3] and specificity toward reactive oxygen and nitrogen species, as well as on the design of novel classes of compounds, may broaden our understanding of oxygen-related metabolism in the living cell and of the protective role of synthetic antioxidants.

The preparation and characterization of the two manganese complexes are described below. One is the water-soluble manganese(III) porhyrin Mn$^{III}$TE-2-PyP$^{5+}$ (Fig. 1B) and the other one is the water-insoluble {Mn$^{III}$BVDME}$_2$ (Fig. 2). Their different lipophilicities may be used to target different *in vitro* and *in vivo* systems. Their equal SOD-like potency, but the inertness of {Mn$^{III}$BVDME}$_2$ toward NO$^{\cdot}$ and H$_2$O$_2$ (Table I), may be employed in targeting different reactive species.

## Experimental

### Preparation of Compounds

*Mn$^{III}$TE-2-PyP$^{5+}$*. Manganese(III) complex, Mn$^{III}$TE-2-PyP$^{5+}$ may be prepared from its non-*N*-alkylated metal-free ligand obtained from Mid-Century Chemicals, Chicago, IL. The *N*-ethylation and the metallation of H$_2$T-2-PyP are done as follows. H$_2$T-2-PyP (500 mg) is *N*-ethylated with 20 ml of ethyl *p*-toluenesulfonate in 100 ml of *N,N*-dimethyl formamide for 24–48 hr at $\geq 100°$. The progressive appearance of mono-, di-, tri-, and tetraethylated species is monitored by thin-layer chromatography (TLC) on Baker-flex silica gel IB using KNO$_3$-saturated H$_2$O–H$_2$O–acetonitrile (1 : 1 : 8, v/v/v). On completion, 200 ml each of water and chloroform are added to the reaction mixture in a separatory funnel and shaken well. The chloroform layer is discarded, and the extraction is repeated several times. The aqueous layer is filtered, and the porphyrin is precipitated by the addition of a concentrated aqueous solution of NH$_4$PF$_6$. The precipitate is thoroughly washed with 2-propanol–diethyl ether (1 : 1, v/v) on a finely

---

[30] C. Quijano, D. Hernandez-Saavedra, L. Castro, J. M. McCord, B. A. Freeman, and R. Radi, *J. Biol. Chem.* **276**, 11631 (2001).

fritted disc and dried briefly *in vacuo*. It is then dissolved in acetone, followed by filtration and precipitation with tetrabutylammonium chloride. The precipitate is washed thoroughly with acetone and dried *in vacuo* at room temperature. The product, $H_2TE$-2-$PyP^{4+}$, is characterized by an absorption maximum in its UV–Vis spectrum at 414.0 nm (log $\varepsilon_{414.0} = 5.33$).[2] The instantaneous insertion of manganese is done in aqueous solution at 20-fold excess $MnCl_2$ over $\sim$1 m$M$ porphyrin after the pH is brought to pH 12.3 to allow the deprotonation of the basic pyrrolic nitrogens.[2,31] The metallation is monitored spectrophotometrically (Soret band at 453.8 nm, log $\varepsilon_{453.8} = 5.14$).[2] On completion, the solution is filtered and the pH dropped to pH $\sim$4 by the addition of diluted HCl. The precipitation of the Mn(III) porphyrin from this solution as $PF_6^-$ salt followed by its isolation from an acetone solution as chloride salt is accomplished as described above for the metal-free ligand. The electrospray mass spectrometry of the 1 : 1 (v/v) aqueous–acetonitrile solution at the declustering potential of 30 V shows the dominant *m/z* at *m/z* 205.74 for $ClMn^{III}TE$-2-$PyP^{4+}$/4, followed by the molecular ion at *m/z* 157.43 ($Mn^{III}TE$-2-$PyP^{5+}$/5).[32]

{$Mn^{III}BVDME$}$_2$. Biliverdin IX dimethyl ester (Frontier Scientific, Logan, UT) is recrystallized from chloroform–petroleum ether (15 : 50, v/v) (35–60°). Fifty milligrams of it is dissolved in a small volume of chloroform and 150 ml of methanol is added. The mixture is heated to $\sim$60° followed by the addition of a 15-fold excess of manganese(II) acetate (0.3 g in 10 ml of methanol). The complex forms immediately, as is evident from the disappearance of the absorption at 666 nm and appearance of a band at 898 nm. (*Caution:* Prolonged air exposure at 60° results in destruction of the compound.) Most of the solvent is evaporated on a rotary evaporator at room temperature. Then 2 ml of methanol is added for solubilization followed by the addition of 200 ml of chloroform and 200 ml of water. The mixture is poured into a separatory funnel, and the metal complex is extracted into the chloroform layer and washed three times with water. Finally, the chloroform layer is shaken with 5 g of $MgSO_4$, followed by filtration and evaporation of the solvent almost to dryness. The residue is transferred to a conical flask and dissolved in 4 ml of dichloromethane. Addition of $\sim$40 ml of petroleum ether gives a precipitate that is collected on a finely fritted glass disc, washed with petroleum ether, and dried overnight *in vacuo* at room temperature. The yield is >80%. The molar absorption coefficients calculated per manganese are log $\varepsilon_{898} = 4.05$, log $\varepsilon_{390} = 4.49$, and log $\varepsilon_{362} = 4.49$.[19] The electrospray mass spectrometry of the methanolic solution gives two major peak groups at *m/z* 1325 and at *m/z* 662, ascribed to oxidized and protonated dimer and oxidized and protonated monomer, respectively.[19]

[31] P. Hambright, Chemistry of water-soluble porphyrins. *In* "The Porphyrin Handbook" (K. M. Kadish, K. M. Smith, and R. Guilard, eds.), Chap. 18. Academic Press, New York, 2000.
[32] I. Batinić-Haberle, R. D. Stevens, and I. Fridovich, *J. Porphyrins Phthalocyanines* **4,** 217 (2000).

*Characterization of Compounds*

*Electrochemistry.* Cyclic voltammetry is performed on 0.5 m$M$ solutions of manganese complexes in either 0.05 $M$ phosphate buffer, pH 7.8 (0.1 $M$ NaCl), or methanol–0.05 $M$ Tris buffer (9 : 1, v/v) (pH 7.8). The three-electrode system is used in a small volume of up to 3 ml, with a 3-mm-diameter glassy carbon button working electrode (Bioanalytical Systems, West Lafayette, IN), Ag/AgCl reference electrode (3 $M$ NaCl; Bioanalytical Systems), and a platinum wire (0.5 mm) as auxiliary electrode. $Mn^{III}TE$-2-$PyP^{5+}$ may be used as internal standard for comparative purposes as well as for expressing $E_{1/2}$ as volts versus NHE.[2]

*Catalysis of $O_2^{\cdot-}$ Dismutation by Cytochrome c Assay.* It has been shown that the stopped-flow technique, pulse radiolysis, and cytochrome $c$ assay are all equally valid in determining the catalytic rate constant for a prospective scavenger of $O_2^{\cdot-}$ (see Spasojević *et al.*[19] and references therein). The cytochrome $c$ assay seems to be the most convenient and widely available method.[33] It is based on competition kinetics using cytochrome $c$ (from horse heart; Fluka, Rokon Koma, NY) as the reference.[19] Xanthine–xanthine oxidase is the source of $O_2^{\cdot-}$ produced at a rate of 1.2 $\mu M$/min, and $1 \times 10^{-5}$ $M$ cytochrome $c$ is the indicating scavenger of $O_2^{\cdot-}$. The reduction of cytochrome $c$ is monitored at 550 nm. Typically 10 $\mu M$ stock solutions of manganese complexes (methanolic stock solutions in the case of {$Mn^{III}BVDME$}$_2$) are diluted into 3 ml of 0.05 $M$ phosphate buffer (pH 7.8) to $10^{-7}$ to $10^{-9}$ $M$. The 1 m$M$ aqueous solution of cytochrome $c$ is diluted to 10 $\mu M$. Catalase (15 $\mu$g/ml) (Boehringer, Mannheim, Germany) is added to eliminate the effect of $H_2O_2$ on the output of the assay. EDTA is added at $10^{-4}$ $M$ to eliminate the interference of "free" metals with the assay. Inhibition of cytochrome $c$ reduction by the compound when plotted as $[(v_0/v_i) - 1]$ versus the concentration of the compound yields a straight line. $v_0$ is the rate of reduction of 10 $\mu M$ cytochrome $c$ by $O_2^{\cdot-}$, and $v_i$ is the rate of reduction of cytochrome $c$ inhibited by catalyst. The concentration that causes 50% of the inhibition of cytochrome $c$ reduction by $O_2^{\cdot-}$ is found at $[(v_0/v_i) - 1] = 1$. At that level of inhibition the rates of the reaction of cytochrome $c$ and the manganese complex with $O_2^{\cdot-}$ are equal, that is, $k_{cat}$ [Mn complex] = $k_{cyt}$ [cytochrome $c$], where $k_{cyt} = 2.6 \times 10^5$ $M^{-1}$ sec$^{-1}$.[34]

*Protection of Superoxide Dismutase-Deficient Escherichia coli.* The $Mn^{III}TE$-2-$PyP^{5+}$ and {$Mn^{III}BVDME$}$_2$, as well as other prospective compounds, may be conveniently tested for their *in vivo* capacity to protect SOD-deficient *E. coli*[2] (obtained from J. A. Imlay[35]) when growing aerobically as compared with SOD-proficient *E. coli*. $Mn^{III}TE$-2-$PyP^{5+}$ and {$Mn^{III}BVDME$}$_2$ offer protection in the range 5 to 25 $\mu M$. Growth of *E. coli* under stringent conditions in minimal medium gives a more sensitive test of SOD-like activity than growth in Casamino acids

---

[33] J. M. McCord and I. Fridovich, *J. Biol. Chem.* **244**, 6049 (1969).
[34] J. Butler, W. H. Koppenol, and E. Margoliash, *J. Biol. Chem.* **257**, 10747 (1982).
[35] J. A. Imlay and S. Linn, *J. Bacteriol.* **169**, 2967 (1987).

medium.[2] The minimal medium contains 0.2% (w/v) glucose, 3 mg/liter each of pantothenic acid and thiamine, and 0.1 g/liter each of leucine, threonine, proline, arginine, and histidine (pH adjusted to pH 7.8).[2] It also contains 0.6% (w/v) $Na_2HPO_4$, 0.3% (w/v) $KH_2PO_4$, 0.1% (w/v) $NH_4Cl$, 0.05% (w/v) NaCl, 2 m$M$ $MgCl_2$, and 20 $\mu M$ $CaCl_2$. The rate of *E. coli* growth is monitored spectrophotometrically as the increase in turbidity of the culture at 700 nm, where manganese porphyrins and related compounds do not absorb. So far the compounds whose potency has been proved in the *E. coli* model are efficacious in rodent models of ischemia–reperfusion as well.[16–18]

*Antioxidant Effect in Oxidation of Rat Brain Homogenate.* Fresh Sprague-Dawley rat brain homogenates (10%, v/v) are made in 0.05 $M$ phosphate buffer, pH 7.4, with the use of a Potter-Elvehjem tissue grinder. The 1-ml aliquots are oxidized with 25 $\mu M$ $FeCl_3$ plus 75 $\mu M$ ascorbate. Samples containing 0.1 to 5 $\mu M$ $Mn^{III}TE-2-PyP^{5+}$ are incubated for 5 min before adding $FeCl_3$ and ascorbate. The reactions are stopped by the addition of 0.1 ml of 60 m$M$ ethanolic solution of butylated hydroxytoluene before analysis of 2-thiobarbituric acid-reactive material as an index of lipid peroxidation.[36,37] The 4 ml of reaction mixture contains 200 $\mu$l of 8.1% (w/v) sodium dodecyl sulfate, 1.5 ml of 20% (w/v) acetic acid (pH 3.5), 1.5 ml of 0.8% (w/v) thiobarbituric acid, and 800 $\mu$l of brain homogenate. The standards are made by diluting 8.2 $\mu$l of malonaldehyde bis(dimethylacetal) into 10 ml of 0.01 $M$ HCl and mixing for 10 min before further diluting to 0.25 to 25 $\mu M$ final concentrations. The samples and the standards are heated in a boiling water bath for 1 hr and then cooled to 25°. Then 1 ml of water and 5 ml of *n*-butanol–pyridine, (15 : 1, v/v) are added to samples, shaken vigorously, and centrifuged at 1000$g$ for 10 min. Absorbance of the organic phase is measured at 532 nm and compared with standards to determine the level of lipid peroxidation. Because of the presence of biological reductants in the brain homogenate[37] the antioxidant effect of $Mn^{II}TE-2-PyP^{4+}$ is observed, with an $IC_{50}$ of 1 $\mu M^{36}$ as compared with an $IC_{50}$ of 15 $\mu M$ for Cu,ZnSOD.[36] The protection by manganese porphyrin, when the ascorbic acid is present, may also be tested in a similar manner on low-density lipoproteins from human plasma.[37]

## Acknowledgments

I am grateful to Prof. Irwin Fridovich for continuous support and guidance as well as for critical reading of the manuscript. I am thankful to Dr. Ivan Spasojević, Dr. Gerardo Ferrer-Sueta, and Dr. Ayako Okado-Matsumoto for helpful suggestions. The financial support from Aeolus/Incara RTP, NC and Duke Comprehensive Cancer Center/NCI Award NCIP30 CA 14236, is greatly appreciated.

[36] B. J. Day, I. Batinić-Haberle, and J. D Crapo, *Free Radic. Biol. Med.* **26**, 730 (1999).
[37] A. Bloodsworth, V. B. O'Donnell, I. Batinić-Haberle, P. H. Chumley, J. B. Hurt, B. J. Day, J. P. Crow, and B. A. Freeman, *Free Radic. Biol. Med.* **28**, 1017 (2000).

## [23] Superoxide Dismutase Mimics: Antioxidative and Adverse Effects

*By* GIDON CZAPSKI, AMRAM SAMUNI, and SARA GOLDSTEIN

Introduction

The superoxide radical anion, which mediates oxidative damage in various biological systems, is formed during normal metabolism as well as in pathophysiological processes through the action of various drugs, poisons, or radiation. Superoxide dismutases (SODs), which are present in all aerobic organisms, provide a defense that is essential for their survival. Such a defense is not complete and $O_2^{·-}/HO_2^·$ plays a role during oxidative stress as in postischemic reperfusion, organ transplantation, and various surgical interventions. Therefore the elevation of the level of SOD may have a therapeutic application. However, the short half-life of exogenously administered SOD, and its poor crossing of blood barriers and permeation into cells, prompted a search for SOD mimics that persist longer in the tissue and more readily enter cells.[1–4]

The detoxification of $O_2^{·-}$ can be achieved catalytically or by reagents that stoichiometrically remove it, and a distinction is generally sought between scavengers of $O_2^{·-}$ and catalysts.[5–7] Scavengers act in a stoichiometric fashion and, therefore, a high flux of $O_2^{·-}$, or even a low flux for an extended time, can rapidly deplete their level. Generally the catalytic process is characterized by [catalyst] ≪ [substrate], yet this condition is not met for the case of SOD and SOD mimics, which are unique catalysts. Unlike most enzyme–substrate systems, in which the substrate concentration far exceeds that of the catalyst, the cellular level of SOD or any SOD mimic exceeds about $10^6$-fold that of the steady-state concentration of its substrate, $O_2^{·-}/HO_2^·$. Proof of a catalytic mechanism is not always readily available, yet it can be demonstrated when the product of (flux of $O_2^{·-}/HO_2^·$) times (duration of the experiment) is significantly greater than the concentration of the catalyst itself. Obviously it is important to understand the mechanism underlying the catalytic activity. Such catalysts, which include native and modified SOD

---

[1] B. Halliwell, *FEBS Lett.* **56**, 34 (1975).
[2] A. Nagele and E. Lengfelder, *Free Radic. Res.* **25**, 109 (1996).
[3] D. P. Riley, *Chem. Rev.* **99**, 2573 (1999).
[4] J. R. Sorenson, *Prog. Med. Chem.* **26**, 437 (1989).
[5] M. C. Krishna, D. A. Grahame, A. Samuni, J. B. Mitchell, and A. Russo, *Proc. Natl. Acad. Sci. U.S.A.* **89**, 5537 (1992).
[6] V. Gadzheva, K. Ichimori, H. Nakazawa, and Z. Raikov, *Free Radic. Res.* **21**, 177 (1994).
[7] M. C. Krishna, A. Russo, J. B. Mitchell, S. Goldstein, H. Dafni, and A. Samuni, *J. Biol. Chem.* **271**, 26026 (1996).

enzymes as well as SOD mimics, greatly differ by their efficiency as antioxidants and by their potential adverse effects. Hence, a selection of a suitable catalyst for removal of $O_2{}^{\cdot-}$ should involve several considerations[8,9]: (1) the redox potential of its reduced and oxidized forms, (2) the kinetics of reaction with $O_2{}^{\cdot-}$, $H_2O_2$, and $O_2$, (3) the reactions with other reagents, (4) the hydrophobic/hydrophilic nature, polarity, and charge, (5) the permeation into cells and distribution among various tissues and cellular compartments, (6) the metabolic half-life in cells and in whole body circulation, (7) the chemical and biological stability as well as the nature of its metabolites, and (8) the characteristics of its dose response and potential adverse effects. Two additional considerations are pertinent to SOD mimics that contain metal: (9) the stability constants of the chelate and (10) the potential formation of ternary complex with cellular components.

## Metal-Based Superoxide Dismutase Mimics

The antioxidative effects of various chelates of manganese,[10–20] copper,[2,21] and iron[1,22–24] have been previously studied and generally attributed to their SOD mimic activity. The structure and physicochemical properties as well as the mechanism and kinetics of reaction of $O_2{}^{\cdot-}$ with metal chelates have been studied.[15] For instance, on the basis of the distinction between SOD mimics and metal chelates

---

[8] G. Czapski and S. Goldstein, *Basic Life Sci.* **49**, 43 (1988).
[9] G. Czapski and S. Goldstein, *Free Radic. Res. Commun.* **4**, 225 (1988).
[10] W. H. Koppenol, F. Levine, T. L. Hatmaker, J. Epp, and J. D. Rush, *Arch. Biochem. Biophys.* **251**, 594 (1986).
[11] H. D. Rabinowitch, C. T. Privalle, and I. Fridovich, *Free Radic. Biol. Med.* **3**, 125 (1987).
[12] H. D. Rabinowitch, G. M. Rosen, and I. Fridovich, *Free Radic. Biol. Med.* **6**, 45 (1989).
[13] W. J. Beyer and I. Fridovich, *Arch. Biochem. Biophys.* **271**, 149 (1989).
[14] W. J. Beyer and I. Fridovich, *Methods Enzymol.* **186**, 242 (1990).
[15] J. D. Rush, Z. Maskos, and W. H. Koppenol, *Arch. Biochem. Biophys.* **289**, 97 (1991).
[16] R. H. Weiss, A. G. Flickinger, W. J. Rivers, M. M. Hardy, K. W. Aston, U. S. Ryan, and D. P. Riley, *J. Biol. Chem.* **268**, 23049 (1993).
[17] M. M. Hardy, A. G. Flickinger, D. P. Riley, R. H. Weiss, and U. S. Ryan, *J. Biol. Chem.* **269**, 18535 (1994).
[18] S. C. Black, C. S. Schasteen, R. H. Weiss, D. P. Riley, E. M. Driscoll, and B. R. Lucchesi, *J. Pharmacol. Exp. Ther.* **270**, 1208 (1994).
[19] R. H. Weiss, D. J. Fretland, D. A. Baron, U. S. Ryan, and D. P. Riley, *J. Biol. Chem.* **271**, 26149 (1996).
[20] D. Salvemini, Z. Q. Wang, J. L. Zweier, A. Samouilov, H. Macarthur, T. P. Misko, M. G. Currie, S. Cuzzocrea, J. A. Sikorski, and D. P. Riley, *Science* **286**, 304 (1999).
[21] S. T. Shuff, P. Chowdhary, M. F. Khan, and J. R. Sorenson, *Biochem. Pharmacol.* **43**, 1601 (1992).
[22] D. Salvemini, D. P. Riley, P. J. Lennon, Z. Q. Wang, M. G. Currie, H. Macarthur, and T. P. Misko, *Br. J. Pharmacol.* **127**, 685 (1999).
[23] B. Halliwell and R. F. Pasternack, *Biochem. Soc. Trans.* **6**, 1342 (1978).
[24] D. Salvemini and D. P. Riley, *Cell. Mol. Life Sci.* **57**, 1489 (2000).

that can reduce $O_2^{\cdot-}$ but are devoid of catalytic activity, it was concluded that an open coordination site is essential but not sufficient to catalyze the dismutation reaction.[10]

## Mechanism of Catalysis of Superoxide Dismutation by Superoxide Dismutase and Superoxide Dismutase Mimics

The mechanism of the catalysis of $O_2^{\cdot-}$ dismutation by SOD as well as by other unbound and chelated transition metals has been suggested to proceed via a "ping–pong mechanism," in which the catalyst oscillates between two oxidation states.[25–28] This mechanism is illustrated here with copper ions. In this case, copper may switch between Cu(II) and Cu(I) or between Cu(II) and Cu(III):

$$Cu(II) + O_2^{\cdot-} \rightleftharpoons Cu(I) + O_2 \quad (1)$$
$$Cu(I) + O_2^{\cdot-} + 2H^+ \to Cu(II) + H_2O_2 \quad (2)$$

or

$$Cu(II) + O_2^{\cdot-} + 2H^+ \to Cu(III) + H_2O_2 \quad (3)$$
$$Cu(III) + O_2^{\cdot-} \to Cu(II) + O_2 \quad (4)$$

The overall reaction (5) is obtained according to both of these mechanisms:

$$O_2^{\cdot-} + O_2^{\cdot-} + 2H^+ \to O_2 + H_2O_2 \quad (5)$$

These two mechanisms cannot be distinguished kinetically, provided reaction (−1) [or reaction (−3)] is negligible. Assuming a steady-state approximation for Cu(I) or Cu(III), rate Eq. (6) is obtained:

$$-d[O_2^{\cdot-}]/dt = k_{cat}[Cu(II)]_0[O_2^{\cdot-}] \quad (6)$$

where

$$k_{cat} = 2k_1k_2/(k_1 + k_2) \quad \text{or} \quad k_{cat} = 2k_3k_4/(k_3 + k_4) \quad (7)$$

The higher the values of $k_1$ (or $k_3$) and $k_2$ (or $k_4$) are, the more efficient is the catalyst. If $k_1$ and $k_2$ differ substantially, $k_{cat}$ approaches the value of the lower rate constant. If reaction (−1) [or reaction (−3)] cannot be neglected, $k_r$, which is defined by Eq. (8),

$$k_r = 2k_1k_2/(k_1 + k_2 + k_{-1}[O_2/[O_2^{\cdot-}]) \quad (8)$$

---

[25] D. Klug-Roth, I. Fridovich, and J. Rabani, *J. Am. Chem. Soc.* **95**, 2786 (1973).
[26] R. Brigelius, R. Spottl, W. Bors, E. Lengfelder, M. Saran, and U. Weser, *FEBS Lett.* **47**, 72 (1974).
[27] S. Goldstein and G. Czapski, *J. Free Radic. Biol. Med.* **2**, 3 (1986).
[28] S. Goldstein and G. Czapski, *J. Am. Chem. Soc.* **112**, 6489 (1990).

## TABLE I
### RELATIVE SOD ACTIVITY AS FUNCTION OF $[O_2^{\cdot-}]$ AND $k_{-1}$. ($[O_2]=0.24$ m$M$)

| | $k_{-1}$ ($M^{-1}$ Sec$^{-1}$) | | | | | |
|---|---|---|---|---|---|---|
| $[O_2^{\cdot-}]$, $M$ | 1 | 10 | $10^2$ | $10^3$ | $10^4$ | $10^5$ |
| $10^{-6}$–$10^{-4}$, pulse radiolysis | 1 | 1 | 1 | 1 | 1 | 1 |
| $10^{-8}$, $\gamma$ radiolysis | 1 | 1 | 1 | 1 | 0.95 | 0.5 |
| $10^{-9}$, xanthine oxidase | 1 | 1 | 1 | 0.9 | 0.5 | 0.1 |
| $10^{-10}$ (in cells) | 1 | 1 | 0.9 | 0.5 | 0.1 | 0.013 |
| $10^{-11}$ (in cells) | 1 | 0.9 | 0.5 | 0.1 | 0.013 | 0.0013 |

replaces $k_{cat}$ in Eq. (6). In the case of bovine Cu,ZnSOD, $k_1 = k_2 = (2\text{–}3) \times 10^9$ $M^{-1}$ Sec$^{-1}$ and $k_{-1} = 0.44$ $M^{-1}$ Sec$^{-1}$.[25] Therefore, this enzyme catalyzes efficiently the dismutation of $O_2^{\cdot-}$. Conversely, SOD mimics, for which $k_{-1}[O_2]/[O_2^{\cdot-}] > (k_1 + k_2)$ and hence $k_r < k_{cat}$, are less efficient catalysts (Table I).

From a thermodynamic point of view, it is required that $\Delta E_1$ (or $\Delta E_3$) $> 0$ and $\Delta E_2$ (or $\Delta E_4$) $> 0$.

$$\Delta E_1 = \Delta E_1^\circ - 0.059 \log \frac{[Cu(I)][O_2]}{[Cu(II)][O_2^{\cdot-}]} > 0 \quad (9)$$

$$\Delta E_2 = \Delta E_2^\circ - 0.059 \log \frac{[Cu(II)][H_2O_2]}{[Cu(I)][O_2^{\cdot-}]} > 0 \quad (10)$$

Therefore, under standard conditions and at pH 7, Eq. 11 is obtained,

$$E^\circ_{(O_2/O_2^{\cdot-})} < E^\circ_{(Cu(II)/Cu(I))} \quad (E^\circ_{(Cu(III)/Cu(II))}) < E^\circ_{(O_2^{\cdot-}, H^+/H_2O_2)} \quad (11)$$

where $E^\circ_{(O_2/O_2^-)} = -0.16$ V and $E^\circ_{(O_2^-, H^+/H_2O_2)} = 0.94$ V.[29] However, in air-saturated solutions, where $[O_2] = 0.24$ m$M$, and under typical *in vivo* concentrations of $O_2^{\cdot-}$ and $H_2O_2$, the range changes (Table II). Hence, the reduction potential of an efficient SOD mimic should vary between 0.28 and 0.64 V, whereas for less efficient catalysts it should vary from 0.16 to 0.76 V (Table II).

### Properties Required for Superoxide Dismutase Mimics Operating *in Vitro* to Work Also *in Vivo*

The SOD activity of a compound can be determined either directly or indirectly. However, the ability of a compound to catalyze $O_2^{\cdot-}$ dismutation *in vitro* does not necessarily indicate that this catalyst will be an efficient SOD mimic *in vivo*. Certain essential properties are required for an efficient catalyst operating *in vitro*

---

[29] D. M. Stanbury, *Adv. Inorg. Chem.* **33**, 69 (1989).

TABLE II
MINIMAL REDOX POTENTIAL$^a$ REQUIRED FOR EFFICIENT SOD MIMICS UNDER VARIOUS CONDITIONS

|  | $k_1/k_2 = 0.01$ | $k_1/k_2 = 1$ | $k_1/k_2 = 100$ |
|---|---|---|---|
| $[O_2] = 2.4 \times 10^{-4} M$<br>$[O_2^{\cdot -}] = 10^{-11} M$ | >0.16 | >0.28 | >0.39 |
| $[O_2] = 2.4 \times 10^{-5} M$<br>$[O_2^{\cdot -}] = 10^{-11} M$ | >0.10 | >0.21 | >0.33 |
| $[O_2^{\cdot -}] = 10^{-11} M$<br>$[H_2O_2] = 10^{-6} M$ | <0.53 | <0.64 | <0.76 |
| $[O_2^{\cdot -}] = 10^{-11} M$<br>$[H_2O_2] = 10^{-8} M$ | <0.65 | <0.76 | <0.88 |
| $[O_2^{\cdot -}] = 10^{-11} M$<br>$[H_2O_2] = 10^{-10} M$ | <0.77 | <0.88 | <1.0 |

$^a$ In volts.

to also act efficiently *in vivo* (Table III). Some of these properties are trivial, such as the compounds should be nontoxic, nonimmunogenic, and inexpensive. In addition, such catalysts should have a long metabolic half-life and should be able to penetrate into the cells. Also, the metal complex should have a high stability constant, otherwise the metal might be sequestered out of the complex by the cell components, which are present at relatively high concentrations.

## Metal-Free Superoxide Dismutase Mimics

Most of the research concerning metal-free SOD mimics has involved stable nitroxide radicals, which were found to catalytically remove superoxide.[30]

TABLE III
PROPERTIES OF SOD AS COMPARED WITH AN OPTIMAL SOD-MIMIC

| Property | Native SOD | Modified SOD | SOD mimic |
|---|---|---|---|
| Immunogenicity | No | No | No |
| Toxicity | No | No | No |
| Metabolic half-life | Short | Long | Long |
| Cell permeability | Poor | Poor | Good |
| Chemical stability | High | High | High |
| Stability constant (for metal chelate) |  |  | High |
| Redox potential (V) | $0.23 \geq E \leq 0.42$ | $0.21 \geq E \leq 0.88$ | $0.21 \geq E \leq 0.88$ |
| Cost | Expensive | Expensive | Inexpensive |
| Reactivity toward $H_2O_2$ and $O_2$ | Low | Low | Low |
| Reactivity toward cell components | Low | Low | Low |

Previously, nitroxides were primarily used as biophysical probes for electron paramagnetic resonance (EPR) spectroscopic studies, particularly of membranes and proteins, for cellular metabolism, and oximetry, and have been tested as contrast agents for *in vivo* nuclear magnetic resonance imaging (MRI). More recently, nitroxides were found to catalyze removal of $O_2^{\cdot-}$ through two different modes. By flip-flopping between two oxidation states, both capable of reacting with $O_2^{\cdot-}$, nitroxides mimic the activity of native SOD.

*Reductive Mode*

The reaction of oxazolidine derivatives, such as OXANO, with $O_2^{\cdot-}$ is a catalytic process characterized by a steady-state distribution of nitroxide and hydroxylamine and a continuous formation of $O_2$ and $H_2O_2$. Thus, the redox couple of 2 ethyl,2,5,5 trimethyl-3-oxazolidine-N-oxyl OXANO acts as an SOD mimic.[30]

$$\text{N}^{\cdot}-\text{O} + O_2^{\cdot-} + H^+ \underset{k_{-12}}{\overset{k_{12}}{\rightleftharpoons}} \text{N}-\text{OH} + O_2 \quad (12)$$

$$\text{N}-\text{OH} + O_2^{\cdot-} + H^+ \xrightarrow{k_{13}} \text{N}^{\cdot}-\text{O} + H_2O_2 \quad (13)$$

*Oxidative Mode*

Piperidine and pyrrolidine nitroxide derivatives, such as TPO, catalyze $O_2^{\cdot-}$ dismutation via an alternative mechanism, which involves an oxoammonium cation (TPO$^+$) intermediate[5]:

$$\text{N}^{\cdot}-\text{O} + O_2^{\cdot-} + 2H^+ \underset{k_{-14}}{\overset{k_{14}}{\rightleftharpoons}} {}^+\text{N}=\text{O} + H_2O_2 \quad (14)$$

$$^+\text{N}=\text{O} + O_2^{\cdot-} \xrightarrow{k_{15}} \text{N}^{\cdot}-\text{O} + O_2 \quad (15)$$

---

[30] A. Samuni, C. M. Krishna, P. Riesz, E. Finkelstein, and A. Russo, *J. Biol. Chem.* **263**, 17921 (1988).

The SOD mimic activity of nitroxides offered a rationale that prompted the investigation of their biological activity. Because they readily enter cells and are neither immunogenic nor cytotoxic, they could substitute for and augment SOD protective activity both extra- and intracellularly. In particular, because nitroxides of various size, charge, and polarity can be synthesized, it seemed promising to develop improved SOD mimics that can selectively function in cellular compartments of the desired lipophilicity.

Additional Modes of Action

Unlike SOD or metal-based SOD mimics that protect by removing $O_2^{\cdot-}$, nitroxides act through several independent pathways (catalytic, pseudocatalytic, and stoichiometric), including oxidation of semiquinone radicals, oxidation of reduced metal ions,[31] procatalase mimic activity and detoxification of hypervalent metals,[32] interruption of radical chain reactions,[33,34] and indirect modulation of NO level. The antioxidative activity of nitroxides is well documented. Nitroxides protect cells, isolated organs, and laboratory animals exposed to oxidative stress. However, it is not yet clear which of the mechanisms is involved, and to what extent it contributes to the protective activity of nitroxides. Therefore, model systems in which $O_2^{\cdot-}/HO_2^{\cdot}$ is directly responsible for the biological damage should serve for evaluation of the SOD mimic effect of nitroxides. For instance, the superoxide-induced inactivation of enzymes, such as papain, having an oxidizable thiol at their active site has been used for this purpose.[35]

Adverse Effects

Pro-oxidative and other adverse effects, which are common to unbound and chelated transition metals, were observed for several SOD mimics as well.[2] For instance, a manganese-containing SOD mimic was found to be capable of removing $O_2^{\cdot-}$ as well as facilitating its formation, depending on the experimental conditions.[36] Even the native enzyme Cu,ZnSOD was found to exert a peroxidative effect upon reaction with $H_2O_2$.[37,38] SOD demonstrates a bell-shaped

---

[31] A. Samuni, D. Godinger, J. Aronovitch, A. Russo, and J. B. Mitchell, *Biochemistry* **30,** 555 (1991).
[32] R. J. Mehlhorn and C. E. Swanson, *Free Radic. Res. Commun.* **17,** 157 (1992).
[33] U. A. Nilsson, L. I. Olsson, G. Carlin, and F. A. Bylund, *J. Biol. Chem.* **264,** 11131 (1989).
[34] V. W. Bowery and K. U. Ingold, *J. Am. Chem. Soc.* **114,** 4992 (1992).
[35] T. Offer, M. Mohsen, and A. Samuni, *Free Radic. Biol. Med.* **25,** 832 (1998).
[36] P. R. Gardner, D. D. Nguyen, and C. W. White, *Arch. Biochem. Biophys.* **325,** 20 (1996).
[37] M. B. Yim, P. B. Chock, and E. R. Stadtman, *Proc. Natl. Acad. Sci. U.S.A.* **87,** 5006 (1990).
[38] S. L. Jewett, A. M. Rocklin, M. Ghanevati, J. M. Abel, and J. A. Marach, *Free Radic. Biol. Med.* **26,** 905 (1999).

dose–response curve in several biological systems.[39–42] A cytotoxic effect due to increased peroxidase activity[43] and lipid peroxidation[42,44] was found in cells. SOD was also reported to amplify sensitivity to ionizing radiation.[45] Apparently, higher than normal levels of SOD, as in transgenic mice or Down syndrome patients, can exert adverse effects or even be lethal.[43] The information concerning potential adverse effects of nitroxide SOD mimics is scarce, yet they were reported to exert mutagenicity in *Salmonella typhimurium*,[46] potentiate the mutagenic effect of $H_2O_2$ in *Salmonella typhimurium* TA 104,[47] cause stress and eventually some necrosis in thymocytes,[48] increase the activity of cytochrome $P_{450}$ oxidase, elevate the production of $O_2^{·-}$ and $H_2O_2$ in human tumor cell lines, and inhibit cell growth possibly by triggering an apoptotic mechanism.[49]

Dose Response

The dose–response relationship of native SOD is not always a monotonous function line but has been found in many cases to exhibit a bell shape.[40,41] In other words, higher doses of SOD lost their protective effects and sometimes exerted even adverse effects.[44,45,50] It is possible that similar behavior characterizes SOD mimics as well.[51] A study with papain as the target for oxidative inactivation demonstrated that nitroxide SOD mimics exhibit a bell-shaped dose–response relationship. These findings offer an additional explanation for the pro-oxidative activity of SOD and SOD mimics, without invoking any dual activity of $O_2^{·-}$ or a combined effect of SOD and $H_2O_2$. The most significant outcome of an increase in SOD level is a decrease in $[O_2^{·-}]_{\text{steady state}}$, rather than any notable elevation of $[H_2O_2]_{\text{steady state}}$. Consequently, the reaction kinetics of the high oxidation state of each catalyst are altered. In the presence of an ultralow $[O_2^{·-}]_{\text{steady state}}$, the oxidized form of SOD

[39] M. Bernier, A. S. Manning, and D. J. Hearse, *Am. J. Physiol.* **256**, 1344 (1989).
[40] B. A. Omar and J. M. McCord, *Free Radic. Biol. Med.* **9**, 473 (1990).
[41] B. A. Omar, N. M. Gad, M. C. Jordan, S. P. Striplin, W. J. Russell, J. M. Downey, and J. M. McCord, *Free Radic. Biol. Med.* **9**, 465 (1990).
[42] J. M. McCord, in "Oxidative Stress in Cancer, AIDS, and Neurodegenerative Diseases" (L. Montagnier, R. Olivier, and C. Pasquier, eds.). Marcel Dekker, New York, 1998.
[43] K. H. Norris and P. Hornsby, *Mutat. Res.* **237**, 95 (1990).
[44] O. Elroy-Stein, Y. Bernstein, and Y. Groner, *EMBO J.* **5**, 615 (1986).
[45] M. D. Scott, S. R. Meshnick, and J. W. Eaton, *J. Biol. Chem.* **264**, 2498 (1989).
[46] B. C. D. Gallez, R. Debuyst, F. Dejehet, and P. Dumont, *Toxicol Lett.* **63**, 35 (1992).
[47] H. Sies and R. Mehlhorn, *Arch. Biochem. Biophys.* **251**, 393 (1986).
[48] A. F. Slater, C. S. Nobel, E. Maellaro, J. Bustamante, M. Kimland, and S. Orrenius, *Biochem. J.* **306**, 771 (1995).
[49] M. B. Gariboldi, S. Lucchi, C. Caserini, R. Supino, C. Oliva, and E. Monti, *Free Radic. Biol. Med.* **24**, 913 (1998).
[50] M. D. Scott, S. R. Meshnick, and J. W. Eaton, *J. Biol. Chem.* **262**, 3640 (1987).
[51] T. Offer, A. Russo, and A. Samuni, *FASEB J.* **14**, 1215 (2000).

[Cu(II),ZnSOD] or nitroxide SOD mimic (oxoammonium cation) does not react with $O_2^{\cdot-}$ but, rather, oxidizes the target molecule that it was supposed to have protected. As a result, these catalysts can exert an anti- or pro-oxidative effect depending on their concentration.

Ternary Complexes

The binding of SOD mimic to components of the biological system might affect its kinetics of reaction with $O_2^{\cdot-}$. This consideration is more relevant to metal chelates that serve as SOD mimics because ternary complexes thus formed might be less efficient catalysts. On the other hand, such binding was found to assist mobilization and distribution of the catalyst in the tissue.[21] No evidence has been collected, so far, indicating formation of such complexes with nitroxide SOD mimics.

Conclusion

Modified SOD enzymes might persist longer and overcome the limitation of a short metabolic half-life, yet their penetration into the cells is not improved. Synthetic low molecular weight SOD mimics can readily enter the cell and, bound to appropriate ligands, may also be directed to selected tissues or cellular compartments. For an SOD mimic to operate effectively in a biological system it should be (1) nontoxic, (2) nonimmunogenic, (3) cell permeable, (4) chemically and metabolically stable, (5) persistent in the circulation, (6) unreactive toward $O_2$ and $H_2O_2$, and (7) unreactive toward other cellular components. These requirements are not easy to meet. Therefore, some efficient SOD mimics *in vitro* may fail to operate *in vivo*, which explains why it is so difficult to find an efficient SOD mimic for *in vivo* conditions.

Acknowledgment

This research was supported by the Israel Science Foundation of the Israel Academy of Sciences (Jerusalem, Israel).

## [24] Superoxide Reductase Activities of Neelaredoxin and Desulfoferrodoxin Metalloproteins

By FRANK RUSNAK, CARLA ASCENSO, ISABEL MOURA, and JOSÉ J. G. MOURA

Superoxide dismutases (SODs) are found in nearly every aerobic organism and represent one form of defense against the toxic effects of reactive oxygen species such as $O_2^-$.[1,2] SOD catalyzes the disproportionation of $O_2^-$ according to Eq. (1):

$$O_2^- + O_2^- + 2H^+ \leftrightarrow H_2O_2 + O_2 \qquad (1)$$

The reaction actually occurs in two, one-electron transfer steps, with one molecule of $O_2^-$ becoming oxidized to oxygen, the other becoming reduced to hydrogen peroxide. SODs contain redox active metal ions, either zinc and copper, iron, manganese, or nickel, which are capable of being either oxidized or reduced by superoxide in a bidirectional electron transfer mechanism.[3-5]

A novel pathway for $O_2^-$ detoxification was revealed by Adams and co-workers[6] in the discovery of an enzyme catalyzing only one of the two redox reactions of SODs, the one-electron reduction of $O_2^-$ to hydrogen peroxide according to Eq. (2):

$$O_2^- + e^- + 2H^+ \leftrightarrow H_2O_2 \qquad (2)$$

Termed superoxide reductase (SOR), biochemical analysis revealed that SOR was a member of a family of iron-containing proteins previously isolated from sulfate-reducing bacteria of the genus *Desulfovibrio*.[7,8] This article summarizes biochemical, spectroscopic, and enzymological characteristics of this newly discovered class of enzymes.

---

[1] J. D. Crapo, J. M. McCord, and I. Fridovich, *Methods Enzymol.* **53**, 382 (1978).
[2] I. Fridovich, *Annu. Rev. Biochem.* **64**, 97 (1995).
[3] J. S. Valentine, L. M. Ellerby, J. A. Graden, C. R. Nishida, and E. B. Grallan, in "Bioinorganic Chemistry: An Inorganic Perspective on Life" (D. P. Kessissoglou, ed.), p. 77. Kluwer Academic Publishers, Dordrecht, 1995.
[4] H.-D. Youn, E.-J. Kim, J.-H. Roe, Y. C. Hah, and S.-O. Kang, *Biochem. J.* **318**, 889 (1996).
[5] H.-D. Youn, H. Youn, J.-W. Lee, Y.-I. Yim, J. K. Lee, Y. C. Hah, and S.-O. Kang, *Arch. Biochem. Biophys.* **334**, 341 (1996).
[6] F. E. Jenney, Jr., M. F. J. M. Verhagen, X. Cui, and M. W. W. Adams, *Science* **286**, 306 (1999).
[7] I. Moura, P. Tavares, J. J. G. Moura, N. Ravi, B. H. Huynh, M.-Y. Liu, and J. Le Gall, *J. Biol. Chem.* **265**, 21596 (1990).
[8] L. Chen, P. Sharma, J. Le Gall, A. M. Mariano, M. Teixeira, and A. V. Xavier, *Eur. J. Biochem.* **226**, 613 (1994).

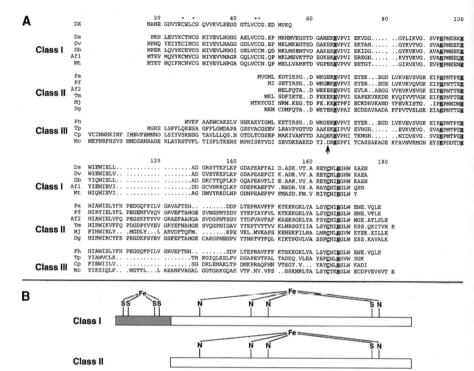

FIG. 1. Multiple sequence alignment of the neelaredoxin and desulfoferrodoxin family of enzymes. (A) The 14 family members listed are separated into three distinct groups on the basis of the presence or absence of an N-terminal domain. All family members contain a conserved C terminus accommodating a mononuclear iron site with $N_4S$ ligation provided by four histidines and a cysteine residue (boldface and underlined). Proteins in class I (desulfoferrodoxins) contain an N-terminal domain that binds a second mononuclear iron site in a distorted tetrahedral $S_4$ coordination, homologous to desulforedoxin (DX). Proteins in class II (neelaredoxins) do not contain an N-terminal domain. Proteins in class III contain an N-terminal extension distinct from the DX domain. The multiple sequence alignment was done with the Wisconsin Package [version 10.0; Genetics Computer Group (GCG), Madison, WI]. Abbreviations are defined as follows: DX, *Desulfovibrio gigas* desulforedoxin [Brumlik *et al., J. Bacteriol.* **172**, 7289 (1990)]; Ds, *D. desulfuricans* desulfoferrodoxin [Devreese *et al., FEBS Lett.* **385**, 138 (1996)]; Dv, *D. vulgaris* desulfoferrodoxin [Brumlik *et al., J. Bacteriol.* **171**, 4996 (1989)]; Db, *Desulfoarculus baarsii* desulfoferrodoxin [Pianzzola *et al., J. Bacteriol.* **178**, 6736 (1996)]; Af1, *Archaeoglobus fulgidus* desulfoferrodoxin [Klenk *et al., Nature (London)* **390**, 364 (1997)]; Mt, *Methanobacterium thermoautotrophicum* desulfoferrodoxin [Smith *et al., J. Bacteriol.* **179**, 7135 (1997)]; Pa, *Pyrococcus abyssi* neelaredoxin; Pf, *P. furiosus* superoxide reductase [Jenney, Jr. *et al., Science* **286**, 306 (1999)]; Af2, *A. fulgidus* neelaredoxin [Klenk *et al., Nature (London)* **390**, 364 (1997)]; Tm, *Thermotoga maritima* neelaredoxin [Nelson *et al., Nature (London)* **399**, 323 (1999)]; Mj, *Methanococcus jannaschii* neelaredoxin [Bult *et al., Science* **273**, 1058 (1996)]; Dg, *D. gigas* neelaredoxin [Silva *et al., Eur. J. Biochem.* **259**, 235 (1999)]; Ph, *P. horikoshii* neelaredoxin [Kawarabayasi *et al., DNA Res.* **5**, 55 (1998); Kawarabayasi *et al., DNA Res.* **5**, 147 (1998)]; Tp, *Treponema pallidum* neelaredoxin [Fraser *et al., Science* **281**, 375 (1998)]; Cp, *Clostridium acetobutylicum* neelaredoxin (http://www.genomecorp.com); Mc, *Magnetococcus* MC-1 (http://www.tigr.org/). The dots indicate cysteine residues that provide

# I. Superoxide Reductase Primary Structure

The first members of this family were isolated from *Desulfovibrio desulfuricans* and *Desulfovibrio vulgaris*. Because the function of this protein was unknown at the time, it was given the descriptive name desulfoferrodoxin (Dfx).[7] Later, a homologous protein termed neelaredoxin (Nlr) was isolated from *Desulfovibrio gigas*.[8] With the advent of genomic sequencing, at least 14 homologs, comprising three distinct protein groups, have been identified (Fig. 1[6,9–24]). All enzymes

---

[9] M. J. Brumlik, G. Leroy, M. Bruschi, and G. Voordouw, *J. Bacteriol.* **172**, 7289 (1990).
[10] B. Devreese, P. Tavares, J. Lampreia, N. Van Damme, J. Le Gall, J. J. G. Moura, J. Van Beeumen, and I. Moura, *FEBS Lett.* **385**, 138 (1996).
[11] M. J. Brumlik and G. Voordouw, *J. Bacteriol.* **171**, 4996 (1989).
[12] M. J. Pianzzola, M. Soubes, and D. Touati, *J. Bacteriol.* **178**, 6736 (1996).
[13] H.-P. Klenk, R. A. Clayton, J.-F. Tomb, O. White, K. E. Nelson, K. A. Ketchum, R. J. Dodson, M. Gwinn, E. K. Hickey, J. D. Peterson, D. L. Richardson, A. R. Kerlavage, D. E. Graham, N. C. Kyrpides, R. J. Fleischmann, J. Quackenbush, N. H. Lee, G. G. Sutton, S. Gill, E. F. Kirkness, B. A. Dougherty, K. McKenney, M. D. Adams, B. Loftus, S. Peterson, C. I. Reich, L. K. McNeil, J. H. Badger, A. Glodek, L. Zhou, R. Overbeek, J. D. Gocayne, J. F. Weidman, L. McDonald, T. Utterback, M. D. Cotton, T. Spriggs, P. Artiach, B. P. Kaine, S. M. Sykes, P. W. Sadow, K. P. D'Andrea, C. Bowman, C. Fujii, S. A. Garland, T. M. Mason, G. J. Olsen, C. M. Fraser, H. O. Smith, C. R. Woese, and J. C. Venter, *Nature (London)* **390**, 364 (1997).
[14] D. R. Smith, L. A. Doucette-Stamm, C. Deloughery, H. Lee, J. Dubois, T. Aldredge, R. Bashirzadeh, D. Blakely, R. Cook, K. Gilbert, D. Harrison, L. Hoang, P. Keagle, W. Lumm, B. Pothier, D. Qui, R. Spadafora, R. Vicaire, Y. Wang, J. Wierzbowski, R. Gibson, N. Jiwani, A. Caruso, D. Bush, H. Safer, D. Patwell, S. Prabhakar, S. McDougall, G. Shimer, A. Goyal, S. Pietrovski, G. M. Church, C. J. Daniels, J.-I. Mao, P. Rice, J. Nölling, and J. N. Reeve, *J. Bacteriol.* **179**, 7135 (1997).
[15] K. E. Nelson, R. A. Clayton, S. R. Gill, M. L. Gwinn, R. J. Dodson, D. H. Haft, E. K. Hickey, J. D. Peterson, W. C. Nelson, K. A. Ketchum, L. McDonald, T. R. Utterback, J. A. Malek, K. D. Linher, M. M. Garrett, A. M. Stewart, M. D. Cotton, M. S. Pratt, C. A. Phillips, D. Richardson, J. Heidelberg, G. G. Sutton, R. D. Fleischmann, J. A. Eisen, O. White, S. L. Salzberg, H. O. Smith, J. C. Venter, and C. M. Fraser, *Nature (London)* **399**, 323 (1999).

---

ligands to the iron ion of center I in desulfoferrodoxins and *D. gigas* desulforedoxin. Residues highlighted as boldface/underlined represent the residues that coordinate the iron ion of center II in *D. desulfuricans* ATCC 27774 desulfoferrodoxin [Coelho *et al., J. Biol. Inorg. Chem.* **2**, 680 (1997)] and *P. furiosus* superoxide reductase [Yeh *et al., Biochemistry* **39**, 2499 (2000)]; a glutamate residue (E14) in *P. furiosus* superoxide reductase is also a ligand in two of four subunits in the unit cell (indicated by an arrow). A search of several unfinished genomes at www.tigr.org (e.g., *Dehalococcoides ethenogenes, Geobacter sulfurreducens,* and *Treponema denticola*) suggests that several more homologs will be discovered when these sequencing projects are eventually completed. (B) Schematic of the domain structure of the three groups of superoxide reductases. All three have a conserved C-terminal domain accommodating the active site iron atom coordinated by four histidine nitrogens (N) and a cysteine sulfur (S). Desulfoferrodoxins are represented in group I because of the presence of an N-terminal domain that binds a second iron atom in a distorted tetrahedral geometry of four cysteine sulfur atoms (gray). Class II enzymes contain only the active site domain. Class III enzymes have unique N termini (black) of unknown function or metal binding capacity. [Adapted with permission from Jovanović *et al., J. Biol. Chem.* **275**, 28439 (2000), copyright © 2000, the American Society for Biochemistry and Molecular Biology; and Ascenso *et al., J. Biol. Inorg. Chem.* **5**, 720 (2000), copyright © 2001, the Society of Biological Inorganic Chemistry.]

in this family share a conserved domain of ~100 amino acids accommodating a mononuclear iron site, now known to be the site of $O_2^-$ reduction. Within this domain, four histidines and one cysteine residue, shown in boldface/underlined type in Fig. 1, are strictly conserved. As is seen below, these residues comprise the metal-binding site. The primary factor distinguishing the three groups is the presence or absence of an N-terminal domain. Enzymes in the first group, represented by Dfx, bind two iron atoms in distinct centers (centers I and II). In Dfx, the active site iron atom resides in center II whereas the iron atom of center I is accommodated within an N-terminal domain homologous to desulforedoxin (Dx), an iron–sulfur protein from *D. gigas* of 36 residues containing a mononuclear iron site coordinated by four conserved cysteine residues in a distorted tetrahedral arrangement.[25] Enzymes in the second group are missing the N-terminal Dx-like domain and are historically referred to as neelaredoxins (Nlrs), in reference to

---

[16] C. J. Bult, O. White, G. J. Olsen, L. Zhou, R. D. Fleischmann, G. G. Sutton, J. A. Blake, L. M. FitzGerald, R. A. Clayton, J. D. Gocayne, A. R. Kerlavage, B. A. Dougherty, J.-F. Tomb, M. D. Adams, C. I. Reich, R. Overbeek, E. F. Kirkness, K. G. Weinstock, J. M. Merrick, A. Glodek, J. L. Scott, N. S. M. Geoghagen, J. F. Weidman, J. L. Fuhrmann, D. Nguyen, T. R. Utterback, J. M. Kelley, J. D. Peterson, P. W. Sadow, M. C. Hanna, M. D. Cotton, K. M. Roberts, M. A. Hurst, B. P. Kaine, M. Borodovsky, H.-P. Klenk, C. M. Fraser, H. O. Smith, C. R. Woese, and J. G. Venter, *Science* **273**, 1058 (1996).

[17] G. Silva, S. Oliveira, C. M. Gomes, I. Pacheco, M. Y. Liu, A. V. Xavier, M. Teixeira, J. LeGall, and C. Rodrigues-Pousada, *Eur. J. Biochem.* **259**, 235 (1999).

[18] Y. Kawarabayasi, M. Sawada, H. Horikawa, Y. Haikawa, Y. Hino, S. Yamamoto, M. Sekine, S.-i. Baba, H. Kosugi, A. Hosoyama, Y. Nagai, M. Sakai, K. Ogura, K. Otsuka, H. Nakazawa, M. Takamiya, Y. Ohfuku, T. Funahashi, T. Tanaka, Y. Kudoh, J. Yamazaki, N. Kushida, A. Oguchi, K.-i. Aoki, T. Yoshizawa, Y. Nakamura, F. T. Robb, K. Horikoshi, Y. Masuchi, H. Shizuya, and H. Kikuchi, *DNA Res.* **5**, 55 (1998).

[19] Y. Kawarabayasi, M. Sawada, H. Horikawa, Y. Haikawa, Y. Hino, S. Yamamoto, M. Sekine, S.-i. Baba, H. Kosugi, A. Hosoyama, Y. Nagai, M. Sakai, K. Ogura, K. Otsuka, H. Nakazawa, M. Takamiya, Y. Ohfuku, T. Funahashi, T. Tanaka, Y. Kudoh, J. Yamazaki, N. Kushida, A. Oguchi, K.-i. Aoki, T. Yoshizawa, Y. Nakamura, F. T. Robb, K. Horikoshi, Y. Masuchi, H. Shizuya, and H. Kikuchi, *DNA Res.* **5**, 147 (1998).

[20] C. M. Fraser, S. J. Norris, G. M. Weinstock, O. White, G. G. Sutton, R. Dodson, M. Gwinn, E. K. Hickey, R. Clayton, K. A. Ketchum, E. Sodergren, J. M. Hardham, M. P. McLeod, S. Salzberg, J. Peterson, H. Khalak, D. Richardson, J. K. Howell, M. Chidambaram, T. Utterback, L. McDonald, P. Artiach, C. Bowman, M. D. Cotton, C. Fujii, S. Garland, B. Hatch, K. Horst, K. Roberts, M. Sandusky, J. Weidman, H. O. Smith, and J. C. Venter, *Science* **281**, 375 (1998).

[21] A. V. Coelho, P. Matias, V. Fülöp, A. Thompson, A. Gonzalez, and M. A. Carrondo, *J. Biol. Inorg. Chem.* **2**, 680 (1997).

[22] A. P. Yeh, Y. Hu, F. E. Jenney, Jr., M. W. W. Adams, and D. C. Rees, *Biochemistry* **39**, 2499 (2000).

[23] T. Jovanović, C. Ascenso, K. R. O. Hazlett, R. Sikkink, C. Krebs, R. Litwiller, L. M. Benson, I. Moura, J. J. G. Moura, J. D. Radolf, B. H. Huynh, S. Naylor, and F. Rusnak, *J. Biol. Chem.* **275**, 28439 (2000).

[24] C. Ascenso, F. Rusnak, I. Cabrito, M. J. Lima, S. Naylor, I. Moura, and J. J. G. Moura, *J. Biol. Inorg. Chem.* **5**, 720 (2000).

[25] J. J. G. Moura, B. J. Goodfellow, M. J. Romão, F. Rusnak, and I. Moura, *Comments Inorg. Chem.* **19**, 47 (1996).

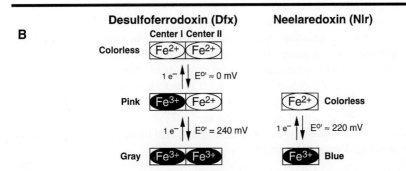

FIG. 2. (A) Schematic of the mononuclear iron sites in desulfoferrodoxin from *D. desulfuricans*, based on the 1.9-Å resolution X-ray structure [Coelho et al., J. Biol. Inorg. Chem. **2**, 680 (1997)] and neelaredoxin from *Pyrococcus furiosus* [Yeh et al., Biochemistry **39**, (2000)]. The amino acid numbering for center II is based on the sequence for the enzyme from *D. desulfuricans*. [Adapted with permission from Ascenso et al., J. Biol. Inorg. Chem. **5**, 720 (2000), copyright © 2000, Society of Biological Inorganic Chemistry. (B) Schematic showing the different oxidation states attainable for centers I and II in Dfx and the iron ion in Nlr.

the prototype from *D. gigas*.[8] The third group contains an N-terminal domain of unknown function. At least in the homolog from *Treponema pallidum*, this domain does not bind a second iron atom.[23,26]

## II. Superoxide Reductase Active Site Structure

The determination of the three-dimensional structures of Dfx[21] and Nlr[22] revealed a novel $N_4S$ coordination to the active site iron atom. In the low-temperature, dithionite-reduced form of Nlr, the geometry is square pyramidal, with four imidazole nitrogens from the conserved histidines coordinating the iron atom in equitorial positions and the conserved cysteine residue forming an axial fifth ligand (Fig. 2A).[21,22,24] All histidines coordinate to the iron atom via $N_\varepsilon$ except one

[26] M. Lombard, D. Touati, M. Fontecave, and V. Nivière, *J. Biol. Chem.* **275**, 27021 (2000).

(His-118; Fig. 2A), which coordinates through $N_\delta$. In the structure of oxidized Nlr at room temperature, two of four subunits in the unit cell had the pentacoordinate geometry shown in Fig. 2A, whereas a glutamate carboxylate provided a sixth ligand to the iron in the other two subunits. At low temperature, all four subunits in the unit cell of the oxidized protein had octahedral geometry: two with an axial glutamate, the other two with a solvent molecule coordinated in the axial position.[22] The presence of this glutamate in the coordination sphere of the iron as well as its conservation in nearly every family member (Fig. 1) implies an important catalytic role that may well involve protonation of a putative peroxo intermediate.[27] Although a pentacoordinate geometry was reported for the ferricyanide-oxidized form of Dfx,[21] a report has suggested that the active site iron ion in that crystal was probably in the ferrous state due to photoreduction by the X-ray source.[27]

In the 1.9-Å structure of Dfx, the 34 amino acids of the N terminus fold into an independent domain (domain I) with a topology superimposed on the structure of Dx with a root-mean-square deviation of 0.59 Å for main-chain atoms.[21,28] The iron center in domain I has the same distorted tetrahedral $FeS_4$ coordination as it does in Dx, with average Fe–S bond lengths of 2.3 Å and a widened S–Fe–S angle of 120 to 126° between vicinyl Cys-28 and Cys-29. The distance between center I and center II iron ions within the Dfx monomer is 22 Å.[21] The independent, modular nature of the two domains of Dfx was demonstrated by recombinant expression of the separate domains.[24]

## III. Overexpression and Purification of Recombinant Forms of Superoxide Reductase

Although superoxide reductases have been purified from native sources,[7,8,17,29–32] this approach requires growing several hundred liters of a bacterial culture in order to obtain enough protein for biochemical and spectroscopic studies. Another approach relies on overexpression of the recombinant protein in *Escherichia coli*.[23,24,26,27,32,33] Recombinant overexpression has the advantage of

---

[27] E. D. Coulter, J. P. Emerson, D. M. Kurtz, Jr., and D. E. Cabelli, *J. Am. Chem. Soc.* **122,** 11555 (2000).
[28] M. Archer, R. Huber, P. Tavares, I. Moura, J. J. G. Moura, M. A. Carrondo, L. C. Sieker, J. LeGall, and M. J. Romão, *J. Mol. Biol.* **251,** 690 (1995).
[29] P. Tavares, N. Ravi, J. J. G. Moura, J. Le Gall, Y.-H. Huang, B. R. Crouse, M. K. Johnson, B. H. Huynh, and I. Moura, *J. Biol. Chem.* **269,** 10504 (1994).
[30] M. F. J. M. Verhagen, W. G. B. Voorhorst, J. A. Kolkman, R. B. G. Wolbert, and W. R. Hagen, *FEBS Lett.* **336,** 13 (1993).
[31] C. V. Romão, M. Y. Liu, J. LeGall, C. M. Gomes, V. Braga, I. Pacheco, A. V. Xavier, and M. Teixeira, *Eur. J. Biochem.* **261,** 438 (1999).
[32] I. A. Abreu, L. M. Saraiva, J. Carita, H. Huber, K. O. Stetter, D. Cabelli, and M. Teixeira, *Mol. Microbiol.* **38,** 322 (2000).
[33] M. Lombard, M. Fontecave, D. Touati, and V. Nivière, *J. Biol. Chem.* **275,** 115 (2000).

producing more protein per liter of culture medium, making large-scale cultures unnecessary and often simplifying the purification procedure. Recombinant overexpression also provides a convenient system in which to perform site-directed mutagenesis of the protein of interest.

A. *Procedure for Cloning Superoxide Reductases*

1. The gene encoding the protein of interest is first amplified by the polymerase chain reaction (PCR), using genomic DNA from the parent organism and a pair of oligonucleotides homologous to the 5' and 3' ends of the gene. Additional DNA sequence representing unique restriction endonuclease recognition sites is added at the 5' end of each primer to facilitate subsequent subcloning into an expression vector.

2. After PCR amplification, the PCR product is purified by electrophoresis in an 8% (w/v) polyacrylamide gel in 90 m$M$ Tris-HCl, 90 m$M$ boric acid, 2.5 m$M$ EDTA buffer. The fragment is excised and the DNA fragment is electroeluted from the gel, using an Irvine Biomedical (Irvine, CA) model UEA unidirectional electroeluter according to the manufacturer instructions.

3. The DNA fragment is precipitated by addition of 2.5 volumes of ethanol. After digestion with the appropriate restriction enzymes and heat inactivation of the restriction enzymes, the fragment is ligated into the plasmid pT7-7,[34] digested similarly. This plasmid contains a multiple cloning site downstream of the T7 promoter.

4. The resulting ligation mix is transformed into a recombination-negative strain of *E. coli* (e.g., DH5-α) and plated on Luria–Bertani (LB) agar containing ampicillin (0.1 mg/ml). After growth at 37°, individual colonies are transferred to 5 ml of LB medium containing ampicillin (0.1 mg/ml) (LB/amp) and grown overnight at 37°.

5. Plasmid DNA from each culture is isolated with the Promega (Madison, WI) Wizard Plus Miniprep DNA purification system.

6. Positive clones are identified by restriction digest analysis of plasmid DNA. The entire gene should be sequenced to ensure that secondary mutations have not been introduced. Plasmids containing the correct sequence can be used for protein overexpression in *E. coli*.

B. *Overexpression and Purification of Recombinant Superoxide Reductases: Treponema pallidum Neelaredoxin*

1. Competent *E. coli* BL21 (DE3) cells (Novagen, Madison, WI) are transformed with the plasmid obtained as described above and plated onto LB agar

[34] S. Tabor, *in* "Current Protocols in Molecular Biology" (F. M. Ausubel, R. Brent, R. E. Kingston, D. D. Moore, J. G. Seidman, J. A. Smith, and K. Struhl, eds.), p. 16.2.1. Greene Publishing and Wiley Interscience, New York, 1990.

plates containing ampicillin (0.1 mg/ml). This strain contains an inducible copy of the T7 RNA polymerase necessary for transcription from the T7 promoter.

2. A colony is used to inoculate 10 ml of LB/amp (0.1 mg/ml) medium. This culture is grown at 37° overnight and the next morning used to inoculate 1.5 liters of LB medium containing ampicillin (0.1 mg/ml) in a 6-liter Erlenmeyer flask. This culture is grown at 37° with shaking until the absorbance of the culture at 595 nm reaches 0.8. At this point, isopropyl-$\beta$-D-thiogalactopyranoside (IPTG) is added to a final concentration of 1 m$M$.

3. After 6 hr, the cells are isolated at 4° by centrifugation at 2500$g$.

4. The cell pellet is resuspended in 50 m$M$ Tris-HCl, pH 7.8, and lysed by three passages through a French pressure cell operating at 15,000 lb/in$^2$ at the orifice. The resulting cell lysate should be centrifuged for at least 60 min at 39,100$g$ to remove unbroken cells and cell debris.

5. The supernatant is subsequently purified by a combination of various chromatographies such as ion-exchange or gel filtration. What follows is a typical purification of recombinant *T. pallidum* neelaredoxin expressed in *E. coli* BL21 (DE3) cells.

6. The supernatant is applied to a column (2.6 × 17.5 cm) containing DEAE-Sepharose CL6B anion-exchange resin equilibrated with 20 m$M$ Tris-HCl, pH 7.8. After loading, the column is washed with the same buffer and the protein is eluted from the column by use of a 0 to 1.0 $M$ NaCl linear gradient in the same buffer. *Treponema pallidum* neelaredoxin elutes as a blue fraction at an NaCl concentration of ~0.1 $M$. Fractions can be analyzed by sodium dodecyl sulfate–polyacrylamide gel electrophoresis (SDS–PAGE) or assayed for superoxide reductase activity as described in Section VI. The best fractions in terms of purity and activity are pooled for the next chromatography step.

7. Pooled fractions are concentrated to ≤10 ml, using an Amicon (Danvers, MA) concentrator equipped with a PM10 membrane, and applied to a column containing Sephadex G-75 gel-filtration resin equilibrated with 50 m$M$ Tris-HCl, 0.3 $M$ NaCl, pH 7.8. The protein is eluted from this column with the same buffer.

## IV. Oxidation States and Redox Properties of Centers I and II of Superoxide Reductases

With only one redox-active iron center in Nlr, there are only two relevant oxidation states, $Fe^{2+}$–Nlr and $Fe^{3+}$–Nlr (Fig. 2B). The redox potential for this center is ~220 mV (Table I[8,23–25,29,30,32,35,36]). A similar redox potential has been

[35] I. Moura, A. V. Xavier, R. Cammack, M. Bruschi, and J. Le Gall, *Biochem. Biophys. Acta* **533**, 156 (1978).
[36] P. Tavares, J. K. Wunderlich, S. G. Lloyd, J. LeGall, J. J. G. Moura, and I. Moura, *Biochem. Biophys. Res. Commun.* **208**, 680 (1995).

TABLE I
REDOX POTENTIALS OF METALLOPROTEINS IMPLICATED IN SUPEROXIDE REDUCTION

| Protein | Organism | Center I[b] | Center II | Ref. |
|---|---|---|---|---|
| Rubredoxin | Several | −50 to 0 | | 25 |
| Desulforedoxin | *Desulfovibrio gigas* | −35 ± 10 | | 35, 36 |
| Dfx N-terminal peptide | *Desulfovibrio vulgaris* | −2 | | 24 |
| Desulfoferrodoxin | *Desulfovibrio desulfuricans* | 4 ± 10 | 240 ± 10 | 29 |
| | *Desulfovibrio vulgaris* | −4 ± 5 | 90 | 30 |
| Neelaredoxin | *Desulfovibrio gigas* | | 190 ± 20 | 8 |
| | *Treponema pallidum* | | 201 ± 4 | 23 |
| | *Archaeoglobus fulgidus* | | 230 | 32 |
| Dfx C-terminal peptide | *Desulfovibrio vulgaris* | | 247 | 24 |

[a] Redox potential relative to normal hydrogen electrode (NHE).
[b] Rubredoxin and desulforedoxin Fe(S–Cys)$_4$ sites listed under the center I category for comparison.

measured for center II of Dfx (Table I). The additional presence of center I in Dfx, which has a midpoint potential near 0 mV (Table I), necessitates that Dfx can exist in three distinct oxidation states: the colorless fully reduced state, with both center I and II in the Fe$^{2+}$ oxidation state; a pink mixed valence form, with center I oxidized (Fe$^{3+}$) and center II reduced (Fe$^{2+}$); and a gray form, in which both centers are in the Fe$^{3+}$ oxidation state (Fig. 2B).

With Dfx, electrons from an as yet unidentified electron donor should be delivered to the active site via center I. Because many biological redox-active compounds and electron transfer proteins have a redox potential more negative than the potential of this center, it is likely that most could, at least thermodynamically, transfer electrons to SOR. Indeed, Lombard *et al.*[33] have demonstrated that unidentified components in cell extracts of *E. coli*, a bacterium that does not encode an SOR within its genome, are capable of donating electrons to Dfx from either NADH or NADPH. These unidentified electron transfer components are presumably also responsible for the ability of SOR to suppress the phenotype of SOD-deficient *E. coli* cells.[12,26,37] A similar phenomenon has been observed with cell extract from *Archaeoglobus fulgidus*.[32]

The electron transfer protein rubredoxin has been shown to be capable of serving as an immediate electron donor to *Pyrococcus furiosus* Nlr.[6] With a midpoint potential close to that observed for Dx (Table I), and a conserved Fe(S–Cys)$_4$ site homologous to center I, it is reasonable to assume that rubredoxin or Dx could

[37] S. I. Liochev and I. Fridovich, *J. Biol. Chem.* **272**, 25573 (1997).

participate in the electron transport chain from reduced pyridine nucleotides, ultimately supplying electrons to Nlr or Dfx for $O_2^-$ reduction.

## A. Preparation of Reduced and Oxidized Forms of Superoxide Reductase: Neelaredoxin

Homogeneous forms of the two oxidation states of Nlr are relatively easy to prepare for biochemical and spectroscopic analysis. As isolated, Nlr is often present as a mixture of ferric and ferrous species. Treatment with a slight excess of either an oxidant or reductant followed by removal of either agent by rapid gel-filtration chromatography can afford the relevant redox species in quantitative yield. Potassium ferricyanide [$K_3Fe(CN)_6$] is a convenient oxidant whereas either sodium ascorbate or sodium dithionite can be used to reduce the enzyme. All yield products with little if any contributions to the optical region of the visible spectrum that might obscure features of superoxide reductase. Nevertheless, after either oxidation or reduction, the ferrous and ferric forms of Nlr are sufficiently stable such that rapid gel-filtration chromatography can be used to desalt each sample.

*Preparation of Oxidized Neelaredoxin.* A sample of oxidized protein is obtained by treating a sample of the purified protein with a slight excess of 15 m$M$ $K_3[Fe(CN)_6]$ in 50 m$M$ Tris-HCl, pH 7.8. A slight excess of $K_3[Fe(CN)_6]$ is defined as the amount of $K_3[Fe(CN)_6]$ added that no longer produces any further increase in the absorbance at 656 nm. The sample is incubated for 10 min at room temperature. The excess $K_3[Fe(CN)_6]$ is removed by passage of the sample over an NAP-25 column containing G25 Sephadex resin (Pharmacia, Uppsala, Sweden) equilibrated in 50 m$M$ Tris-HCl (pH 7.8) buffer. Blue-colored fractions containing the protein are pooled and used immediately.

*Preparation of Reduced Neelaredoxin.* The reduced protein is obtain by treating a sample of the protein with a slight excess of a freshly prepared solution of 15 m$M$ sodium ascorbate in 50 mM Tris-HCl, pH 7.8. A slight excess is defined the amount of sodium ascorbate added that is sufficient to eliminate the absorbance of the oxidized protein at 656 nm. Excess reductant is removed with an NAP-25 column as described above, except that fractions are colorless and are analyzed for protein either by their absorbance at 280 nm or by a protein dye method.

## B. Preparation of Reduced and Oxidized Forms of Superoxide Reductase: Desulfoferrodoxin

Generation of reduced Dfx can be accomplished after treatment of the enzyme with sodium dithionite, whereas either potassium ferricyanide or superoxide can be used to oxidize center II to the gray form as described above for Nlr. Poising Dfx in the intermediate mixed valence state may pose some challenges. Recombinant forms of Dfx are often isolated in the pink mixed valent state, thus representing one manner of producing this intermediate oxidation state. At least in one case, the pink

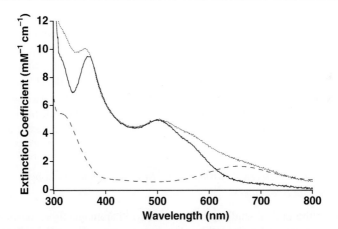

FIG. 3. Visible spectra of *D. vulgaris* desulfoferrodoxin and *T. pallidum* neelaredoxin. Solid line, *D. vulgaris* Dfx pink form; dotted line, *D. vulgaris* Dfx gray form; dashed line, *T. pallidum* Nlr oxidized (blue) form. The spectrum of *T. pallidum* Nlr is adapted from Jovanović *et al., J. Biol. Chem.* **275,** 28439 (2000) with permission, copyright © 2000, the American Society for Biochemistry and Molecular Biology.]

and gray forms of Dfx from *D. desulfuricans* could be resolved by ion-exchange chromatography on a high-performance liquid chromatograph (HPLC) using a Protein-Pack column-DEAE SPW (Waters, Milford, MA), presumably reflecting slight differences in isoelectric point between the fully oxidized gray and mixed-valence pink forms.[29] Furthermore, because of the 200-mV separation between midpoint potentials for centers I and II in *D. desulfuricans* Dfx, it was possible to generate pink Dfx by treatment with ascorbate, which reduces center II but leaves center I oxidized. Methods for generating the pink form may differ depending on the source of the enzyme (recombinant versus native; bacterial isoform, etc.). The investigator may need to try alternative methods, possibly even redox titrations, to poise the enzyme in this intermediate oxidation state.

## V. Spectroscopic Properties of Superoxide Reductases

### A. UV–Visible Spectroscopy

Optical studies of Dfx resolved distinct spectra for centers I and II.[7,25,29] When both center I and II are in the $Fe^{2+}$ oxidation state, the protein is colorless and exhibits a featureless optical spectrum for $\lambda \geq 350$ nm. The pink form, with center I oxidized and center II reduced, exhibits $\lambda_{max}$ at 370 and 495 nm and a shoulder at ~590 nm (Fig. 3,[23] solid line). This spectrum is identical to the optical spectrum of $Fe^{3+}$–Dx.[38–40]

[38] I. Moura, M. Bruschi, J. Le Gall, J. J. G. Moura, and A. V. Xavier, *Biochem. Biophys. Res. Commun.* **75,** 1037 (1977).

Further oxidation produces the gray form, in which both centers are in the $Fe^{3+}$ oxidation state, yielding contributions to the optical spectrum from both centers I and II (Fig. 3, dotted line).[29] UV–visible difference spectroscopy between the gray and pink forms of Dfx revealed that the oxidized form of center II exhibits absorption maxima at 635 and 335 nm, which resonance Raman studies assigned to a sulfur-to-iron charge transfer band originating from at least one cysteine ligand.[29] These optical features are observed in the blue, oxidized form of Nlr (Fig. 3, dashed line) and represent one of the distinguishing features of this active site center.[8,23,26,33]

B. *Mössbauer and Electron Paramagnetic Resonance Spectroscopy of Superoxide Reductases*

As with the optical studies, distinct electron paramagnetic resonance (EPR) spectra are observed for the oxidized states of centers I and II of Dfx.[7,24,29,31] The pink form of Dfx, with center II reduced, affords an EPR spectrum with contributions provided solely by center I. Thus, the EPR spectrum of pink Dfx has $g$ values at 7.7, 5.7, and 4.1, characteristic of a high-spin $Fe^{3+}$ ion with a positive zero-field splitting term, $D$, and rhombicity, $E/D$, approximately 0.08.[7,29,31] As expected, the EPR signal from oxidized center I is nearly identical to $Fe^{3+}-Dx$[39–41] and the recombinant N-terminal domain of Dfx (Fig. 4B).[23,24] The EPR spectrum of $Fe^{3+}$–center II in Dfx (e.g., from gray Dfx) is obscured by the signal from center I. Nevertheless, careful spectral analysis and deconvolution has revealed the EPR spectrum of center II to be more rhombic, with signals observed at $g = 9.5$–9.8 and 4.3.[29] A better representation of the EPR spectrum of center II can be observed with Nlr or the recombinant C-terminal domain of Dfx, neither of which contains a Dx-like N-terminal domain. EPR spectra of *T. pallidum* Nlr[23,26] and the recombinant C-terminal domain of Dfx are shown in Fig. 4.[23,24] The $g$ values of 9.6–9.7 and ~4.3 are representative of rhombic high-spin $Fe^{3+}$ species with $E/D \approx 1/3$.

The resemblance of center I in Dfx to the $Fe(S-Cys)_4$ site in Dx was further demonstrated by Mössbauer spectroscopy.[7,29] The Mössbauer parameters of oxidized and reduced forms of center I were characteristic of tetrahedral sulfur coordination and compared well with those observed for *D. gigas* Dx.[39] In both the pink and dithionite-reduced forms of Dfx, the active site center II iron ion exhibited Mössbauer parameters ($\delta = 1.04$–1.07 mm/sec and $\Delta E_Q = 2.8$–2.9 mm/sec) consistent with a high-spin $Fe^{2+}$ ion. Similar values were observed for the active

---

[39] I. Moura, B. H. Huynh, R. P. Hausinger, J. Le Gall, A. V. Xavier, and E. Münck, *J. Biol. Chem.* **255**, 2493 (1980).
[40] C. Czaja, R. Litwiller, A. J. Tomlinson, S. Naylor, P. Tavares, J. LeGall, J. J. G. Moura, I. Moura, and F. Rusnak, *J. Biol. Chem.* **270**, 20273 (1995).
[41] L. Yu, M. Kennedy, C. Czaja, P. Tavares, J. J. G. Moura, I. Moura, and F. Rusnak, *Biochem. Biophys. Res. Commun.* **231**, 679 (1997).

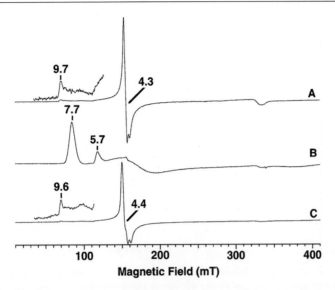

FIG. 4. EPR spectra of *T. pallidum* neelaredoxin and recombinant N- and C-terminal polypeptides of *D. vulgaris* desulfoferrodoxin. (A) EPR spectrum (4.3 K) of the C-terminal domain of *D. vulgaris* Dfx; (B) EPR spectrum (4.2 K) of the N-terminal domain of Dfx; (C) EPR spectrum (3.6 K) of *T. pallidum* Nlr. The EPR spectrum of oxidized (gray) Dfx consists of a sum of spectra (A) and (B). [Adapted with permission from Jovanović *et al., J. Biol. Chem.* **275**, 28439 (2000), copyright © 2000, the American Society for Biochemistry and Molecular Biology; and Ascenso *et al., J. Biol. Inorg. Chem.* **5**, 720 (2000), copyright © 2000, Society of Biological Inorganic Chemistry.]

site iron ion of Nlr, which had $\delta = 1.02$ mm/sec and $\Delta E_Q = 2.80$ mm/sec.[23] The isomer shifts of 1.02 to 1.07 are much lower than the $\delta \approx 1.3$ mm/sec observed for octahedral N/O-coordinated $Fe^{2+}$ and suggest either a lower coordination number or increased covalency due to sulfur coordination or both. These isomer shifts are consistent with a pentacoordinate geometry for the reduced enzyme.[27]

## VI. Superoxide Reductase Assays and Activity Measurements

Although both Dfx and Nlr were originally characterized as SODs,[17,31] Adams and colleagues were the first to clarify the fundamental differences between the reactions catalyzed by SODs and SORs [cf. Eq. (1) vs. Eq. (2)]. Indeed, although *P. furiosus* SOR exhibited a high SOD activity (∼4000 U/mg) in the classic cytochrome *c* reduction assay, the SOD activity of SOR was significantly lower when acetylated cytochrome *c* was used or in three other SOD assays, demonstrating that SORs have little if any SOD activity. Similar results have been demonstrated for Dfx from *D. desulfuricans*[31] and *Desulfoarculus baarsii*,[33] Nlrs from *T. pallidum*[23,26] and *A. fulgidus*,[32] and the C-terminal domain of *D. vulgaris* Dfx,[24] all of which have measurable but catalytically inefficient SOD activities of

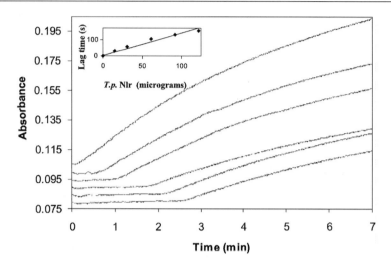

FIG. 5. The effect of *T. pallidum* neelaredoxin on the rate of cytochrome $c$ reduction by superoxide. The reduction of cytochrome $c$ is monitored at 550 nm as a function of time in the presence of increasing amounts of neelaredoxin. At $t = 0$, xanthine oxidase is added to initiate the reaction. Each curve has been offset to facilitate comparisons between curves. From top to bottom, 0, 15, 30, 61, 91, and 121 μg of neelaredoxin have been included in each assay. *Inset*: A plot of the lag time of the initial phase as a function of the neelaredoxin concentration. [Used with permission from Jovanović *et al.*, *J. Biol. Chem.* **275**, 28439 (2000), copyright © 2000, the American Society for Biochemistry and Molecular Biology.]

10–70 U/mg (or $10^6\ M^{-1}\ \text{sec}^{-1}$ as measured by pulsed radiolysis, which is $\leq 1\%$ the second-order rate constant for canonical SODs.[32]

*Distinguishing Superoxide Dismutase and Superoxide Reductase Activities*

That these enzymes exhibit both SOD and SOR activity is nicely demonstrated by use of the classic cytochrome $c$ reduction assay as shown in Fig. 5.[23,42] In this assay, a steady-state concentration of $O_2^-$ generated by xanthine–xanthine oxidase reduces cytochrome $c$, thereby leading to an increase in absorbance at 550 nm (top trace; Fig. 5).

1. The assay is performed at 25° in a 1-ml quartz optical cuvette containing 50 m$M$ potassium phosphate, pH 7.8, containing 100 μ$M$ EDTA, 10 μ$M$ horse heart cytochrome $c$, and 7.5 m$M$ xanthine.

2. An amount of xanthine oxidase that gives an initial rate of $\Delta A_{550}$ of ~0.025 absorbance unit per minute is added. Various amounts of Nlr or Dfx are added and the absorbance at 550 nm is monitored over time. One unit of SOD activity is defined as the amount of protein that inhibits the rate of cytochrome $c$ reduction by 50%.

[42] L. Flohé and F. Ötting, *Methods Enzymol.* **105**, 93 (1984).

In the presence of SOR, the reduction of cytochrome $c$ by $O_2^-$ occurs in two distinct phases. In the initial lag phase, in which no increase in $A_{550}$ occurs, the $O_2^-$ generated by xanthine–xanthine oxidase is immediately reduced by the SOR present in solution [cf. Eq. (2)]. The length of this initial phase is proportional to the amount of SOR present in solution (inset; Fig. 5), as expected for this bimolecular reaction. That the reduced form of SOR is responsible for the lag phase can be determined using the different oxidation states of either Nlr or Dfx as described above in Sections IV,A and IV,B. When all the SOR is oxidized, the second phase commences, corresponding roughly to a linear rate of increase in $A_{550}$ as $O_2^-$ reduces cytochrome $c$. At high enough concentrations of SOR ($\geq 15$ $\mu$g), its low but measurable SOD activity results in a slight inhibition of cytochrome $c$ reduction, as evidenced by a decrease in slope of the $A_{550}$-versus-time graph. From this observed change in slope of these curves, an SOD activity can be computed ($10 \pm 5$ U/mg).[23]

Additional studies have shown that $O_2^-$ is capable of oxidizing SORs to the $Fe^{3+}$ state, but not the opposite, consistent with the reaction of Eq. (2).[23,24,26] The second-order rate constant for the reaction of reduced SORs with $O_2^-$ has been measured by monitoring the increase in absorbance due to the ferric form of the active site as a function of time in the presence or absence of SODs, yielding values of $10^8$–$10^9$ $M^{-1}$ sec$^{-1}$.[23,26,33] Kurtz and colleagues have used rapid kinetic techniques to monitor the reaction of SOR with $O_2^-$ and have proposed a minimal three-step reaction mechanism.[27] In the first step, $O_2^-$ binds to $Fe^{2+}$–SOR at a nearly diffusion-controlled rate of $1.5 \times 10^9$ $M^{-1}$ cm$^{-1}$ to form an intermediate that has $\lambda_{max}$ centered at 600 nm. On the basis of spectroscopic studies of model compounds, this intermediate is proposed to be a ferric-(hydro)peroxo species. In the second step, the glutamate seen coordinating the iron ion in X-ray structures of the oxidized enzyme displaces hydrogen peroxide from the active site iron ion after proton transfer, thus forming the octahedral $Fe^{3+}$–SOR species. The reaction cycle is completed after reduction of the iron and displacement of the carboxylate to form the trigonal bipyramidyl species capable of binding another $O_2^-$ molecule for another round of catalysis. There is much support for this simplified mechanism and it is hoped that additional studies will reveal whether other intermediates can be isolated and characterized.

Summary

Superoxide reductases have now been well characterized from several organisms. Unique biochemical features include the ability of the reduced enzyme to react with $O_2^-$ but not dioxygen (reduced SORs are stable in an aerobic atmosphere for hours). Future biochemical assays that measure the reaction of SOR with $O_2^-$ should take into account the difficulties of assaying $O_2^-$ directly and the myriad of redox reactions that can take place between components in the assay, for example, direct electron transfer between cytochrome c and Dfx.[24] Future prospects include further delineation of the reaction mechanism, characterization of the putative (hydro)peroxo intermediate, and studies that uncover the components between reduced pyridine nucleotides and SOR in the metabolic pathway responsible for $O_2^-$ detoxification.

## [25] Purification and Preparation of Prion Protein: Synaptic Superoxide Dismutase

*By* MAKI DANIELS and DAVID R. BROWN

The prion protein ($PrP^c$) has become widely known because of the association of an abnormal isoform of the protein ($PrP^{Sc}$) with prion diseases, which include Creutzfeld–Jakob disease, scrapie, and bovine spongiform encephalopathy.[1] This form of the protein is identical to the normal isoform in terms of its primary sequence but is misfolded, has altered secondary structural characteristics, readily aggregates into fibrils, and may be infectious. In contrast to this isoform associated with mammalian neurodegenerative disease, the normal isoform is expressed by all vertebrates and is emerging as an important brain protein.[2] Understanding this brain protein is important for two reasons. First, $PrP^c$ is highly concentrated and localized in synapses[3] and therefore is important to synaptic activity. Second, expression of $PrP^c$ is necessary if not sufficient for the development of prion disease.[4] Mice lacking $PrP^c$ expression due to genetic ablation cannot be infected with prion disease. This is because the host must express $PrP^c$ in order for it to be converted to $PrP^{Sc}$. Therefore the events that cause the conformational change of $PrP^c$ to $PrP^{Sc}$ are the key to understanding these diseases. Defining $PrP^c$ functionally is therefore a prerequisite to understanding this conversion and its consequence for cell biology. $PrP^c$ is now accepted to be a copper-binding protein.[5] Evidence from cell culture and biochemistry studies point to a role of $PrP^c$ in two kinds of cellular activity. First, expression of $PrP^c$ can alter the cellular distribution of copper, suggesting its expression may alter copper uptake and release.[5,6] Second,

[1] S. B. Prusiner, *Proc. Natl. Acad. Sci. U.S.A* **95,** 13363 (1998).
[2] D. R. Brown, *Trends Neurosci.* **24,** 85 (2001).
[3] N. Sàles, K. Rodolfo, R. Hässig, B. Faucheux, L. Di Giamberdino, and K. L. Moya, *Eur. J. Neurosci.* **10,** 2464 (1998).
[4] H. Büeler, A. Aguzzi, A. Sailer, R. A. Greiner, P. Autenried, M. Aguet, and C. Weissmann, *Cell* **73,** 1339 (1993).
[5] D. R. Brown, K. Qin, J. W. Herms, A. Madlung, J. Manson, R. Strome, P. E. Fraser, T. Kruck, A. von Bohlen, W. Schulz-Schaeffer, A. Giese, D. Westaway, and H. Kretzschmar, *Nature (London)* **390,** 684 (1997).
[6] D. R. Brown, *J. Neurosci. Res.* **58,** 717 (1999).

the majority of evidence suggests that $PrP^c$ is an antioxidant protein with an activity like that of superoxide dismutase.[7]

The superoxide dismutase (SOD) activity of $PrP^c$ can be detected with standard spectrophotometric assays. The difficulty in studying this protein is not in performing the assays but in preparing the $PrP^c$ protein. We have used two methods to prepare $PrP^c$ for study. The first of these is the more difficult but was necessary to verify our findings. Native $PrP^c$ was isolated from the brains of mice, using an affinity isolation procedure employing an antibody with high affinity for $PrP^c$ coupled to cyanogen bromide-activated Sepharose beads.[8]

The second and more versatile method employs recombinant protein technology.[9] Standard methods are used to prepare the protein. However, as outline below, once pure protein is obtained the key to gaining active protein is correct incorporation of copper into the protein molecule. Incorrect copper association with the protein[10] can cause it to switch conformation to an inactive $\beta$-sheet-rich isoform that readily aggregates into protofilaments. However, by denaturing the protein with an agent such as urea and gradually removing the urea in the presence of a fixed concentration of copper, correct incorporation of copper or incorporation of substitutes such as manganese can be achieved. Assessment of this can be achieved by copper-detecting methods such as mass spectroscopy[11] or spectrophotometic assays.[12,13] In addition, the effect of the copper incorporation of the secondary structure of the protein can be easily monitored by circular dichroism.[14] Recombinant technology also allows the production of mutant proteins. Proteins lacking the copper-binding domain serve as useful controls. The one disadvantage of this expression system is that the protein produced lacks the glycosylation associated with the native form.

## Protein Preparation

The coding frames for many prion proteins from different species have now been cloned. The models used in our system are the mouse prion protein or the

---

[7] D. R. Brown, B.-S. Wong, F. Hafiz, C. Clive, S. J. Haswell, and I. M. Jones, *Biochem J.* **344**, 1 (1999).
[8] D. R. Brown, C. Clive, and S. J. Haswell, *J. Neurochem* **76**, 69 (2001).
[9] B. S. Wong, H. Wang, D. R. Brown, and I. M. Jones, *Biochem. Biophys. Res. Commun.* **279**, 352 (1999).
[10] K. Qin, D.-S. Yang, Y. Yang, M. A. Christi, L.-J. Meng, C. M. Yip, P. E. Fraser, and D. Westaway, *J. Biol. Chem.* **275**, 19121 (2000).
[11] R. M. Whittal, H. L. Ball, F. E. Cohen, A. L. Burlingame, S. B. Prusiner, and M. A. Baldwin, *Protein Sci.* **9**, 332 (2000).
[12] A. J. Brenner and E. D. Harris, *Anal. Biochem.* **226**, 80 (1995).
[13] O. C. J. Gabler and M. Lauey, *J. Biol. Chem.* **215**, 510 (1953).
[14] B.-S. Wong, C. Vénien-Bryan, R. A. Williamson, D. R. Burton, P. Gambetti, M.-S. Sy, D. R. Brown, and I. M. Jones, *Biochem. Biophys. Res. Commun.* **276**, 1217 (2000).

chicken homolog and the techniques described are based on these proteins. However, others can be used. The prion protein contains two signal peptides. Amplification, by polymerase chain reaction (PCR), of the coding region is carried out so that the expression of the signal peptides is excluded and only the mature protein is generated.

For our purposes the coding region of mouse or chicken PrP$^c$ is cloned into a commercial plasmid such as pET (Novagen, Madison, WI) or pTrkHis (Invitrogen, Carlsbad, CA). These plasmids are then transformed into strains of *Escherichia coli* bacteria optimized for protein expression. Expression of the protein is driven by a T7 promoter that is induced by the addition of 1 m$M$ isopropyl-$\beta$-D-thiogalactopyranoside (IPTG). Culture volumes of 2 liters are grown after addition of 50 ml of an overnight inoculum. After 3–4 hr the cultures will have reached the late log phase of growth. At this time the IPTG should be added and the cultures left to grow for a further 2 hr. After collecting the bacteria by centrifugation, the pellets can either be stored or used directly for extraction.

At this point two possibilities exist for extraction of the protein from the bacterial pellets. Most forms of PrP$^c$ are insoluble. However, some mutants we have studied are soluble and therefore the protein will not remain in the inclusion bodies after further treatment.

*Soluble Protein.* For soluble protein we add 50 ml of urea buffer (8 $M$ urea, 200 m$M$ NaCl, 20 m$M$ Tris-HCl, pH.80) to the pellets and sonicate the pellets with a metal-tip sonicator until the pellets are completely broken up. After centrifugation at 12,000$g$ the supernatant is sonicated again briefly and centrifuged again. The supernatant is filtered through a 0.45-$\mu$m pore size filter before passing it through an immobilized metal affinity chromatography (IMAC) column.

*Insoluble Protein.* For insoluble proteins inclusion body extraction is carried out before addition of urea buffer. The technique we use is a standard enzymatic lysis procedure with lysozyme and deoxycholic acid. Bacterial pellets from 2 liters of cells are resuspended in lysis buffer (50 m$M$ Tris-HCl, 1 m$M$ EDTA, 100 m$M$ NaCl, pH.7.4), using 3 ml for every gram of pellet. To this is added, at 4 $\mu$l/ml, phenylmethylsulfonyl fluoride (PMSF, 100 m$M$) and, at 30 $\mu$l/ml, lysozyme (10 mg/ml). The solution is mixed at room temperature for 20 min. Two milligrams of deoxycholic acid per milliliter of lysis buffer is then added and the mixture is incubated at 37° until the lysate becomes viscous. At this point 20 $\mu$g of DNase I is added for each milliliter of lysis buffer and is mixed at room temperature until viscosity is lost. The mixture is centrifuged at 12,000 rpm at room temperature for 10 min. The pellet is then resuspended in lysis buffer (10 ml/mg of original pellet) and centrifuged again. This wash procedure is repeated three times. After the final wash step the purified inclusion bodies are resuspended in urea buffer as for soluble protein and sonicated as described above. After filtering the supernatant through a 0.45-$\mu$m pore size filter the protein extract is applied to an IMAC column.

The protein prepared by this method has a hexahistidine tag at either the N or C terminus. Attachment of a tag at the N terminus has proved to be optimal for purification in that it is easier to cleave the tag from the protein once the purified protein has been refolded out of the buffer containing 8 $M$ urea. However, for many experiments removal of the histidine tag is not necessary. Other tagging methods may also work effectively. Indeed, the histidine tag method has the disadvantage of adding an additional metal-binding site to the protein. As studies of $PrP^c$ involve analysis of its metal-binding ability this method can be a disadvantage. However, for our studies this did not prove to be the case. For some of our studies using deletion mutants lacking the octameric repeat region associated with metal binding the hexahistidine tag provided a site for addition of copper that did not endow the protein with SOD activity. This control helped demonstrate the enzymatic nature of the SOD activity of $PrP^c$.

The IMAC column is run at room temperature. A 4- to 6-ml bed volume of chelating Sepharose (Pharmacia, Uppsala, Sweden) is prepared and washed with 5 bed volumes of distilled water. The column is then washed with 0.3 $M$ $NiCl_2$ (10 ml). The column is then equilibrated with urea buffer. The flow rate is adjusted to 2 ml/min. The protein sample is then loaded. The column is washed with 50 ml of urea buffer followed by urea buffer with the addition of 20 m$M$ imidazole. The eluant is monitored at 280 nm with a spectrophotometer. Once the absorbance reaches zero the column is washed with an additional 20 ml of urea buffer. Elution of the bound protein is achieved by addition of urea buffer containing 400 m$M$ imidazole. One-milliliter fractions are collected. The prion protein normally elutes in fractions 2–6.

It should be noted that prion protein will also bind to an IMAC column via the octameric repeat region. IMAC purification has been used to purify native protein from hamster brain.[15] However, the technique is difficult and the yield is small in comparison with the affinity technique mentioned above. This is probably because $PrP^c$ already has copper bound to it *in vivo*. However, it is also possible to adapt the IMAC procedure to purify recombinant $PrP^c$ that is not histidine tagged. In this case it is better to charge the IMC column with copper rather than nickel. After washing as described above the protein can be eluted with 50 m$M$ EDTA.

Copper Incorporation

After elution of the protein from the IMAC column the concentration of protein should be between 1 and 2 mg/ml in approximately 4–5 ml. The protein in 8 $M$ urea and salts will also contain substantial amounts of imidazole, which must

---

[15] K.-M. Pan, N. Stahl, and S. B. Prusiner, *Protein Sci.* **1,** 1343 (1992).

be removed before the addition of copper. Furthermore, dialysis of the protein to remove the urea and imidazole, will cause precipitation of the protein. These problems are overcome by using sequential dilution and concentration. However, once the concentration of urea is reduced to less than 1 $M$ dialysis can be used to remove the urea completely.

1. Refolding without the presence of copper: If the protein is to be used in experiments that do not require incorporation of copper the protein is diluted gradually. Assuming a starting volume of 1 ml, 100 $\mu$l of water is added 10 times with thorough mixing. From this point water is added 50 $\mu$l at a time, followed by thorough mixing. Once the volume reaches approximately 5 ml the protein can be concentrated with spin concentrators with a cutoff of 10 kDa (Millipore, Bedford, MA) at 12,000 rpm until the initial 1-ml volume is restored. The protein can then be concentrated further or dialyzed.

2. Refolding to incorporate copper: The protein (1 ml) is diluted to 5 ml with urea buffer (8 $M$ urea, 200 m$M$ NaCl, 20 m$M$ Tris-HCl, pH.8). The protein is concentrated with spin concentrators (10-kDa cutoff) to 0.5 ml and then diluted with urea buffer to 5 ml. At this point $CuSO_4$ is added to make a final concentration of 1 m$M$. The protein is again concentrated to 0.5 ml. To this is added 50 $\mu$l of water containing 1 m$M$ $CuSO_4$. The solution is mixed thoroughly and the addition of 50-$\mu$l volumes with mixing is repeated until the volume is 2 ml. Then 1 m$M$ $CuSO_4$ in water is added in 100-$\mu$l amounts (10 times) followed by thorough mixing. Then 1 m$M$ $CuSO_4$ in water is added in 500-$\mu$l amounts followed by thorough mixing until the volume is 5 ml or greater. At this point the protein can be concentrated again to 1 ml, using a spin concentrator. At this point the urea concentration will be less than 0.5 $M$. The protein is diluted with 3 ml of water and then concentrated again. At this point the protein can be dialyzed to remove residual urea and unbound copper.

Once preparation of the protein is complete to this step the hexahistidine tag can be cleaved off. This is especially necessary if the protein is to be used for studies of metal binding. In using commercial constructs such as the pTrkHis plasmid, enterokinase cleaves the tag from the protein. Commercial kits are available to remove the enterokinase or, alternatively, the peptidase can be coupled to beads and removed after digestion. When removal of the hexahistidine tag is necessary, the major dialysis steps should follow the enterokinase cleavage. Small amounts of urea and copper do not interfere markedly with the digestion. However, we have found that imidazole has a strong inhibitory effect above 5 m$M$. In addition, we have found complete digestion to be difficult to achieve. Digestion of PrP with concentrations above 2 units/ml leads to degradation of the protein. Underdigestion is therefore preferred to protein loss. The fidelity of the digestion can be checked by analysis with a an antibody against hexahistidine.

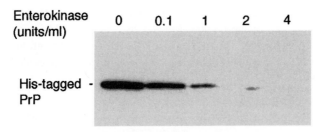

FIG. 1. The prion protein after purification was digested with enterokinase to remove the histidine-tag. After digestion, equal amounts of protein were loaded onto a polyacrylamide gel and electrophoresed. After Western blotting and immunodetection with an anti-His tag antibody only the uncut protein retains the histidine tag.

Figure 1 shows Western blots of copper-refolded protein from which the histidine tag has been effectively removed. When digestion is incomplete the tagged protein can be separated from the cleaved protein by mixing the protein with nickel-loaded chelating Sepharose (10 μg/ml, assuming a protein concentration of 1 mg/ml). and pelleting the bound material. The supernatant will then contain pure cleaved protein.

## Analysis by Circular Dichroism

For analysis of the change in structure of the protein after copper incorporation care must be taken to remove all urea from the preparation. Dialysis into water is sufficient to remove the urea. Figure 2 shows a comparison of chicken prion protein refolded into water or refolded into water with copper present. The comparison

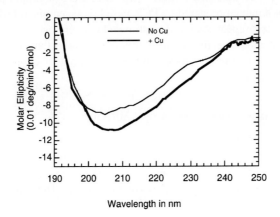

FIG. 2. Far-UV circular dichroism spectra of chicken prion protein shows the change in the spectra after incorporation of copper in the molecule. The protein was measured in water at pH.7. Shown is the average of five spectra, each measured at 2 sec/0.5 nm.

was made in the absence of salts, as it has been found that salts can change the secondary structure of the protein. Metal salts, especially those of copper, can cause $PrP^c$ to aggregate. This is especially the case for $PrP^c$ refolded in water. Addition of copper during refolding appears to stabilize the molecule and make it more soluble.

### Analysis of Copper Content

It is essential to ensure that the refolding technique leads to correct incorporation of copper into the molecule. There are two ways to determine whether copper has been incorporated into the protein. The first involves the use of spectroscopy techniques such as total X-ray reflection fluorescence spectroscopy or mass spectroscopy. The second involves simpler alternatives based on spectrophotometric assay. The disadvantage of the first is that it requires expensive equipment. The disadvantages of the second are reduced sensitivity and that the assays will determine only copper content and not other metals. One assay we have used is a modification of the bicinchoninic acid assay often used for protein assays.[13,16] An alternative is an assay using diethyl dithiocarbamate.[12] There is also a version of this assay that can be used in-gel assay.[17]

Prepare diethyl dithiocarbamate (DDC) at 200 m$M$ in water. Prepare copper standards from copper sulfate to give a standard curve for copper from 0.1 to 10 $\mu$g/ml. The assay is most effective for $PrP^c$ at relatively high concentrations of 100–10 $\mu$g/ml. Mix the DDC solution with an equal volume of either the standards or the test samples. Allow the reaction to stand at room temperature for 0.5 hr and then measure absorbance at 458 nm in a spectrophotometer. The values measured for chicken $PrP^c$ refolded with copper were 12 $\pm$ 2 $\mu$g of copper per milligram of protein. For a mutant prepared by PCR-based mutagenesis to lack the copper-binding hexameric repeat region, the estimated copper content was found to be 0.5 $\pm$ 0.5 $\mu$g of copper per milligram of protein, indicating that the refolding led to the successful incorporation of approximately four atoms of copper per molecule of wild-type chicken $PrP^c$.

### Analysis of Superoxide Dismutase Activity

The activity of the prion protein has been characterized using several different techniques. In principle, none of these techniques are novel and virtually any known SOD-detecting technique can be used. At present there is no assay to distingish $PrP^c$ activity from that of other SODs in tissue samples. The activity of $PrP^c$ can

[16] D. R. Brown, I. K. Iorndanova, B.-S. Wong, C. Vénien-Bryan, F. Hafiz, L. L. Glassmith, M.-S. Sy, P. Gambetti, I. M. Jones, C. Clive, and S. J. Haswell, *Eur. J. Biochem.* **267**, 2452 (2000).
[17] S. L. Jewett and A. M. Rocklin, *Anal. Biochem.* **217**, 236 (1994).

be inhibited by diethyl dithiocarbamate, which also inhibits Cu,ZnSOD. However, the activity cannot be inhibited by potassium cyanide, indicating a possible way in which the activity can be distinguished from, for example, MnSOD. In addition, immunodepletion has been used successfully to distinguish how much of the total cellular SOD can be related to PrP$^c$ activity.[18]

The specific activity of PrP$^c$ has been determined to be one-tenth that of Cu,Zn SOD.[7] Therefore 1 unit of PrP$^c$ activity is approximately 0.1 U of Cu,ZnSOD. However, this is dependent on the amount of copper incorporated into the PrP$^c$ molecule. This estimate is based on four atoms of copper per molecule. However, in vivo PrP$^c$ may bind less and then the activity is correspondingly less.

Spectrophotometric techniques are sufficient to detect the activity.[19–21] Other techniques such as stopped flow may also be used, but the equipment and expense needed for such studies make such analysis superfluous in the face of adequate simpler alternatives. An alternative assay for SOD is the in-gel assay or the zymogen assay.[22]

*In-Gel Assay*

The in-gel assay is basically as described by other authors. The protein (5–20 μg) is electrophoresed on a 7% (w/v) polyacrylamide gel without sodium dodecyl sulfate (SDS) or reducing agents. After electrophoresis the gel is soaked in a solution of 5 m$M$ nitroblue tetrazolium at room temperature with rocking for 20 min. The gel is then rinsed briefly with distilled water and a developing solution (30 μ$M$ riboflavin, 30 m$M$ tetramethylethylenediamine, 40 m$M$ potassium phosphate, pH 7.8) for 15 min. At this point the gel is exposed to light until a uniform blue color covers the gel. Protein with superoxide dismutase activity leaves the gel transparent. However, if the reaction is allowed to proceed for too long the contrast between these regions is lost. As can be seen in Fig. 3 the prion protein runs close to the top of the gel, unlike Cu,ZnSOD.

*Spectrophotometric Assay*

Many different spectrophotometric assays can be used. The one used most often in our laboratory is the assay based on nitroblue tetrazolium (NBT) conversion to a purple formazan product by the presence of superoxide. Inhibition of the formation of this product corresponds to superoxide dismutase activity. The agent used to

---

[18] B.-S. Wongs, T. Pan, T. Liu, R. Li, P. Gambetti, and M.-S. Sy, *Biochem. Biophys. Res. Commun.* **273**, 136 (2000).
[19] L. W. Oberley and D. R. Spitz, *Methods Enzymol.* **105**, 457 (1984).
[20] S. Marklund, *J. Biol. Chem.* **251**, 7504 (1976).
[21] L. Flohé and F. Ötting, *Methods Enzymol.* **105**, 93 (1984).
[22] C. Beauchamp and I. Fridovich, *Anal. Biochem.* **44**, 276 (1971).

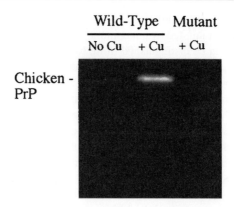

FIG. 3. In-gel assay of the activity of chicken prion protein. Samples were run on a native polyacrylamide gel and stained as described in text. Wild-type chicken prion protein either with copper incorporated (+Cu) or without (no Cu) was run on the gel. In addition, a mutant protein lacking the hexameric repeat region, refolded in the presence of copper, was run in parallel to show that the activity is related to specific binding of copper to the this region.

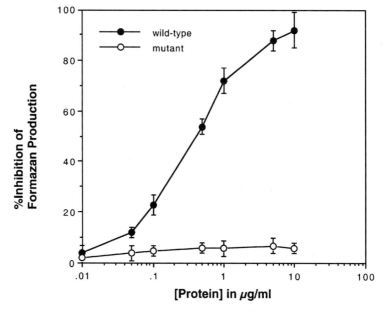

FIG. 4. Analysis of chicken prion protein superoxide dismutase activity, using a spectrophotometric assay based on the the conversion of nitroblue tetrazolium to a colored formazan product. Protein refolded in the presence of copper was tested for activity in the assay. SOD activity is expressed as inhibition of formazan production. Wild-type protein (●) showed high activity in comparison with mutant protein (○), in which the hexameric repeat region had been deleted by mutagenesis. Shown are the means and SE of four separate experiments.

form superoxide is xanthine oxidase, and the assay is performed in the presence of catalase to break down hydrogen peroxide. Although chelators such as EDTA can be added to the reaction we have found that they make no difference to the measured readings.

Prepare a master solution containing 0.5 m$M$ xanthine, catalase (40 U/ml), and 10 m$M$ NaCl in 100 m$M$ sodium phosphate, pH 7. The solution should be filtered through a 0.2-$\mu$m pore size filter. A stock of xanthine should have been prepared previously by dissolving it in water at a higher concentration by adding a few drops of 10 N NaOH. NBT can be prepared as a 100× concentrated stock in dimethyl formamide. Xanthine oxidase should be diluted to an appropriate concentration ($\sim$20 mU/ml) in 100 m$M$ sodium phosphate–10 m$M$ NaCl, pH 7. Into 1-ml aliquots of master solution add amounts of prion protein of up to 10 $\mu$g/ml.

All solutions should be at room temperature before starting the assay. Ideally, the spectrophotometer should contain a temperature-regulating chamber to keep the assay solution at room temperature during the assay. Begin the assay by assaying the increased absorption at 560 nm of a 1-ml control solution with no added protein. Xanthine oxidase solution (100 $\mu$l) is added to the master solution. The change in absorption over 5 min relative to the control sample is monitored to ensure linearity in the reaction. After this, protein samples are assayed by the addition of 100 $\mu$l of xanthine oxidase solution. The change in absorption per minute is determined. An example of the results of such an assay is shown in Fig. 4.

Acknowledgment

The authors thank Ian Jones and Boon Seng Wong for the original chicken PrP$^c$-expressing construct and the basic method for protein production.

# Section IV
## *In Vivo* Sources, Cell Signaling

## [26] Mitochondrial Superoxide Anion Production and Release into Intermembrane Space

By DERICK HAN, FERNANDO ANTUNES, FRANCESCA DANERI, and ENRIQUE CADENAS

The mitochondrial electron transport chain is the major cellular source of superoxide anion ($O_2^{·-}$),[1] largely originating from autoxidation of ubisemiquinone.[2–4] $O_2^{·-}$, generated in this manner, can be vectorially released toward the mitochondrial matrix or the intermembrane space, for ubisemiquinone formation occurs on the inner membrane near the matrix ($Q_i$ site) and in the intermembrane space ($Q_o$ site).[5] The vectorial release of $O_2^{·-}$ into the matrix space is a well-established notion supported by the following observations: (1) submitochondrial particles, in which the inner membrane is inverted, generate $O_2^{·-}$ into the medium as monitored by cytochrome $c$- or adrenaline-based assays[3,6]; and (2) the mitochondrial matrix contains high levels of manganese superoxide dismutase ($1.1 \times 10^{-5}$ $M$),[7] and the expectedly high rate of $O_2^{·-}$ production in the mitochondrial inner membrane would be in a functional relationship with the localization of this enzyme. Conversely, the release of $O_2^{·-}$ toward the intermembrane space has been difficult to assess because of the localized environment in this compartment given by (1) the high levels of cytochrome $c$, which reacts with $O_2^{·-}$ at considerable rates[8] and interferes with the classic detection method for $O_2^{·-}$, that is, superoxide dismutase-sensitive cytochrome $c$ reduction, and (2) the occurrence of a Cu,Zn superoxide dismutase isoform.[9]

The detection of $O_2^{·-}$ in the intermembrane space can best be accomplished by a combination of approaches involving, on the one hand, mitoplasts (mitochondria devoid of portions of the outer membrane) and, on the other, spin-trapping electron paramagnetic resonance (EPR) or redox-sensitive fluorescent dyes as detection methodologies. The current methods to obtain mitoplasts not only remove portions of the outer membrane, but also remove cytochrome $c$ and

---

[1] B. Chance, H. Sies, and A. Boveris, *Physiol. Rev.* **59**, 527 (1979).
[2] A. Boveris, E. Cadenas, and A. O. M. Stoppani, *Biochem. J.* **156**, 435 (1976).
[3] A. Boveris and E. Cadenas, *FEBS Lett.* **54**, 311 (1975).
[4] E. Cadenas, A. Boveris, C. I. Ragan, and A. O. M. Stoppani, *Arch. Biochem. Biophys.* **180**, 248 (1977).
[5] S. de Vries, *J. Bioenerg. Biomembr.* **18**, 196 (1986).
[6] O. Dionisi, T. Galeotti, T. Terranova, and A. Azzi, *Biochim. Biophys. Acta* **403**, 292 (1975).
[7] D. D. Tyler, *Biochem. J.* **147**, 493 (1975).
[8] J. Butler, W. Koppenol, and E. Margoliash, *J. Biol. Chem.* **257**, 10747 (1982).
[9] P. J. Iñarrea, personal communication (2000).

intermembrane superoxide dismutase that can interfere with $O_2^{\cdot-}$ determination.[10] EPR spin trapping or redox-sensitive fluorescent dyes offer sensitive methods for $O_2^{\cdot-}$ determination that can also be validated by using superoxide dismutase.

## Preparation of Mitoplasts

Mitoplasts are prepared from rat liver and heart mitochondria. Liver mitochondria are isolated from adult male Wistar rats by differential centrifugation as previously described.[11] Rat livers are excised, washed with 0.25 $M$ sucrose, and homogenized in an ice-cold isolation buffer consisting of 210 m$M$ mannitol, 70 m$M$ sucrose, 2 m$M$ HEPES, and 0.05% (w/v) bovine serum albumin with 10 strokes of a loose-fitting Potter-Elvehjem Teflon pestle. The homogenate is centrifuged at 800$g$ for 8 min at 4°, the pellet is removed, and the centrifugation process is repeated. The resulting supernatant is centrifuged at 8000$g$ for 10 min at 4° and washed with isolation buffer, and the centrifugation is repeated. Mitochondria are resuspended in isolation buffer and left on ice for the duration of the experiments.

Heart mitochondria are isolated from adult male Wistar rats by differential centrifugation, using Nagarse.[2,4] Rat hearts are excised, washed, and chopped finely. Chopped hearts are suspended in an isolation buffer consisting of 230 m$M$ mannitol, 70 m$M$ sucrose, 1 m$M$ EDTA, and 5 m$M$ Trizma-HCl buffer, pH 7.4. The chopped heart is treated with Nagarse (1 mg/heart) for 5 min and then homogenized by 10 strong strokes of a loose-fitting Potter-Elvehjem teflon pestle. The homogenate is centrifuged at 800$g$ for 8 min at 4°, the pellet is removed, and the centrifugation process is repeated. The resulting supernatant is centrifuged at 8000$g$ for 10 min at 4° and washed with isolation buffer, and the centrifugation is repeated.

The outer and inner membranes of mitochondria differ in lipid composition and membrane integrity. The outer membrane is considered more fragile than the inner membrane,[12] and thus portions of the outer membrane can be removed with little damage to the inner membrane. The disruption of the outer membrane to generate mitoplasts is generally accomplished by (1) physical shearing with a French press,[13] (2) exposure to detergents such as digitonin,[11] and (3) swelling of mitochondria in hypo-osmotic buffer.[14] Although the shearing force of a French press is effective in removing the outer membrane, the French press is not readily available in most laboratories.

---

[10] D. Han, E. Williams, and E. Cadenas, *Biochem. J.* **353**, 411 (2000).
[11] P. L. Pedersen, J. W. Greenawalt, B. Reynafarje, J. Hullihen, G. L. Decker, J. W. Soper, and E. Bustamante, *Methods Cell Biol.* **20**, 411 (1978).
[12] R. A. Weisiger and I. Fridovich, *J. Biol. Chem.* **248**, 4793 (1973).
[13] G. L. Decker and J. W. Greenawalt, *J. Ultrastruct. Res.* **59**, 44 (1977).
[14] B. Burnette and P. P. Batra, *Anal. Biochem.* **145**, 80 (1985).

## Digitonin Treatment

Mitoplasts are prepared from isolated rat liver mitochondria by digitonin treatment.[11] Digitonin (10%, w/v) is dissolved in water and heated at 95° until solubilized. Rat liver mitochondria (30 mg/ml) in isolation buffer are treated with various concentrations of digitonin and stirred for 15 min on ice. Samples are diluted 20-fold with isolation buffer and spun down at 8000$g$ for 10 min at 4°. The pellet is washed once with isolation buffer and the suspension is centrifuged a final time at 8000$g$ for 10 min at 4°.

## Hypo-osmotic Treatment

The placement of mitochondria in hypo-osmotic buffer results in the swelling of mitochondria, which ruptures the outer membrane with little damage to the inner membrane. If swelling is carried out in a series of steps involving hypo-osmotic KCl, cytochrome $c$ can be extracted from the intermembrane space.[15,16] Although the hypo-osmotic treatment is not effective in removing significant portions of the outer membrane (unless other agents are added[14]), mitoplasts generated by this procedure have several experimental uses. Mitoplasts prepared from isolated rat heart mitochondria are suspended in cold hypotonic medium (10 m$M$ KCl and 2 m$M$ HEPES, pH 7.4) for 20 min at a concentration of 1 mg/ml.[16] The suspension is centrifuged at 8000$g$ for 10 min at 4° and resuspended at 1 mg/ml in a solution containing 150 m$M$ KCl and 2 m$M$ HEPES, pH 7.2. The suspension is centrifuged at 8000$g$ for 10 min at 4° and redissolved at 1 mg/ml in 150 m$M$ KCl solution. After a final centrifugation, the pellet is dissolved in isolation buffer at a concentration of 8 mg/ml.

## Mitoplast Purity and Functional Integrity

Mitochondrial functional integrity is assessed in terms of the respiratory control ratio (RCR). Oxygen uptake is measured polarographically with a Clark-type electrode (Hanstech, Norfolk, UK). Oxygen consumption by mitochondria and mitoplasts is determined in a respiration buffer consisting of 230 m$M$ mannitol, 70 m$M$ sucrose, 30 m$M$ Tris-HCl, 2 m$M$ MgCl$_2$, 5 m$M$ KH$_2$PO$_4$, 1 m$M$ EDTA, and 0.1% (w/v) bovine serum albumin, pH 7.4, using succinate (7.5 m$M$) as respiratory substrate. The respiratory control ratio is defined as the state 3/state 4 ratio. For rat liver mitoplasts both state 3 and state 4 measurements are carried out in the presence of 2 $\mu M$ cytochrome $c$, because a portion of cytochrome $c$ is lost after digitonin treatment. Monoamine oxidase activity is employed as a marker of

---

[15] E. E. Jacobs and D. R. Sanadi, *J. Biol. Chem.* **235**, 531 (1960).
[16] J. F. Turrens, A. Alexandre, and A. L. Lehninger, *Arch. Biochem. Biophys.* **237**, 408 (1985).

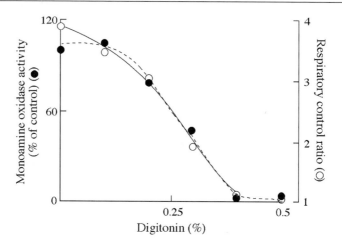

FIG. 1. Effect of digitonin on the respiratory control ratio and monoamine oxidase activity of isolated rat liver mitochondria. Assay conditions are described in text.

outer membrane activity and it is determined by measuring $H_2O_2$ production in the presence of tyramine.[17]

The outer membrane contains a higher level of cholesterol than the inner membrane,[18] and thus is more susceptible to digitonin treatment. Digitonin will remove in a dose-dependent manner portions of the outer membrane as measured by monoamine oxidase activity, an outer membrane marker (Fig. 1). As levels of digitonin increase, the damage to the inner membrane also increases as demonstrated by changes in the RCR (Fig. 1). The mitochondrial outer membrane cannot be completely removed with digitonin without mitochondria becoming uncoupled. At a digitonin concentration of 0.2% (w/v), mitoplasts remain coupled whereas portions of the outer membrane are removed (Fig. 1) along with some of the cytochrome $c$.[10] Overall, these mitoplasts retain respiratory parameters similar to those of mitochondria, albeit with a decrease in RCR values, which is usually observed with mitoplasts preparations.[19,20]

Oxygen measurements of rat heart mitoplasts obtained by hypo-osmotic treatment are carried out in respiration buffer containing 230 m$M$ mannitol, 70 m$M$ sucrose, 30 m$M$ Tris-HCl, 5 m$M$ KH$_2$PO$_4$, 1 m$M$ ETDA, and 0.1% (w/v) bovine serum albumin, pH 7.4. Respiration by heart mitochondria or mitoplasts proceeds more efficiently when magnesium is excluded from the assay medium. Hypotonic

---

[17] N. Hauptmann, J. Grimsby, J. C. Shih, and E. Cadenas, *Arch. Biochem. Biophys.* **335**, 295 (1996).
[18] L. Ernster and G. Schatz, *J. Cell Biol.* **91**, 227s (1981).
[19] C. L. Hoppel, J. Kerner, P. Turkaly, J. Turkaly, and B. Tandler, *J. Biol. Chem.* **273**, 23495 (1998).
[20] J. W. Greenawalt, *Methods Enzymol.* **55**, 88 (1979).

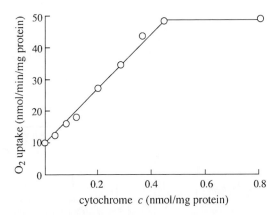

FIG. 2. Dependence of oxygen consumption by heart mitoplasts on cytochrome $c$ concentration. Assay conditions are described in text.

solutions to generate mitoplasts are generally not effective in removing significant portions of the outer membrane unless other agents such as digitonin are added.[14] Mitoplasts generated with hypotonic solutions show a decreased respiration rate due to loss of cytochrome $c$ (Fig. 2). The addition of exogenous cytochrome $c$ restores respiration in a dose-dependent manner.

## Detection of $O_2^{\cdot-}$ Formation by Liver and Heart Mitoplasts

### Spin-Trapping Electron Paramagnetic Resonance

The use of EPR for $O_2^{\cdot-}$ detection is advantageous over monitoring cytochrome $c$ reduction for several reasons: first, EPR in conjunction with the spin trap 5,5-dimethyl-1-pyrroline-$N$-oxide (DMPO) has been estimated to be 20-fold more sensitive than cytochrome $c$ reduction.[21] As with the cytochrome $c$ assay, spin trapping of $O_2^{\cdot-}$ needs to be validated by its sensitivity to superoxide dismutase. Second, the use of the cytochrome $c$ assay in mitochondrial membranes is complicated by the effective oxidation of reduced cytochrome $c$ by cytochrome oxidase. This cannot be overcome by inhibition of cytochrome oxidase by KCN, a condition that—together with antimycin A—leads to suppression of $O_2^{\cdot-}$ by preventing ubisemiquinone formation on inhibition of electron transfer from the Rieske iron sulfur center to cytochrome $c_1$.[4] Adenylated cytochrome $c$, a less efficient substrate for cytochrome oxidase,[22,23] cannot be applied to isolated liver mitoplasts to detect $O_2^{\cdot-}$ for similar reasons.

---

[21] S. P. Sanders, S. J. Harrison, P. Kuppusamy, J. T. Sylvester, and J. L. Zweier, *Free Radic. Biol. Med.* **16**, 753 (1994).

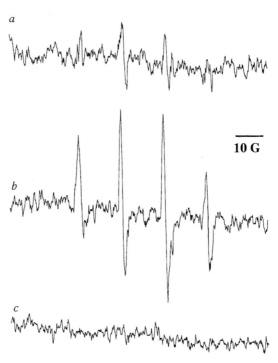

FIG. 3. Spin-trapping EPR signals of rat liver mitochondria and mitoplasts. (a) Rat liver mitochondria plus antimycin A. (b) Rat liver mitoplasts plus antimycin. (c) As in (b) plus superoxide dismutase.

For EPR analysis, mitoplasts are placed in buffer consisting of 230 m$M$ mannitol, 70 m$M$ sucrose, 20 m$M$ Tris [the pH is adjusted to pH 7.4 with morpholinepropanesulfonic acid (MOPS)] in the absence or presence of mitochondrial respiratory substrates or inhibitors. DMPO is present at a concentration of 160 m$M$. EPR spectra are recorded on a Bruker ECS106 spectrometer with the following settings: receiver gain, $5 \times 10^4$; microwave power, 20 mW; microwave frequency, 9.81 GHz; modulation amplitude, 0.505 G; time constant, 1.3 sec; scan time, 167.7 sec; scan width, 100 G. The EPR traces are a summation of five scans.

Isolated rat liver mitochondria generate a small DMPO–OH adduct signal in the presence of antimycin and DMPO; this signal originates from the decomposition of the $O_2^{\cdot-}$ adduct of DMPO (DMPO–OOH) (Fig. 3a). When mitoplasts generated with 0.2% (w/v) digitonin treatment are treated with antimycin

---

[22] A. Azzi, C. Montecucco, and C. Richter, *Biochem. Biophys. Res. Commun.* **65,** 597 (1975).
[23] A. Boveris, *Methods Enzymol.* **105,** 429 (1984).

(a complex III inhibitor that increases the steady-state levels of ubisemiquinone) and DMPO, a clear DMPO–OH signal is observed (Fig. 3b). The DMPO–OH signal has the following characteristics: (1) it is suppressed by superoxide dismutase, thus confirming $O_2^{\cdot-}$ as the source of the DMPO–OH signal (Fig. 3c); (2) it is decreased in a dose-dependent manner by ferricytochrome $c$ (Fig. 4a–c), which reacts with $O_2^{\cdot-}$ [reaction (1); $k_1 = 2.5 \times 10^5 \, M^{-1} \, \text{sec}^{-1}$][8] at a rate considerably faster that DMPO [reaction (2); $k_2 = 10 \, M^{-1} \, \text{sec}^{-1}$].

$$O_2^{\cdot-} + \text{cyt}\, c^{3+} \rightarrow O_2 + \text{cyt}\, c^{2+} \quad (1)$$

$$O_2^{\cdot-} + \text{DMPO} \rightarrow \text{DMPO–OOH} \quad (2)$$

The decrease in EPR signal intensity by cytochrome $c$ may be sustained by the competitive mechanisms shown in Eq. (3).

$$\%\text{Inhibition} = k_1\,[\text{cytochrome}\, c^{3+}]/(k_1[\text{cytochrome}\, c^{3+}] + k_2[\text{DMPO}]) \quad (3)$$

This effect also suggests that DMPO reacts with $O_2^{\cdot-}$ at the cytosolic side of the inner membrane; and (3) the EPR signal is broadened by the membrane-impermeable spin trap-broadening agent chromium trioxalate[24,25] (Fig. 4d and e): this strengthens further the localizatin of the spin adduct on the cytosolic side of the inner membrane.

EPR offers a sensitive and qualitative method for the detection of $O_2^{\cdot-}$. However, the spin adducts can be reduced to undetectable hydroxylamines by the respiratory chain.[26]

*Hydroethidine Oxidation*

$O_2^{\cdot-}$ has been shown to specifically oxidize hydroethidine (HE) to the fluorescent ethidium ($E^+$).[27] Although hydroethidine is not oxidized by $H_2O_2$, HOCl, or $ONOO^-$,[28] ethidium fluorescence may be affected by changes in mitochondrial membrane potential,[29] making $O_2^{\cdot-}$-mediated ethidium fluorescence a qualitative assay. $E^+$ fluorescence ($\lambda_{\text{exc}} = 470$ nm; $\lambda_{\text{em}} = 590$ nm) may be monitored as a measure of superoxide anion.[28] Heart mitoplasts suspended in 230 m$M$ mannitol, 70 m$M$ sucrose, and 20 m$M$ Tris–HCl (buffered to pH 7.4 with MOPS) are supplemented with respiratory substrates and 3 $\mu M$ hydroethidine and fluorescence is monitored with the settings described above. Fluorescence intensity depends

---

[24] S. P. Berg and D. M. Nesbitt, *Biochim. Biophys. Acta* **548**, 608 (1979).
[25] C. S. Lai, W. Froncisz, and L. E. Hopwood, *Biophys. J.* **52**, 625 (1987).
[26] A. T. Quintanilha and L. Packer, *Proc. Natl. Acad. Sci. U.S.A.* **74**, 570 (1977).
[27] L. Benov, L. Sztejnberg, and I. Fridovich, *Free Radic. Biol. Med.* **25**, 826 (1998).
[28] V. P. Bindokas, J. Jordán, C. C. Lee, and R. J. Miller, *J. Neurosci.* **16**, 1324 (1996).
[29] S. L. Budd, R. F. Castilho, and D. G. Nicholls, *FEBS Lett.* **415**, 21 (1997).

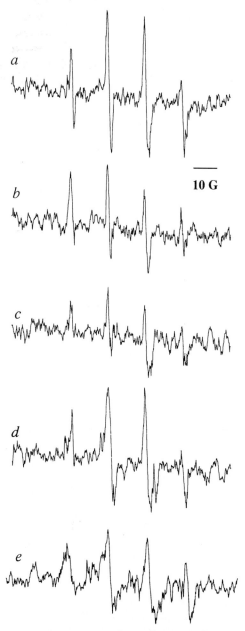

FIG. 4. Effect of cytochrome $c$ and chromium trioxalate on the EPR signal intensity of antimycin-treated mitoplasts. (a) No treatment. (b) Plus 5 $\mu M$ cytochrome $c$. (c) Plus 10 $\mu M$ cytochrome $c$. (d) Plus 2.5 m$M$ chromium trioxalate. (e) Plus 5 m$M$ chromium trioxalate.

## TABLE I
### $O_2^{·-}$ AND $H_2O_2$ PRODUCTION BY HEART MITOPLASTS[a]

|  | Ethidium fluorescence (arbitrary units) | $H_2O_2$ production (nmol/min/mg protein) |
|---|---|---|
| Succinate | 22.1 ± 13.1 | 0.08 ± 0.06 |
| Succinate + antimycin | 382.0 ± 80.1 | 0.32 ± 0.14 |
| Succinate + antimycin + SOD | 0 | 1.25 ± 0.25 |
| Succinate + antimycin + SOD + myxothiazol | 0 | 0.25 ± 0.13 |

[a] Assay conditions: Heart mitoplasts (0.1 mg) were treated with 3 $\mu M$ hydroethidine and incubated in a buffer containing 230 m$M$ mannitol, 70 m$M$ sucrose, 20 m$M$ Tris-HCl (adjusted to pH 7.4 with MOPS) at 37°. Ethidium fluorescence was monitored as a measure $O_2^{·-}$ as described in text. For $H_2O_2$ measurements, heart mitoplasts (0.2 mg) were incubated in the aforementioned buffer at 37° with horseradish peroxidase (10 U) and $p$-hydroxyphenylacetate (1 m$M$). Exogenous cytochrome $c$ (30 pmol/mg) was added for maximum ethidium fluorescence and $H_2O_2$ production by mitoplasts. When present, the concentrations of succinate, antimycin, SOD, and myoxthiazol were 7.5 m$M$, 1 $\mu$g/mg of protein, 400 U/mg, and 1 $\mu M$, respectively.

on the availability of respiratory substrates, is enhanced by antimycin A, and is abolished by superoxide dismutase (Table I).

## Hydrogen Peroxide Formation by Mitoplasts

$O_2^{·-}$, generated by the mitochondrial respiratory chain, is the stoichiometric precursor of mitochondrial $H_2O_2$; hence, the latter reflects $O_2^{·-}$ production. Conventionally, $H_2O_2$ is measured with peroxidase-based assays[23] in conjunction with redox-sensitive fluorescence dyes, such as scopoletin and $p$-hydroxyphenylacetate; the values of $H_2O_2$ measured with horseradish peroxidase are usually lower than those measured with cytochrome $c$ peroxidase, because of the competition between the fluorescent dyes and endogenous fluorescence donors for the horseradish peroxidase–$H_2O_2$ complex. Heart mitoplasts supplemented with respiratory substrates, antimycin A, and cytochrome $c$ generate $H_2O_2$ measured by monitoring $p$-hydroxyphenylacetate fluorescence increase[30] ($\lambda_{exc}$ = 320 nm; $\lambda_{em}$ = 400 nm) on a PerkinElmer (Norwalk, CT) LS-5 spectrofluorometer equipped with a thermally controlled and magnetic stirring sample compartment. Cytochrome $c$ is required to restore electron transfer and, thereby, $H_2O_2$ production by mitoplasts,[16] although the heme protein interferes with $p$-hydroxyphenylacetate fluorescence. The rate of $H_2O_2$ formation by mitoplasts is increased 4-fold by superoxide dismutase and decreased by myxothiazol, which inhibits ubisemiquinone formation (Table I). The rate of $O_2^{·-}$ production by mitoplasts can be calculated

[30] P. A. Hyslop and L. A. Sklar, *Anal. Biochem.* **141**, 280 (1984).

by multiplying by two the difference between the rate of $H_2O_2$ formation in the presence of superoxide dismutase and that measured in the absence of superoxide dismutase.

Summary

The topological distribution of ubiquinone in the mitochondrial respiratory chain suggests that both ubiquinone pools may participate in $O_2^{\cdot-}$ production and, hence, are vectorially released into the matrix and intermembrane space. Mitoplasts, obtained by either digitonin or hypotonic KCl treatment, are a suitable experimental model for measuring $O_2^{\cdot-}$ in the intermembrane space. The use of membrane-impermeable spin-broadening agents strengthens the notion that part of the $O_2^{\cdot-}$ generated by the respiratory chain may be released into the intermembrane space. This, together with the putative occurrence of a Cu,Zn-superoxide dismutase in this compartment may account for part of $H_2O_2$ released by mitochondria and contributing to a cytosolic steady-state level of this species in cytosol.

# [27] Measurement of Superoxide Radical and Hydrogen Peroxide Production in Isolated Cells and Subcellular Organelles

By ALBERTO BOVERIS, SILVIA ALVAREZ, JUANITA BUSTAMANTE, and LAURA VALDEZ

Superoxide radicals and hydrogen peroxide are constantly produced in aerobic cells. The early recognition that both species of molecules are able to initiate free radical reactions harmful to cells and tissues is now complemented by the concept that both $O_2^-$ [1] and $H_2O_2$ [2] are carefully regulated metabolites capable of signaling the regulatory devices of the biochemical and genetic systems of the cell. In addition, both chemical species are likely to have important roles in intercellular signaling and communication. The redox regulation of gene expression and intercellular communication is just starting to be understood as a vital mechanism in health and disease. Quantitatively precise measurements of $O_2^-$ and $H_2O_2$ production rates and steady-state concentrations are needed to enhance knowledge of signaling by these two intermediates of the partial reduction of oxygen. For instance, intracellular $H_2O_2$ steady-state levels regulate Jurkat-T cell function within narrow

[1] J. M. McCord, *Am. J. Med.* **108,** 652 (2000).
[2] A. Boveris and E. Cadenas, *IUBMB Life* **50,** 245 (2000).

limits; below 0.7 $\mu M$ H$_2$O$_2$, cells are in a proliferative state; at 1.0–3.0 $\mu M$ H$_2$O$_2$, cells develop programmed cell death; and at levels higher than 3.0 $\mu M$ H$_2$O$_2$, cells undergo necrosis.[3] Regulated intracellular steady-state concentrations of O$_2^-$ and H$_2$O$_2$ seem critical for cell physiology, much as it is the case, for instance, for K$^+$ and Na$^+$, or for Ca$^{2+}$ and Mg$^{2+}$.

Among the various intracellular organelles, the production of O$_2^-$ and H$_2$O$_2$ has been most studied and well documented in mitochondria, accounting for 1% (O$_2^-$) and 0.5% (H$_2$O$_2$), respectively, of organ O$_2$ uptake in rat liver and heart under physiological conditions, considering the pO$_2$ of the *in situ* organs.[4] In the liver, the cells of which have well-developed endoplasmic reticulum and peroxisomes, total tissue O$_2^-$ and H$_2$O$_2$ production can be estimated at about 3% of total liver O$_2$ uptake for both metabolites. In the heart, with almost no oxidases other than cytochrome oxidase, tissue O$_2^-$ and H$_2$O$_2$ production account for 1% and 0.5%, respectively, of organ oxygen uptake, as stated above. Intermediate situations are expected for other organs, such as lung, kidney, intestine, and brain, which have either considerable endoplasmic reticulum (lung), or peroxisomes (kidney), or other oxidases (as, e.g., intestinal xanthine oxidase and brain monoamino-oxidase). An early review, still valid in almost its entirety, provides a general outline of O$_2^-$ and H$_2$O$_2$ metabolism in mammalian organs.[5] A detailed description of assays for the measurement of O$_2^-$ and H$_2$O$_2$ production in mitochondria and submitochondrial particles was provided in a previous contribution to this series.[6]

## Superoxide Radicals

Superoxide radicals are produced at significant levels by (1) submitochondrial particles (SMPs) obtained by sonication of mitochondria (normally heart mitochondria; other types of mitochondria do not yield reproducible or well-characterized preparations); (2) fragments of mitochondrial membranes (any type of mitochondria), usually obtained by freezing and thawing mitochondrial suspensions and homogenizing them by passage through a tuberculin needle; (3) endoplasmic reticulum preparations from liver and lung; and (4) specialized cells of the immune system (neutrophils, macrophages, and lymphocytes). In the latter case, the white blood cells respond to stimulation by phorbol myristate acetate (PMA) or *N*-formyl-methionine-leucine-phenylalanine (fMLP) as described in a previous contribution to this series.[7] The preparations described in categories

---

[3] F. Antunes and E. Cadenas, *FEBS Lett.* **475**, 121 (2000).
[4] A. Boveris, L. E. Costa, and E. Cadenas, in "Understanding the Process of Aging" (E. Cadenas and L. Pacher, eds.), p. 1. Marcel Dekker, New York, 1999.
[5] B. Chance, H. Sies, and A. Boveris, *Physiol. Rev.* **59**, 527 (1979).
[6] A. Boveris, *Methods Enzymol.* **105**, 429 (1984).
[7] M. C. Carreras, J. J. Poderoso, E. Cadenas, and A. Boveris, *Methods Enzymol.* **269**, 65 (1996).

(1) to (4) can be assayed for $O_2^-$ generation by the techniques described in the following sections. Isolated cells, other than the immunocompetent cells described previously, do not produce $O_2^-$ in the extracellular space at important rates (in the 1–3% range of total oxygen uptake, as mentioned previously). However, plasma membranes have an NADH oxidase that vectorially releases $O_2^-$ into the extracellular space.[5] It has long been understood that mitochondria do not release $O_2^-$ to the reaction medium or to the cytosolic space.[5] However, it has been claimed more recently that rat liver mitoplasts release $O_2^-$ to the reaction medium on the basis of the broadening of the electron paramagnetic resonance (EPR) signal of the $O_2^-$-derived DMPO–OH adduct.[8]

*Quantitation of $O_2^-$ Production*

Isolated organelles, suborganellar preparations, and isolated cells require a reaction medium to be assayed for $O_2^-$ generation. Usually, mitochondria, submitochondrial particles, and fragments of mitochondrial membranes are suspended in 0.23 $M$ mannitol, 0.07 $M$ sucrose, 20–30 m$M$ Tris-HCl or Tris–morpholinepropanesulfonic acid (MOPS) buffer, pH 7.4.[6] For other subcellular structures, such as microsomes and peroxisomes, a reaction medium resembling cytosol in ionic composition, that is, 135 m$M$ KCl–20 m$M$ phosphate buffer, pH 7.4, is advisable.[9] For the case of isolated cells of any cell type, and certainly neutrophils, macrophages and lymphocytes, a suspending medium resembling extracellular fluid in ionic composition, such as 140 m$M$ NaCl–5 m$M$ KCl–20 m$M$ phosphate buffer, pH 7.4, is a common choice.[7,10] The rates of $O_2^-$ generation are best assessed by spectrophotometric determination. A highly sensitive spectrophotometer, such as a double-beam dual-wavelength model [SLM (Aminco, Lake Forest, CA); 356 (PerkinElmer, Norwalk, CT), etc.] or a modern single-beam instrument that discounts initial absorption, is required. Basically, adrenochrome formation and acetylated cytochrome $c$ reduction are the options. The formation of the EPR-detectable adduct DMPO–$O_2^-$ (or other similar adducts) provides qualitative evidence of $O_2^-$ generation,[8–11] but does not give a quantitative measurement of $O_2^-$ production.

*Adrenochrome Formation.* A description of this assay, showing a spectrophotometric trace of $O_2^-$-dependent adrenochrome formation by SMPs, was provided earlier in this series.[6] The reaction medium is supplemented with 1–2 m$M$ epinephrine (adrenaline) and adrenochrome formation is monitored spectrophotometrically. When the measurements are carried out in a dual-wavelength mode, 480–475 nm is used as the wavelength setting, with 480 nm as the active band,

---

[8] D. Han, E. Williams, and E. Cadenas, *Biochem. J.* **353**, 411 (2001).
[9] A. Boveris, N. Oshino, and B. Chance, *Biochem. J.* **128**, 617 (1972).
[10] L. Valdez and A. Boveris, *Antiox. Redox Signal.* **3**, 505 (2001).
[11] C. Giulivi, A. Boveris, and E. Cadenas, *Arch. Biochem. Biophys.* **316**, 909 (1995).

575 nm as the reference, and $E = 2.96$ m$M^{-1}$ cm$^{-1}$. In the case of a single-wavelength instrument, 480 nm and $E = 4.0$ m$M^{-1}$ cm$^{-1}$ are used.

The addition of 0.1–0.3 $\mu M$ Cu,Zn-superoxide dismutase provides the rate of the SOD-sensitive adrenochrome formation, which is taken stoichiometrically as the rate of $O_2^-$ production. The assay gives reliable information at rates of 1–5 $\mu M$ $O_2^-$/min. Epinephrine autoxidizes at alkaline pH in and SOD-sensitive process[12]; control of the baseline during the assay is essential. Catalase (0.1–0.3 $\mu M$) and 1 m$M$ EDTA or 0.2 m$M$ DETAPAC are advisable additions to prevent $H_2O_2$- and $Fe^{2+}$-dependent side reactions that lead to overestimation of $O_2^-$ generation rates.

*Acetylated Cytochrome c Reduction.* A description of the assay was provided earlier in this series.[6] Acetylated cytochrome $c$ reacts rapidly with $O_2^-$ ($k = 1 \times 10^5 \, M^{-1}$ cm$^{-1}$), is not efficiently reduced by the cytochrome $c$ reductase activities of the inner mitochondrial membrane and of the plasma membrane, and is not efficiently oxidized by cytochrome oxidase.[13] Ferricytochrome $c$ (50 mg) is acetylated by dissolving it in 50% saturated sodium acetate solution in an ice bath and adding 70 $\mu$l of pure acetic anhydride. The reaction mixture is kept in the ice bath for 30 min and then dialyzed against 50 m$M$ phosphate buffer, pH 7.2.[13] Succinylated cytochrome $c$ can be prepared similarly by replacing acetic anhydride by succinic anhydride. The reduction of 20–30 $\mu M$ modified cytochrome $c$ is monitored spectrophotometrically at 550–540 nm in a dual-wavelength spectrophotometer ($E = 19$ m$M^{-1}$ cm$^{-1}$), or at 550 nm with a sensitive single-beam instrument ($E = 21$ m$M^{-1}$ cm$^{-1}$). The addition of superoxide dismutase (0.1–0.3 $\mu M$ Cu,ZnSOD) gives the SOD-sensitive rate of acetylated cytochrome $c$ reduction, which gives the stoichiometric rate of $O_2^-$ generation. The assay provides reliable information at rates of 0.2–5 $\mu M$ $O_2^-$/min.

## Hydrogen Peroxide

Hydrogen peroxide is produced in all aerobic cells and is highly diffusible across biological membranes. Consequently, $H_2O_2$ is to be considered an intercellular metabolite: it will diffuse out of cells provided the steady-state concentration in the surrounding or extracellular medium, or in the neighboring cell, is lower than in the generating cell. Hydrogen peroxide diffusion out of hepatocytes,[5] thymocytes,[14] and MCF-7 breast tumor cells,[15] among others, has been reported. The determination of $H_2O_2$ produced at the plasma membrane of neutrophils and other cells of the immune system, and its release to the suspending medium, is a

[12] H. Misra and I. Fridovich, *J. Biol. Chem.* **247**, 3170 (1972).
[13] A. Azzi, C. Montecucco, and C. Richter, *Biochem. Biophys. Res. Commun.* **65**, 597 (1975).
[14] J. Bustamante, A. Tovar-Baraglia, G. Montero, and A. Boveris, *Arch. Biochem. Biophys.* **337**, 121 (1997).
[15] J. Bustamante, M. Galleano, E. Medrano, and A. Boveris, *Breast Cancer Res. Treat.* **17**, 145 (1990).

common functional parameter of phagocytic cells.[7,10] Interestingly enough, $H_2O_2$ diffusion out of plant organs, such as soybean embryonic axes,[16] and out of the gill of small (2–4 g) live fishes[17] has been reported.

*Quantitation of $H_2O_2$ Production and Steady-State Concentrations*

The assays suitable for determining $H_2O_2$ production rates and steady-state levels are based almost exclusively on the use of peroxidases. Chemical assays based on $I^-$ oxidation or $KMnO_4$ reduction are no longer utilized. Horseradish peroxidase and yeast cytochrome $c$ peroxidase are commonly used and their proper use was described earlier in this series.[6] Yeast cytochrome $c$ peroxidase (CCP) reacts with $H_2O_2$, forming an enzyme–substrate complex that in the absence of cytochrome $c$, the hydrogen donor, is an excellent indicator for quantitating $H_2O_2$. However, CCP is not commercially available and its use is restricted. Horseradish peroxidase (HRP) reacts with $H_2O_2$ and various hydrogen donors or reductants ($AH_2$) according to reactions (1) and (2):

$$HRP + H_2O_2 \rightarrow HRP-H_2O_2 \qquad (1)$$
$$HRP-H_2O_2 + AH_2 \rightarrow HRP + 2H_2O + A \qquad (2)$$

The reactions have a series of one-electron intermediates that are disregarded in Eqs. (1) and (2). The enzyme–substrate complex, $HRP-H_2O_2$, unfortunately, reacts with a variety of hydrogen donors that may be present in biological preparations[6] and also with tyrosine residues of its own HRP polypeptide chain, which leads to underestimation of the $H_2O_2$ measurements.[18] Calibration of the HRP-based assay with known rates of $H_2O_2$ production by glucose oxidase or uricase are highly advisable.[6] Fluorometric measurement of the coupled oxidation of hydrogen donors ($AH_2$) is usually performed. $p$-Hydroxyphenylacetic acid ($p$-HPA), the oxidized form of which (A) yields a fluorescent tetramer, is the hydrogen donor most often used. Others are diacetyldihydrodichlorofluorescein and scopoletin, the latter being fluorescent in its reduced form. The fluorometric determination, with excitation at 317 nm and emission at 414 nm, largely eliminates the problem of the light scattering by cell and organelle suspensions. The assay is calibrated, using equivalence of fluorescence units equivalent to $H_2O_2$ concentrations, with a standard $H_2O_2$ solution whose concentration is determined at 240 nm ($E = 43.6 \, M^{-1} \, cm^{-1}$).

*$H_2O_2$ Production Rates.* Organelles or cells are suspended at 0.1–0.5 mg of protein/ml or $1-5 \times 10^5$ cells/ml in the reaction medium outlined above, to which is added HRP (1–3 U/ml, types VI and VI-A; Sigma, St. Louis, MO) and 10–40 $\mu M$

---

[16] S. Puntarulo, R. A. Sanchez, and A. Boveris, *Plant Physiol.* **86**, 626 (1988).
[17] D. Wilhem-Filho, B. Gonzalez-Flecha, and A. Boveris, *Braz. J. Med. Biol. Res.* **27**, 2879 (1994).
[18] E. Cadenas and A. Boveris, *Biochem. J.* **188**, 31 (1980).

$p$-HPA, supplemented with the corresponding substrates, and the changes in fluorescence are recorded. The slopes of the fluorescence increases are converted to rates of $H_2O_2$ production, using fluorescence units equivalent to $H_2O_2$ concentration (micromolar).

$H_2O_2$ *Steady-State Concentrations.* Cells release $H_2O_2$ to the suspending medium until the $H_2O_2$ concentration in the extracellular medium reaches a diffusion equilibrium with the internal $H_2O_2$ steady state concentration.[14,15] The same principle has been applied to white blood cells,[10] soybean embryonic axes,[16] and swimming fishes.[17] Basically, 0.5- to 2-ml aliquots of the suspending medium are taken at periods of about 3–5 min and analyzed for $H_2O_2$ concentration. If the cells are in suspension, a fast centrifugation at about 300$g$ for 5 min at 4° provides a clear supernatant for analysis. If the cells are in a monolayer, an aliquot of the overlaying medium is the sample to be analyzed. Figure 1 shows the procedure applied to mononuclear cells. Medium aliquots of 1.5 ml are centrifuged to obtain a clear supernatant; the aliquot is divided into two parallel samples of 0.6 ml (measurement and control; Fig. 1), both added to 1 ml of medium containing HRP and $p$-HPA in 0.1 $M$ phosphate buffer, pH 7.4, being the control sample pretreated with 0.1–0.3 $\mu M$ catalase. The differential fluorescence readings (measurement minus control) are translated into $H_2O_2$ concentrations by the calibration relationship described previously. Figure 2A shows the kinetics of $H_2O_2$ release by human

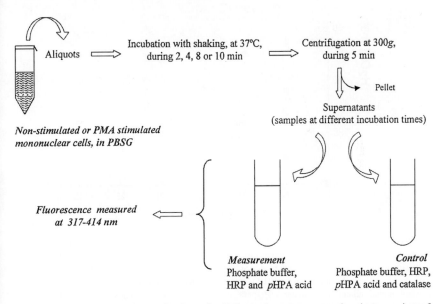

FIG. 1. Experimental design used to determine $H_2O_2$ steady-state concentrations in suspensions of mononuclear cells. The assay is suitable for use with other types of cells in suspension or cultured cells adhered in monolayers. PBSG, PBS containing 7.5 m$M$ glucose.

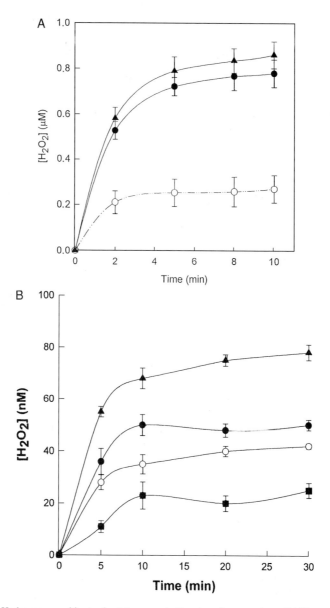

FIG. 2. Hydrogen peroxide steady-state concentrations in cell suspensions. (A) Human mononuclear cells ($2 \times 10^6$ cells/ml) suspended in 140 m$M$ NaCl, 5 m$M$ KCl, 20 m$M$ phosphate buffer, pH 7.4, with a final volume of 8.0 ml, were incubated at 37° in a regulated shaking water bath. Aliquots (1.5 ml) were taken at the indicated times and $H_2O_2$ concentrations were analyzed as described in text. (○) Control, nonstimulated mononuclear cells; (●) mononuclear cells stimulated by PMA (0.1 μg/ml); (▲) same as (●) but supplemented with 1 μ$M$ SOD. (B) Rat thymocytes ($25 \times 10^6$ cells/ml) suspended in 130 m$M$ NaCl, 10 m$M$ KCl, 20 m$M$ phosphate buffer, pH 7.4. Independent samples from which 1.5-ml aliquots were taken were incubated. (○) Control thymocytes; (●) thymocytes plus 10 μ$M$ methyl prednisolone; (▲) thymocytes plus 5 μ$M$ thapsigargin; (■) thymocytes plus 10 μ$M$ etoposide.

mononuclear cells, nonstimulated and stimulated by PMA, and also shows the superoxide dimutase effect. Nonstimulated mononuclear cells reach an equilibrium with external $H_2O_2$ at about 0.25 $\mu M$ $H_2O_2$, and after stimulation by PMA the level is increased to about 0.80 $\mu M$ $H_2O_2$. Figure 2B shows the kinetics of $H_2O_2$ release from thymocytes. The basal situation indicates an intracellular concentration of about 35 n$M$ $H_2O_2$, which is increased to about 45 and 75 nM $H_2O_2$ after apoptosis stimulation by methylprednisole and thapsigargin, respectively. The addition of superoxide dismutase to the reaction medium usually has no significant effect on cells other than the phagocytic cells, which release significant amounts of $O_2^{·-}$ to the suspending medium.

# [28] Biochemical Assay of Superoxide Dismutase Activity in Drosophila

By ROBIN J. MOCKETT, ANNE-CÉCILE V. BAYNE, BARBARA H. SOHAL, and RAJINDAR S. SOHAL

## Introduction

A variety of indirect assays of superoxide dismutase (SOD) activity have been described.[1] These assays use chemical or enzymatic systems to generate superoxide anion radicals ($O_2^{·-}$), which then react with an indicator molecule. Superoxide dismutase activity decreases the supply of $O_2^{·-}$, and is measured on the basis of the degree of inhibition of the indicator reaction. Validation of these methods requires the exclusion of numerous potential interferences. Potential sources of error include direct inhibition or amplification of either $O_2^{·-}$ production or the indicator reaction by other enzymes, metal ions, or other substances in tissue samples.

One convenient method uses xanthine–xanthine oxidase to generate $O_2^{·-}$, and reduction of nitroblue tetrazolium (NBT) to a blue formazan as the indicator reaction. The NBT assay was originally used to determine SOD activity either spectrophotometrically or in polyacrylamide gels.[2] The method was later modified to suppress various interferences occurring in crude mammalian tissue homogenates.[3] The introduction of metal chelators that also inhibit the electron transport chain was particularly advantageous. In our laboratory, this assay has been adapted slightly for measurement of SOD activity in Drosophila homogenates.

[1] L. Flohé and F. Ötting, *Methods Enzymol.* **105**, 93 (1984).
[2] C. Beauchamp and I. Fridovich, *Anal. Biochem.* **44**, 276 (1971).
[3] D. R. Spitz and L. W. Oberley, *Anal. Biochem.* **179**, 8 (1989).

## Superoxide Dismutase Activity Assay

*Principle*

Xanthine–xanthine oxidase is used to generate $O_2^{·-}$, which reduces NBT to a blue formazan. Inhibition of this reaction by SOD is monitored spectrophotometrically at 560 nm. Treatment of samples with cyanide permits direct measurement of MnSOD activity.[3] Pretreatment with sodium dodecyl sulfate (SDS) permits direct measurement of Cu,ZnSOD.[4] Total SOD activity is measured in untreated samples.

*Reagents*

Potassium phosphate buffer (50 m$M$), pH 7.8: All other reagents are dissolved in this buffer except SDS and KCl.
Xanthine (1.8 m$M$): NaOH is added to dissolve the xanthine.
Nitroblue tetrazolium (NBT; 2.24 m$M$): Prepare fresh daily and shield from light.
Diethylenetriaminepentaacetic acid (DTPA; 1.33 m$M$).
Catalase (40 units/ml)
Bathocuproine disulfonic acid, disodium salt (BCS; 10 m$M$).
NaCN (0.33 $M$): Prepare fresh daily.
Bovine serum albumin (BSA, defatted; 0.20 mg/ml) in potassium phosphate buffer containing 1.33 m$M$ DTPA: Referred to in working solution as potassium phosphate–DTPA–BSA
Xanthine oxidase (about 0.025 units/ml) in potassium phosphate buffer containing 1.33 m$M$ DTPA and 0.20 mg/ml BSA; the amount is adjusted until the absorbance change in working solution is 0.02–0.03/min in the absence of SOD.
Sodium dodecyl sulfate (SDS; 10%, w/v).
KCl (3 $M$)
Working solution: The reagents listed above are mixed in the following ratio and shielded from light: 132 ml of potassium phosphate–DTPA–BSA, 5 ml of catalase, 5 ml of NBT, 17 ml of xanthine, and 1 ml of BCS (for assay of MnSOD, add 3 ml of NaCN and subtract 3 ml of potassium phosphate–DTPA–BSA).

*Sample Preparation*

Whole-body homogenates (5%, w/v) of fresh or frozen flies are prepared in 50 m$M$ potassium phosphate buffer, pH 7.8, and centrifuged 10 min at 10,000$g$, 4°. For total SOD activity, the supernatants are diluted 8 : 11 with the same buffer (80 $\mu$l

---

[4] B. L. Geller and D. R. Winge, *Anal. Biochem.* **128**, 86 (1983).

of supernatant is combined with 30 µl of buffer), and recentrifuged for 10 min, 20,000g, 4°. For MnSOD activity, the 20,000g supernatant is then incubated with 5 m$M$ NaCN (final concentration) for at least 45 min at room temperature. For Cu,ZnSOD activity, the 10,000g supernatant is mixed with 10% (w/v) SDS (20 µl of SDS is combined with 80 µl of supernatant), incubated for 30 min at 37°, and then chilled for 5 min on ice. The mixture is treated next with 3 $M$ KCl (10 µl KCl per 100 µl), chilled for 30 min on ice, and then centrifuged for 10 min, 20,000g, 4°. The supernatants are retained for measurement of activity.

*Procedure*

1. The absorbance change is initially measured in one cuvette, containing 800 µl of working solution, 100 µl of potassium phosphate, and 100 µl of xanthine oxidase, for 4–6 min at 560 nm, 25°, to check the dilution of xanthine oxidase and the linearity of the reaction. If the reaction becomes nonlinear after less than 4 min, fresh xanthine oxidase should be prepared.

2. Measurements of SOD-containing samples are made simultaneously in six cuvettes. Mix 800 µl of working solution, 100 µl of potassium phosphate plus diluted sample, and 100 µl of xanthine oxidase per cuvette. Include one cuvette with no sample to obtain the maximum, uninhibited rate. Add varying amounts of samples to the other cuvettes. A typical sample is diluted about 8× for measurement of total SOD, 6× for Cu,ZnSOD, and 3× for MnSOD. Typical sample volumes with these dilutions are 15, 20, 30, 40, and 60 µl. Ideally, the dilution is such that one SOD sample inhibits the reaction by less than 50%, another inhibits by about 50%, and the other three inhibit by more than 50%. Some deviation from this guideline is permissible.

3. A blank reaction with the maximum concentration of sample and either no xanthine or no xanthine oxidase will normally have a rate close to 0, but this assumption should be checked occasionally.

4. The protein concentration is determined separately for the samples with and without SDS pretreatment. The Lowry and bicinchoninic acid (BCA) assays are both suitable for this purpose.

*Calculation of Activity*

The unit of SOD activity is defined as the amount that inhibits NBT reduction half-maximally under the conditions described above. Note that the unit is conventionally defined as half-maximal inhibition in a 3-ml reaction volume, whereas the reaction volume has been reduced to 1 ml here, so a conversion factor of 3 is introduced in comparisons with some other work.

The slope divided by the intercept (ordinate) of double-reciprocal plots, that is, plots of 1/% inhibition of NBT reduction vs. 1/protein content, gives the amount of sample protein that contains 1 unit of SOD activity.[5] The reciprocal of this number is the number of units of SOD activity per unit protein. Although this calculation gives a reasonable approximation of the SOD activity in most samples, it also tends to amplify the impact of pipetting errors or residual interferences in the homogenates. More consistent activity results are obtained by determining the reciprocal of the $IC_{50}$, using Eq. (1) (provided in the GraFit 4 software package[6]):

$$\text{NBT reduction rate} = \frac{\text{range}}{1 + \frac{(\text{protein content})^s}{(IC_{50})^s}} + \text{background} \qquad (1)$$

Use of Eq. (1) requires the choice of reasonable starting estimates for the four parameters in the estimatrix. The estimatrix value of the background is almost always zero, except in the special case of flies with Cu,ZnSOD gene mutations, discussed below. The range is estimated as the absorbance change in the uninhibited reaction, minus any background. The starting estimate for the slope factor, $s$, is normally 1.0, and the estimate for the $IC_{50}$ depends on the expected level of SOD activity in the sample. If the initial estimate differs from the final result by more than 1 unit, the procedure can be repeated with new starting values in the estimatrix to ensure that the iterative calculation converges on a reasonable final result. The reasonableness of the result is inferred from the fit of the curve to the data points in the plot of NBT reduction rate versus protein content (protein content is plotted on a logarithmic scale).

Activity is determined three or four times for both total SOD and either MnSOD or Cu,ZnSOD (via NaCN treatment or SDS pretreatment, respectively). The remaining quantity may then be calculated by subtraction of the other two, but for some samples this is less accurate than direct measurement of all three quantities.

Discussion

To confirm that NBT reduction by *Drosophila* tissue samples is due to SOD activity, homogenates prepared as described for the spectrophotometric assay were also separated on native polyacrylamide gels. These gels were soaked in NBT, followed by riboflavin–$N,N,N',N'$-tetramethylethylenediamine (TEMED), and SOD activity was detected on the basis of its inhibition of $O_2^{\cdot-}$-mediated NBT photoreduction.[2] Using homogenates of flies containing a wild-type complement of SOD alleles prepared for the assay of total SOD, three bands were observed. The uppermost band, identified putatively as MnSOD, was always eliminated by pretreatment of samples with SDS. The two lower bands, identified putatively as Cu,ZnSOD, were both totally absent in samples from SOD *x39*/SOD *x16* flies,

[5] J. E. Bell and E. T. Bell, "Proteins and Enzymes." Prentice-Hall, Englewood Cliffs, New Jersey, 1988.
[6] R. J. Leatherbarrow, "GraFit," version 4.0. Erithacus Software, Staines, U.K., 1998.

which contain deletions overlapping part of the coding region of both copies of the Cu,ZnSOD gene.[7] Bovine erythrocyte SOD, used as a positive control, also migrated as a double band on native gels, an observation that has also been made with purified SOD from other sources.[1,2]

These identifications were confirmed immunologically, using rabbit anti-*Drosophila* MnSOD and Cu,ZnSOD antibodies.[8] After SDS-PAGE and immunoblot analysis, strong signals were detected from wild-type samples with both antibodies. The SDS pretreatment essentially abolished the signal detected with the MnSOD antibody, without decreasing the signal with the Cu,ZnSOD antibody. Conversely, in the SOD *x39*/SOD *x16* sample, there was no reaction with the Cu,ZnSOD antibody, but a strong signal was detected with the MnSOD antibody. Pretreatment of SOD *x39*/SOD *x16* samples with SDS abolished SOD activity on activity gels, and abolished the reaction of the denatured sample with MnSOD antibody in immunoblot analysis. Regrettably, these antibodies are unreactive with the native proteins, and SOD activity is lost on SDS-containing gels, so a direct link cannot be made between the bands on activity gels and immunoblots.

In homogenates of *Drosophila* samples with wild-type or higher levels of Cu,ZnSOD and MnSOD activity, the sum of the SOD activities after SDS pretreatment (Cu,ZnSOD) and NaCN treatment (MnSOD) approximates the total SOD activity fairly. The background level of NBT reduction in the $IC_{50}$ calculation is zero, that is, half-maximal inhibition of NBT reduction is the same as 50% inhibition. In the case of flies with Cu,ZnSOD gene deletions, the $IC_{50}$ background is still zero in the calculation of total SOD activity, because of the presence of MnSOD. However, after pretreatment of such samples with SDS, where no Cu,ZnSOD activity is detectable on activity gels, there remains a low level of inhibition of NBT reduction in the spectrophotometric assay. The $IC_{50}$ calculation does not then converge on a reasonable result, that is, there is no fit to the data points, unless up to 65% of the NBT reduction is treated as background in the estimatrix. The absorbance maximum at the end of the reaction is shifted from 545–550 to 520–525 nm, and there is a plateau in NBT reduction, so that the maximum percent inhibition is 35–60%, not 100%. With large volumes of undiluted samples, the plateau is followed eventually by a further decrease in NBT reduction, possibly through direct inhibition of $O_2^{\cdot -}$ production.

It is likely that most or all of the "activity" detected in the SDS-pretreated, Cu,ZnSOD null samples is spurious, reflecting residual interferences in the assay that are drowned out in samples with wild-type levels of SOD activity. Owing to the high background in the $IC_{50}$ calculation, there is a large discrepancy between 50% inhibition and half-maximal inhibition of NBT reduction. For such samples,

---

[7] J. P. Phillips, J. A. Tainer, E. D. Getzoff, G. L. Boulianne, K. Kirby, and A. J. Hilliker, *Proc. Natl. Acad. Sci. U.S.A.* **92**, 8574 (1995).
[8] V. I. Klichko, S. N. Radyuk, and W. C. Orr, *Neurobiol. Aging* **20**, 537 (1999).

an alternative calculation can be made by taking the point closest to 50% inhibition, calculating the actual protein content and percent inhibition, and then making a linear extrapolation to 50% inhibition. Although this estimate is not accurate, because the assumption of a linear relationship between protein content and percent inhibition is false, the activity gel results are much more consistent with this crude approximation than with the $IC_{50}$ calculation. The results of the two calculations are almost identical when the estimatrix background is zero, but they diverge by up to 5-fold when both copies of the Cu,ZnSOD gene contain deletions. When only one Cu,ZnSOD gene is expressed, as a transgene in a Cu,ZnSOD null background, the estimatrix background may be zero or there may be a low background and a divergence between the calculations, depending on the level of transgene expression.

## Conclusions

The conclusions drawn from these experiments are that *Drosophila* Cu,ZnSOD and MnSOD inhibit the reduction of NBT by $O_2^{\cdot-}$, and that, under most conditions, most or all of the activity detected is attributable to the two SOD enzymes. The Cu,ZnSOD activity can be measured by pretreating the samples with SDS and the MnSOD activity can be measured by treatment with NaCN. Where the level of SOD activity is low, residual interferences in the assay become apparent, and the activity can be estimated only on the basis of rough approximations.

## Acknowledgments

The authors thank S. N. Radyuk and W. C. Orr for the *Drosophila* Cu,ZnSOD and MnSOD antibodies and L. W. Oberley for advice about the SOD assay.

## [29] Transcriptional Regulation and Environmental Induction of Gene Encoding Copper- and Zinc-Containing Superoxide Dismutase

By MUN SEOG CHANG, HAE YONG YOO, and HYUNE MO RHO

### Introduction

Superoxide dismutase (SOD) constitutes the first coordinated unit of defense against reactive oxygen species (ROS), which have been implicated both in the aging process and in degenerative diseases, including arthritis and cancer.[1,2] Cu,Zn superoxide dismutase (SOD1), which catalyzes the dismutation of superoxide radicals ($O_2^-$) to oxygen and hydrogen peroxide, is a key enzyme in the metabolism of oxygen free radicals.[3] The human and rat *SOD1* gene consists of five exons and four introns spanning about 11 and 6 kb of genome, respectively.[4,5] The physiological significance of the induction and regulation of SOD has been assessed. One report suggested that the overexpression of SOD1 and catalase could increase the average life span of a fly.[6] Because Cu,ZnSOD is not easily taken up by cells, the regulation and induction mechanism of the *SOD1* gene would be of great interest.

Stress conditions, including heat shock and oxidative stresses, are deleterious to normal cellular function. To survive these and other environmental and physiological stresses, all organisms possess specialized defense mechanisms to protect themselves. In this study, we report fundamental insights into the transcription of the *SOD1* gene in normal and induced states, using molecular biological approaches. Also, we examine whether the *SOD1* gene could be induced by environmental factors such as heavy metals, xenobiotics (dioxin), peroxisome proliferator (arachidonic acid), oxidative stress ($H_2O_2$ and paraquat), and heat shock through their corresponding *cis* elements.

### Transcriptional Regulation of *SOD1* Gene

*Overall Approach*

For the analysis of the mechanisms of *SOD1* gene transcription, the promoter structure, especially *cis* elements in the 5'-flanking region, must be analyzed. We

---

[1] J. Duschesne, *J. Theor. Biol.* **66**, 137 (1977).
[2] L. W. Oberley, *in* "Superoxide Dismutase" (L. W. Oberley, ed.), Vol. 2, p. 127. CRC Press, Boca Raton, Florida, 1983.
[3] I. Fridovich, *Annu. Rev. Biochem.* **44**, 147 (1975).
[4] D. Levanon, J. Lieman-Hurwitz, N. Dafni, M. Wigderson, L. Sherman, Y. Bernstein, Z. Laver-Rudich, E. Danciger, O. Stein, and Y. Groner, *EMBO J.* **4**, 77 (1985).
[5] Y. H. Kim, H. Y. Yoo, G. Jung, J. Kim, and H. M. Rho, *Gene* **133**, 267 (1993).

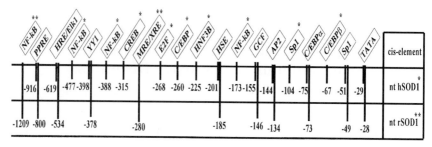

FIG. 1. Comparison of transcription factor-binding sites between rat and human *SOD1* promoter. Asterisks represent transcription factor-binding sites present only in humans (*) or rats (**). The numbers in the names of transcription factor-binding sites represent the 5' end point of corresponding *cis*-element sequence.

have isolated rat and human *SOD1* genes and cloned the promoter regions.[5,7] Subsequently, the promoter sequence was analyzed by the FASTA program with TFD, an EMBL database of transcription factors. The sequence motifs and consensus sequences for the putative transcription factor-binding site were mapped as shown in Fig. 1. To identify responsible *cis* elements and corresponding *trans*-acting factors, we employed several methods: deletion analyses, mutations, mobility shift assays, and transcription efficiency tests in the heterologous promoter and natural context systems.

*Cloning of Promoters*

In general, the major *cis* elements are located within −2 kb of the TATA box. Nucleotide (nt) sequences for the rat *SOD1* gene and 1.7 kb of the 5'-flanking region[5] have been determined and the 1.7-kb *Bam*HI–*Sma*I fragment (nt −1633 to +85) from the rat *SOD1* gene has been inserted into pBLCAT2, which contains the bacterial chloramphenicol acetyltransferase structural gene.[8] Plasmid phSOD-1500, which contains the promoter region of human *SOD1* gene, has been prepared by polymerase chain reaction using synthetic oligonucleotide primers (5'-upstream primer, 5'-GGATCTCCCTTGGCAAGTTTACAATGAACT-3'; 3'-downstream primer, 5'-CCGCTCGAGCGGAAACCCAGACGCTGC-3'), which amplify the genomic sequence from nt −1500 to −27 of the promoter. The cloned fragment has been inserted into the *Bam*HI and *Xho*I sites of the multiple cloning site of pBLCAT2.[8]

---

[6] W. C. Orr and R. S. Sohal, *Science* **263**, 1128 (1994).
[7] H. T. Kim, Y. H. Kim, J. H. Nam, H. J. Lee, H. M. Rho, and G. Jung, *Biochem. Biophys. Res. Commun.* **201**, 1526 (1994).
[8] B. Luckow and G. Schütz, *Nucleic Acids Res.* **15**, 5490 (1987).

## Construction of Serial Deletion Mutants of Promoter

Primer extension analysis demonstrates that a transcription start point (tsp) exists 93 bp upstream of the ATG start codon.[5] In the promoter region of the *SOD1* gene, the consensus TATA box is found at nt −28 from the tsp. Farther upstream from the TATA box, putative AP2, CREB, and HSF-binding sites have been shown by sequence homology analysis, using the FASTA program with TFD, an EMBL database of transcription factors. To obtain a series of deletions in the 5′-flanking region of rat *SOD1* gene, deletion by ExoIII (Promega, Madison, WI) is performed as follows. The 1.7-kb rat *SOD1* upstream region is taken out with *Sph*I and *Xba*I. This linear DNA contains a 3′ overhang that is resistant to ExoIII digestion, and a 5′-overhang that is susceptible to ExoIII. Five micrograms of the DNA is dissolved in 60 $\mu$l of ExoIII buffer [66 m$M$ Tris-HCl (pH 8.0) and 660 $\mu M$ MgCl$_2$]. Digestion proceeds at about 450 bases/min at 37°. There is a 20- to 30-sec lag before the start of the reaction. After adding ExoIII, each sample (2.5 $\mu$l) is collected at 30-sec intervals into S1 tubes on ice [40 m$M$ potassium acetate (pH 4.6), 340 m$M$ NaCl, 1.4 m$M$ ZnSO$_4$, 6.8% (v/v) glycerol, and 8 units of S1 nuclease]. After all the samples have been taken, the tubes are allowed to stand at room temperature for 30 min. The S1 nuclease is inactivated by adding 1 $\mu$l of stop buffer (0.3 $M$ Tris, 0.05 $M$ EDTA) to each sample and heating at 70° for 10 min. The samples from each time point are transferred to 37° and 1 $\mu$l of Klenow mix [20 m$M$ Tris-HCl (pH 8.0), 100 m$M$ MgCl$_2$, and 0.1–0.2 unit of Klenow] is added to each sample. Each sample is incubated for 3 min and then 1 $\mu$l of the dNTP mix (0.125 m$M$ dNTP) is added. After an additional 5 min at 37°, the samples are transferred to room temperature and 40 $\mu$l of ligase mix [50 m$M$ Tris-HCl (pH 7.6), 10 m$M$ MgCl$_2$, 1 m$M$ ATP, 5% (w/v) polyethylene glycol (PEG), 1 m$M$ dithiothreitol (DTT), and 0.05 unit of T4 DNA ligase] is addèd. After 1 hr at room temperature, each ligation mix is introduced into *Escherichia coli* HB101 by the CaCl$_2$ transformation method.[9] Among a nested set of clones, we choose about six to nine deletion mutants depending on the TFB data and restriction enzyme recognition sites on the *SOD1* promoter.[10]

## Chloramphenicol Acetyltransferase Assays for Determination of Promoter Strength of Deletion Mutants

*Transfection.* The following protocol is optimized for 60-mm culture dishes and all volumes are for single transfection. Exponentially growing cells are divided into 60-mm culture dishes (5 × 10$^5$ cells per dish) no more than 24 hr before the transfection. These cells should be grown overnight in 10 ml of Dulbecco's

---

[9] J. Sambrook, E. F. Fritsch, and T. Maniatis, "Molecular Cloning: A Laboratory Manual," 2nd Ed. Cold Spring Harbor Laboratory Press, Cold Spring Harbor, New York, 1989.
[10] Y. H. Kim, K. H. Park, and H. M. Rho, *J. Biol. Chem.* **271,** 24539 (1996).

modified Eagle's medium (DMEM) to a confluency of approximately 20–40%. Five micrograms of supercoiled DNA with 1 μg of RSV-$\beta$-Gal DNA is diluted with $H_2O$ to 225 μl. Sequentially, 25 μl of 2.5 $M$ $CaCl_2$ and 250 μl of 2× BBS [50 m$M$ $N$,$N$-bis-(2-hydroxyethyl)-2-aminoethanesulfonic acid (BES), 280 m$M$ NaCl, 1.5 m$M$ $Na_2HPO_4$ (pH 6.96)] are added to the DNA mixture, and further incubated for 10–20 min at room temperature. The precipitate is gently added to the culture dropwise and the plate is swirled gently to distribute evenly. Cells are then incubated for 12 to 24 hr. In the case of HepG2 cells, the incubation time is 18 hr. The $CO_2$ concentration is adjusted to 3.0%, which improves transfection efficiency. After 18 hr of incubation, the medium is removed and the cells are rinsed twice with phosphate-buffered saline (PBS). After adding fresh complete medium, the cells are allowed to incubate for 24 hr under 5% $CO_2$ concentration.

*Cell Extract Preparation.* Cells are harvested after rinsing twice with PBS, and then resuspended in a minimal volume of PBS (1 ml) by scraping them off the dish surface with a rubber policeman. Harvested cells (at 4°) are resuspended in 100 μl of 0.25 $M$ Tris-HCl (pH 7.8). Cell disruption is accomplished by a freezing and thawing method involving switching the tube to an ethanol–dry ice bath for 5 min and thawing in a 37° water bath for 5 min. After repeating this process three times, cell extracts are collected by centrifugation with 13,000 rpm for 5 min at 4°. The protein content (mg/ml) of each sample is determined by the optical density of the reactions at a wavelength of 595 nm, using the standard Bradford assay[9] and normalized to the same amount of protein.

*$\beta$-Galactosidase Assay.* The *E. coli* $\beta$-galactosidase gene is used as an internal reference in the transfection study. The construct that carries the $\beta$-galactosidase gene downstream from the long terminal repeat of Rous sarcoma virus[11] is used. The protocol for the $\beta$-galactosidase assay is as follows. Each sample is assayed in the assay mixture [1 m$M$ $MgCl_2$, $O$-nitrophenyl-$\beta$-D-galactopyranoside (ONPG, 1 mg/ml), cell extract, 33 m$M$ sodium phosphate (pH 7.5)]. The reactions are incubated at 37° for 1 hr or until a faint yellow color develops. The reactions are stopped by adding 500 μl of 1 $M$ $Na_2CO_3$ solution. The optical density of the reaction is determined at a wavelength of 420 nm.

*Chloramphenicol Acetyltransferase Assay.* The chloramphenicol acetyltransferase (CAT) assay is performed as follows. Each extract is assayed in the assay mixture [0.5 $M$ Tris-HCl (pH 7.8), 0.5 m$M$ acetyl-CoA, 50 μl of cell extract, and 5 μl of [$^{14}$C chloramphenicol]. After 1 hr of incubation, chloramphenicol is extracted with 1 ml of ethyl acetate by vortexing for 30 sec. After centrifugation in a microcentrifuge for 30 sec at room temperature, the top phase is carefully collected and ethyl acetate is evaporated with a Speed-Vac (Savant, Hicksville, NY). Samples are dissolved in 20 μl of ethyl acetate and spotted on a silica gel thin-layer

[11] C. M. Gorman, G. T. Merlino, M. C. Willingham, I. Pastan, and B. H. Howard, *Proc. Natl. Acad. Sci. U.S.A.* **79**, 6777 (1982).

chromatography (TLC) plate. The TLC plate is placed in an equilibrated tank containing chloroform–methanol (95 : 5, v/v) for separation. After air drying, the TLC plate is placed in a cassette containing X-ray film and exposed for an appropriate length of time.

*Confirmation of Binding of Protein to Cis Elements*

For confirmation of the biological role of the *cis* element and its corresponding transcription factor in a given tissue, binding tests should be carried out by gel mobility shift assays.

*Nuclear Extract Preparation.* Nuclear extracts are prepared by a modified procedure of Andrews and Faller.[12] This method is typically applied to between $5 \times 10^5$ and $1 \times 10^7$ cells in a 100-mm dish. Adherent cells are scraped into 1.5 ml of cold PBS. The cell suspension is then transferred to a microcentrifuge tube. Cells are pelleted for 10 sec and resuspended in 400 µl of cold buffer A [10 m$M$ HEPES, (pH 7.9), 1.5 m$M$ MgCl$_2$, 10 m$M$ KCl, 0.5 m$M$ DTT, and 0.2 m$M$ phenylmethylsulfonyl fluoride (PMSF)] by flicking the tube. The cells are allowed to swell on ice for 20 min and then sonicated for a short period (3 sec). Samples are centrifuged for 10 sec and the supernatant is discarded. The pellet is resuspended in 100 µl of cold buffer C [20 m$M$ HEPES (pH 7.9), 25% (v/v) glycerol, 420 m$M$ NaCl, 1.5 m$M$ MgCl$_2$, 0.2 m$M$ EDTA, 0.5 m$M$ DTT, and 0.2 m$M$ PMSF] and incubated on ice for 20 min for high-salt extraction. Cellular debris is removed by centrifugation for 2 min at 4° and the supernatant fraction is stored at −70°.

*Gel Mobility Shift Assay.* The double-stranded oligonucleotides corresponding to transcription factor-binding sites have been synthesized by GIBCO–BRL (Gaithersburg, MD). The oligonucleotide are labeled with [$\gamma$-$^{32}$P]ATP and polynucleotide kinase.[9] Nuclear extracts are prepared as follows. DNA-binding reactions are carried out in a 15-µl volume that typically contains 10 m$M$ HEPES (pH 7.9), 60 m$M$ KCl, 5 m$M$ MgCl$_2$, 1 m$M$ dithiothreitol, 1 m$M$ EDTA, 10% (v/v) glycerol, and 500 ng of poly(dI-dC) (Sigma, St. Louis, MO). Before the reaction with DNA, an equal amount (10 µl) of nuclear extract prepared from HepG2 cells is mixed and incubated for 20 min. In the supershift assay, the specific polyclonal antibody is added to the reaction mixture after the binding reaction. DNA binding is started by adding 10,000 cpm of probe to the preincubated reaction mixture followed by an additional incubation for 15 min at room temperature. For competition assays, the binding reaction is performed with an unlabeled probe or competitor DNA fragments. Samples are loaded on a 4% (w/v) polyacrylamide gel [acrylamide–bisacrylamide, 49 : 1 (w/w)] in 0.5× TBE (44 m$M$ Tris, 44 m$M$ boric acid, and 1 m$M$ EDTA). After electrophoresis, the gels are dried and exposed to X-ray film. All experiments are repeated at least three times.

[12] N. C. Andrews and D. V. Faller, *Nucleic Acids Res.* **19**, 2499 (1991).

*Functional Analysis*

From the deletion analysis data and gel mobility shift assays, we could determine which *cis* element is responsible for promoter strength and for the given inducer. For the development of a functional test, we usually adapted heterologous promoter and mutation analyses systems of the responsible *cis* elements.

*Heterologous Promoter System.* For the construction of a heterologous promoter, the synthetic oligonucleotide corresponding to the responsible *cis* element of the *SOD1* promoter[5] is cloned into the *Bam*HI site of pBLCAT2Δ, which is derived from pBLCAT2. The plasmid pBLCAT2Δ has a minimal region (nt −80 to +51) of the herpes simplex virus thymidine kinase (*tk*) promoter.[8] Three or four copies of the oligonucleotides for wild type or mutant are introduced, respectively. The insertion of consensus oligonucleotide and mutant sequences is confirmed by DNA sequencing. These plasmids, containing the various *cis* elements of the *SOD1* gene fused to the *tk* minimal promoter and *CAT* gene, are transfected into HepG2 cells and we measure functional activity by CAT assays as described above.

*Transcriptional Regulation of SOD1 Gene in Normal State*

Serial deletions from nt −1633 to −55 revealed at least three important *cis*-acting elements in the promoter region of the *SOD1* gene.[13] A DNA fragment from nt −576 to −412 (positive regulatory element, PRE) was found to enhance CAT activity by 6- to 7-fold. In contrast, DNA fragments from nt −412 to −305 (negative regulatory element, NRE) reduced CAT activities by about 7-fold, implying that the silencer element is involved in the transcription of the *SOD1* gene in HepG2 cell.[13] The CCAAT element-binding proteins (C/EBPα) are located at nt −73 and −75 of the rat and human *SOD1* genes, respectively (Fig. 1). When expression was examined in the presence of pMSV-C/EBPα with pRSP-412, the level of expression was dramatically increased (Table I[10,13,14,19–21,23]). We tested the direct binding activity of C/EBPα to this site by mobility shift assay (Fig. 2A). These results showed that C/EBPα could activate the transcription of the *SOD1* promoter in HepG2 cell.[14] To further characterize the elements that were responsible for a specific binding of the PRE and NRE, various oligonucleotide competitors were synthesized. The specificity of PRE binding was confirmed by using synthetic competitors (Fig. 2B). A functional test of Elk1, using transfection of mutant plasmid mElk1tkCAT, showed little or no change in promoter activity, whereas the CAT activity of Elk1tkCAT increased about 5-fold (Table I). The shifted DNA–protein complex was observed to contain only the YY1-binding site, which was critical for the DNA-binding activity detected by mutation and supershift assay with YY1 antibody (Fig. 2C). YY site-mediated repression of tk promoter activity was less than 10–20% relative to control vector (pBLCAT2) promoter activity (Table I), whereas the mutant

---

[13] M. S. Chang, H. Y. Yoo, and H. M. Rho, *Biochem. J.* **339**, 335 (1999).
[14] Y. H. Kim, H. Y. Yoo, M. S. Chang, G. Jung, and H. M. Rho, *FEBS Lett.* **401**, 267 (1997).

## TABLE I
### Transcriptional Regulation and Induction of SOD1 by Environmental Factors

| Status[a] | Transcription factor[b] / inducer[c] | Cis element[d] | Activation[e] (fold) | Ref. |
|---|---|---|---|---|
| Normal | C/EBPα | $_{-73}$TAGCGATTGGTTCCCTGCC$_{-55}$ | 27.0[f] | 14 |
| | YY1 | $_{-376}$AGCATCCATCTTGGCTCAC$_{-358}$ | 0.1[g] | 13 |
| | Elk1 | $_{-533}$TGCCTAGGAAGCGC$_{-520}$ | 4.5[g] | 13 |
| Induced | AP2/Ginsenoside-Rb$_2$ | $_{-134}$CCCCGCCC$_{-127}$,$_{-118}$CCCCGCGG$_{-111}$ | 5.0[h] | 10 |
| Environmental stress[a] | PPREBP/AA | $_{-797}$AGGTCAGAGGCA$_{-786}$ | 4.0[g] | 19 |
| | HREBP/H$_2$O$_2$ | $_{-533}$TGCCTAGGAAGCGC$_{-520}$ | 4.0[g] | 20 |
| | MREBP/Cu, Zn | $_{-273}$GCGCGCA$_{-267}$ | 4.0[g] | 21 |
| | XREBP/β-NF, TCDD | $_{-268}$GCACGCA$_{-262}$ | 4.0[g] | 23 |
| | HSE/42°, PQ | $_{-184}$ATTCTGGAACTTTC$_{-171}$ | 5.0[g] | 20 |

[a] Various conditions involved in *SOD1* transcription.
[b] Transcription factor involved in *SOD1* promoter transcription.
[c] Environmental stresses and ginsenoside-Rb$_2$ were used in this study.
[d] *Cis*-element sequence and location for corresponding *trans*-acting factors present in *SOD1* promoter.
[e] Relative CAT activity determined under indicated conditions.
[f] Relative CAT activity of pRSP-412 plasmid cotransfected with or without MSV-C/EBPα.
[g] CAT activity determined by synthetic oligonucleotide in the heterologous promoter system.
[h] Effect of ginsenoside-Rb$_2$ on the synthetic AP2 in the heterologous promoter system.

having a mutation in the YY-binding site was not affected and relieved the repression from YY1 function.[13]

## Induction of *SOD1* Gene by Ginsenoside-Rb$_2$

*Panax ginseng* C. A. Mayer (Araliaceae) is one of the most popular natural tonics in Asian countries. We identified the active fraction of ginseng saponin and further characterized ginsenoside-Rb$_2$ (Rb$_2$) as a strong inducer of the *SOD1* gene (Table I). To identify the target sequence of Rb$_2$ in the upstream region of the *SOD1* gene, deletion mutants were prepared and transfected into HepG2 cell with or without Rb$_2$ in the medium. We demonstrated that transcription factor AP2 was increased by Rb$_2$ treatment, and its binding site was in the promoter region of the *SOD1* gene (Fig. 2D). This was also confirmed by DNA sequence analysis in which two sites of AP2 were located between nt $-134$ and $-111$ ($_{-134}$CCCCGCCC$_{-127}$ and $_{-118}$CCCCGCGG$_{-111}$).[10]

## Environmental Induction of SOD1

### Background

The production of superoxides has been observed subsequent to treatment with peroxisome proliferator, metal ions, xenobiotics, and heat shock at the cellular and

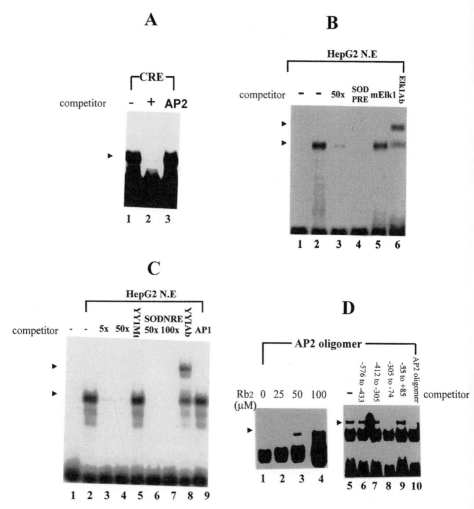

FIG. 2. Identification of transcription factor-binding sites involved in *SOD1* transcription in normal state and ginsenoside activation. (A) Direct binding of C/EBPα to the CAAT box sequence of the rat *SOD1* promoter. $^{32}$P-labeled double-stranded oligonucleotide corresponding to the CP2 site ($-74$ to $-55$ region in the *SOD1* promoter) was incubated with 100 ng of C/EBPα expressed in *E. coli*. (B) Binding of transcription factor Elk1 to its binding site in the PRE region. Nucleotides from positions $-537$ to $-516$ of the *SOD1* promoter were used. Specificity of binding was determined by the competition reaction. (C) Mobility shift assay with the $^{32}$P-labeled YY1 oligonucleotide probe encompassing region $-378$ to $-359$ of the *SOD1* promoter. Arrowheads indicate the position of the major DNA–protein complex and of the supershifted band by YY1-specific antibody. (D) Transcription factor AP2 induced by ginsenoside-Rb$_2$. MSA using the AP2 oligonucleotide with Rb$_2$-treated HepG2 nuclear extract was done. The increasing amount of AP2–DNA complex was observed by the increasing amount of Rb$_2$.

the organismic levels.[15–18] To survive these and other environmental and physiological stresses, all organisms possess specialized defense mechanisms to protect themselves. The diversity of *SOD1* regulatory *cis* elements including environmental factor response *cis* elements suggests that multiple antioxidative mechanisms and mechanisms by which cells receive and respond to environmental stresses should exist.

*SOD1* expression is highly induced during environmental stresses. However, little is known about how the environmental signal is transmitted to a transcriptional regulator in cells during oxidative stress such as that caused by hydrogen peroxide, paraquat (PQ), and heat shock. The regulatory regions of rat and human *SOD1* genes contain five environmental factor response *cis* elements for peroxisome proliferator, hydrogen peroxide, heavy metals, xenobiotics, and heat shock. Analyses of these elements were carried out by mobility shift assay (MSA), mutations, and heterologous promoter systems as described above.

*Environmental Factors*

*Chemical Treatment of Transfected Cells.* After transfection with the appropriate vector containing the responsible *cis* element, the cells are washed twice with PBS. The cells are then treated with the appropriate metal ion, xenobiotics, or peroxisome proliferators for 30 min in PBS [for the observation of environmental factors, arachidonic acid (AA), 50 $\mu M$; $H_2O_2$, 200 $\mu M$; $Cu^{2+}$, 400 $\mu M$; $\beta$-naphthoflavone ($\beta$-NF), 50 $\mu M$; 2,3,7,8-tetrachlorodibenzo-*p*-dioxin (TCDD), 10 n$M$; heat shock (HS), 42° for 2 hr; paraquat (PQ), 50 $\mu M$]. After a 30-min incubation, we remove the PBS and add fresh complete medium, and the cells are allowed to incubate for 2–3 hr under optimized conditions. Subsequently, cells are harvested and analyzed by CAT assay. All chemicals are prepared as 1000-fold concentrated solutions in appropriate solvent just before use. An equal volume of solvent is added to the control experiment. The various chemicals are evaluated for cytotoxicity over a concentration range of 1 to 500 $\mu M$ by monitoring cell death. Nuclear extracts are prepared for the MSA and the functional analysis is performed as described above.

*Induction of SOD1 Gene by Environmental Factors*

The *cis* elements for the peroxisome proliferator response element (PPRE), metal response element (MRE), xenobiotics response element (XRE), hydrogen peroxide response element (HRE), and heat shock element (HSE) are located

[15] P. H. Chan and R. A. Fishman, *J. Neurochem.* **35**, 1004 (1980).
[16] V. C. Culotta and D. H. Hamer, *Mol. Cell. Biol.* **9**, 1376 (1989).
[17] C. C. Winterbourn, G. F. Vile, and H. P. Monteiro, *Free Radic. Res. Commun.* **12–13**, 107 (1991).
[18] R. Omar and M. Pappolla, *Eur. Arch. Psychiatry Clin. Neurosci.* **242**, 262 (1993).

upstream of the *SOD1* gene (Fig. 1). The effect of environmental factors on the induction of the *SOD1* gene is examined as follows. pRSP-1633, which is the rat *SOD1* upstream region fused to the bacterial chloramphenicol acetyltransferase (CAT) plasmid, is introduced into HepG2 cells. Thirty-six hours after transfection, arachidonic acid (AA), copper ion ($Cu^{2+}$), $\beta$-naphthoflavone ($\beta$-NF), and TCDD are added separately to the transfected cells at appropriate concentrations. After 22 hr, the CAT activity of each transfected cell is determined. In the case of oxidative stresses, the CAT activity is determined after 2 hr. These treatments increase SOD1 activity about 5-fold (Table I).

The PPRE is identified as nt $-797$ to $-792$ in the *SOD1* promoter.[19] This PPRE contains two tandem repeats of the AGGTCA motif with one intervening nucleotide (G). When cells are transfected with the pPPREtk plasmid, which has three copies of the PPRE oligonucleotide linked to the heterologous thymidine kinase (*tk*) promoter, CAT activity is induced by AA (Table I). The CAT activity of the plasmid pmPPREtk containing a mutant PPRE is not affected by AA. These results indicate that induction of the *SOD1* gene by peroxisome proliferator is mediated through the PPRE site (Fig. 3A).

When the plasmid pRSP-576 is transfected into HepG2 cells and exposed to $H_2O_2$, SOD1 induction is observed only in pRSP-576.[20] The sequence Elk1 (nt $-533$ to $-520$), named the hydrogen peroxide response element (HRE), is tested for SOD1 induction ability by $H_2O_2$. The CAT activity is increased in cells transfected with pHREtk by 5-fold (Table I), but not with pmHREtk containing a mutated HRE site. These results confirm that HRE is responsible for the induction by $H_2O_2$ (Fig. 3B).

The MRE is located at nt $-273$ to $-267$ upstream of the *SOD1* gene. The *SOD1* gene is induced about 3- to 4-fold by various heavy metal ions, including $Cd^{2+}$, $Zn^{2+}$, and $Cu^{2+}$.[21] The CAT activity of pMREtk is induced about 4-fold (Table I) by copper but the CAT activity of pmMREtk, containing a mutated MRE site, is not induced by copper. These findings confirm that the MRE is a response element for copper ion (Fig. 3C).

The dioxin–Ah receptor complex binds to the core sequence TWGCGTG.[22] The complementary sequence of the xenobiotic response element (XRE) is located in a reverse orientation at nt $-268$ to $-262$ of the *SOD1* promoter. The CAT activity of pRSP-412 is also activated about 5-fold by treatment with *tert*-butylhydroquinone (tBHQ) and iodoacetamide (IA).[23] Almost identical results are observed by administration of $\beta$-NF and TCDD to the pXREtk plasmid, which

---

[19] H. Y. Yoo, M. S. Chang, and H. M. Rho, *Gene* **234**, 87 (1999).
[20] H. Y. Yoo, M. S. Chang, and H. M. Rho, *J. Biol. Chem.* **274**, 23887 (1999).
[21] H. Y. Yoo, M. S. Chang, and H. M. Rho, *Mol. Gen. Genet.* **262**, 310 (1999).
[22] H. Reyes, S. Reisz-Porszasz, and O. Hankinson, *Science* **256**, 1193 (1992).
[23] H. Y. Yoo, M. S. Chang, and H. M. Rho, *Biochem. Biophys. Res. Commun.* **256**, 133 (1999).

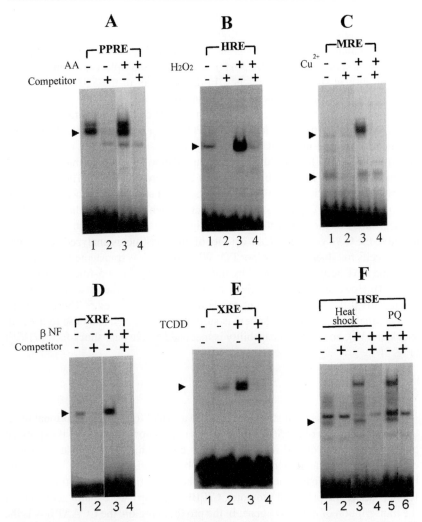

FIG. 3. Binding and induction of transcription factors involved in environmental induction of the *SOD1* gene. (A) Binding and induction of the PPRE-binding protein. MSA was performed using nuclear extracts prepared from AA (50 $\mu M$) treated cells. The intensity of the retarded DNA–protein complex was increased when the cells were treated with AA. (B) HRE-binding protein and increased binding activity due to $H_2O_2$. The DNA–protein complex disappeared with unlabeled competitor, and increased binding activity of HRE-binding protein *in vivo* was observed. (C) Binding and induction of MRE-binding transcription factors. When HepG2 cells were treated with $Cu^{2+}$, a strong DNA–protein complex appeared. (D) Binding and induction of XRE-binding protein *in vivo*. The intensity of the retarded DNA–protein complex was increased when the cells were treated with $\beta$NF. (E) Induction of XRE binding by TCDD. The intensity of the retarded DNA–protein complex was increased when the cells were treated with TCDD. (F) The change of binding of HSE-binding protein (HSF) by PQ and heat shock. Using nuclear extracts prepared from heat- and PQ-treated cells, a far-shifted DNA–protein complex appeared.

has three copies of the XRE linked to a heterologous *tk* promoter (Table I). These findings strongly confirm that induction of the *SOD1* gene by xenobiotics such as dioxin is mediated through the XRE site (Fig. 3D and E).

The heat shock element (NGAAN) is found in multiple copies in the promoter of all heat shock protein genes. In the promoter of the *SOD1* gene, three tandem repeats of the heat shock element are observed at region $-185$ to $-171$. When pRSP-305 is transiently expressed in HepG2 cells, about 4-fold induction of CAT activity by heat shock (42°) and PQ is observed (Fig. 3F). Heat shock treatment of murine macrophages results in an enhanced capacity to release superoxide anion ($O_2^-$).[24] In addition, Omar and Pappolla[25] showed that heat shock protein synthesis is induced by high levels of superoxide anion. Therefore, it is assumed that induction of the *SOD1* gene by PQ, a superoxide-generating agent, may be mediated through the same heat shock element. To confirm this possibility, the heat shock element is synthesized and ligated to the heterologous *tk* promoter linked to the CAT gene. The CAT activity of this construct is analyzed after exposure of cells to either heat shock or PQ. When pHSEtk is transfected into HepG2 cells, the CAT activity is induced by heat shock or PQ about 5-fold, respectively (Table I). However, the CAT activity is not induced by either heat shock or PQ when pmHSEtk containing a mutated HSE site is transfected.[20] These findings indicate that the heat shock element is sufficient to induce a response to the heat shock and PQ, and that this capability is mediated by the heat shock factor (HSF) (Fig. 3F).

## Conclusions

From the results of our study, we propose a model for the transcriptional regulation and induction of the *SOD1* gene (Fig. 4). It is known that *SOD1* is expressed at high levels in liver and its distribution is consistent with the relative protein concentrations and activity of SOD. These observations revealed the relationship between the tissue distribution of SOD1 and the role of tissue-specific transcription factors. Under the proximal promoter, C/EBPα and Sp1 were major activators of the transcription of the *SOD1* gene. In the proximal region, the CCAAT box is the major activator in the basal transcription of the rate *SOD1* gene. In the upstream region, transcriptional activation and repression of the *SOD1* gene by Elk1 and YY1 are elicited by their direct binding to their binding sites in the PRE and NRE of the *SOD1* promoter, respectively. The transcriptional activation by PRE was increased about 6-fold even in the presence of the NRE, implying that the

---

[24] M. V. Reddy and P. R. Gangadharam, *Infect. Immun.* **60,** 2386 (1992).
[25] R. Omar and M. Pappolla, *Eur. Arch. Psychiatry Clin. Neurosci.* **242,** 262 (1993).

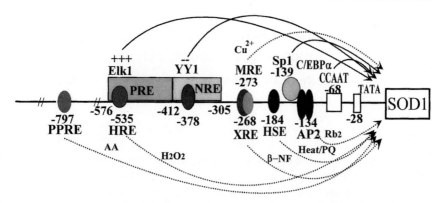

FIG. 4. Model for the transcriptional regulation and induction of the *SOD1* gene. Solid and dotted lines represent the regulation of the *SOD1* gene in the normal state and induced state, respectively. PPRE, peroxisome-proliferating activator response element; HRE, hydrogen peroxide response element; PRE, positive regulatory element; NRE, negative regulatory element; MRE, metal-binding element; XRE, xenobiotic response element; HSE, heat shock response element; CCAAT, CAT box; TATA, TATA box.

strength of the PRE overrode the effect of the NRE. The *SOD1* gene was induced by nontoxic ginsenoside-$Rb_2$ through the AP2 site, which appeared to be important because SOD1 is not easily taken up by cells. SOD1 also could be induced by various environmental factors through their corresponding elements. These types of SOD1 inductions may prevent hazardous effects resulting from oxidative and environmental stresses. Also, the diversity of SOD1 inducers implies that there are multiple regulatory elements to achieve proper adjustment to various conditions. The observations described above support the idea that the *SOD1* gene promoter has the potential to respond to normal and changing circumstances.

Acknowledgments

We recognize the contributions of Dr. Young Ho Kim and Jin Sik Cho to some aspects of these studies. This work was supported in part by the BK21 Foundation, the Ministry of Education, and the Korea Science and Engineering Foundation (KOSEF).

## [30] Transcription Regulation of Human Manganese Superoxide Dismutase Gene

By DARET K. ST. CLAIR, SUREERUT PORNTADAVITY, YONG XU, and KELLEY KININGHAM

### Introduction

Manganese superoxide dismutase (MnSOD) is a member of the family of SODs that control the cellular levels of superoxide radicals.[1] The human MnSOD gene is located on chromosome 6q25.[2,3] It is a single-copy gene consisting of five exons separated by four introns with typical splice junctions.[4] The basal promoter region is GC rich, and contains a cluster of several SP-1- and AP-2-binding sites without TATA or CATT boxes.[5] The gene is transcribed and processed into two MnSOD mRNA species of approximately 1 and 4 kb in size.

To study transcriptional regulation of the gene, *in vivo* and *in vitro* mapping strategies as well as sequence analyses by computer searches are used to probe regulatory sequences located within the promoter of a given gene. Analysis of DNase I-hypersensitive sites in cultured cells can provide information on when and how proteins occupy a given regulatory region of DNA under physiological conditions. However, *in vivo* footprinting of a single-copy regulatory region in large genomes is technically challenging because of the high-level signal-to-noise ratio. Although *in vivo* mapping coupled to sequence inspection may provide clues to the identity of proteins bound to a DNA region, they do not provide an accurate map of overlapping sites. Thus, *in vitro* mapping on naked DNA with purified proteins will be needed to establish the exact binding sequences. Because of the GC-rich nature of the human MnSOD promoter, the computer-predicted *cis*-acting DNA-binding elements in the basal promoter region can differ greatly from those determined by footprinting using purified transcription factors.[6] For example, sequence analyses by computer predicted five SP-1 and one AP-2 sites for the human SOD promoter from nucleotides −210 to +40, whereas *in vitro* footprints using purified SP-1 and AP-2 proteins identified eight SP-1 and nine AP-2 overlapping sites.[6]

The fact that the basal promoter contains nearly 80% GC makes it difficult to amplify and sequence by standard protocols, thus limiting success in obtaining a

---

[1] I. Fridovich, *Annu. Rev. Biochem.* **64,** 97 (1995).
[2] R. Cregan, J. Tischfield, F. Ricciuti, and F. H. Ruddle, *Humangenetik* **20,** 203 (1973).
[3] S. L. Church, J. W. Grant, E. U. Meese, and J. M. Trent, *Genomics* **14,** 823 (1992).
[4] X. S. Wan, M. N. Devalaraja, and D. K. St. Clair, *DNA Cell Biol.* **13,** 1127 (1994).
[5] Y. Xu, K. K. Kiningham, M. N. Devalaraja, C. C. Yeh, H. Majima, E. J. Kasarskis, and D. K. St. Clair, *DNA Cell Biol.* **18,** 709 (1999).
[6] C. C. Yeh, X. S. Wan, and D. K. St. Clair, *DNA Cell Biol.* **17,** 921 (1998).

promoter region for gene regulation studies. In the following sections, we describe an optimized protocol that has been used successfully in our hands to amplify this promoter from genomic DNA.

## Amplification of Human MnSOD Promoter by Polymerase Chain Reaction

Genomic DNA isolated from cultured cells or human blood by standard methods can be used in this reaction.

*Oligonucleotide sequences:* The oligonucleotide sequences of the primers designed to amplify a fragment from nt $-321$ to $+70$ in the 5' flanking region of the human MnSOD gene are 5'-GCCTTCGGGCCGTACCAACTCCAA-3' (forward primer) and 5'-CTAGTGCTGGTGCTACCGCTGATGC-3' (reverse primer).

*PCR mixture:* The polymerase chain reaction (PCR) mixture contains 20 m$M$ Tris-HCl (pH 8.8), 2 m$M$ MgSO$_4$, 10 m$M$ KCl, 10 m$M$ (NH$_4$)$_2$SO$_4$, 0.1% (v/v) Triton X-100, nuclease-free bovine serum albumin (BSA, 1 mg/ml) dATP, dCTP, and dTTP (160 $\mu M$ each) 120 $\mu M$ 7-deaza-2'-dGTP (C7 dGTP; Boehringer Mannheim, Indianapolis, IN) and 40 $\mu M$ dGTP (3:1 C7 dGTP:dGTP instead of 160 $\mu M$ dGTP in the dNTP mixture), a 0.2 p$M$ concentration of each primer, 6% (v/v) dimethyl sulfoxide (DMSO), 1 $\mu$g of genomic DNA, and 2.5 U of DNA polymerase in 50 $\mu$l.

*Thermal cycling settings:* Initial denaturation is at 98° for 10 min, 5 cycles of 95° for 45 sec, and 72° for 1 min for primer annealing, followed by 35 cycles consisting of denaturation at 95° for 45 sec, annealing at 60° for 1 min, and extension at 72° for 1.5 min (automatic segment extension of 2 sec per cycle), and final extension at 72° for 10 min.

*Product analysis:* The PCR products can be analyzed on a 1% (w/v) agarose gel in Tris–acetate buffer with ethidium bromide staining and the products can be purified by an MC membrane (Millipore, Bedford, MA) according to the manufacturer's recommendations.

Using the PCR conditions described above, we have found a set of three unique mutations in the basal promoter region of the human MnSOD gene from several cell lines derived from colon cancer patients. These mutations lead to an increase in one AP-2-binding site in addition to those found in the noncancerous promoter. The mutated promoter also has significantly lower transcription activity compared with the wild-type promoter.[7] However, the prevalence and significance of these mutations in human cancer remain to be further investigated. These findings may be important for our understanding of how transcription of the human MnSOD gene can be regulated, and may also have pathological implications.

[7] Y. Xu, A. Krisnan, X. S. Wan, H. Majima, C. C. Yeh, G. Ludewig, E. J. Karsaksis, and D. K. St. Clair, *Oncogene* **18**, 93 (1999).

## Trancriptional Regulation

To study the transcriptional regulation of the human MnSOD gene, both direct assay of the abundance of the MnSOD mRNA and indirect assay of reporter constructs expressing reporter gene products under the control of MnSOD promoter-enhancer elements can be used to identify *cis*-acting regulatory elements responsive to stimulus-dependent increases in MnSOD mRNA. Direct assay of the mRNA abundance has the advantage of avoiding potential effects of stimuli on translation and posttranslational events. However, this method can be wearisome when mRNA for many mutant constructs must be purified and analyzed. Thus, the more practical method is to link the *cis*-acting elements to a reporter gene that encodes a novel enzymatic activity (e.g., chloramphenicol acetyltransferase or firefly luciferase) and assay the amount of enzymatic activity of the reporter gene product. However, these assays are extremely indirect and care must be taken to monitor for transcription initiated at other sites in the vector. Furthermore, in most studies in which transient transfection is used, a coreporter is usually included to monitor transfection efficiency. In this case, it is important to take into consideration the *trans* effects between promoters on cotransfected plasmids and the potential effect of activating agents on the expression of the coreporter. These potential problems can be minimized by selecting appropriate pairs of experimental and control reporter vectors. For studying the effect of activating agents, however, it is advisable to do batch transfection and subsequently divide the transfected cells into multiple dishes for treatments.

In the case of cotransfection with vectors that will express a protein of interest, it is important to verify that the proper protein of interest is actually expressed in the cells. We have used the reporter gene assay to demonstrate that SP-1 sites and the transcription factor SP-1 are needed for basal promoter activity. In contrast, binding of the AP-2 protein to the overlapping AP-2/SP-1 sites may allow for competition, resulting in a negative effect on transcription.[6] However, it will be important to directly verify that expression of AP-2 protein plays a negative role in the basal expression of the human MnSOD gene. To accomplish this, we transfected an AP-2 expression vector into human hepatocellular carcinoma cells, which have no detectable endogenous AP-2 protein. As shown in Fig. 1A, transfection of an AP-2 expression vector, pRSV-AP-2 (a generous gift from W. Trevor, Yale University, New Haven, CT), results in a concentration-dependent decrease in transcription activity. In contrast, the levels of AP-2 protein were increased in a concentration-dependent manner (Fig. 1B). Moreover, the expressed AP-2 was capable of binding to the consensus AP-2 DNA sequence (Fig. 1C).

## MnSOD Induction

The composition of MnSOD promoter from all mammalian sources examined to date appears to have typical characteristics of housekeeping gene promoters;

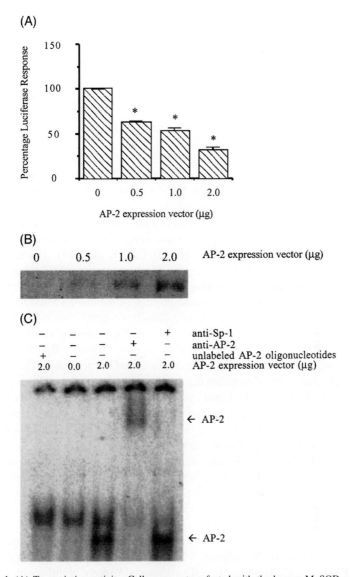

FIG. 1. (A) Transcription activity. Cells were cotransfected with the human MnSOD promoter-driven luciferase vector (PGL3/P7) and an AP-2 expression vector (pRSV-AP-2). The effect of AP-2 on transcription of the reporter gene was normalized to that in cells transfected with vector alone. Asterisks ($p < 0.01$) indicate significant differences between control and tests. (B) Western analysis of AP-2 protein. Proteins from nuclear extracts of cells transfected with vector alone or AP-2 expression vectors were separated on a 10% (w/v) SDS–polyacrylamide gel, transferred to nitrocellulose membrane, and probed with antibody against AP-2 protein. (C) Electrophoretic mobility shift assay. Nuclear extract from cells transfected with AP-2 expression vectors was incubated with $^{32}$P-labeled consensus AP-2 oligonucleotides. The specificity of binding was probed by adding unlabeled competing oligonucleotides. The presence of SP-1 or AP-2 protein on the labeled DNA is identified by supershift with specific antibody against each protein.

however, the expression of the MnSOD gene is highly responsive to both endogenous and exogenous stimuli. The 5' flanking region of the human MnSOD gene contains several potential enhancer binding sites including NF-$\kappa$B, AP-1, and ARE. These elements were not responsive to numerous agents that are known to induce MnSOD, including tumor necrosis factor $\alpha$ (TNF-$\alpha$), interleukin 1$\beta$ (IL-1$\beta$), and phorbol 12-myristate 13-acetate (TPA) in simian virus 40 (SV40)-transformed human lung fibroblast (VA13) cells. However, a CREB-1/ATF-1 like element between nt $-1292$ and $-1202$ was responsive to TPA-induced transcription activation in human lung carcinoma (A549) cells.[8] These results suggested that transcriptional activation of the human MnSOD gene may be cell-type dependent.

In addition to elements in the promoter region, we have identified in the second intron of the human MnSOD gene several potential enhancer elements that are responsive to induction by TNF-$\alpha$, IL-1$\beta$, and TPA.[5] These elements are potential binding sites for transcription factors NF-$\kappa$B, C/EBP, and NF-1. Transcriptional activation of the human MnSOD gene by cytokines such as TNF-$\alpha$ and IL-1$\beta$ requires a complex interaction among multiple enhancer elements and transcription factors. For example, increased transcription of the human MnSOD gene by TNF-$\alpha$ requires the presence of intronic NF-$\kappa$B, C/EBP$\beta$, and NF-1 for high-level expression of the gene.[5] Whereas mutation of the intronic NF-$\kappa$B site completely abolished the activation of MnSOD gene by TNF-$\alpha$ or IL-1$\beta$, mutations in the intronic C/EBP and NF-1 sites merely reduced the activation. Treatment of cells with multiple agents together can have a synergistic effect. The synergistic activation of MnSOD by TPA and cytokine is due to the ability of each agent to activate different NF-$\kappa$B-dependent pathways.[9] Together, these studies suggest activation of multiple transcription factors and pathways leading to increased NF-$\kappa$B-mediated mechanisms for synergistic induction of MnSOD.

Accumulated evidence based on nuclear run-on transcription assays and assays of steady-state MnSOD mRNA in the presence of actinomycin D suggests that in many cases, the increased MnSOD mRNA is due to increased *de novo* transcription.[10–14] Increases in transcription of the human MnSOD gene is usually translated into increased active protein. The data in Fig. 2 demonstrate that the induction of MnSOD in response to TNF-$\alpha$ was a result of TNF-$\alpha$-induced increased biosynthesis of the enzyme. A Northern blot (Fig. 2A) shows an increase in both the 1- and 4-kb human MnSOD transcripts. Western blotting with specific

---

[8] H. P. Kim, J. H. Roe, B. Chock, and M. B. Yim, *J. Biol. Chem.* **274**, 37455 (1999).
[9] K. K. Kiningham, Y. Xu, C. Daosukho, B. Popova, and D. K. St. Clair, *Biochem. J.* **353**, 147 (2001).
[10] G. H. W. Wong and D. V. Goeddel, *Science* **242**, 941 (1988).
[11] G. A. Visner, W. C. Dougall, J. M. Wilson, I. A. Burr, and H. S. Nick, *J. Biol. Chem.* **265**, 2856 (1990).
[12] T. Yoshioka, T. Homma, B. Meyrick, M. Takeda, T. Moore-Jarrett, V. Kon, and I. Ichikawa, *Kidney Int.* **46**, 405 (1994).
[13] L. A. H. Borg, E. Cagliero, S. Sandler, N. Welsh, and D. L. Eizirik, *Endocrinology* **130**, 2851 (1992).
[14] J. E. White and F. Tsan, *Am. J. Physiol.* **266**, L664 (1994).

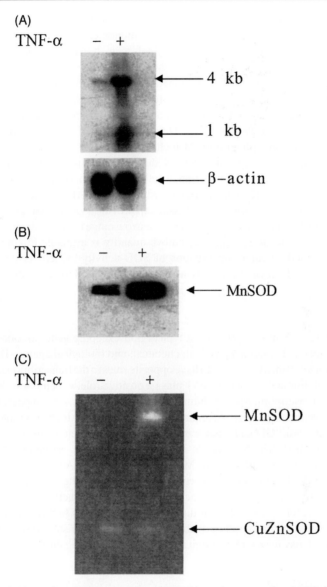

FIG. 2. (A) Expression of human MnSOD mRNA. Thirty micrograms of total RNA from a neuroblastoma cell line, SK-N-SH, treated with TNF-$\alpha$ (200 U/ml) for 6 hr was electrophoresed and transferred. The blot was hybridized with a $^{32}$P-labeled MnSOD cDNA probe. RNA loading was verified with a $^{32}$P-labeled $\beta$-actin cDNA probe. (B) Western analysis of MnSOD protein. Proteins (50 $\mu$g) were separated on a 12.5% (w/v) SDS–polyacrylamide gel, transferred to nitrocellulose, and immunoblotted with anti-human MnSOD (1 : 5000). (C) Native gel assay for SOD activity. Proteins (100 $\mu$g) were separated on native polyacrylamide gels and stained for SOD activity by the photoinduced NBT reaction. Samples were treated with TNF-$\alpha$ (200 U/ml) for 24 hr.

antibody against purified human MnSOD protein indicates an increase in the 22-kDa band corresponding to the predicted monomer molecular weight of the human MnSOD protein (Fig. 2B). Activity gels (Fig. 2C) demonstrate an increase in the level of active proteins.

Northern analysis of the MnSOD mRNA is straightforward and so far is the most reliable method for quantitation of steady-state mRNA levels. So long as intact mRNAs are available, standard procedures for Northern analysis can be easily applied for the study of human MnSOD mRNA. Assay of the protein by Western analysis, immunoprecipitation, or histochemical staining could be a problem if the antibody used is nonspecific. Some of the antibodies used in the literature and available commercially do not recognize authentic human MnSOD protein.[15] The antibody used in Fig. 2B was a generous gift from L. Oberley of the University of Iowa (Iowa City, IA). The activity gel assay can provide a visual means to verify the authenticity of the protein. However, it has a narrow range for accurate quantitation. Thus, unless authentic protein with known quantity is available to concurrently generate a standard curve, quantitation of SOD activity by activity gels can be problematic, and enzymatic activity assays should be used to verify the result.

## Summary

The human MnSOD gene has a typical housekeeping gene promoter, but is highly inducible by various physical, chemical, and biological agents. Transcription factors SP-1 and AP-2 seem to have opposite roles in the transcriptional activity of the basal promoter. Whereas SP-1 plays a positive role, which is absolutely essential for transcription from the human MnSOD promoter, AP-2 appears to play a negative role in this process. An enhancer element is found in the promoter region of the human MnSOD gene. Several important enhancer elements are located in the second intron. The NF-$\kappa$B site in the second intron is essential but not sufficient for high-level induction of MnSOD by cytokines. Although mutations in the regulatory elements may be partially responsible for the lack of induction of MnSOD in some cell types, differences in the degree of induction exist that cannot be accounted for by the defect in the DNA sequence. It is highly likely that this difference is due to the presence or absence of coactivator or suppressor proteins in the cells and may have a physiological role in the defense against oxidative stress.

## Acknowledgments

This work was supported by the National Institutes of Health Grant RO1 CA 49797-11 and by a research career development award HL-03544 to D. K. St. Clair.

---

[15] T. Yan, X. Jiang, H. J. Zhang, S. Li, and L. W. Oberley, *Free Radic. Biol. Med.* **25,** 688 (1998).

# [31] Assessment of Oxidants in Mitogen-Activated Protein Kinase Activation

By LANCE S. TERADA and RHONDA F. SOUZA

Metazoan cells respond to a wide variety of extracellular signals and environmental stresses by activating various gene expression programs, which in turn determine whether the cell proliferates and invades local tissue, or, conversely, differentiates, develops specialized functions, and/or undergoes programmed cell death. The mitogen-activated or microtubule-associated protein kinase (MAPK) cascades constitute a signaling network critical to the organized activation of these gene sets. The three most intensely studied metazoan MAPKs, the ERK, JNK, and p38 families, are homologous to the yeast MPK1, KSS/Fus3, and HOG1 kinases, which, not surprisingly, are also activated by both extracellular signals (pheromones) and environmental stresses (osmolar changes, starvation). These MAPK families are distinguished by their hallmark dual phosphorylation motifs, TEY, TPY, and TGY, respectively.

Each MAPK signaling pathway appears to be preferentially involved in certain cellular responses. For instance, the ERK pathway is most potently activated by mitogenic stimuli such as growth factors. Indeed, protooncogene products such as Ras and Raf are common upstream mediators of this pathway, and interruption of the ERK pathway can block the proliferative response to a variety of stimuli.[1,2] In contrast, JNK and p38 appear to be preferentially activated by environmental stimuli such as UV radiation or heat shock,[3] cytokines,[4] and inflammatory lipids.[5] Activation of these pathways instead correlates strongly with cell cycle arrest and programmed cell death. However, fixed assignment of given signaling pathways to specific cell responses is not always warranted. In B cells, for instance, ligation of CD40 activates JNK and rescues cells from apoptosis, whereas engagement of surface IgM activates ERK and initiates apoptosis.[6] In fact, as with the ERK pathway, protooncogene products such as Rac1 and Cdc42 act upstream from JNK,[7,8] suggesting a role for JNK in control of proliferation under certain conditions.

[1] M. W. Renshaw, X. D. Ren, and M. A. Schwartz, *EMBO J.* **16,** 5592 (1997).
[2] L. S. Mulcahy, M. R. Smith, and D. W. Stacey, *Nature (London)* **313,** 241 (1985).
[3] M. Hibi, A. Lin, T. Smeal, A. Minden, and M. Karin, *Genes Dev.* **7,** 2135 (1993).
[4] G. Natoli, A. Costanzo, A. Ianni, D. J. Templeton, J. R. Woodgett, C. Balsano, and M. Levrero, *Science* **275,** 200 (1997).
[5] J. K. Westwick, A. E. Bielawska, G. Dbaibo, Y. A. Hannun, and D. A. Brenner, *J. Biol. Chem.* **270,** 22689 (1995).
[6] N. Sakata, H. R. Patel, N. Terada, A. Aruffo, G. L. Johnson, and E. W. Gelfand, *J. Biol. Chem.* **270,** 30823 (1995).

Endogenously produced superoxide and secondary oxidants have been recognized as physiologic signaling elements, specifically in the MAPK system. Platelet-derived growth factor (PDGF)-stimulated tyrosine phosphorylation, ERK activation, and DNA synthesis, for instance, are accompanied by a transient burst of $H_2O_2$,[9] and chemical or enzymatic scavengers diminish both ERK activation and DNA synthesis. Likewise, endogenous or exogenous oxidants activate ERK at or upstream of $p21^{ras}$ GTP loading.[10,11] Similarly, scavengers or oxidase inhibitors suppress interleukin $1\beta$ (IL-$1\beta$) and tumor necrosis factor $\alpha$ (TNF-$\alpha$)-stimulated activation of JNK or ERK2 kinase activation, whereas exogenous oxidants stimulate kinase activation.[12,13] Oxidants may also mediate the effects of TGF-$\beta$[14] and insulin,[15] which activate MAPK cassettes. Superoxide specifically has been noted to increase after stimulation of fibroblasts with cytokines.[16] Potential targets for superoxide relevant to the MAPK cascade are the protein tyrosine phosphatases. The catalytic Cys-215 of PTP-1B, for instance, appears to be reversibly oxidized by superoxide in an efficient reaction facilitated by an active site cationic charge trap.[17]

Although much is known of the proteins that mediate upstream MAPK signaling, the source, regulation, and mechanism of oxidant involvement are poorly defined at present. The mainstay of studies that implicate the involvement of oxidants in MAPK signaling has been the correlation of MAPK activation and oxidant levels in the presence of various interventions. Methods for the two end points employed in this approach are outlined below.

## Mitogen-Activated Protein Kinase Activation

A number of methods for assessment of MAPK activity have been published; of these, the MAPK phosphorylation, immunoprecipitation (IP) kinase, and IP active kinase approaches are the most approachable and are generally reliable.

---

[7] O. A. Coso, M. Chiariello, J. C. Yu, H. Teramoto, P. Crespo, N. Xu, T. Miki, and J. S. Gutkind, *Cell* **81**, 1137 (1995).
[8] A. Minden, A. Lin, F. X. Claret, A. Abo, and M. Karin, *Cell* **81**, 1147 (1995).
[9] M. Sundaresan, Z. X. Yu, V. J. Ferrans, K. Irani, and T. Finkel, *Science* **270**, 296 (1995).
[10] A. K. Bhunia, H. Han, A. Snowden, and S. Chatterjee, *J. Biol. Chem.* **272**, 15642 (1997).
[11] H. M. Lander, J. S. Ogiste, K. K. Teng, and A. Novogrodsky, *J. Biol. Chem.* **270**, 21195 (1995).
[12] W. A. Wilmer, L. C. Tan, J. A. Dickerson, M. Danne, and B. H. Rovin, *J. Biol. Chem.* **272**, 10877 (1997).
[13] Y. Y. C. Lo, J. M. S. Wong, and T. F. Cruz, *J. Biol. Chem.* **271**, 15703 (1996).
[14] M. Ohba, M. Shibanuma, T. Kuroki, and K. Nose, *J. Cell Biol.* **126**, 1079 (1994).
[15] D. Heffetz and Y. Zick, *J. Biol. Chem.* **264**, 10126 (1989).
[16] B. Meier, H. H. Radeke, S. Selle, M. Younes, H. Sies, K. Resch, and G. G. Habermehl, *Biochem. J.* **263**, 539 (1989).
[17] W. C. Barrett, J. P. DeGnore, Y. F. Keng, Z. Y. Zhang, M. B. Yim, and P. B. Chock, *J. Biol. Chem.* **274**, 34543 (1999).

## Mitogen-Activated Protein Kinase Phosphorylation

*Considerations.* Because dual Thr/Tyr phosphorylation is in essence necessary and sufficient for MAPK activation, the phosphorylated TXY epitope has been used as a surrogate marker for MAPK activation. Commercial availability of "phosphospecific" antibodies has popularized this approach. The advantages of this method include the ease and speed of assay (2 days), the simultaneous assessment of different phosphorylated MAPK subtypes separable by molecular mass (e.g., $p42^{ERK2}$ and $p44^{ERK1}$), the absence of radioactivity, and the option of normalizing the blot for total MAPK. The principal disadvantage of this method is that it does not measure activity directly, and therefore in theory may not detect the effects of pharmacologic or endogenous active site inhibitors.

*Method*

1. Generally one 75-cm$^2$ flask (3–8 × 10$^6$ cells) is sufficient to yield a strong signal.
2. Cells are washed twice with phosphate-buffered saline (PBS). Adherent cells are mechanically harvested without trypsin, and pelleted.
3. The cell pellet is resuspended in 100 μl of Laemmli buffer, and sonicated briefly, and boiled, and 20 μl is loaded on a 10% (w/v) sodium dodecyl sulfate (SDS)–polyacrylamide minigel.
4. The gel is immunoblotted by standard techniques, using phosphospecific antisera for ERK, JNK, or p38 [currently available from Cell Signaling Technology (Beverly, MA), Santa Cruz Biotechnology (Santa Cruz, CA), Upstate Biotechnology (Lake Placid, NY), and others]. Visualization by a chemiluminescent method affords maximum sensitivity.
5. As a loading control, blots should be reprobed with antibodies recognizing total (phosphorylated and nonphosphorylated) MAPK. Before reprobing, blots are stripped by incubation in 62.5 m$M$ Tris (pH 6.7), 0.1 $M$ 2-mercaptoethanol (280 μl/40 ml), and 2% (w/v) SDS at 50° for 30 min, with occasional agitation.

## Immunoprecipitation Kinase

*Considerations.* Perhaps the most widely employed approach, the IP kinase method directly assays activity *in vitro*. Its chief disadvantages are the requirement for an efficient and specific immunoprecipitating antibody and the use of radioactive isotopes. The efficiency and specificity of the IP antibody for different MAPK subfamily members in a given study system should be assessed at least once with an immunoblot of the immunoprecipitate, using panspecific antisera (e.g., JNK1/2/3 specific).

*Method*

1. One or two 75-cm$^2$ dishes per assay is adequate, depending on the strength of the signal. Care should be taken to standardize the degree of cell confluency.

For proliferative signals 16–24 hr of serum starvation [0.1–0.5% (v/v) serum] will decrease unstimulated baseline MAPK (particularly ERK) activity.

2. After stimulation with agonist, cells are washed once with ice-cold PBS and 1 ml of ice-cold radioimmunoprecipitation assay (RIPA) buffer [50 m$M$ Tris (pH 7.2), 150 m$M$ NaCl, 0.1% (w/v) SDS, 0.5% (v/v) sodium deoxycholate, 1% (v/v) Triton X-100, 10 m$M$ sodium pyrophosphate, 25 m$M$ $\beta$-glycerophosphate, 2 m$M$ sodium vanadate, aprotinin (10 $\mu$g/ml), leupeptin (10 $\mu$g/ml), and 1 m$M$ phenylmethylsulfonyl fluoride (PMSF)] is added. Cells are scraped with a rubber policeman, transferred to a microcentrifuge tube, and disrupted by three draws through a 23-gauge needle.

3. After a 10-min spin at 4000$g$, 4°, the supernatant is immunoprecipitated with 20 $\mu$l of 50% (v/v) protein G-agarose (protein A-Sepharose for rabbit antisera) and 10 $\mu$l of antibody at 4° for 90 min. For JNK1, commercially available antibodies work well (e.g., Santa Cruz sc-474).

4. The precipitate is washed three times with ice-cold RIPA buffer and once in 1 ml of 1× kinase buffer (see below). Prior to the final spin, an aliquot (250 $\mu$l) is removed for immunoblot with anti-MAPK to assess the equivalence of immunoprecipitation across different samples.

5. As much of the supernatant as possible is drawn off, and the precipitate is resuspended in 30 $\mu$l of a master mix solution. The master mix contains 10× kinase buffer [250 m$M$ HEPES (pH 7.6), 200 m$M$ $\beta$-glycerophosphate, 200 m$M$ MgCl$_2$, 20 m$M$ dithiothreitol (DTT), 1 m$M$ Na$_3$VO$_4$, 200 m$M$ $p$-nitrophenyl phosphate] in deionized H$_2$O, freshly added 20 $\mu$$M$ cold ATP, and 5 $\mu$Ci of [$\gamma$-$^{32}$P]ATP (7000 Ci/mmol) per reaction. The 10× kinase buffer should be replaced every 1–2 weeks.

6. The reaction is started by addition of 2 $\mu$g of target peptide (2 $\mu$l of a 1 $\mu$g/ml stock in 1× kinase buffer). Myelin basic protein or Elk-1 is used for ERK, c-Jun for JNK, and ATF-2 for p38. The mixture is incubated for 30 min at 30° with frequent trituration to resuspend the beads, and stopped by addition of 10 $\mu$l of Laemmli buffer.

7. Samples are boiled, pulsed, and run on a 12% (w/v) SDS–polyacrylamide minigel. The gel is dried on 3MM paper and subjected to autoradiography at $-80°$ for 2–24 hr (Fig. 1).

*Immunoprecipitation Active Kinase*

*Considerations.* The method also assesses kinase activity *in vitro*. Again, the relative success of this method depends on availability of an efficient and specific antibody that does not interfere with kinase activity. Two phosphospecific IP antibodies have been released in "kit" form for ERK1/2 and p38 kinases (Cell Signaling Technology). For assessment of MAPK activity in both tissue biopsies and cultured cells, we have found these kits to be reliable, sensitive, and without the need for radioactive isotopes. A third kit requiring capture of JNK

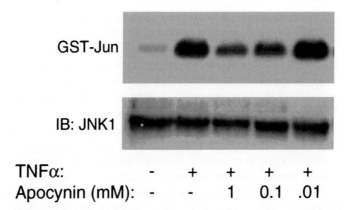

FIG. 1. Typical JNK1 activity experiment. ECV-304 cells were treated with TNF-$\alpha$ (100 ng/ml) for 15 min and assayed as described. *Top:* Phosphorylation of target fusion protein GST–Jun. *Bottom:* IB of JNK1 immunoprecipitate with anti-pan-JNK, demonstrating equivalent capture of JNK1. The NADPH oxidase antagonist apocynin decreased JNK activation.

with immobilized c-Jun has proved less sensitive than the traditional IP kinase method (described above) in our hands. A potential disadvantage of this method is that it does not allow for normalization on the basis of immunoprecipitation efficiency, because only the active kinase is precipitated. Normalization can be accomplished by assaying equivalent amounts of tissue (weight), protein, or cell numbers. Alternatively, aliquots of lysates may be taken before immunoprecipitation and immunoblotted for total MAPK.

*Method.* The lysis, immunoprecipitation, and kinase methods are performed essentially as described above except that phosphospecific antibodies to ERK or p38 (Cell Signaling Technology) are used for immunoprecipitation. In addition, the commercial kit substitutes immunoblot of phosphorylated target peptide (phospho-Elk-1 or phospho-ATF-2) for $[\gamma\text{-}^{32}P]ATP$ incorporation, obviating radionuclide use.

For assaying MAPK activity in cultured cells, one 75-$cm^2$ dish is adequate. For tissues, fresh biopsy fragments weighing approximately 4–10 mg are blotted gently on filter paper and placed into preweighed tubes containing 1 ml of iced lysis buffer. Tubes are weighed again to calculate tissue weight, and then samples are finely minced with surgical scissors for 15 sec and sonicated four times on ice for 5 sec each. The 18,000$g$ supernatants are then immunoprecipitated and assayed as above.

## Oxidant Production

### Considerations

Correlation of MAPK activation with production of superoxide or other oxidants is a significant challenge because the levels of endogenous oxidants produced after ligand binding appear to be limited. The production of superoxide can be specifically detected by using the spin trap 5,5-dimethyl-1-pyrroline-$N$-oxide (DMPO),[16] but electron paramagnetic resonance (EPR) is a less sensitive detection method than many others. Other methods rely on the addition of radical-specific scavengers [e.g., superoxide dismutase (SOD)] to implicate specific radicals.

The oxidation of the fluorochrome 2′,7′-dichlorodihydrofluorescein diacetate ($H_2$-DCFDA) has been widely used to detect oxidants because of its sensitivity and ease of use. The nonfluorescent, membrane-permeant compound is deacetylated by endogenous esterases, leaving a less permeant anionic compound. $H_2$-DCF becomes oxidized to the fluorescent 2′,7′-dichlorofluorescein. The compound appears to be more readily oxidized by peroxides such as $H_2O_2$ than superoxide, and requires cellular peroxidases or other redox-active factors. The oxidized product can be detected by either flow cytometry or quantitative microscopy (Fig. 2). In either case, the loading concentration of $H_2$-DCFDA must be titrated for the particular application to optimize the signal-to-noise ratio. Lower concentrations of $H_2$-DCFDA are tolerated for microscopy (2–4 $\mu M$), as we have found that the compound photooxidizes and therefore will increase in fluorescence intensity if the microscopic field is left static. At higher concentrations, leakage of the

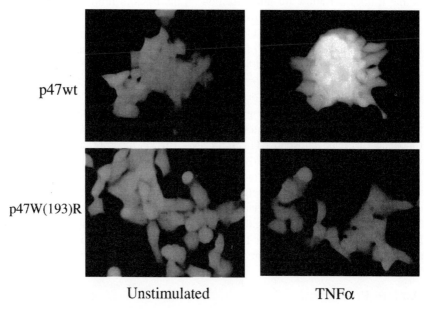

FIG. 2. Dichlorofluorescein fluorescence demonstrating oxidant production. ECV-304 cells were transfected with either wild-type p47$^{phox}$ or mutant p47$^{phox}$ defective in its first SH3 domain, loaded with $H_2$-DCFDA as described, and then stimulated with TNF-$\alpha$ (100 ng/ml) for 15 min.

deacetylated compound into the extracellular medium can be noted within 20 sec, further confounding microscopic quantification.

*Method*

1. For flow cytometry, 10- to 25-cm$^2$ dishes are adequate for each assay. Cells must be washed twice in Hanks' balanced salt solution (HBSS) to remove serum-associated esterases before loading with H$_2$-DCFDA. A convenient 20 m$M$ stock solution can be made on the day of use by weighing 5–10 mg of H$_2$-DCFDA into a preweighed tube. Absolute ethanol is used to dissolve the compound (milliliters of ethanol equals milligrams of H$_2$-DCFDA divided 9.74), which is kept on ice in foil. The compound may be more stable in dimethyl sulfoxide (DMSO) but we have not found this solution to yield as strong a signal. Cells are generally loaded with 5 $\mu M$ H$_2$-DCFDA in HBSS for 15 min at 37°. Pretreatment with inhibitors may be performed during this step.

2. Cells are washed once with medium and stimulated with agonist. Cells in suspension may be repetitively assayed by flow cytometry. Adherent cells must be washed twice with Ca$^{2+}$, Mg$^{2+}$-free HBSS, trypsinized, and studied. For signals requiring the presence of extracellular Ca$^{2+}$, cells are stimulated while adherent in relevant Ca$^{2+}$-containing medium for an appropriate length of time before washing and trypsinization. To minimize clumping, cells are not washed after trypsinization; cells may be diluted with 1–2 ml of Ca$^{2+}$, Mg$^{2+}$-free HBSS. Adherent cells are drawn through a Pasteur pipette several times immediately before flow cytometric analysis and therefore an autoloader is not recommended. Excitation and emission wavelengths are set at 488 and 525 nm [or FL1 on the Becton Dickinson (Mountain View, CA) FACS Caliber]. The forward–side scatter plot as well as a quick microscopic check should reveal a relatively homogeneous distribution of single cells without large clumps.

Interventions

Implicating a specific oxidase in transduction of a given signal can be difficult. The oxidase should be shown to exist in significant levels in the system studied, such as through assessment of activity of xanthine oxidase[18] or NAD(P)H oxidase.[19] If the oxidase is present, biochemical inhibitors provide the principal strategy for linking a specific oxidase with MAPK signaling. Superoxide scavengers (covered elsewhere in Section V) provide further evidence of involvement.

Xanthine oxidase is inhibited with relatively high specificity by oxypurinol (10–50 $\mu M$) or allopurinol (50–100 $\mu M$), although these compounds appear to have nonspecific scavenging properties at higher concentrations. These compounds

[18] L. S. Terada, J. A. Leff, and J. E. Repine, *Methods Enzymol.* **186,** 651 (1990).
[19] K. M. Mohazzab, P. M. Kaminski, and M. S. Wolin, *Am. J. Physiol.* **266,** H2568 (1994).

must first be dissolved in 1 $N$ NaOH and the pH of the buffer is readjusted after addition. Alternatively, inactive xanthine oxidase will replace the active form by growth of cells in 10 ppm sodium tungstate.[20] Arachidonic acid metabolism proceeding through cyclooxygenase (COX) or 5-lipoxygenase (LO) can produce peroxides. Phenidone (100 $\mu M$) inhibits both COX and LO, whereas ibuprofen (10 $\mu M$) and indocin (1–10 $\mu M$) inhibit COX-1 and COX-2. NS398 (10 $\mu M$) selectively inhibits COX-2.[21] LO is inhibited by nordihydroguaiaretic acid (masoprocol, 1–10 $\mu M$) or esculetin (1 $\mu M$).[22] Mitochondrial respiration is also known to produce both superoxide and $H_2O_2$. Because oxidants are thought to "leak" at both complexes I and II, inhibitors of complex I (rotenone, 1–25 $\mu M$; amytal, 400 $\mu M$) or complex II (thenoyltrifluoroacetone, 15–250 $\mu M$) have been shown to decrease oxidant production.[23] Conversely, complex III inhibitors (antimycin A) are thought to increase oxidant leak. Cytochrome P450s (CYP) are also known to be a metabolic source of oxidants. Because of the diverse isoforms, a combination of inhibitors may be required to investigate CYP-related oxidant production. Ketoconazole (2.5–30 $\mu M$) and furafylline (3–5 $\mu M$) are prototype CYP1A2 and 3A4/5 inhibitors, whereas quinidine (0.1–10 $\mu M$) is a potent inhibitor of CYP2D6. Other inhibitors include omeprazole (10 $\mu M$, CYP2C19), fluvoxamine (10–30 $\mu M$, CYP1A2, 2C19, and 3A4), 8-methoxypsoralen (2.5–10 $\mu M$, CYP2A6), and $\alpha$-naphthoflavone (10 $\mu M$, CYP1).[24–27] Finally, NADPH oxidase is strongly inhibited by diphenylene iodonium (1–10 $\mu M$); however, other flavoenzymes may be inhibited. Apocynin (acetovanillone, 1–5 m$M$) may be a more specific antagonist, apparently blocking oxidase assembly[28] (Fig. 1).

---

[20] L. S. Terada, D. M. Guidot, J. A. Leff, I. R. Willingham, M. E. Hanley, D. Piermattei, and J. E. Repine, *Proc. Natl. Acad. Sci. U.S.A.* **89,** 3362 (1992).
[21] R. F. Souza, K. Shewmake, D. G. Beer, B. Cryer, and S. J. Spechler, *Cancer Res.* **60,** 5767 (2000).
[22] F. A. Attiga, P. M. Fernandez, A. T. Weeraratna, M. J. Manyak, and S. R. Patierno, *Cancer Res.* **60,** 4629 (2000).
[23] K. Schulze-Osthoff, A. C. Bakker, B. Vanhaesebroeck, R. Beyaert, W. A. Jacob, and W. Fiers, *J. Biol. Chem.* **267,** 5317 (1992).
[24] P. Taavitsainen, M. Anttila, L. Nyman, H. Karnani, J. S. Salonen, and O. Pelkonen, *Pharmacol. Toxicol.* **86,** 215 (2000).
[25] T. Niwa, T. Shiraga, Y. Mitani, M. Terakawa, Y. Tokuma, and A. Kagayama, *Drug Metab. Disposition* **28,** 1128 (2000).
[26] R. Abbas, C. P. Chow, N. J. Browder, D. Thacker, S. L. Bramer, C. J. Fu, W. Forbes, M. Odomi, and D. A. Flockhart, *Hum. Exp. Toxicol.* **19,** 178 (2000).
[27] L. L. Koenigs, R. M. Peter, S. J. Thompson, A. E. Rettie, and W. F. Trager, *Drug Metab. Disposition* **25,** 1407 (1997).
[28] J. Stolk, T. J. N. Hiltermann, J. H. Dijkman, and A. J. Verhoeven, *Am. J. Respir. Cell Mol. Biol.* **11,** 95 (1994).

## [32] Assaying Binding Capacity of Cu,ZnSOD and MnSOD: Demonstration of Their Localization in Cells and Tissues

By INGRID EMERIT, PAULO FILIPE, JOAO FREITAS, ALFONSO FERNANDES, FRÉDÉRIC GARBAN, and JANY VASSY

### Introduction

The concept that superoxide dismutase (SOD) exerts its antiinflammatory action according to its documented catalytic function on superoxide anion radicals released into the extracellular space appeared doubtful, because there was no correlation between the level of circulating exogenous SOD and the antiinflammatory effects.[1] It was therefore suggested that the antiinflammatory action of exogenous SOD was due to attachment of the enzyme to the cell membrane.[2] This was supported by observations from our laboratory, demonstrating prevention of perinuclear halo formation in UVA-exposed fibroblast cultures by pretreatment with exogenous SOD, even after rinsing of the cells and resuspension in fresh, SOD-free medium before irradiation.[3] Also, observations of anticlastogenic effects in SOD-pretreated and washed lymphocyte cultures suggested cellular binding of the enzyme.[4] Binding to membranes, penetration, and intracellular localization of the enzyme were studied, using fluorescently labeled SOD.[5] Differences between cell types, as well as between Cu,Zn- and MnSODs, were noted. It could also be shown that this protein can enter the upper layers of the epidermis despite its molecular weight of 30,000.[6]

### Materials and Methods

*Preparation of Fluorescently Labeled Enzymes*

All SODs are studied before use for their specific activity and purity. The activity is assayed through the reduction of cytochrome $c$ by superoxide generated with a xanthine–xanthine oxidase reaction. The protein content of the samples

---

[1] A. Baret, G. Jadot, and K. Puget, *Biochem. Pharmacol.* **33**, 2755 (1984).
[2] A. M. Michelson, K. Puget, and G. Jadot, *Free Radic. Res. Commun.* **2**, 43 (1986).
[3] I. Emerit, A. M. Michelson, E. Martin, and J. Emerit, *Dermatologica* **163**, 295 (1981).
[4] I. Emerit, F. Garban, J. Vassy, A. Levy, P. Filipe, and J. Freitas, *Proc. Natl. Acad. Sci. U.S.A.* **93**, 12799 (1996).
[5] P. Filipe, I. Emerit, J. Vassy, A. Levy, V. Huang, and J. Freitas, *Mol. Med.* **5**, 517 (1999).
[6] P. Filipe, I. Emerit, J. Vassy, J. P. Rigaut, R. Martin, J. Freitas, and A. Fernandes, *Exp. Dermatol.* **6**, 116 (1997).

is measured according to the Lowry method. The purity of the enzymes is studied by high-performance liquid chromatography (HPLC) and polyacrylamide gel electrophoresis (PAGE).

Fluorescein isothiocyanate (FITC; Sigma, St. Louis, MO) is covalently conjugated to the various SODs by constant stirring in 50 m$M$ Na$_2$CO$_3$ at 4° for 1 hr. The FITC–SODs are purified by Sephadex G-25 chromatography with 10 m$M$ potassium phosphate buffer (pH 7.4) as the eluent. The enzymatic activity of FITC-modified SODs is determined with the cytochrome $c$ assay in comparison with the respective unlabeled enzymes.

*Preparation of Isolated Cells*

Venous blood from healthy blood donors is anticoagulated with heparin. Polymorphonuclear neutrophils (PMNs) and mononuclear cells are separated by dextran sedimentation and Ficoll-Hypaque density gradient centrifugation. Pure monocytes are prepared by adherence on plastic dishes. Human skin fibroblasts from semiconfluent cultures are detached from the Falcon plastics with a cell scraper. Trypsinization is avoided, because contact with trypsin perturbs membrane integrity and inhibits membrane binding of the SOD. Single-cell suspensions are obtained by successive passages through needles of decreasing diameter.

*Flow Cytometry*

Cells (1.5 × 10$^6$) are incubated in phosphate-buffered saline (PBS) at 37° in the presence of FITC–SOD at various concentrations and exposure times. To avoid nonspecific binding, the cells are pretreated with 5% (v/v) fetal calf serum for 20 min before exposure to FITC–SOD. After incubation, cells are washed three times in PBS to eliminate all residual free SOD. FITC–goat anti-mouse immunoglobulin fragments are used as a negative control.

Analysis is performed on a FACScan (Becton Dickinson, Mountain View, CA). In the mononuclear cell suspension, monocytes and lymphocytes are gated according to forward scatter and side scatter. Large cells are monocytes (<99% CD14$^+$), small cells are lymphocytes (>90% CD3$^+$ or CD19$^+$). PMNs are distinguished from contaminating red blood cells as large granular cells according to forward and side scatter. FITC fluorescence is detected at 520 nm. Data are collected and analyzed by Lysis II software (Becton Dickinson).[4]

*High-Resolution Confocal Scanning Laser Microscopy*

Fixation of cells is realized with 4% (v/v) formol for 30 min. Nuclei are stained with chromomycin A3 (Sigma) for 30 min at room temperature in the dark. After rinsing with PBS, the cell layer is mounted in Fluoprep (Bio-Merieux, Marcy l'Etoile, France). Confocal fluorescent images are obtained with an MRC-600 confocal scanning laser microscope (Bio-Rad, Hercules, CA) equipped with a 25-mW multiline argon ion laser (Ion Laser Technology, Salt Lake City, UT), a $z$-stepping motor, an 80386/87 MS/DOS Nimbus microcomputer (Research Machines, Oxford, UK), and SOM software (Bio-Rad). The Optiphot (Nikon, Tokyo, Japan) is equipped with a ×60 plan apochromate (Nikon) objective with a 1.4 numerical aperture. Immersion oil with a 1.515 refraction index is used. The ($x$, $y$) scanning is usually done with a 1.0 electronic zoom. To reduce noise, a Kalman filter recording an average of 10 images/sec is used for confocal imaging. Excitation at 488 nm and emission at LP 515 nm are the wavelengths for detection of FITC–SOD. For imaging of the chromomycin-stained DNA of nuclei, a combination of two filters is used: one filter for excitation at 458 nm and a high-pass filter for emission at 550 nm.[5,6]

The CA3 stain results in a red color, whereas FITC yields green fluorescence.

*Preparation of Samples for Histologic Studies*

For human skin as well as for mouse skin, the fluorescent SOD is applied twice daily for 10 days in a carbopol gel preparation. For a concentration of 1%, each application represents 160 units/cm$^2$. The prolonged treatment is based on previous studies, indicating a correlation between duration of treatment and protective effects.[7] The last application is given in the evening, and biopsies are taken the next morning after extensive rinsing of the skin with distilled water. A gel without SOD is applied on the opposite side for control. A biopsy is taken from both areas. Part of the specimen is used for conventional histology studies, and the sections are stained with hematoxylin–eosin. The remainder is immediately frozen in liquid nitrogen, and cryosections, 6 $\mu$m thick, are fixed in acetone for 10 min and stored at $-30°$ until microscopy is performed. Fixed slides are stained for DNA with chromomycin A3 (CA3; Sigma) for 30 min and mounted in Fluoprep (Bio-Mérieux). With a ×20 objective and an electronic zoom of 1.0, a magnification to 1.25 pixels/$\mu$m is obtained. With the ×60 objective and an electronic zoom of 1 or 2, the magnification is 3.6 and 7.3 pixels/$\mu$m, respectively. Autofluorescence of the tissues is bypassed using a filter with an excitation wavelength at 514 nm and emission band pass 540 DF of 30 nm. For imaging of the CA3-stained nuclei the same filter is used as for isolated cells.

*Evaluation of Protective Effects*

The superoxide production by PMNs and monocytes is studied by the classic cytochrome $c$ assay after stimulation with the tumor promoter phorbol myristate acetate (PMA; Chemicals for Cancer Research, Eden Prairie, MN).[4,5]

---

[7] A. Alaoui-Youseffi, I. Emerit, and J. Feingold, *Eur. J. Dermatol.* **4**, 389 (1994).

The inhibition of chromosome-damaging effects of chemical clastogens by SOD are studied by cytogenetic methods. For anticlastogenesis, FITC–SOD at a final concentration of 150 cytochrome $c$ units/ml medium is added to the cultures for a 30-min period. The cells are then washed and resuspended in fresh medium, which is free of residual SOD as monitored spectrophotometrically with the cytochrome $c$ assay. The cells are exposed to the clastogen for the entire cultivation period of 48 hr.[4]

The antiinflammatory effect of topically applied FITC–SOD is studied by a psoralen plus UVA light (PUVA)-mediated skin reaction[6]

## Results

### Flow Cytometry

The experiments with flow cytometry showed cellularly bound FITC–SOD. At a constant incubation time of 1 hr before rinsing, the fluorescent labeling increased with the SOD concentration. Monocytes regularly showed more labeling than lymphocytes and PMNs, for which only the highest concentration of 1500 units/ml yielded significant results. The degree of labeling was not only concentration dependent, but also time dependent. No fluorescence labeling was observed on erythrocytes in four independent experiments. The contamination with erythrocytes did not influence the results for PMNs. Labeling of lymphocytes from various individuals was not improved for activated cells after stimulation with phytohemagglutinin (1 $\mu$g/ml) for 1, 4, 24, and 48 hr before incubation with FITC–SOD.

Fibroblasts showed significant FITC fluorescence when trypsin was not used for detachment of cells. Cytochalasin B, an inhibitor of endocytosis, diminished FITC-related fluorescence, but did not suppress it completely.

### High-Resolution Confocal Laser Microscopy

*Isolated Cells.* In agreement with the results obtained with flow cytometry, lymphocytes from different donors did not show any fluorescence even after an incubation time of 3 hr with FITC–SOD doses of 1500 units/ml. This was true also after PHA stimulation of 48 hr. Only rarely were PMNs labeled with FITC–SOD, despite the use of a maximal incubation time of 3 hr and the highest SOD concentration of 1500 units/ml. In contrast to lymphocytes and PMNs, monocytes regularly had a diffuse fluorescence pattern, which was seen after FITC treatment only, but not in untreated monocyte layers.

Monocytes were chosen for the comparison of different SODs. At identical activity, the intracellular distribution of the fluorescently labeled SOD was similar for bovine, equine, and human Cu,ZnSOD. The intensity of fluorescence was time dependent. As in the experiments with flow cytometry, fluorescent labeling was regularly visible on the membrane after 15 min of incubation with FITC–SOD. A small number of cells was already labeled after 5 min. Up to 1 hr, the fluorescence remained limited to the membrane. In the following, fluorescence was present inside the cell and accumulated around the nucleus. The remaining fluorescence on the membrane and in the more peripheral cytoplasmic areas suggested a unidirectional intracellular migration from the cellular to the nuclear membrane, which appeared to be a barrier for further progression. As ascertained by focusing through the cell, the fluorescence was never located at the nuclear level. Compared with the bovine and equine SODs, the human recombinant Cu,ZnSOD penetrated slower; however, the difference was rather quantitative than qualitative. On the other hand, with MnSOD from *Escherichia coli* and recombinant human MnSOD (rhMnSOD), the number of labeled cells was low. Therefore the intracellular distribution of fluorescence and its pattern could not be studied. The rare labeled cells showed a single intracellular focal spot different from the diffuse punctuate distribution pattern of the Cu,Zn enzymes.

*Skin.* Because all tissue autofluorescence was eliminated by the filter, the fluorescence visible on the tissue section from SOD-treated human and mouse skin corresponded to FITC–SOD. In agreement herewith, specimens from vehicle-treated skin did not show any fluorescence. With a magnification of 1.2 pixels/$\mu$m, a green fluorescent film was visible in the stratum corneum, indicating the presence of FITC–SOD despite careful rinsing of the skin before the specimen was excised. With a higher magnification (3.6 pixels/$\mu$m), it could be seen that the fluorescence reaches the granulosa with decreasing intensity. The openings of the appendages and the lumena of the hair canals were filled with green fluorescent material. The observations were similar for human skin. With gray-level intensity measurements, taken along a line at 20-pixel thickness, a gradient could be demonstrated, suggesting progressive penetration of the FITC–SOD through the upper epidermis.

For fluorescence presentations see Emerit *et al.*[4] and Filipe *et al.*[5]

*Demonstration of Protective Effects*

*Phorbol Myristate Acetate-Stimulated Superoxide Production.* Equal numbers of monocytes were pretreated for 2 hr with different SODs at a concentration of 300 units/ml. After removal of all residual extracellular SOD with three washes, the cells were stimulated with PMA as indicated in Materials and Methods. Superoxide production, measured with the cytochrome *c* assay, was significantly reduced after pretreatment with bovine, equine, and rhCu,ZnSODs compared with nonpretreated cells from the same donor. MnSODs slightly reduced the superoxide production, but the difference was not significant. This was true as well for the human as for the *E. coli* enzyme. No difference was noted between native and FITC-modified SOD at equal concentrations. Cytochrome *c* reduction was also inhibited in three experiments in which the exposure time to SOD was only 5 min. As already mentioned, SOD is fixed on the cellular membrane at that time, but has not yet penetrated into the cytoplasm.

*Inhibition of Inosine Triphosphate-Induced Clastogenic Effects by Pretreatment with Fluorescein Isothiocyanate-Labeled Superoxide Dismutase.* The clastogen chosen for these experiments was inosine triphosphate (ITP), shown to act via superoxide-mediated mechanisms. It was dissolved in distilled water to a final concentration of 20 $\mu M$. This concentration induced 22.2% chromosomal breaks per 100 cells. The aberration rates were significantly reduced in the cell cultures, which had been set up with cells pretreated with SOD. The measurements of SOD activitiy in the cell supernatants showed a reduction of 150 to 12 units/ml after the first wash. No activity was detectable after the second wash.

*Protective Effects in Psoralen Plus Ultraviolet A-Induced Inflammatory Skin Reactions.* Macroscopically, the area pretreated with the vehicle showed an inflammatory reaction after exposure to PUVA, whereas this was not the case for the SOD-treated area (see Filipe et al.[6]). Histologically, the vehicle-treated skin revealed PUVA-induced necrosis that did not occur in the SOD-treated skin.

## Conclusions and Comments

The experiments with fluorescently labeled SOD clearly showed that the enzyme binds to cellular membranes, but that the intensity of binding varies according to cell type. This was found with both technical procedures, flow cytometry as well as fluorescence microscopy.

Among blood cells, monocytes were regularly more fluorescent than the other blood cells. The preferential binding of exogenous SOD to monocytes compared with neutrophils may explain why therapeutic effects of SOD have been reported mainly for chronic inflammatory diseases, in particular rheumatoid arthritis, in which the monocytes are in an activated state and exhibit increased superoxide production.

Whereas membrane binding was rapid and detectable after a 5-min incubation time, passage of the enzyme through the membrane appeared to be relatively slow and fluorescence was seen after 1 hr only. Because biological effects of exogenous SOD such as reduced superoxide production by stimulated cells could be observed after short incubation times not sufficient for intracellular penetration according to confocal microscopy, one may conclude that intracellular uptake of SOD is not required for its protective action. This was true also for anticlastogenic effects of SOD, which were seen after a 15-min preincubation in the presence of the enzyme and that were preserved even after extensive washing of the cells. The experiments demonstrate that the capacity of exogenous SOD to prevent chromosome damage is not dependent on its presence in the cell nucleus, because no FITC fluorescence was detectable at the nuclear level on focusing through the cell with a confocal microscope. This confirms earlier findings of others, who studied the uptake of radioactive SOD in different cellular fractions and found only minor quantities in the nuclear fraction.[8] Superoxide-mediated clastogenesis[4] occurs via indirect action mechanisms leading to the production of secondary, more long-lived clastogenic species. These comprise metabolites from the arachidonic acid cascade and other mediators, in particular tumor necrosis factor, released as a consequence of increased superoxide production by competent cells. Binding of SOD to the cell surface was accompanied by diminished superoxide production by stimulatd cells and consequently by decreased formation of clastogenic factors. These findings confirm earlier observations with radioactive SOD.[8]

Use of fluorescence techniques not only provided interesting data for isolated cells, but gave evidence that topically applied SOD can penetrate into human and mouse epidermis, either through the unbroken stratum corneum or through the follicular appendages, with progressively decreasing intensity of the fluorescence in the tissue. FITC fluorescence could not be detected with our techniques in the dermis, where the inflammatory reaction was prevented in the SOD-treated skin. It is possible that the SOD reaches the dermis, but that its concentration is not sufficient for detection by our methods. Another explanation would be that the enzyme located in the epidermis scavenges superoxide radicals, which otherwise would generate secondary products, in particular lipid peroxidation products and cytokines, responsible for inflammatory reactions in the dermis.

The differences between Cu,Zn- and MnSODs with respect to membrane binding were striking. The reasons for these discrepancies between SODs with different active sites are not clear. Differences in electrostatical charges may be responsible. Because the differences observed for binding were concomitant with decreased inhibition of stimulated superoxide production, they may be of importance also for the therapeutic efficacy of these enzymes.

---

[8] A. Petkau, W. S. Chelack, K. Kelly, and H. K. Friesen, in "Pathology of Oxygen" (A. Autor, ed.), p. 223. Academic Press, New York, 1982.

# Section V

# Superoxide Dismutase in Aging and Disease Therapy

# [33] Superoxide Dismutase in Aging and Disease: An Overview

*By* JOE M. MCCORD

## Introduction

The most striking observation regarding the role of oxidative stress in disease is that it appears to be nearly universal. As our understanding of free radical pathology has progressed, we have been continually surprised as disease after disease has been found to involve free radicals and related oxidants and the systems designed to keep them in check. There are several common biochemical mechanisms that underlie these relationships between oxidative stress and disease, as well as a number of novel and unique ones. The purpose of this review is not to focus on any particular oxidant-related disease, but rather to provide a framework for categorizing them as to the nature of the biochemical disturbance.

Not all oxidants are bad, and not all antioxidants are good. Rather, we live in a redox-based world where oxidation and reduction form the basis of our ability to power the reactions on which our lives depend. Because our cells are replete with reducing substances and oxidizing substances, unwanted reactions must be minimized by kinetic barriers or by physical compartmentalization. For example, evolution has carefully designed one of our most potent reductants, NADH, such that it does not spontaneously react with molecular oxygen, one of our most potent oxidizing substances, even though the reaction is favorable thermodynamically. Thus, we have multiple pools of compounds coexisting within our cells, both oxidants and reductants, that have great potential for reaction. They do not react, however, until or unless we want them to do so, because enzymes maintain kinetic control over which compounds may react. Further control may be provided by subcellular structure and physical compartmentalization. When cells are sick or injured, these kinetic or structural barriers may be breached. Active transport pumps may fail to keep metabolites in their proper compartments. Osmotic imbalance may cause swelling and stretching of delicate cellular architecture, especially that of the mitochondria. Figure 1 diagrams a general scheme whereby disease or trauma leads to increased production of cellular oxidants. There are various checks and balances in place, but these also represent sites vulnerable to damage, and therefore capable of upsetting the overall balance. It is important to remember that the superoxide radical appears to have important roles in normal cellular metabolism as a terminator of lipid peroxidation, as a signaling molecule, and as a molecule that can affect the concentration of nitric oxide, another free radical with important regulatory roles in the maintenance of homeostasis. The production of too much superoxide is bad, but the complete elimination

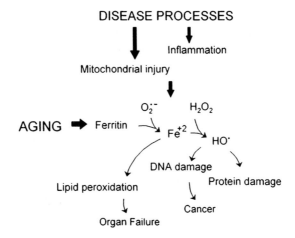

FIG. 1. A general scheme depicting ways in which disease and aging may upset oxidative balance. Production of $O_2^{·-}$ and $H_2O_2$ may be increased in damaged mitochondria or by the NADPH oxidase of inflammatory cells. Aging often results in accumulation of tissue ferritin. Damage may be increased as more stored iron is liberated by $O_2^{·-}$ for subsequent redox participation. Damage may be moderated by the availability of dietary antioxidants and phytochemicals capable of scavenging the reactive oxygen species.

of superoxide may be worse. Therapeutic efforts aimed at diseases involving oxidative stress must deal with this daunting paradox. It is *balance* that must be preserved or restored.

## Mitochondrial Injury

The most fundamental need of every cell in the human body is to generate the energy it requires to function. Our bodies do not have a centralized facility for the generation and distribution of power—every cell is responsible for its own needs. This puts a substantial burden on the cell, because the chemistry involved in the aerobic combustion of foodstuffs and the generation of ATP via mitochondrial electron transport is potentially nasty chemistry requiring careful containment to ensure that things do not get out of hand. In a healthy tissue the situation is well contained, and the low but unavoidable production of noxious by-products by oxidative metabolism is offset by the antioxidant enzymes [superoxide dismutases (SODs), catalases, and glutathione peroxidases] of the cell and its intake of antioxidant vitamins and related reducing substances. When conditions become stressful or injurious, however, the power generation system may become a liability. If the structural integrity of the cell is compromised or altered by physical trauma, metabolic stress, or by energy starvation, then adequate containment of the electron transport system becomes problematic and increased production of noxious by-products such as superoxide and hydrogen peroxide may exceed the detoxification capacity of the cell. The superoxide-mediated release of redox-active transition metal ions serves to exacerbate the situation, and uncontrolled oxidative processes such as lipid peroxidation, protein oxidation, and DNA damage may ensue. This damage may initiate the self-destruct mechanism of apoptosis, or may be sufficient to cause direct cell death via necrosis. Thus, a healthy cell is actually in a precarious steady-state balancing act, and it may be vulnerable to seemingly innocuous perturbations of that steady state. Insults that seem relatively minor (such as transient interruption of fuel supply) may precipitate a catastrophic failure of the power generation system from which the cell may not recover. If an automobile runs out of gas it simply stops until it is refueled. If an ocean liner runs out of fuel, however, and its bilge pumps cease pumping, it is in a more vulnerable situation. Soon, the engine room may become flooded, and attempts to restart the power generators may be disastrous.

Oxidative stress is now thought to make a significant contribution to all diseases that involve transient interruptions of blood supply, and hence of energy generation. These include ischemic diseases involving virtually any organ, but especially heart disease,[1] stroke,[2] intestinal ischemia,[3] and organ preservation associated with transplantation.[4]

Inflammation

The second major reason for the common involvement of free radicals in human disease stems from the reliance of our immune systems on the generation of superoxide by phagocyte NADPH oxidase as a means of keeping us free from microbial infection. The importance of this role of superoxide is difficult to overemphasize. It was first proposed by B. Babior and co-workers in 1973.[5] The cells of the immune system are ever vigilant and hyperreactive by nature. If they fail to recognize an invasion by a microbe, the mistake may be fatal. If they respond inappropriately when no real threat of infection exists, however, as in the case of a minor contusion, for example, the only price to be paid is a transiently sore muscle due to the additional tissue injury caused by the barrage of free radical production by inflammatory cells. More serious "false-positive" reactions by the surveillance team do

[1] B. A. Omar and J. M. McCord, *J. Mol. Cell. Cardiol.* **23,** 149 (1991).
[2] K. Baker, C. B. Marcus, K. Huffman, H. Kruk, B. Malfroy, and S. R. Doctrow, *J. Pharmacol. Exp. Ther.* **284,** 215 (1998).
[3] D. A. Parks, G. B. Bulkley, D. N. Granger, S. R. Hamilton, and J. M. McCord, *Gastroenterology* **82,** 9 (1982).
[4] F. Biasi, M. Bosco, I. Chiappino, E. Chiarpotto, G. Lanfranco, A. Ottobrelli, G. Massano, P. P. Donadio, M. Gay, E. Andorno, M. Rizzetto, M. Salizzoni, and G. Poli, *Free Radic. Biol. Med.* **19,** 311 (1995).
[5] B. M. Babior, R. S. Kipnes, and J. T. Curnutte, *J. Clin. Invest.* **52,** 741 (1973).

occur, and may result in chronic autoimmune diseases such as rheumatoid arthritis or systemic lupus erythematosus. Even so, the threat to our continued existence is less than that of a single undetected infection. Thus, the system reacts by producing superoxide not only to foreign antigens that may (infection) or may not (allergy) represent a real threat, but also to internal signals indicating tissue injury that may correlate with and presage a vulnerability to infection. This hyperresponsiveness of the immune system, coupled with the fact that its ability to discriminate self from nonself is not perfect, leads to the result that immune system-generated superoxide is often involved in a broad spectrum of human diseases, extending far beyond infectious diseases. Oxidative stress is now thought to make a significant contribution to arthritis, vasculitis, glomerulonephritis, lupus erythematosus, adult respiratory distress syndrome,[6] scleroderma,[7] and many other diseases characterized by local accumulations of activated inflammatory cells. Indeed, most diseases attributable to a disruption of the power-generating system (e.g., ischemia–reperfusion) will involve as well an inflammatory response initiated by factors signaling that tissue injury has occurred. Within a few hours of an ischemia-induced heart attack, the infarcted area is surrounded by activated neutrophils.

### Balance between Superoxide and Superoxide Dismutases

The possibility of a balance between superoxide and superoxide dismutases was suggested early on by the work of Y. Groner and co-workers, who produced transfected cells and transgenic mice[8,9] overexpressing SOD1, the cytosolic Cu,ZnSOD.[9] To the surprise of many, too much SOD turned out to be problematic. Transfected cells showed higher rates of lipid peroxidation, not lower. Patients with Down syndrome, whose cells contain 50% more SOD1 due to the gene dosage effect of having a third copy of chromosome 21, similarly show increased rates of lipid peroxidation. Could it be that superoxide is actually somehow beneficial? Further demonstration of the phenomenon came with studies of isolated perfused hearts. In these studies the organ was injured by a period of no-flow ischemia. On reperfusion, the rate of superoxide production exceeded the ability of endogenous SODs to maintain an optimal level of the radical. Adding exogenous SOD to the perfusate provided substantially improved recovery of function. Adding more SOD beyond the optimal concentration, however, resulted in an ever-increasing failure to recover, with the damage finally exceeding that observed when no exogenous SOD

---

[6] P. K. Gonzalez, J. Zhuang, S. R. Doctrow, B. Malfroy, P. F. Benson, M. J. Menconi, and M. P. Fink, *Shock* **6**(Suppl. 1), S23 (1996).

[7] C. M. Stein, S. B. Tanner, J. A. Awad, L. J. Roberts, and J. D. Morrow, *Arthritis Rheum.* **39**, 1146 (1996).

[8] O. Elroy-Stein, Y. Bernstein, and Y. Groner, *EMBO J.* **5**, 615 (1986).

[9] C. J. Epstein, K. B. Avraham, M. Lovett, S. Smith, O. Elroy-Stein, G. Rotman, C. Bry, and Y. Groner, *Proc. Natl. Acad. Sci. U.S.A.* **84**, 8044 (1987).

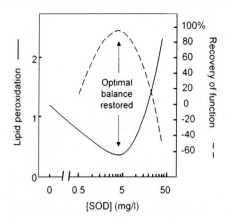

FIG. 2. Relationship between net lipid peroxidation and functional recovery of isolated rabbit hearts, in response to increasing dosages of superoxide dismutase. Recovery of developed pressure correlates inversely with net lipid peroxidation. Maximal protection is seen at a dosage of 5 mg/liter in the coronary perfusate. [Data are replotted from S. K. Nelson, S. K. Bose, and J. M. McCord, *Free Radic. Biol. Med.* **16**, 195 (1994).]

at all was added to the perfusate.[10] That is, bell-shaped dose–response curves were observed. In subsequent studies we found that lipid peroxidation was inversely related to the bell-shaped dose response, seen in Fig. 2.[11] Functional recovery was maximal (and nearly completely restored) at a single concentration of SOD, and lipid peroxidation was minimal at this point. Lipid peroxidation increased when SOD concentration in the perfusate was either increased or decreased from this optimal value.[11] The explanation for this rather bizarre observation, we believe, lies in the fact that lipid peroxidation is a free radical chain reaction. It must be initiated by a free radical (or by a one-electron redox event), and can be terminated only by a free radical (or by a one-electron redox event). Superoxide can act, indirectly at least, to initiate the process by reducing and liberating iron from its storage protein ferritin (see Fig. 1). This ferrous ion can cause reductive lysis of a lipid hydroperoxide molecule (LOOH), generating an alkoxy radical (LO·) capable of initiating a new chain reaction. Alternatively, superoxide can drive Fenton chemistry to produce hydroxyl radical, also capable of initiating a new chain by abstracting an allylic hydrogen from a lipid molecule. At the same time, superoxide is capable of terminating the free radical chain reaction via a radical–radical annihilation with the propagating lipid dioxyl radical (LOO·), reducing it to a lipid hydroperoxide. The termination of lipid peroxidation by superoxide *in vivo* appears to be a significant

---

[10] B. A. Omar, N. M. Gad, M. C. Jordan, S. P. Striplin, W. J. Russell, J. M. Downey, and J. M. McCord, *Free Radic. Biol. Med.* **9**, 465 (1990).
[11] S. K. Nelson, S. K. Bose, and J. M. McCord, *Free Radic. Biol. Med.* **16**, 195 (1994).

physiological function of the radical, and explains the puzzling observation that cells from individuals with Down syndrome show increased rates of lipid peroxidation despite their 50% higher content of SOD.[8] The single, optimal concentration of SOD in any system, then, depends on the rate at which superoxide is being produced and results from the best compromise between maintaining good termination of lipid peroxidation without excessive initiation of the process.[11] Unfortunately, the concentration of SOD required to maintain optimal balance may be rapidly changing as the pathological insult evolves. This presents a daunting challenge for therapeutic intervention with scavengers of superoxide, undoubtedly explaining the failure of many clinical trials. It presents an equally daunting challenge to the cell as it tries to maintain oxidative balance. One might predict that this optimal concentration of SOD would also be a function of the concentrations of other inhibitors of lipid peroxidation in the system, such as vitamin E.

### Can Superoxide Dismutase Activity Be Rapidly Modulated to Maintain Balance?

If indeed a delicate balance exists between superoxide and SOD activity, it is clear that some of the pathological situations considered above may cause sudden increases in rates of superoxide production that would necessitate rapid compensatory changes in SOD activity to avoid excessive amounts of lipid peroxidation. Can a cell rapidly adjust its SOD activity in the face of changing stress? Little research has focused on this question. All three SOD genes are inducible, although the *SOD1* gene is often considered to be constitutively expressed. The most dramatically inducible is the *SOD2* gene encoding the mitochondrial MnSOD, which may be induced more than 10-fold by tumor necrosis factor.[12] Induction of mammalian genes, however, is a relatively sluggish process usually requiring 24 to 48 hr. The pathological increases in superoxide production may have a much more rapid onset. An interesting study by Yamashita *et al.*[13] strongly suggests the existence of a different mechanism for the regulation of SOD2 activity. Rats were intensively exercised for 20 min. Thirty minutes after the exercise ended there was a 60% increase in myocardial *SOD2* activity with no accompanying increase in immunoprecipitable protein. By 3 hr the activity returned to normal. At 48 hr a second increase in SOD2 activity was seen, this time accompanied by a corresponding increase in SOD2 protein, reflecting induced gene expression. By 72 hr things returned once again to normal. Both increases in SOD2 activity were abolished if animals were pretreated with anti-tumor necrosis factor $\alpha$ (TNF-$\alpha$) and anti-interleukin 1$\beta$ (IL-1$\beta$) antibodies. Thus, there appears to be a cytokine-dependent

---

[12] G. H. Wong and D. V. Goeddel, *Science* **242**, 941 (1988).
[13] N. Yamashita, S. Hoshida, K. Otsu, M. Asahi, T. Kuzuya, and M. Hori, *J. Exp. Med.* **189**, 1699 (1999).

mechanism whereby mitochondrial SOD can be rapidly and reversibly activated. The biochemical basis for this remains to be determined.

## Iron Overload Exacerbates Superoxide-Dependent Pathology

Hemochromatosis is a disease of abnormally high iron absorption, and is associated with a common Cys282Tyr mutation in the *HFE* gene.[14] Homozygotes become severely iron loaded, and heterozygotes (about 14% of the population) show moderate degrees of iron overload. Heterozygotes are twice as likely to develop acute myocardial infarction[15] or type 2 diabetes.[16] Why does excess iron increase the risk of superoxide-related diseases? The most generally destructive action of superoxide may be the bringing about of the reductive release of iron from its storage protein, ferritin.[17] Because of its redox activity, iron is handled carefully by cells. In the healthy state, there is no appreciable concentration of "free" iron. Free Fe(II) is chelated by compounds such as citrate or ADP, but these complexes readily participate in redox reactions, catalyzing the formation of ·OH, which can initiate lipid peroxidation or cause DNA strand breaks.[18] Most small chelators of iron can accommodate the coordination geometry of either Fe(II) or Fe(III), providing little hindrance to its cyclical reduction or oxidation. Many forms of oxidative stress are ameliorated by iron-chelating agents such as desferrioxamine. The macromolecular chelators of iron such as transferrin and ferritin, as well as drugs such as desferrioxamine, provide binding sites of such rigid specificity that Fe(III) is bound extremely tightly, but Fe(II) is not bound at all. Because of kinetic restrictions as well as the thermodynamics of binding, the iron in transferrin and ferritin is difficult to reduce by most cellular reductants, and is thus shielded from release. Abnormally high levels of superoxide, however, can reductively liberate this iron. As ferritin concentration increases within a cell, it presents, in effect, a bigger target for the superoxide available and will capture a larger fraction of that superoxide, with the deleterious result of releasing catalytically active iron. When faced with the same oxidative insult, iron-loaded tissues suffer greater damage and loss of function than tissues with normal iron stores.[19] Certain other iron-containing proteins (such as aconitase or iron-responsive element-binding protein) within the

[14] J. N. Feder, A. Gnirke, W. Thomas, Z. Tsuchihashi, D. A. Ruddy, A. Basava, F. Dormishian, R. Domingo, Jr., M. C. Ellis, A. Fullan, L. M. Hinton, N. L. Jones, B. E. Kimmel, G. S. Kronmal, P. Lauer, V. K. Lee, D. B. Loeb, F. A. Mapa, E. McClelland, N. C. Meyer, G. A. Mintier, N. Moeller, T. Moore, E. Morikang, C. E. Prass, L. Quintana, S. M. Starnes, R. C. Schatzman, K. J. Brunke, D. T. Drayna, N. J. Risch, B. R. Bacon, and R. K. Wolff, *Nat. Genet.* **13**, 399 (1996).
[15] T. P. Tuomainen, K. Kontula, K. Nyyssonen, T. A. Lakka, and J. T. Salonen, *Circulation* **98**(Suppl.), abstract 2417 (1998).
[16] T. Kwan, B. Leber, S. Ahuja, R. Carter, and H. C. Gerstein, *Clin. Invest. Med.* **21**, 251 (1998).
[17] P. Biemond, H. G. van Eijk, A. J. G. Swaak, and J. F. Koster, *J. Clin. Invest.* **73**, 1576 (1984).
[18] B. Halliwell and J. M. C. Gutteridge, *FEBS Lett.* **307**, 108 (1992).
[19] A. M. M. Van der Kraaij, L. L. Mostert, H. G. van Eijk, and J. F. Koster, *Circulation* **78**, 442 (1988).

cell are also vulnerable to superoxide.[20,21] These proteins, however, contain iron in functional or catalytic settings and serve as enzymes or regulatory proteins rather than storage proteins. Quantitatively, they represent much less iron than is typically present in ferritin, and their concentrations are largely independent of iron status. Their reaction with superoxide may adversely affect cellular metabolism, however, because of their loss of function. In healthy cells, stored iron appears to pose little problem because superoxide concentration is low. Even so, one might predict that the iron-dependent "background" damage due to unscavenged superoxide would be proportionately higher in iron-overloaded individuals, perhaps contributing to the accumulation of random damage associated with aging.

### Iron Stores and Aging: An Underappreciated Relationship

A peculiar aspect of iron metabolism is the fact that we have a specific mechanism for its absorption (regulated to a degree) but we have no mechanism to eliminate excess iron. In fact, loss of blood is the primary way our bodies lose iron. As a result, the iron stores of American males increase almost linearly with age.[22] Females, because of menstruation, are in relatively good iron balance until the age of menopause, after which time they begin to accumulate iron at rates comparable to that of males. Sullivan[23] has suggested that this difference in amounts of stored iron accounts for the gender difference observed in the mortality statistics for ischemic heart disease. Ferritin stores provide no know benefit other than to provide iron for new hemoglobin synthesis in the event of a substantial, nonreplaced blood loss, something that rarely happens in modern society.

Human clinical data support the concept that high iron stores impose additional risk from many types of disease. Salonen et al.[24] identified high levels of stored iron as a clear risk factor for heart disease. In a study of patients with small cell carcinoma of the lung, Milman et al.[25] found that patients with the lowest serum ferritin levels at the time of diagnosis had significantly longer survival times. Because, as already mentioned, oxidative stress is associated with nearly all forms of disease, one might expect that iron stores would become a liability to almost all critically ill patients. Because iron stores increase with age, one might then expect that the ravages of disease will be amplified in the elderly, relative to the young. Ironically, iron stores are easily monitored and readily lowered by blood donation and the avoidance of iron supplementation, yet few physicians pay much attention to this important factor. It is not difficult to find breakfast cereals that provide, with

[20] P. R. Gardner, I. Raineri, L. B. Epstein, and C. W. White, *J. Biol. Chem.* **270**, 13399 (1995).
[21] C. Bouton, M. Raveau, and J. C. Drapier, *J. Biol. Chem.* **271**, 2300 (1996).
[22] J. D. Cook, C. A. Finch, and N. J. Smith, *Blood* **48**, 449 (1976).
[23] J. L. Sullivan, *Lancet* **1**, 1293 (1981).
[24] J. T. Salonen, K. Nyyssonen, H. Korpela, J. Tuomilehto, R. Seppanen, and R. Salonen, *Circulation* **86**, 803 (1992).
[25] N. Milman, H. Sengelov, and P. Dombernowsky, *Br. J. Cancer.* **64**, 895 (1991).

a degree of pride, 100% of your recommended daily allowance of iron (12–15 mg) in a single serving.

## Superoxide Dismutase Deficiency Diseases

One of the more interesting and obvious possibilities for oxidative stress-related disease is the underproduction of SOD activity. Mammals have three known genes encoding SODs: a Cu,ZnSOD found in the cytosol (*SOD1*), an MnSOD in the mitochondria (*SOD2*), and a secreted extracellular SOD (*SOD3*). Knockout mice have now been produced for each with surprising results. SOD1 knockouts and SOD3 knockouts get along well, especially under nonstressful conditions. SOD1 knockouts are more vulnerable to ischemia–reperfusion injury[26] and they display a motor axonopathy and vulnerability to axonal injury.[27,28] SOD3 knockouts are more sensitive to hyperoxic lung injury,[29] to ischemia–reperfusion injury,[30] and to alloxan-induced diabetes.[31] Perhaps the most interesting feature of SOD3 deficiency, however, is impairment in spatial learning and radial maze performance.[32] These characteristics are also seen in SOD3 transgenics that overexpress the activity, underscoring once again the importance of balance in free radical metabolism. In this case, extracellular $O_2^{\cdot-}$ is thought to modulate the level of nitric oxide, a neurotransmitter.

A polymorphism (Arg213Gly) in SOD3 has been found in 3–5% of the population. This arginine-to-glycine replacement is in the hydrophilic C-terminal "tail" of the enzyme, which is responsible for the binding of SOD3 to negatively charged endothelial surfaces and macromolecules such as collagen fibrils in the extracellular spaces. The mutation weakens the binding of SOD3 to polyanionic surfaces, and results in about 10-fold more SOD3 activity circulating in the plasma. The frequency of the mutation is about twice as high in patients undergoing renal dialysis as in the population at large, suggesting that the diminished adherence of SOD3 to cell surfaces may be a risk factor for renal failure.[33]

---

[26] T. Yoshida, N. Maulik, R. M. Engelman, Y. S. Ho, and D. K. Das, *Circ. Res.* **86,** 264 (2000).
[27] J. M. Shefner, A. G. Reaume, D. G. Flood, R. W. Scott, N. W. Kowall, R. J. Ferrante, D. F. Siwek, M. Upton-Rice, and R. H. J. Brown, *Neurology* **53,** 1239 (1999).
[28] A. G. Reaume, J. L. Elliott, E. K. Hoffman, N. W. Kowall, R. J. Ferrante, D. F. Siwek, H. M. Wilcox, D. G. Flood, M. F. Beal, R. H. Brown, Jr., R. W. Scott, and W. D. Snider, *Nat. Genet.* **13,** 43 (1996).
[29] L. M. Carlsson, J. Jonsson, T. Edlund, and S. L. Marklund, *Proc. Natl. Acad. Sci. U.S.A.* **92,** 6264 (1995).
[30] H. Sheng, T. C. Brady, R. D. Pearlstein, J. D. Crapo, and D. S. Warner, *Neurosci. Lett.* **267,** 13 (1999).
[31] M. L. Sentman, L. M. Jonsson, and S. L. Marklund, *Free Radic. Biol. Med.* **27,** 790 (1999).
[32] E. D. Levin, T. C. Brady, E. C. Hochrein, T. D. Oury, L. M. Jonsson, S. L. Marklund, and J. D. Crapo, *Behav. Genet.* **28,** 381 (1998).
[33] H. Yamada, Y. Yamada, T. Adachi, H. Goto, N. Ogasawara, A. Futenma, M. Kitano, K. Hirano, and K. Kato, *Jpn. J. Hum. Genet.* **40,** 177 (1995).

The complete elimination of SOD2 activity is lethal shortly after birth,[34] illustrating the profound importance of mitochondrial SOD. The animals die of neurodegeneration and dilated cardiomyopathy.[35] Heterozygotes expressing about half the normal amount of MnSOD, however, appear normal and do not show increased sensitivity to hyperoxic injury[36] like the SOD3 knockouts, showing the importance of compartmentalization.

### Oxidative Stress and Malignancy

The natural inclination of a cell is to divide. Reining in the biological imperative of a cell to proliferate is no small feat. Indeed, it may require more sophisticated cellular engineering to prevent proliferation than to promote it. For a normal postmitotic cell to become malignantly transformed, several conditions may have to be met. It may be necessary to relieve certain evolutionary constraints that tell the cell not to enter into mitosis. It may be necessary to provide mild oxidative stress, which seems to be a nearly universal signal for proliferation. It may even be necessary to disable yet another set of evolutionary constraints designed to prevent cells from running amok by triggering apoptosis.[37] This latter set of constraints can, in fact, be triggered by oxidative stress per se. Whereas wild proliferation may be a mark of success for a bacterium, it is dangerous in higher organisms. The entire organism can be brought down by what begins as a single errant cell that has broken free of its evolutionary constraints. Thus, we have evolved a failsafe system, apotosis, which can detect out-of-control proliferation, and can cause programmed self destruction of any cell showing this behavior.

How might an initiated cell, one on its way to becoming cancerous, achieve and maintain the condition of mild oxidative stress necessary to drive its proliferation? It has been shown that many types of human cancer cells have reduced SOD2 activity.[38] In most cases, the reduced activity has been assumed to be due to defective expression of the gene (i.e., changes in the promoter region of the gene).[39] Oberley *et al.* and St. Clair *et al.*[38,39] have observed in numerous studies that transfection with the human *SOD2* gene can reverse the malignant phenotype of tumor

---

[34] Y. Li, T. T. Huang, E. J. Carlson, S. Melov, P. C. Ursell, J. L. Olson, L. J. Nobel, M. P. Yoshimura, C. Berger, P. H. Chan, D. C. Wallace, and C. J. Epstein, *Nat. Genet.* **11,** 376 (1995).
[35] R. M. Lebovitz, H. Zhang, H. Vogel, J. Cartwright, Jr., L. Dionne, N. Lu, S. Huang, and M. M. Matzuk, *Proc. Natl. Acad. Sci. U.S.A.* **93,** 9782 (1996).
[36] M. F. Tsan, J. E. White, B. Caska, C. J. Epstein, and C. Y. Lee, *Am. J. Respir. Cell Mol. Biol.* **19,** 114 (1998).
[37] J. M. McCord and S. C. Flores, *in* "Oxidative Processes and Antioxidants" (R. Paoletti, ed.), p. 13. Raven Health Care Communications, New York, 1994.
[38] L. W. Oberley and G. R. Buettner, *Cancer Res.* **39,** 1141 (1979).
[39] D. K. St. Clair and J. C. Holland, *Cancer Res.* **51,** 939 (1991).

cells, suggesting that mitochondrial SOD functions as a tumor suppressor.[40] Xu et al.[41] reported finding a variant sequence containing a cluster of three mutations in the promoter regions of the MnSOD genes from 5 of 14 human cancer cell lines examined. All five cell lines were heterozygous for the variant sequence. The mutations change the binding pattern of transcription factor AP-2 and cause a marked diminution in the efficiency of the promoter, using a luciferase reporter assay system. Whereas SOD2 is inducible by interleukin $1\beta$ or by tumor necrosis factor $\alpha$, there are no recognition sequences in the promoter for the expected involvement of transcriptional factor NF-$\kappa$B or C/EBP-$\beta$. Such recognition sites do exist, however, in an enhancer region located in intron 2 of the gene.[42] Thus, it is interesting to speculate that serious, tumor-promoting oxidative stress may result in a cell that is unable to respond to cytokine signals, even though its basal expression of SOD2 may be perfectly normal. An alternative is that mutations in the coding region of SOD2 may adversely affect catalytic efficiency or the stability of the protein. In a preliminary examination of genomic DNA isolated from lung and prostate cancers, we have found *SOD2* genes containing mutations in all the categories described above. Impaired expression of MnSOD may be a nearly universal characteristic of transformed cells.

[40] W. Zhong, L. W. Oberley, T. D. Oberley, T. Yan, F. E. Domann, and D. K. St. Clair, *Cell Growth Differ.* **7**, 1175 (1996).
[41] Y. Xu, A. Krishnan, X. S. Wan, H. Majima, C. C. Yeh, G. Ludewig, E. J. Kasarskis, and D. K. St. Clair, *Oncogene* **18**, 93 (1999).
[42] P. L. Jones, D. Ping, and J. M. Boss, *Mol. Cell. Biol.* **17**, 6970 (1997).


# [34] Tissue-Specific Mitochondrial Production of $H_2O_2$: Its Dependence on Substrates and Sensitivity to Inhibitors

*By* LINDA K. KWONG *and* RAJINDAR S. SOHAL

## Introduction

It is commonly accepted that reactive oxygen species (ROS) contribute to a variety of pathophysiological conditions ranging from neurodegenerative diseases to aging.[1,2] However, they also function as second-messenger molecules in cellular

[1] N. Hogg, *Semin. Reprod. Endocrinol.* **16**, 241 (1998).
[2] R. S. Sohal and W. C. Orr, *Aging (Milano)* **10**, 149 (1998).

signal transduction pathways.[3] Thus, an imbalance between ROS generation and antioxidative defense may result in a change from physiological to pathophysiological state.

The major intracellular site of ROS production is the mitochondria. It is estimated that ~95% of the oxygen consumed by an eukaryotic organism is utilized in the mitochondria for oxidative phosphorylation and that ~1–2% of the oxygen is leaked from the respiratory chain and forms superoxide ion ($O_2^{\cdot -}$).[4] There are two main sites of $O_2^{\cdot -}$ generation in the mitochondria, NADH dehydrogenase at complex I and ubiquinone at complex III. Measurement of the $O_2^{\cdot -}$ generation rate from these sites in the intact mitochondria is hampered by the inability of $O_2^{\cdot -}$ to pass outward through the inner mitochondrial membrane, the presence of superoxide dismutase in the mitochondrial matrix, the spontaneous dismutation of $O_2^{\cdot -}$ to hydrogen peroxide ($H_2O_2$), and the inability of detection reagents to enter the mitochondria.[5] Thus, most laboratories use submitochondrial particles (SMPs) to measure the rate of $O_2^{\cdot -}$ generation. However, there is still the possibility that the measured rate in SMPs may not be indicative of the rate in the intact mitochondria due to alteration or damage to the membrane during isolation. Therefore, a need remains for a reliable method to measure the $O_2^{\cdot -}$ generation in intact mitochondria.

Our method of measuring mitochondrial ROS generation takes advantage of the quantitative stoichiometric dismutation of $O_2^{\cdot -}$ to $H_2O_2$ and the ease with which $H_2O_2$ is able to permeate the mitochondrial membrane.[6] Thus, the rate of $H_2O_2$ released from the mitochondria reflects the rate of intramitochondrial generation of $O_2^{\cdot -}$. The released $H_2O_2$ is quantitated by a coupled reaction in which horseradish peroxidase (HRP) is used to catalyze the oxidation of $p$-hydroxyphenylacetate (PHPA), forming a fluorescent dimeric product, 2,2′-dihydroxy-biphenyl-5,5′-diacetate [(PHPA)$_2$] in the presence of $H_2O_2$.[7] The observed increase in fluorescence is catalase sensitive, demonstrating the specificity of this assay.[8] The high concentration of HRP used in our assay and the relative difference in rate constants between HRP and mitochondrial glutathione peroxidase and catalase strongly favor the reaction of $H_2O_2$ with HRP. Therefore, the increase in fluorescence is directly proportional to the amount of $H_2O_2$ generated in the mitochondria, and the rate of $H_2O_2$ generation can be measured by monitoring the rate of increase in fluorescence.

[3] K. Hensley, K. A. Robinson, S. P. Gabbita, S. Salsman, and R. A. Floyd, *Free Radic. Biol. Med.* **28**, 1456 (2000).
[4] B. Chance, H. Sies, and A. Boveris, *Physiol. Rev.* **59**, 527 (1979).
[5] J. F. Turrens and J. M. McCord, in "Free Radicals, Lipoproteins, and Membrane Lipids" (A. C. Paulet, L. Douste-Blazy, and R. Paoletti, eds.), p. 203. Plenum Press, New York, 1990.
[6] E. Cadenas and K. J. Davies, *Free Radic. Biol. Med.* **29**, 222 (2000).
[7] P. A. Hyslop and L. A. Sklar, *Anal. Biochem.* **141**, 280 (1984).
[8] L. K. Kwong and R. S. Sohal, *Arch. Biochem. Biophys.* **350**, 118 (1998).

## Assay Method

### Reagents

Incubation buffer (buffer B): 154 m$M$ potassium chloride, 10 m$M$ potassium phosphate, 3 m$M$ MgCl$_2$ and 0.1 m$M$ EGTA; adjust to pH 7.4 with KOH. Make fresh weekly or dispense 100-ml aliquots and store at $-20°$.
PHPA: 10 mg/ml in buffer B, adjust to pH 7.4 with KOH; dispense 1-ml aliquots and store at $-20°$.
HRP (EC 1.11.1.7, type VI): 100-U/ml stock solution in buffer B; dispense 800-$\mu$l aliquots and store at $-20°$.
Substrates: 100× stock solutions [FADH linked: succinate, $\alpha$-glycerophosphate, and $\beta$-hydroxybutyrate (0.7 $M$); NADH-linked: pyruvate (0.5 $M$)/malate (0.1 $M$)] in buffer B, adjust to pH 7.4 with KOH; dispense 500-$\mu$l aliquots and store at $-20°$.
Inhibitors: 10-ml stock solutions of antimycin (complex III inhibitor; 1 mg/ml) in 100% ethanol and rotenone (complex I inhibitor; 1 mg/ml) in chloroform; dispense 500-$\mu$l aliquots and store at $-20°$. On the day of assay, dilute 165 $\mu$l of antimycin stock solution and 100 $\mu$l of rotenone stock solution to 1 ml with ethanol to make working solutions. Rotenone is light sensitive; therefore cover tubes of stock and working solution with aluminum foil.
Hydrogen peroxide standards: Add 1 $\mu$l of 30% (v/v) H$_2$O$_2$ to 1 ml of buffer B to make a stock solution. Blank the spectrophotometer with 500 $\mu$l of buffer B in a 1-ml quartz cuvette at 230 nm. Add 500 $\mu$l of diluted H$_2$O$_2$ and mix well by inversion. Measure the absorbance of the solution in the cuvette. Calculate the [H$_2$O$_2$] in the stock solution by dividing the absorbance by 0.071, the millimolar extinction coefficient for H$_2$O$_2$, and multiply by 2 to compensate for the dilution in the cuvette. Determine the amount of stock solution needed to make 1 ml of 1 m$M$ solution. Dilute the 1 m$M$ solution 1:500 with buffer B to make the standard solution of 2 pmol/$\mu$l. The stock solution may be stored at 4° for 3 days, but diluted standard solution should be made on the day of assay.

### Equipment

Optimal equipment is a computer-controlled spectrofluorometer equipped with a thermally controlled and magnetic stirring sample compartment, and kinetic analysis software. A regular spectrofluorometer equipped with a chart record may also be used (set chart speed to 1 cm/min). If the sample compartment does not have magnetic stirring capability, remove the cuvette from the sample compartment after addition of substrate or inhibitor, cover the cuvette top with Parafilm, invert to mix the contents well, and place the cuvette back into the sample compartment. The crucial requirement for any type of spectrofluorometer used is the temperature-controlled sample compartment. Set the excitation wavelength to 320 nm and the emission wavelength to 400 nm, slit to 10, and temperature to 37° for mammalian mitochondria and 30° for insect mitochondria.

*Procedure*

1. Warm buffer B to the required temperature and place thawed aliquots of stock solutions on ice. Pipette buffer B, 50 μl of PHPA, 40 μl of HRP, and mitochondria solution containing 50–200 μg of protein (final volume, 2.970 ml) into a plastic 4.5-ml clear-sided disposal methacrylate cuvette containing a stir bar designed for spectrophotometer cells (Cell Spinbar). Cover the cuvette top with Parafilm, invert to mix the contents well, and place the cuvette into the sample compartment. Initiate data acquisition or chart recording. Monitor the reaction for 1 min to determine baseline endogenous rate.

2. Add 30 μl of substrate [heart (FADH linked, succinate; NADH linked, low rate); brain (FADH linked, succinate α-glycerophosphate, or β-hydroxybutyrate; NADH linked, pyruvate/malate); kidney (FADH linked, succinate, α-glycerophosphate, or β-hydroxybutyrate; NADH linked, pyruvate/malate); skeletal muscle (FADH linked, succinate or α-glycerophosphate)] to the cuvette and continue to monitor for an additional 2–3 min. This represents the substrate-supported $H_2O_2$ generation during state 4 respiration, during which the mitochondria are highly reduced.

3. Add 5–15 μl of inhibitor [heart (0.5 $\mu M$ antimycin or rotenone); brain and skeletal muscle (1 $\mu M$ antimycin or rotenone); and kidney (1.5 $\mu M$ antimycin or rotenone)] and continue to monitor for another 2–3 min to determine the rate in the presence of inhibitor. Respiratory inhibitors block electron flow through the mitochondria, producing a full reduction of all components located on the substrate side of the inhibition site. If the mitochondrial site of $O_2^{·-}$ production is located between the substrate and the inhibition site, the rate of $H_2O_2$ would be stimulated.

4. Add 5–15 μl of second inhibitor (if antimycin is added in step 3, add rotenone at this step, and vice versa) and continue to monitor for another 2–3 min. Addition of rotenone after antimycin eliminates the back flow of electrons from complex III to complex I, and the addition of antimycin after rotenone gives an indication of electron passage beyond the rotenone inhibition site.

5. Standard curve: Add 10, 20, 40, 60, 80, and 100 μl of $H_2O_2$ standard solution (20–200 pmol) to cuvettes containing buffer B, PHPA, HRP, and specific substrate or inhibitor so that the total volume is 3 ml. Determine the changes in fluorescence units relative to the blank cuvette and plot picomoles of $H_2O_2$ versus fluorescence units to generate the standard curve. The fluorescence response is rapid, requiring less than 30 sec to stabilize.

*Calculation*

Measure the rate of fluorescence increase under each condition, using computer kinetic analysis software or the delta change in height per minute on the chart.

Determine the picomoles of $H_2O_2$ generated per minute, using the standard curve and divide the value by the milligrams of mitochondrial protein used in the assay. This represents the picomoles of $H_2O_2$ generated per minute per milligram protein.

Comments

Mitochondria may be isolated by using various published methods. However, the final pellet should be rinsed free of any fatty acids and any light or loosely packed damaged mitochondria before resuspending gently in a minimal volume of buffer, preferably >2 mg/ml. We successfully use four passes, without grinding motion, of a PTEE pestle in a glass Potter-Elvehjem homogenizer to obtain a homogeneous suspension of coupled mitochondria. Moreover, buffer containing calcium should be avoided.

We have successfully used the method described above to study the rate and sites of $H_2O_2$ generation in mitochondria isolated from flight muscles of *Musca* and *Drosophila,* and from different tissues of C57/BL mice. The assay is highly sensitive, requiring only ~50 μg of mitochondrial protein, except for kidney mitochondria, where ~150 μg of protein is needed for a reliable state 4 $H_2O_2$ generation rate. Intra- and interassay variations are usually within 15% when the rates are at least 200 pmol of $H_2O_2$ generated/min/mg protein. A major contributor to the intraassay variation is the deterioration of the mitochondria after isolation. Therefore, mitochondrial preparations should be studied within 1 hr of isolation. Interassay variation in fluorescence responses to known amounts of $H_2O_2$ is minimal (Fig. 1). The concentrations of HRP (1.33 U/ml) and PHPA (1.1 m$M$) used

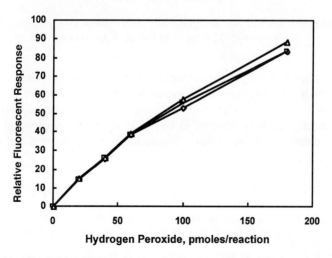

FIG. 1. Interassay variation of fluorescent responses to known amounts of hydrogen peroxide. Three standard curves were done separately over a 1-week time period under identical conditions and with different concentrated stock solutions.

in this method circumvented interference by glutathione reductase/catalase and endogenous hydrogen donors, respectively.

There are tissue-specific responses to the substrate as well as inhibitor used in the $H_2O_2$ generation assay. Mitochondria isolated from different mouse organs have specific substrate preferences.[8] For example, $\alpha$-glycerophosphate did not support $H_2O_2$ generation in heart mitochondria, but was preferred over succinate by skeletal muscle mitochondria. In addition, there are distinct responses by the mitochondria from each tissue to respiratory inhibitors. Heart mitochondria were most sensitive and responsive to antimycin. Brain mitochondria are the most sensitive and responsive to rotenone, whereas kidney mitochondria were nonresponsive.

## Conclusion

The method described in this article, using a coupled fluorescence reaction to measure the rate of mitochondrial $H_2O_2$ generation, is specific, quantitative, and reproducible. As little as 50 $\mu$g of mitochondrial protein can be used and a minimum rate of 100 pmol of $H_2O_2$ generated/min/mg protein can be determined reliably. However, for each tissue and organism studied, the preferred substrate(s) and inhibitor(s) must be determined empirically for maximal $H_2O_2$ generation.

## [35] Targeting Superoxide Dismutase to Critical Sites of Action

By MASAYASU INOUE, EISUKE SATO, MANABU NISHIKAWA, AH-MEE PARK, KENSAKU MAEDA, and EMIKO KASAHARA

Because reactive oxygen species rapidly react with various molecules and interfere with cellular functions, these species have been postulated to play critical roles in the pathogenesis of various diseases. Thus, protection of tissues from oxygen toxicity has been believed to be one of the major prerequisites to aerobic life.[1] Studies[2-5] revealed that a cross-talk of reactive oxygen and nitrogen

[1] H. Sies, "Oxidative Stress." Academic Press, New York, 1985.
[2] Y. Takehara, H. Nakahara, S. Okada, K. Yamaoka, K. Hamazaki, A. Yamazato, M. Inoue, and K. Utsumi, *Free Radic. Res.* **30**, 287 (1999).
[3] M. Nishikawa, E. Sato, K. Utsumi, and M. Inoue, *Cancer Res.* **56**, 4535 (1996).
[4] M. Inoue, M. Nishikawa, E. Sato, A. Park, M. Kashiba, Y. Takehara, and K. Utsumi, *Free Radic. Res.* **31**, 251 (1999).
[5] M. Inoue, E. Sato, A. Park, M. Nishikawa, E. Kasahara, M. Miyoshi, A. Ochi, and K. Utsumi, *Free Radic. Res.* **33**, 757 (2000).

species constitutes a supersystem that regulates circulation, energy metabolism, reproduction, and remodeling of tissues, and functions as a major defense system for pathogens. Hence, to minimize oxidative tissue injury, these reactive species should be scavenged selectively at the site(s) of their hazardous action.

Although Cu,Zn-type superoxide dismutase (SOD) and catalase have been used successfully to inhibit oxygen toxicity *in vitro*,[6] intravenously administered SOD and catalase rapidly disappear from the circulation because of renal filtration[7,8] and extraction by reticuloendothelial cells, respectively. Thus, it has been difficult to use these enzymes as tools for *in vivo* study of the pathophysiological roles of the reactive oxygen species or as therapeutics. To minimize oxidative stress caused by superoxide and related metabolites, these enzymes and related antioxidants should be targeted specifically to the site of tissue injury. Thus, various types of chemically modified SODs with a prolonged *in vivo* half-life have been synthesized by conjugating the enzyme with macromolecules, such as polyethylene glycol, dextran, and albumin.[7] However, targeting of these enzymes to an injured site cannot be achieved simply by conjugating with such polymers.

## Targeting Long-Acting Superoxide Dismutase to Tissue with Decreased pH

Most hydrophobic anions, such as fatty acids, bilirubin, and warfarin, circulate bound to albumin, thereby escaping from renal filtration and nonspecific distribution to tissues.[9] When the affinity of such compounds for albumin is sufficiently high, these ligands and their protein conjugates also circulate bound to this plasma protein with prolonged *in vivo* lifetimes. On the basis of this concept, we synthesized various types of Cu,ZnSOD derivatives (SM-SOD) by covalently linking half-butyl esterified co-poly(styrenemaleic acid) (SM, MW = 1600) to either Cys−111 or a lysyl residue of the enzyme (Fig. 1). SM-SODs circulate bound to albumin with half-lives longer than 6 hr.[8,10,11]

Covalent modification of lysyl amino groups of Cu,ZnSOD can be achieved by using SM having an anhydride group within the ligand.[8] The incubation medium contains, in a final volume of 1 ml, 0.5 $M$ sodium bicarbonate, 30 mg of SOD, and 2.2 m$M$ SM anhydride (SMA; Fig. 1a). The reaction is initiated by adding

---

[6] I. Emerit, L. Packer, and C. Auclair (eds.), "Antioxidants in Therapy and Preventive Medicine." Plenum Press, New York, 1990.
[7] M. Inoue, "Role of Reactive Oxygen Species in Diseases." Gakkai Shutsupan Center, Tokyo, Japan, 1992.
[8] T. Ogino, M. Inoue, Y. Ando, M. Awai, and Y. Morino, *Int. J. Protein Peptide Res.* **32**, 464 (1988).
[9] M. Inoue, *Hepatology* **5**, 892 (1985).
[10] M. Inoue, I. Ebashi, and N. Watanabe, *Biochemistry* **28**, 6619 (1989).
[11] N. Watanabe, M. Inoue, and Y. Morino, *Biochem. Pharmacol.* **38**, 3477 (1989).

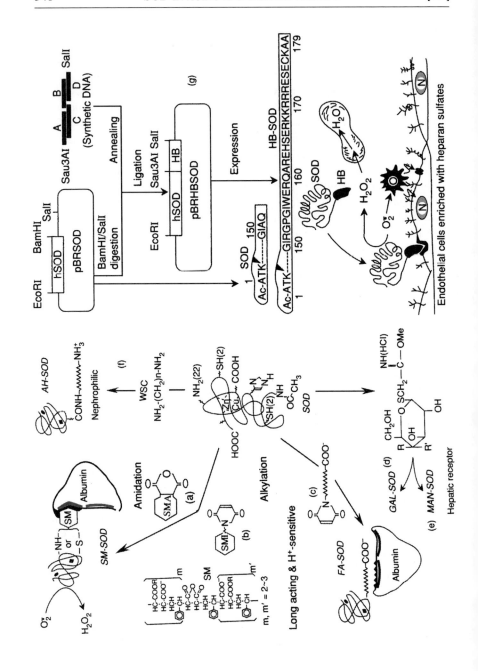

SM anhydride dissolved in dimethyl sulfoxide; the final concentration of dimethyl sulfoxide is less than 10%. During incubation at 37°, the enzyme activity and trinitrobenzene sulfonate-titrable amino groups are determined. After 6 hr, the incubated mixture is passed over a Sephadex G-100 column (2 × 30 cm) equilibrated with 20 m$M$ phosphate-buffered saline. Under these conditions, 1 mol of SM covalently incorporates in the monomeric subunit of SOD; the specific activity of the modified SOD is similar (2700 units/mg) to that of native enzyme (3000 units/mg). After lyophilization of SM-SOD, the enzyme can be stored at −20° for at least 6 months.

Covalent modification of the Cys−111 residue of Cu,ZnSOD can be achieved by using a maleimide derivative of SM that specifically reacts with free SH groups.[10] The incubation medium contains, in a final volume of 1 ml, 0.1 $M$ Tris-HCl buffer (pH 8.0), 0.1 m$M$ EDTA, 30 mg of SOD, and 10 m$M$ $\alpha$-4-{[6-($N$-maleimide)hexanoyloxymethyl]cumyl}half-butyl-esterified copoly(styrenemaleic acid) (SMI; Fig. 1b). Under a nitrogen stream, the reaction is initiated by adding SMI dissolved in dimethyl sulfoxide; the final concentration of dimethyl sulfoxide is 10%. During incubation at 37°, the enzyme activity and dithiobis(nitrobenzoic acid)-titrable free thiol groups are determined. After 6 hr, the incubated mixture is passed over a Sephadex G-100 column (2 × 30 cm) equilibrated with 20 m$M$ phosphate-buffered saline. Physicochemical and pharmacokinetic properties of SMI-linked SOD are similiar to those of SMA-linked SOD.

The SM moiety of SM-SOD can be replaced with other amphipathic anions having similar physicochemical properties. $N$-Maleimide derivatives of $\omega$-aminocarboxylic acids with various carbon chain lengths [$NH_2$-$(CH_2)_n$-COOH, FA, $n = 17$–$21$] are such examples. FA is linked covalently to the Cys−111 of human SOD (Fig. 1c), essentially as described for the synthesis of SMI-linked SOD.[10] The incubation medium contains, in a final volume of 5 ml, 0.1 $M$ Tris-HCl buffer (pH 8.0), 0.1 m$M$ EDTA, 150 mg of SOD, and 5 m$M$ FA. Under a nitrogen stream, the reaction is initiated by adding FA dissolved in dimethyl sulfoxide; the final concentration of dimethyl sulfoxide is 10%. During incubation at 37°, the enzyme activity and dithiobis(nitrobenzoic acid)-titrable free thiol groups are determined. After 12 hr, the incubated mixture is passed over a Sephadex G-100 column equilibrated with 20 m$M$ phosphate-buffered saline. Physicochemical and pharmacokinetic properties of FA-SOD are similar to those of SM-SODs.

---

FIG. 1. Strategy for targeting SOD. (a) SMA, anhydride of half-butyl esterified co-poly(styrenemaleic acid); (b) $\alpha$-4-{[6-($N$-maleimide)hexanoyloxymethyl]cumyl}-derivative of SM; SM-SOD, SM-conjugated SOD; (c) FA, $N$-maleimide-derivative of fatty acids; FA-SOD, FA-conjugated SOD; (d) GAL-SOD, galactose-conjugated SOD; (e) MAN-SOD, mannose-conjugated SOD; R and R', OH or H; (f) AH-SOD, diaminohexane-conjugated SOD; WSC, water-soluble carbodiimide; (g) HB-SOD with high affinity for heparan sulfate on vascular endothelial cells; A–D, Synthetic DNA.

When the carbonyl groups of SM and FA moieties of these SOD derivatives are protonated in tissues with decreased pH, the enzymes are released from albumin and accumulate in cell surface membrane/lipid bilayers chiefly because of the amphipathic nature of the ligands.[8,10] SOD derivatives that circulate bound to albumin with prolonged *in vivo* half-life and accumulate in tissues with decreased pH permit studies on the role of superoxide radicals in and around ischemic and hypoxic tissues.[7,11]

### Targeting Superoxide Dismutase to Hepatic Constituent Cells

Hepatocytes and Kupffer cells possess surface receptors specific for terminal galactose and mannose residues of glycoproteins, respectively.[12,13] To target Cu,ZnSOD specifically to these cells, galactose-conjugated enzyme (GAL-SOD; Fig. 1d) or mannose-conjugated enzyme (MAN-SOD; Fig. 1e) are synthesized by reacting the enzyme with activated galactose or mannose derivatives, respectively.

After incubation of 10 m$M$ sodium methoxide with 1 mmol of either cyanomethyl 1-thiogalactoside or 1-thiomannoside[14] at room temperature for 24 hr, the mixture is evaporated *in vacuo*. The resulting syrup of 2-imino-2-methoxyethyl derivative of either 1-thiogalactoside and 1-thiomannoside is incubated with 10 mmol of SOD dissolved in 15 ml of 50 m$M$ borate buffer, pH 10, at room temperature for 5 hr. The incubated mixture is concentrated with a Diaflow membrane followed by gel-filtration chromatography on a Sephadex G-25 column equilibrated with 0.1 $M$ acetate buffer, pH 6.0. GAL-SOD and MAN-SOD thus prepared bind to peanut lectin-Sepharose and ConA-Sepharose columns, respectively.[15] When injected intravenously in rats, GAL-SOD and MAN-SOD rapidly disappear from the circulation; more than 95% of GAL-SOD and MAN-SOD selectively accumulate in hepatocytes and Kupffer cells, respectively. The two SOD derivatives permit studies of the pathophysiological roles of the superoxide radical in these hepatic constituent cells. Because reticuloendothelial cells in other tissues also have mannose receptors, some fractions of intravenously administered MAN-SOD also accumulate in these cells and tissues. Interestingly, administration of MAN-SOD rapidly induces a marked hypotension in normal animals, suggesting that the mannose receptor-containing cells are responsible for the maintenance of physiological blood pressure.

---

[12] G. Ashwell and A. Morell, *Adv. Enzymol.* **47,** 99 (1974).
[13] T. Brown, L. Henderson, S. Thorpe, and J. Baynes, *Arch. Biochem. Biophys.* **188,** 418 (197).
[14] Y. Lee, C. Stowell, and M. Krantz, *Biochemistry* **15,** 3956 (1976).
[15] T. Fujita, M. Nishikawa, C. Tamaki, Y. Takakura, M. Hashida, and H. Sezaki, *J. Pharmacol. Exp. Ther.* **263,** 971 (1993).

## Targeting Superoxide Dismutase to Renal Proximal Tubule Cells

To test the possible involvement of superoxide and related metabolites in the pathogenesis of renal diseases, we synthesized an SOD derivative that rapidly accumulates in renal proximal tubule cells (Fig. 1f). Incubation medium contains, in a final volume of 5 ml, 0.5 $M$ hexamethylenediamine (AH, pH 8.0) and 500 mg of SOD. The reaction is started by adding 2 mmol of water-soluble carbodiimide (WSC) at 0°. After incubation at 4° for 4 hr, the incubated mixture is dialyzed against 5 liters of 20 m$M$ phosphate buffer, pH 7.4, containing 0.15 $M$ NaCl. The specific activity of AH-SOD is 2000 units/mg protein. Intravenously administered AH-SOD rapidly undergoes glomerular filtration, binds to lumenal membranes of renal proximal tubules, enters into these cells, and stays within the cells in a catalytically active form for a fairly long time (Fig. 2).[16] AH-SOD is useful for studying the role of superoxide radicals in and around renal tubule cells.

Although cisplatin is a potent anticancer drug, its clinical use is highly limited because of its strong side effects. Kinetic analysis revealed that mitochondria in renal proximal tubule cells are one of the potential targets of cisplatin. In fact, cisplatin increases the generation of the superoxide radical in and around renal proximal tubules, impairs mitochondrial function, and oxidatively injures their DNA, thereby inducing apoptosis of these cells. Intravenous administration of AH-SOD effectively inhibits the renal injury induced by cisplatin without affecting its anticancer action and increases the survival of tumor-bearing animals (Fig. 3).[16,17]

## Targeting Superoxide Dismutase to Vascular Endothelial Cells

Vascular endothelial cells are one of the critical targets for reactive oxygen species.[18] To scavenge superoxide radicals in and around vascular endothelial cells, a fusion gene encoding a hybrid SOD (HB-SOD) consisting of human Cu,ZnSOD and a C-terminal basic peptide with high affinity for heparan sulfates is constructed (Fig. 1g).[19] A full-length cDNA encoding the enzyme is obtained from the human placental cDNA library. An *Eco*RI restriction site is made before the initiation codon by site-directed mutagenesis. A 450-base pair *Eco*RI–*Sau*3AI fragment of the SOD gene is subcloned into *Eco*RI–*Bam*HI sites of pBR322. The constructed plasmid is designated pBRSOD1.

---

[16] M. Inoue, M. Nishikawa, E. F. Sato, K. Matsuno, and J. Sasaki, *Arch. Biochem. Biophys.* **368**, 354 (1999).
[17] M. Nishikawa, H. Nagatomi, B. Chang, E. Sato, and M. Inoue, *Arch. Biochem. Biophys.* **387**, 78 (2001).
[18] R. Kunitomo, Y. Miyauchi, and M. Inoue, *J. Biol. Chem.* **267**, 8732 (1992).
[19] M. Inoue, N. Watanabe, J. Sasaki, K. Matsuno, Y. Tanaka, H. Hatanaka, and T. Amachi, *J. Biol. Chem.* **266**, 16409 (1991).

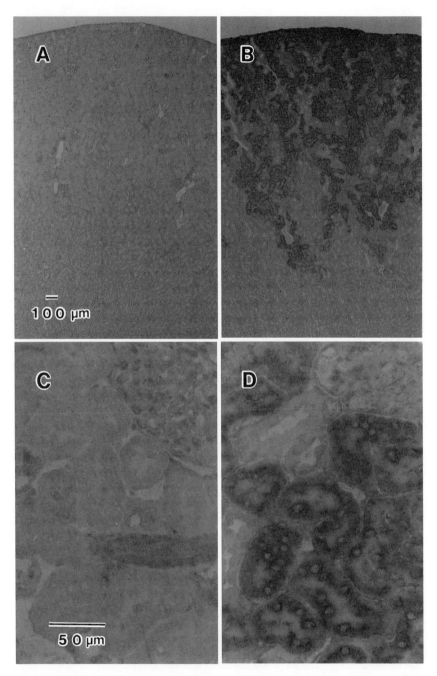

FIG. 2. Targeting SOD to renal proximal tubule cells. Using anti-human SOD antibody, rat kidney was immunohistochemically examined 30 min after intravenous administration of either saline (A and C) or 10 mg of AH-SOD (B and D).

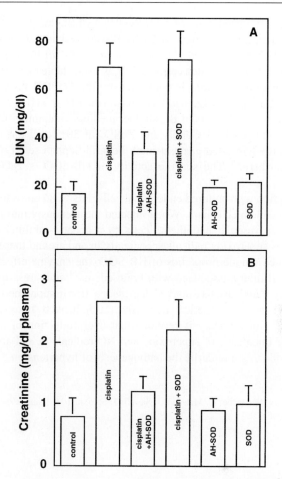

FIG. 3. AH-SOD inhibits cisplatin-induced renal injury. Rats were intraperitoneally administered cisplatin (5 mg/kg). At the same time, 0.2 ml of saline or either SOD or AH-SOD (5 mg/kg) was intravenously injected. On day 5, plasma levels of BUN and creatinine were determined.

A nucleotide fragment encoding a heparin-binding domain similar to that of extracellular SOD[20] with a stop codon and restriction sites of Saw3AI and SalI is constructed by annealing four synthetic DNAs: 5'-GATCTGCGGGCCCGGGCTCT GGGAGCGCCAGGCGCGGGAGCACT-3' (A), 5'-TCTTGCGCTCTGAGTGC TCCCGCGCCTGGCGCTCCCAGAGCCCGGGCCCGCA-3' (B), 5'-CAGAGC GCAAGAAGCGGCGGCGCGAGAGCGAGTGCAAGGCCGCCTGAG-3' (C), and 5'-TCGACTCAGGCGGCCTTGCACTCGCTCTCGCGCCGCCGTC-3' (D).

[20] M. Inoue, N. Watanabe, J. Sasaki, Y. Tanaka, and T. Amachi, *FEBS Lett.* **269**, 89 (1990).

The fragments are ligated with a 4.5-kb BamHI–SalI fragment of pBRSOD1, and the constructed plasmid is designated pBRHBSOD. A 0.549-kb EcoRI–SalI fragment of pBRHBSOD is ligated with an 8.0-kb EcoRI–SalI fragment in yeast expression vector pYHBS1, which carries *TRP1* as a selectable marker for yeast. The HB-SOD gene in the plasmid is controlled by the G3PDH promoter in yeast. Yeast strain EH13-15 (*Matα, trp1*) is transformed by pYHBS1. The transformants are cultured in 30 liters of Burkholder medium containing 0.5% Casamino acids and 0.1 m$M$ each of $CuSO_4$ and $ZnSO_4$ at 30° for 2 days. HB-SOD is purified from the harvested yeast, using a heparin-Sepharose CL-6B column as described elsewhere.[19] The specific activity of HB-SOD is 2800 units/mg of protein.

On incubation with cultured endothelial cells, HB-SOD binds to the cells by a heparin-inhibitable mechanism. When injected intravenously into rats, HB-SOD rapidly binds to vascular endothelial cells by a heparin-inhibitable mechanism and dismutates superoxide radicals specifically in and around these cells *in vivo*. Because of such unique properties of HB-SOD, the enzyme effectively inhibits endothelial cell injury associated with brain edema[20] and stress-induced gastric mucosal injury,[20] and reveals a marked depressor action in spontaneously hypertensive rats as well as in nongenetic hypertensive rats induced by deoxycorticosterone and sodium.[21] Kinetic analysis revealed that overproduction of the superoxide radical and/or imbalance of superoxide and NO radicals in and around vascular endothelial cells might underlie the pathogenesis of hypertension.

[21] K. Nakazono, N. Watanabe, K. Matsuno, J. Sasaki, T. Sato, and M. Inoue, *Proc. Natl. Acad. Sci. U.S.A.* **88**, 10045 (1991).

# [36] *In Vitro* Quantitation of Biological Superoxide and Hydrogen Peroxide Generation

*By* KEVIN R. MESSNER and JAMES A. IMLAY

A central issue in understanding the role of superoxide in biological systems is the elucidation of sources of this species. Indeed, one of the initial objections to McCord and Fridovich's original proposal for a biological role for $O_2^-$, and for superoxide dismutase as a protective agent,[1] was that monovalently reduced oxygen could not be formed in significant quantities within the constraints of

[1] J. M. McCord and I. Fridovich, *J. Biol. Chem.* **244**, 6049 (1969).

biological chemistry. Whereas the relevance of $O_2^-$ in biology has been well established, progress in establishing its sources in enzymes, cells, and tissues has been slow. One difficulty arises from the inability of $O_2^-$ to cross membranes at physiological pH. Because of this physical limitation, detector reagents outside the membrane-enclosed space will not successfully detect internally generated $O_2^-$. In such cases, measurement of hydrogen peroxide, which arises as the product of $O_2^-$ dismutation, can be used as a proxy for the latter species. Peroxide is much more stable than $O_2^-$; hence it can accumulate to readily detectable levels. In addition, $H_2O_2$ diffuses readily through lipid and can provide an estimate of reactive oxygen generation, where the site of generation is within a membrane-enclosed compartment. These aspects can make $H_2O_2$ detection a more facile means of evaluating oxidant generation from whole cell systems. Also, because some sources of $O_2^-$ in addition directly generate $H_2O_2$, assay of both species can be of use in mechanistic investigations of these sources.

## Measurement of Superoxide

Progress in evaluation of $O_2^-$ generation has also been substantially hindered by technical difficulties with many of the detection systems employed to assay $O_2^-$. This is particularly true when quantitative rigor in the rate of production is needed. The lucigenin chemiluminescence method has been the subject of considerable controversy, with several reports that this compound is unreliable as an indicator of $O_2^-$. Lucigenin must first be reduced monovalently (by the experimental sample) to a radical species, which is then able to react with $O_2^-$, producing light.[2,3] Thus the experimental material must be able to reduce lucigenin directly to allow $O_2^-$ detection to occur. Moreover, this radical species is also reactive with molecular oxygen, producing some artifactual $O_2^-$. Use of nitroblue tetrazolium as an indicator of $O_2^-$ is similarly problematic, because this compound is also reduced in two monovalent steps, and the intermediate radical leads to spurious $O_2^-$ generation.[4] Hence both assays can lead to $O_2^-$ generation where it would not otherwise occur, and hence they are less useful when the $O_2^-$ flux from the experimental sample is small.

The compound hydroethidine fluoresces on oxidation by $O_2^-$ and has been employed as a nonquantitative detector of $O_2^-$ production *in vitro* and *in vivo*. The utility of hydroethidine for precise quantitation is complicated by its ability to catalyze the dismutation of $O_2^-$.[5] Estimation of $O_2^-$ by epinephrine oxidation to adrenochrome[6] generally correlates in a predictable manner with other methods,

---

[2] S. I. Liochev and I. Fridovich, *Arch. Biochem. Biophys.* **337,** 115 (1997).
[3] S. I. Liochev and I. Fridovich, *Free Radic. Biol. Med.* **25,** 926 (1998).
[4] S. I. Liochev and I. Fridovich, *Arch. Biochem. Biophys.* **318,** 408 (1995).
[5] L. Benov, L. Sztejnberg, and I. Fridovich, *Free Radic. Biol. Med.* **25,** 826 (1998).
[6] H. P. Misra and I. Fridovich, *J. Biol. Chem.* **247,** 3170 (1972).

such as cytochrome $c$ reduction. Some caution is still necessary, as in some cases the rate of epinephrine oxidation disagrees with the rates seen using other assays. The four-electron oxidation occurs by a complex, pH-dependent chain reaction involving $O_2^-$; a second oxidation process, independent of $O_2^-$, may account for some of the variations seen.[6]

Spin traps such as 5,5-dimethyl-1-pyrroline-$N$-oxide (DMPO)[7] have been used with considerable success, although precautions are necessary.[8,9] High concentrations (often 100 m$M$) of DMPO are necessary to detect $O_2^-$ due to the low rate constant of adduct formation; hence, effects of the spin trap on the experimental material must be considered. Further, nitrones such as DMPO can be readily reduced or oxidized, particularly in the presence of metal ions, leading to false signals. The half-life of the DMPO–OOH adduct is ~60 sec at pH 7, limiting the quantitative utility of the method. The limited availability and cumbersomeness of electron paramagnetic resonance (EPR) instrumentation may also pose practical impediments to use of this technique for extensive quantitative analysis.

Assay of Superoxide by Cytochrome $c$

Because of these concerns with various assay methods we have limited our consideration of superoxide generation mainly to *in vitro* systems, where the interaction of the detection system with cellular components is more controllable. Modified versions of the cytochrome $c$ method,[1] previously considered in this series,[10,11] remain the most facile and quantitatively reliable assays of *in vitro* $O_2^-$ formation. Superoxide reduces ferricytochrome $c$ to give the intensely red-colored ferrocytochrome $c$, providing a straightforward means to detect $O_2^-$ generation continuously in many experimental systems. Because enzymes that transfer electrons to $O_2$ also tend to transfer them directly to other acceptors, sometimes including cytochrome $c$, saturating amounts of SOD are used in a reference reaction to determine any such $O_2^-$-independent, direct reduction of cytochrome $c$ by the sample. This method has several advantages that make it the first choice in investigations of $O_2^-$ generation. The ferrocytochrome product is sufficiently stable for the purposes of a typical steady-state experiment. Any SOD-independent, direct cytochrome $c$ reduction can be minimized by reducing the concentration of the indicator, without sacrificing accuracy. Moreover, superoxide is not generated by the assay system, so there is a direct 1 : 1 relationship between $O_2^-$ produced by the experimental sample and the amount of cytochrome $c$ reduced.

[7] E. Finkelstein, G. M. Rosen, and E. J. Rauckmann, *J. Am. Chem. Soc.* **102,** 4994 (1980).
[8] G. M. Rosen and E. Rauckmann, *Methods Enzymol.* **105,** 198 (1984).
[9] G. R. Buettner and R. P. Mason, *Method Enzymol.* **186,** 127 (1990).
[10] M. J. Green and H. A. O. Hill, *Methods Enzymol.* **105,** 3 (1984).
[11] A. Boveris, *Methods Enzymol.* **105,** 429 (1984).

*Assay*

The assay is typically carried out in 50 m$M$ potassium phosphate buffer at pH 7.8, although other buffers can readily be used. A 1 m$M$ stock solution of bovine ferricytochrome $c$ (Sigma, St. Louis, MO) is prepared in the same buffer. This stock solution can be frozen in aliquots and stored at $-70°$. The precise concentration in the stock solution can be determined by adding a few grains of sodium hydrosulfite (dithionite) to 10 $\mu$l of the stock in 1 ml of buffer. Using a $\Delta\varepsilon_{red-ox}$ of 21,000 $M^{-1}$ cm$^{-1}$, the actual concentration of oxidized cytochrome $c$ in the stock is determined and an appropriate volume used to give 10 $\mu M$ in the reaction mixture. A total reaction volume of 1 or 3 ml is convenient. A parallel reaction is set up to contain SOD at 8 U/ml. The reaction is started by adding the experimental sample and its substrate, and the increase in absorbance at 550 nm is recorded continuously, generally for 3–10 min. The rate of cytochrome $c$ reduction for the two reactions is calculated by using the above-described extinction coefficient. The signal from the SOD-containing reaction is subtracted from that of the reaction without SOD to obtain the rate of cytochrome $c$ reduction due to $O_2^-$ (Fig. 1). Rates of superoxide generation as low as 20 nmol/min can be detected by the method.

*Precautions and Suggestions*

It must be stressed that because $O_2^-$ does not effectively cross biological membranes, this method is not useful for measurement of $O_2^-$ generated inside intact cells or membrane-bound fractions. For the cytochrome $c$ assay, assurance must be provided that the sample is devoid of SOD activity if an accurate estimation of $O_2^-$ is to be made. If the sample contains cytochrome $c$ reductase or oxidase activities, they must be inhibited or otherwise accounted for. Measurement of $O_2^-$ formation by xanthine–xanthine oxidase,[1] in the presence and absence of the experimental sample, will reveal such interfering activities. As in all methods of $O_2^-$ detection, the SOD control is vital, as in some cases the rate of direct reduction of cytochrome $c$ is considerable or even predominant. The reaction of $O_2^-$ with ferricytochrome $c$ has a high rate constant[12] and is not highly dependent on the cytochrome $c$ concentration, whereas direct reduction of cytochrome $c$ by the sample is typically directly proportional to the concentration of cytochrome $c$. Hence, if direct reduction is a substantial portion of the total, this background may be minimized by reducing the concentration of detector molecule to 5 or 2 $\mu M$. Acetylated cytochrome $c$ can also be prepared and used where direct reduction of native cytochrome $c$ remains problematic.[13] Failing this, alternative methods such as $Fe^{3+}$-ADP reduction, visualized by ferrene, must be employed.[14] The use of

---

[12] E. J. Land and A. J. Swallow, *Arch. Biochem. Biophys.* **145**, 365 (1971).
[13] A. Azzi, C. Montecucco, and C. Richter, *Biochem. Biophys. Res. Commun.* **65**, 597 (1975).
[14] K. R. Messner and J. A. Imlay, *J. Biol. Chem.* **274**, 10119 (1999).

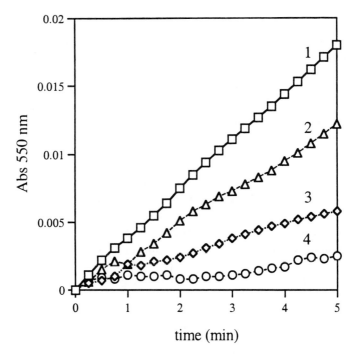

FIG. 1. Time course of cytochrome $c$ reduction by $E.$ $coli$ NADH dehydrogenase II. The rate of $O_2^-$ generation is determined by subtraction of the rate when SOD is present from the rate without SOD. 1, Complete reaction without SOD; 2, calculated cyt $c$ reduction due to $O_2^-$; 3, complete reaction with SOD present; 4, reaction lacking NADH dehydrogenase II.

the respiratory inhibitor cyanide in the assay is limited to lower temperatures; at 37°, cyanide rapidly modifies the cytochrome $c$ and effectively removes it from participation in the assay. Where cyanide is used, MnSOD (Sigma) should be used in the control reactions as Cu,ZnSOD is also inhibited by this compound.

Measurement of Hydrogen Peroxide

Several methods of assay for $H_2O_2$ have been used, many using horseradish peroxidase (HRP) and one of several dye compounds. These dyes are oxidized by HRP with concomitant reduction of $H_2O_2$ to water. The oxidized dye has an altered absorbance or fluorescence that can be monitored spectrophotometrically. The utility of these dyes in measurement of $H_2O_2$ varies considerably. The dye $o$-dianisidine[15] is often convenient, with little autoxidation and a moderately stable increase in absorbance at 460 nm on oxidation. One drawback of the method is its

[15] H. P. Misra and I. Fridovich, $Anal.$ $Biochem.$ **79,** 553 (1977).

sensitivity, which is limited to about 200 n$M$ H$_2$O$_2$ under ideal conditions. In addition, background absorbance and light scattering from the experimental material can impose limits on the accuracy of the method. Consequently, fluorescent dyes with greater sensitivity and fewer background restrictions are often employed. Scopoletin is probably the most widely used of the HRP indicator compounds, its oxidation leading to a decrease in fluorescence.[16] We have noted, however, that this compound spontaneously generates some H$_2$O$_2$, as revealed by a control reaction containing catalase. An additional problem is that the assay monitors a loss of fluorescence from a relatively high background. Hence, the sensitivity and range of the assay are still limited. The compound dichlorodihydrofluorescein[17] gives an increase in fluorescence on oxidation, but its use is also problematic because it is also a source of H$_2$O$_2$ due to spontaneous oxidation.[18] In addition, the dye is oxidized univalently by HRP, and this resulting radical can further react with O$_2$ to give artifactual O$_2^-$ generation and H$_2$O$_2$ on dismutation.

## Assay of Hydrogen Peroxide by Horseradish Peroxidase–Amplex Red

The compound Amplex Red ($N$-acetyl-3, 7-dihydroxyphenoxazine) is oxidized to resorufin in the presence of HRP–H$_2$O$_2$, giving a fluorescence peak of 587 nm when excited at 563 nm. This is the basis of a sensitive H$_2$O$_2$ assay.[19] Because the assay measures an increase in fluorescence, a better signal-to-noise ratio can be achieved than with scopoletin. Autoxidation of the compound is reasonably slow and little H$_2$O$_2$ generation by the assay system is seen.

*Assay*

Amplex Red (Molecular Probes, Eugene, OR) is dissolved in dimethyl sulfoxide (DMSO) to give a concentration of 1.25 mg/ml. This stock can be frozen in aliquots and stored at $-70°$ for several months. A 0.8-ml aliquot of the stock is mixed with 18 ml of 50 m$M$ potassium phosphate, pH 7.8, and kept shielded from light; this preparation is good for several hours. Horseradish peroxidase (type II; Sigma) is made in 18 ml of the same buffer to 20 $\mu$g/ml and kept on ice. Stock H$_2$O$_2$ is diluted in the same buffer to give approximately 10 m$M$; the absorbance of this solution is measured at 240 nm and the exact concentration determined with an $\varepsilon$ of 43.6 $M^{-1}$ cm$^{-1}$. For the standard curve, 1-ml aliquots of H$_2$O$_2$ standards are mixed with 0.5 ml of the Amplex Red solution and 0.5 ml of HRP. The fluorescence of this mixture can be read within 30 sec. Prolonged coincubation of the dye and HRP leads to autoxidation of Amplex Red, as noted for other dyes; hence, the two

[16] A. Boveris, E. Martino, and A. O. M. Stoppani, *Anal. Biochem.* **80,** 145 (1977).
[17] A. S. Keston and R. Brandt, *Anal. Biochem.* **11,** 1 (1965).
[18] C. Rota, C. F. Chignell, and R. P. Mason, *Free Radic. Biol. Med.* **27,** 873 (1999).
[19] M. Zhou, Z. Diwu, N. Panchuk-Voloshina, and R. Haugland, *Anal. Biochem.* **253,** 162 (1997).

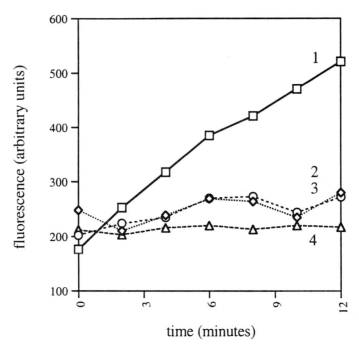

FIG. 2. Time course of $H_2O_2$ generation by *E. coli* fumarate reductase, determined by fluorescence of Amplex Red–horseradish peroxidase. The assay detects $H_2O_2$ generated by dismutation of $O_2^-$, as well as $H_2O_2$ made directly by the source material. 1, Complete reaction without catalase; 2, reaction lacking succinate; 3, reaction lacking fumarate reductase; 4, complete reaction with catalase present.

solutions are kept separate from each other until they are mixed with the $H_2O_2$ sample. The concentration of HRP is kept to a minimum for the same reason. We use a Shimadzu RF-Mini 150 fluorometer fitted with fluorescence filters with an excitation cutoff of <520 nm and emission cutoff of <620 nm. The standard curve should be linear up to approximately 5 $\mu M$ $H_2O_2$. The detection limit using this method is approximately 50 n$M$.

For peroxide-producing reactions, the subject materials and substrates are mixed as needed. SOD (8 units) should be included in all reactions to ensure complete detection of superoxide-derived peroxide; in addition, some dyes are affected by the presence of $O_2^-$.[20] At intervals 1-ml samples are removed from this reaction mixture and assayed for $H_2O_2$ as described above. At least two time points are needed, and the difference in fluorescence is measured to subtract any background from the reaction mixture. Figure 2 shows an assay of production of

[20] A. J. Kettle, A. C. Carr, and C. C. Winterbourn, *Free Radic. Biol. Med.* **17,** 161 (1994).

$H_2O_2$ from fumarate reductase (FRD) of *E. coli*, using succinate as the enzyme substrate. The change in peroxide content is determined by reference of the reaction fluorescence to the standard curve.

*Precautions*

Our laboratory has used this assay successfully for isolated enzymes, respiratory membrane vesicles, and whole cells. As with the presence of SOD in assays of superoxide, peroxide-degrading activities such as catalase and peroxidases must be absent from the experimental material, whether by biochemical or genetic means. The absence of such activities should be confirmed by demonstrating the quantitative detection of a spike of $H_2O_2$ of a similar quantity to that being measured. The separation of the assay protocol into $H_2O_2$ accumulation and detection steps is necessary; in all HRP–dye assays, if these detection components are present during peroxide generation the observed $H_2O_2$ generation rate can be misestimated, compared with a discontinuous assay.[21] Care must be taken that experimental substrates or their products do not compete with the dye in the peroxidase reaction. Both NADH and urate (the product of the xanthine oxidase reaction) present this difficulty. NADH can be removed from samples by adding catalytic amounts of respiratory membranes.[14] In addition, low levels of NADH or NAD ($\sim 10\ \mu M$) in assay samples give rise to artifactual $H_2O_2$ signals. These signals increase with the time of dye development. It is recommended that if NADH is used as an experimental substrate, measurements be made at several concentrations of NADH to detect such artifacts.

## Parallel Estimation of Superoxide and Peroxide

It is sometimes useful to distinguish between $H_2O_2$ generated as a result of $O_2^-$ dismutation, and $H_2O_2$ generated directly from the experimental material. If quantitative assays of both can be accomplished under like conditions, the rate of "direct" $H_2O_2$ generation should be equal to the total measured rate of $H_2O_2$ generation (with SOD present) minus one-half the rate of $O_2^-$ generation.

## Acknowledgment

This work was supported by GM49640 from the National Institutes of Health.

---

[21] K. Stanick and H. Nohl, *Biochim. Biophys. Acta* **1413**, 70 (1999).

# Author Index

Numbers in parentheses are footnote reference numbers and indicate that an author's work is referred to although the name is not cited in the text.

## A

Aasa, R., 75
Abbas, R., 320
Abdul, T. H., 5
Abdul-Tehrani, H., 5
Abel, J. M., 240
Abernethy, J. L., 49, 50(4)
Abo, A., 313(8), 314
Abramsson, L., 74
Abreu, I. A., 248, 250(32), 251(32), 255(32), 256(32)
Abushamaa, A. M., 182
Adachi, T., 339
Adams, M. D., 244(13; 16), 245, 245(16), 246
Adams, M. W. W., 124, 125(10), 245(22), 246, 247(22), 248(22)
Aguet, M., 258, 261(4)
Aguzzi, A., 258, 261(4)
Ahuja, S., 337
Ajello, M., 165
Akiyama, M., 131
Alaoui-Youseffi, A., 323
Aldredge, T., 244(14), 245
Alexander, D. D., 181, 182(4), 192(12), 193
Alexander, J. E., 148
Alexandre, A., 273
Allen, J. M., 191
Allen, R. M., 10, 11(8)
Allison, S. A., 45, 46, 46(15), 47(23)
Altschul, S. F., 82(13), 83
Alvarez, B., 23, 24, 25(8), 32, 33(35), 36(8)
Alvarez, S., 280
Amachi, T., 351, 353, 354(19; 20)
Ammendolia, M. G., 165
Andersen, P. M., 74
Anderson, A. J., 148

Anderson, K. L., 213
Anderson, S. D., 213
Ando, Y., 347, 350(8)
Andorno, E., 333
Andrew, P. W., 148
Andrew, P. Y., 124, 125(10)
Andrews, B., 49, 59(6)
Andrews, N. C., 297
Andrews, S. C., 5
Angerer, L. M., 204
Angerer, R. C., 204
Angerhofer, A., 66, 71(26)
Anscher, M. S., 226, 233(17b)
Antonelli, M., 189
Anttila, M., 320
Antunes, F., 271, 281
Aoki, K.-i., 244(18; 19), 245(18; 19), 246
Appella, E., 130, 139(7)
Ara, J., 24, 34(7)
Arceneaux, J. E., 84
Archer, M., 248
Archibald, A. L., 191
Archibald, F. S., 80(7), 81, 155
Arias, J., 180
Arnelle, D. R., 102
Arnold, L. D., 49, 59(6)
Arnold, R. S., 21
Aronovitch, J., 240
Artiach, P., 244(13; 20), 245, 245(20), 246
Aruffo, A., 313
Asahi, M., 336
Asayama, K., 114
Ascenso, C., 124, 243, 245(23; 24), 246, 247(23; 24), 248(23; 24), 250(23; 24), 251(23), 254(23; 24), 255(23; 24), 256(23), 257(23; 24)
Ashwell, G., 350
Aston, K. W., 223, 235
Attiga, F. A., 320
Auclair, C., 347

Augusto, O., 102
Aust, S. D., 3
Autenried, P., 258, 261(4)
Avraham, K. B., 181, 182(3), 192, 197, 334
Awad, J. A., 334
Awai, M., 347, 350(8)
Axelrod, A. E., 112
Azizova, O. A., 3, 6
Azzi, A., 271, 275(22), 276, 283, 357

## B

Baba, S.-i., 244(18; 19), 245(18; 19), 246
Babior, B. M., 333
Bacquet, R. J., 46, 47(23)
Badger, J. H., 244(13), 245
Bagg, A., 3
Baker, E. N., 69, 80, 83(6), 87, 89
Baker, H. M., 80, 83(6)
Baker, K., 333
Bakker, A. C., 320
Baldino, F., 206
Baldwin, M. A., 259
Ball, H. L., 259
Balsano, C., 313
Banci, L., 47, 64
Banemann, A., 145(5), 146
Bannister, J. V., 38, 70, 81, 83(9; 10)
Bannister, W. H., 38, 70, 81, 83(9; 10)
Barbeito, L., 24, 54
Baret, A., 321
Baron, D. A., 235
Barra, D., 81, 83(9), 116
Barrett, W. C., 314
Barth, G., 116
Basava, A., 337
Bashirzadeh, R., 244(14), 245
Batinic-Haberle, I., 25, 26, 27, 27(21), 28, 28(15; 21), 223, 224(2; 3), 225, 225(2; 3), 226, 226(2; 14), 227, 227(2), 228, 228(2; 3; 19; 22; 33), 229, 229(2; 3; 20), 230(2; 3), 231, 231(2; 19), 232, 232(2; 19), 233, 233(2; 16–18)
Batra, P. P., 272, 275(14)
Batson, D., 188
Battistoni, A., 44, 48, 48(10), 49, 49(12), 50, 51, 51(17), 53(12; 16), 54(12; 16), 55(12), 56(12), 59, 59(12; 16), 60(16), 61(16), 84,

115, 155, 157, 158(18), 159(18), 161, 161(18), 165
Baudhuin, J., 106, 107(4), 113(4), 114(4)
Bauer, A. W., 162
Baumler, A. J., 155
Bay, E. D., 64
Bayne, A.-C. V., 216, 287
Baynes, J., 350
Beal, M. F., 129, 339
Bearson, S., 155
Beauchamp, C., 62, 92, 159, 265, 287, 292(2)
Beckman, J. S., 10, 24, 25, 26, 27(21a), 32, 32(4), 37, 54, 101, 103, 104, 115
Beckman, T. W., 10, 24, 101
Beebe, A. M., 194
Beem, K. M., 50
Beer, D. G., 320
Behar, D., 62
Behringer, R. R., 191
Beinert, H., 13
Bell, E. T., 290
Bell, J. E., 290
Benes, J. J., 214
Bennett, B., 12
Benov, L., 26, 27, 27(21), 28(21), 151, 155, 160(3), 161(3), 223, 224(2), 225(2), 226(2), 227(2), 228(2), 229(2), 230(2), 231(2), 232(2), 233(2), 277, 355
Bensch, K. G., 214
Benson, L. M., 124, 245(23), 246, 247(23), 248(23), 250(23), 251(23), 254(23), 255(23), 256(23), 257(23)
Benson, P. F., 334
Benz, W. K., 216
Berche, P., 81, 82(11)
Berg, S. P., 277
Berger, C., 130, 211, 340
Berman, H. M., 45
Bernier, M., 241
Bernstein, Y., 241, 293, 334, 336(8)
Beyaert, R., 320
Beyer, W. F., 33, 64, 68(13), 72
Beyer, W. F., Jr., 87, 162, 165(34)
Beyer, W. J., 235
Bezkorovzainy, A., 3
Bhat, T. N., 45
Bhattacharjee, J. K., 168
Bhunia, A. K., 314
Bianchi, A., 140

Biasi, F., 333
Bibb, M. J., 91
Bielawska, A. W., 313
Bielski, B. H. J., 40
Biemond, P., 337
Biliaderis, C. G., 53(25), 54
Bilinski, T., 168
Bill, E., 5
Bindokas, V. P., 277
Bissinger, P. H., 170
Bjelle, A., 74
Bjerrum, M. J., 70
Black, S. C., 235
Blake, C. C. F., 64, 81, 83(8; 9), 90
Blake, J. A., 244(16), 245(16), 246
Blakely, D., 244(14), 245
Blondeau, R., 91
Bloodsworth, A., 233
Boggess, K. A., 74
Boiron, P., 91
Bolognesi, M., 38, 45, 48, 50, 59
Bomford, A., 3
Bonaccorsi di Patti, M. C., 49, 50, 50(10), 53(10; 16), 54(16), 59(16), 60(16), 61(16)
Bonini, M. G., 102
Bonner, F. T., 105
Borchelt, D. R., 192(10), 193
Borders, C. L., 70, 110
Bordo, D., 38, 48, 50, 59
Borg, L. A. H., 310
Borgstahl, G. E. O., 64, 68(13), 69
Borodovsky, M., 244(16), 245(16), 246
Bors, W., 236
Bosco, M., 333
Bose, K., 91, 93(11), 95(11), 96(11), 98(11), 99(11)
Bose, S. K., 85, 335, 336(11)
Boss, J. M., 341
Bossa, F., 81, 83(9), 116
Bossi, L., 155
Botstein, D., 194
Bottaro, G., 48
Bouchard, L., 4, 150
Boulet, A. M., 214
Boulianne, G. L., 214, 218, 291
Bounds, P. L., 115, 117, 122
Bourassa, J. L., 26, 27(21b)
Bourne, P. E., 45
Bourne, Y., 49(11), 50, 53(11), 59(11)

Bouton, C., 338
Boveris, A., 271, 275(4; 23), 276, 279(23), 280, 281, 282, 282(5; 6), 283, 283(5), 284, 284(6), 285(10; 14–17), 342, 356, 359
Bowery, V. W., 240
Bowman, C., 244(13; 20), 245, 245(20), 246
Boyle, J. A., 64, 139
Bradford, M. M., 117
Bradley, B., 226, 233(17a)
Brady, T. C., 339
Braga, V., 248, 254(31), 255(31)
Bramer, S. L., 320
Brandt, R., 359
Brandts, J. F., 115, 116
Bratt, P. J., 66, 71(26)
Bray, R. C., 38, 72
Bredesen, D. E., 115
Brenner, A. J., 259, 264(12)
Brenner, D. A., 313
Brenner, M. K., 140
Brenowitz, M., 80
Breslauer, K. J., 49, 50(7), 53(7), 54(7), 59(7)
Breuer, W., 4
Brigelius, R., 236
Brinster, R. L., 191
Brito, C., 24, 34(6)
Britt, R. D., 90
Briviba, K., 102
Bronson, R. T., 193
Browder, N. J., 320
Brown, D. R., 258, 259, 261(2; 5; 6), 264, 265(7)
Brown, R. H., Jr., 129, 185, 339
Brown, T., 350
Bruijn, L. I., 213
Brumlik, M. J., 244(9; 11), 245
Bruschi, M., 244(9), 245, 250, 251(35), 253
Brusov, O. S., 106, 107(3)
Bruton, C. J., 91
Bry, C., 181, 182(3), 192, 334
Buchmeier, N. A., 164
Budd, S. L., 277
Büeler, H., 258, 261(4)
Buettner, G. R., 131, 340, 356
Bugnon, P., 24, 102
Bulkley, G. B., 333
Bull, C., 62, 63, 65(9), 67(9; 19), 68(9), 73(19), 89

Bult, C. J., 244(16), 245(16), 246
Burger, A. R., 116
Burke, R. E., 189
Burlingame, A. L., 259
Burnette, B., 272, 275(14)
Burr, I. A., 130, 310
Burr, I. M., 114
Burton, D. R., 259
Bush, D., 244(14), 245
Bush, K. M., 24
Bustamante, J., 241, 272, 273(11), 280, 283, 285(14; 15)
Butler, A., 49, 50(7), 53(7), 54(7), 59(7)
Butler, J., 39, 41(4), 232, 271
Buyukafsar, K., 104
Byers, B. R., 84
Bylund, F. A., 240

## C

Cabantchik, Z. I., 4
Cabelli, D. E., 40, 47, 51, 64, 66, 72, 72(27), 73(42), 91, 93(11), 95(11), 96(11), 98(11), 99(11), 224, 229, 248, 250(32), 251(32), 255(27; 32), 256(32), 257(27)
Cabrito, I., 245(24), 246, 247(24), 248(24), 250(24), 251(24), 254(24), 255(24), 257(24)
Cadenas, E., 271, 272, 274, 274(10), 275(4), 280, 281, 282, 282(6), 284, 284(6), 342
Cadet, J. L., 182, 183, 185(10)
Café, C., 115
Cagliero, E., 310
Calabrese, L., 38, 44, 45(9), 47, 49, 50(10), 53(10), 59, 72, 116
Calagui, B., 192
Caliendo, J., 181, 182(4), 192(12), 193
Calson, E., 183, 185(10)
Calvi, C. C., 226, 233(18)
Cambria, M. T., 44, 48(10)
Cammack, R., 250, 251(35)
Campbell, G., 155
Campbell, K. A., 90
Canvin, J. R., 156, 163(14)
Cao, J., 193
Caparon, M. G., 148
Capo, C., 44, 45(9), 59, 161

Carbone, G. M., 199
Card, T. D., 74(13), 75
Carita, J., 248, 250(32), 251(32), 255(32), 256(32)
Carlin, G., 240
Carlini, P., 49, 50(10), 53(10)
Carlioz, A., 86, 145(6), 146, 149(6), 150(6), 162
Carlson, E. J., 129(2), 130, 182, 187, 192, 193, 194, 196, 199(20), 211, 340
Carlsson, L. M., 74, 75, 75(3), 77(3), 130, 339
Carr, A. C., 360
Carri, M. T., 48, 50, 51, 53(16), 54(16), 59, 59(16), 60(16), 61(16), 84, 115
Carrie, K., 66
Carrondo, M. A., 245(21), 246, 247(21), 248, 248(21)
Carter, R., 337
Cartwright, J., Jr., 130, 193, 211(18), 340
Caruso, A., 244(14), 245
Casareno, R. L., 167
Caserini, C., 241
Caska, B., 340
Castilho, R. F., 277
Castro, L., 11(23), 12, 13, 13(23), 22(27), 25, 32(13), 34(13), 36(13), 37(13), 230
Cathraw, S., 153
Caugant, D. A., 200
Cayota, A. M., 13, 22(27), 24, 34(6)
Cervoni, L., 49(12), 50, 53(12), 54(12), 55(12), 56(12), 59(12)
Chan, P. H., 129(2), 130, 192, 193, 194, 211, 301, 340
Chan, P. K., 130
Chance, B., 271, 281, 282, 282(5), 283(5), 342
Chang, B., 351
Chang, E., 168
Chang, M. S., 293, 298, 299(13; 14; 19–21; 23), 302
Chang, Y. S., 5
Charloteaux-Waters, M., 3
Charon, M. H., 91
Chater, K. F., 91
Chatterjee, S., 314
Chelack, W. S., 327
Chen, C., 180
Chen, F. C., 26, 27(21b)
Chen, J., 10, 24, 37, 101, 158, 159, 159(21)
Chen, L., 243, 245(8), 247(8), 248(8), 250(8), 251(8), 254(8)

Chen, S. F., 129(2), 130, 192, 193, 194
Chen, T., 158, 159(21)
Chen, W., 181, 182(4), 192(12), 193
Chen, X., 66
Chen, Y., 199
Cheng, I. C., 159
Cheng, S. H., 26, 27(21b)
Chesselet, M. F., 206
Chiapotto, E., 333
Chiappino, I., 333
Chiaraluce, R., 49
Chiariello, M., 313(7), 314
Chidambaram, M., 244(20), 245(20), 246
Chignell, C. F., 359
Chin, S. M., 3(17), 4
Chiu, A. Y., 181, 182(4)
Cho, Y.-H., 100
Chock, B., 310
Chock, P. B., 130, 240, 314
Choi, J.-H., 100
Choi, T., 191
Choudhury, S. B., 91, 93(11), 95(11), 96(11), 98(11), 99(11)
Chow, C. P., 320
Chowdhary, P., 235, 242(21)
Christi, M. A., 259
Christianson, D. W., 26
Christodoulou, D., 101, 104(2), 105(2)
Chui, A. Y., 192(12), 193
Chumley, P. H., 226, 233, 233(16)
Chun, J., 91
Chung, A. B., 21
Chung, H.-J., 92, 98(19), 99(19), 100, 100(19)
Church, G. M., 244(14), 245
Church, S. L., 306
Claire Kennedy, M., 229
Claret, F. X., 313(8), 314
Clark, A. J., 191
Clark, T. W., 44
Clayton, P. E., 223
Clayton, R. A., 244(13; 15; 16; 20), 245, 245(16; 20), 246
Clements, M. O., 153
Cleveland, D. W., 192(10), 193, 213
Clive, C., 259, 264, 265(7)
Cockle, S. A., 72
Coelho, A. V., 124, 245(21), 246, 247(21), 248(21)
Coffman, R. L., 194

Cohen, F. E., 259
Commack, R., 106, 107(5)
Conrad, C. C., 19
Consalvi, R., 49
Cook, J. A., 101, 104(2), 105(2)
Cook, J. D., 338
Cook, R., 244(14), 245
Cooper, C. E., 102
Copeland, N. G., 192(10), 193
Copin, J. C., 192
Cordingley, F. T., 140
Coronado, M. A., 124
Coskun, P., 19
Coso, O. A., 313(7), 314
Costa, L. E., 281
Costantino, G., 10, 12(18), 13(18), 22, 22(18)
Costanzo, A., 313
Cotton, M. D., 244(13; 15; 16; 20), 245, 245(16; 20), 246
Cottrell, B., 19
Coulter, E. D., 229, 248, 255(27), 257(27)
Cousens, L. S., 51
Cox, K. H., 204
Crane, D., 108
Crapo, J. D., 74, 130, 182, 193, 200, 226, 233, 233(18), 243, 256(1), 339
Cregan, R., 306
Crespo, P., 313(7), 314
Crichton, R. R., 3
Cross, C. E., 25
Crouse, B. R., 124, 248, 250(29), 251(29), 253(29), 254(29)
Crow, J. P., 24, 25, 32, 37, 54, 115, 229, 233
Crumbliss, A. L., 26, 27(21), 28(21), 223, 224(2), 225(2), 226(2), 227(2), 228(2), 229(2), 230(2), 231(2), 232(2), 233(2)
Cruz, C., 226, 233(17a)
Cruz, T. F., 314
Cryer, B., 320
Culotta, V. C., 167, 168(7), 169, 171, 172(11), 301
Cunningham, D. D., 102, 103(5), 106(5)
Cunningham, K. W., 167
Curnutte, J. T., 333
Currie, M. G., 235
Curtis, J. F., 25
Cuzzocrea, S., 235
Czaja, C., 253(40), 254
Czapski, G., 24, 62, 234, 235, 236
Czuczman, M., 112

## D

Dacheux, D., 153
Dafni, H., 234
Dafni, N., 203, 293
Da Gai, R., 50, 53(16), 54(16), 59(16), 60(16), 61(16)
Dal Canto, M. C., 181, 182(4), 192(12), 193
Danciger, E., 293
D'Andrea, K. P., 244(13), 245
Daneri, F., 271
Daniels, C. J., 244(14), 245
Daniels, M., 258
Danne, M., 314
Daosukho, C., 310
D'Ari, R., 150
Darvasi, A., 194
Das, D. K., 339
Davidson, G., 91, 93(11), 95(11), 96(11), 98(11), 99(11)
Davidson, J. F., 170
Davies, K. J., 342
Davies, K. M., 163
Davies, S., 44, 45(9)
Day, B. J., 21, 226, 229, 233, 233(16; 18)
Day, C. E., 139
Day, E. D., Jr., 139
Dbaibo, G., 313
De Bilbao, F., 183, 184(9)
Debuyst, R., 241
Decker, G. L., 272, 273(11)
DeGnore, J. P., 314
DeGraff, W. G., 101, 104(2), 105(2)
De Groote, M. A., 153, 155, 162, 163(33)
Dejehet, F., 241
De Koster, B. M., 105
DeLeon, D. V., 204
Delic-Attree, I., 153
Deloughery, C., 244(14), 245
De Martino, A., 84
Deng, H. X., 115, 181, 182(4), 192(12), 193
Denicola, A., 24, 25(8), 36(8)
Desai, M., 102
DeShazer, D., 145(4), 146
Desideri, A., 38, 44, 45, 47, 47(18), 48, 48(10), 49(12), 50, 51, 53(12), 54(12; 22), 55(12), 56(12), 59, 59(12), 84, 115, 161
Despied, S., 4, 150
Deters, D., 104
Deutsch, H. F., 131

Devalaraja, M. N., 306, 310(5)
Devreese, B., 244(10), 245
de Vries, S., 271
Dewhirst, M. W., 226, 233(17b)
Dey, M. S., 193
Dickerson, J. A., 314
Dickinson, D., 214
Dickinson, J. H., 153
Diekert, K., 171, 172(11)
Dieterle, C. S., 74(12), 75
Dighiero, G., 24, 34(6)
Di Giamberdino, L., 258, 261(3)
Dijkman, J. H., 320
Di Mascio, P., 102
Dinauer, M. C., 153, 162, 163(33)
Dionisi, O., 271
Dionne, L., 130, 193, 211(18), 340
Diven, W. F., 112
Diwu, Z., 359
Dixon, M. M., 64, 67(17), 83
Dizdarolgu, M., 19
Djerassi, C., 116
Djinovic, K., 45
Djinovic-Carugo, K., 48, 50, 59
D'Mello, R. A., 148, 156, 160(16)
Dobashi, J. K., 114
Doctrow, S. R., 223, 333, 334
Dodson, R. J., 244(13; 15; 20), 245, 245(20), 246
Doetschman, T., 192
Domann, F. E., 341
Dombernowsky, P., 338
Domingo, R., Jr., 337
Donadio, P. P., 333
Donald, C. E., 105
Donaldson, D., 185
Dong, Z., 49
Donnarumma, G., 161, 165
Dorfman, L. M., 62
Dormishian, F., 337
Douce, R., 22
Doucette-Stamm, L. A., 244(14), 245
Dougall, W. C., 130, 310
Dougan, G., 156, 163(14)
Dougherty, B. A., 244(13; 16), 245, 245(16), 246
Downey, J. M., 241, 335
Drapier, J.-C., 10, 16(19), 22(19), 338
Dreyer, J.-C., 13
Driscoll, E. M., 235
Dubois, J., 244(14), 245
Dubois-Dauphin, M., 183, 184(9)

Dumont, P., 241
Dunlap, P. V., 157
Dunn, K. L., 155
Duo, C. F., 151
Duong, D. K., 67, 70(31)
Duong, M. N., 155
Duschesne, J., 293

# E

Earnshow, A., 26
Eaton, J. W., 241
Ebashi, I., 347, 350(10)
Edlund, A., 74, 75(9; 10), 77
Edlund, T., 74, 75, 75(2; 3; 9; 10), 76(18), 77, 77(3), 130, 339
Edwards, R. A., 69, 80, 83(6), 87, 89
Eggerding, F., 191
Eisen, J. A., 244(15), 245
Eiserich, J. P., 25
Eizirik, D. L., 310
Elferink, J. G., 105
Elia, A. J., 214
Eling, T. E., 25
Ellerby, L. M., 115, 224, 243
Elliott, J. L., 129, 339
Ellis, M. C., 337
Elmquist, L.-G., 74
Elroy-Stein, O., 181, 182(3), 192, 241, 334, 336(8)
Emerit, I., 323, 347
Emerit, J., 321, 322(4), 323(4–6), 324(6), 325(4; 5), 327(4)
Emerson, J. P., 229, 248, 255(27), 257(27)
Emptage, M. H., 4, 10, 12(11), 13, 13(11), 17(11), 127, 149
Endo, Y., 134, 138
Engel, P. C., 49
Engelman, R. M., 339
Engels, W. R., 216
Enghild, J. J., 74(12), 75
Engström, Å., 74, 75(2)
Epp, J., 235, 236(10)
Epstein, C. J., 19, 129(2), 130, 181, 182, 182(3), 183, 185(10), 187, 191, 192, 193, 194, 196, 199(20), 211, 334, 340
Epstein, L. B., 13, 18(25), 19(25), 20(25), 22(25), 338
Epsztejn, S., 4

Ermler, U., 91
Ernster, L., 274
Espey, M. G., 102, 105(12)
Estevez, A. G., 24, 54

# F

Faber, J. C., 3
Fahn, S., 183, 185(10), 186
Falconi, M., 38, 45, 47, 48, 51, 59
Faller, D. V., 297
Fang, F. C., 153, 155, 162, 163(33)
Faraggi, M., 225
Farr, S. B., 150
Farrant, J. L., 155, 156, 163(14)
Faucheux, B., 258, 261(3)
Faulkner, K. M., 27
Feder, J. N., 337
Fee, J. A., 62, 63, 64, 65, 65(9), 67(9; 17; 19), 68(9), 71, 73(19), 83, 87, 89, 89(32), 90, 97, 117
Feelisch, M., 101, 102, 103(5), 104(2), 105, 105(2; 12), 106(5)
Feingold, J., 323
Feng, Z., 45
Ferber, D. M., 152, 157, 163(17)
Fernandes, A., 321, 323(6), 324(6)
Fernandez, P. M., 320
Ferrans, V. J., 314
Ferrante, R. J., 129, 339
Ferrario, M., 45, 47(18)
Ferreira, M., 186
Ferrer-Sueta, G., 23, 24, 25, 25(8), 27(27), 28, 28(15), 32, 33(35), 36(8), 228, 229(20)
Fielden, E. M., 38, 65, 67(20), 72
Fiers, W., 320
Figueroa-Bossi, N., 155
Filipe, P., 321, 322(4), 323(4–6), 324(6), 325(4; 5), 327(4)
Finazzi Agró, A., 51, 54(20), 116
Finch, C. A., 338
Finch, C. E., 21
Fine, R., 45, 46(17), 47
Fink, G. R., 167
Fink, M. P., 334
Finkel, T., 314
Finkelstein, E., 238(30), 239, 356
Fisher, C. L., 47, 64
Fishman, R. A., 301

FitzGerald, L. M., 244(16), 245(16), 246
Fleischmann, R. D., 244(13–16), 245, 245(16), 246
Fleming, J. E., 214
Flickinger, A. G., 235
Flint, D. H., 4, 10, 11(8), 12(11), 13(11), 17(11), 127, 149
Flockhart, D. A., 320
Flohé, L., 33, 256, 265, 287, 292(1)
Flood, D. G., 129, 339
Flores, S. C., 85, 226, 233(17a), 340
Floris, R., 25
Floyd, R. A., 342
Folcarelli, S., 48, 49(12), 50, 51, 53(12), 54(12), 55(12), 56(12), 59(12), 165
Folz, R. J., 182
Fong, N. M., 51
Fontecave, M., 123, 124, 124(2), 125(2), 127(2), 247, 248, 248(26), 251(26; 33), 254(26; 33), 255(26; 33), 257(26; 33)
Fontecilla-Camps, J. C., 91
Forbes, W., 320
Fork, D. C., 148, 153
Forman, H. J., 110
Forman, H. S., 49, 50(2)
Forsgren, L., 74
Forsman, C., 70
Foster, J. W., 155
Foster, S. J., 153
Fox, R. A., 65, 67(20)
Frank, L., 199
Fraser, C. M., 244(13; 15; 16; 20), 245, 245(16; 20), 246
Fraser, P. E., 258, 259, 261(5)
Freeman, B. A., 10, 24, 25, 32, 32(13), 33(35), 34(13), 36(13), 37(13), 101, 226, 230, 233, 233(16)
Freitas, J., 321, 322(4), 323(4–6), 324(6), 325(4; 5), 327(4)
Frenkel, K., 102
Fretland, D. J., 235
Frey, H. E., 49, 53(9), 54(9), 59(9)
Frey, M., 91
Fridovich, I., 4, 9, 10, 12, 12(10), 13, 13(10), 14(1), 16, 17(10; 31), 18(24), 19(20; 24), 20, 20(24), 21, 21(14), 22, 22(10), 25, 26, 27, 27(21), 28, 28(15; 21), 33, 49, 50(2), 62, 65, 72, 80, 86, 87, 90, 92, 92(22), 93, 97, 103, 106, 107(1; 2; 6), 108, 109(20), 110, 116, 123, 149, 150(10), 151, 152(21),
155, 159, 160(3), 161, 161(3), 162, 163, 165(34), 180(2), 181, 200, 223, 224, 224(2), 225, 225(2), 226, 226(2; 14), 227, 227(2), 228, 228(2; 19), 229, 229(2; 20), 230(2), 231, 231(2; 19), 232(2; 19), 233(2; 16; 17b; 18), 235, 236, 237(25), 243, 251, 256(1), 265, 272, 277, 283, 287, 292(2), 293, 306, 354, 355, 358
Friedman, R. L., 145(4), 146
Friesen, A. D., 53(25), 54
Friesen, H. K., 327
Fritsch, E. F., 295, 296(9)
Fritz, H., 139
Froncisz, W., 277
Fu, C. J., 320
Fuhrmann, J. L., 244(16), 245(16), 246
Fujii, C., 244(13; 20), 245, 245(20), 246
Fujii, J., 130, 131, 138
Fujimura, T., 32, 34(32)
Fujita, T., 350
Fukuto, J. M., 101, 102, 104(2), 105(2; 12)
Fullan, A., 337
Fülöp, V., 124, 245(21), 246, 247(21), 248(21)
Funahashi, T., 244(18; 19), 245(18; 19), 246
Futenma, A., 339

## G

Gabbianelli, R., 50, 53(16), 54(16), 59(16), 60(16), 61(16), 84, 161
Gabbita, S. P., 342
Gabler, O. C. J., 259, 264(13)
Gad, N. M., 241, 335
Gadzheva, V., 234
Gaillard, J., 22
Galeotti, T., 271
Galleano, M., 283, 285(15)
Gallegos, C., 51
Gallez, B. C. D., 241
Galtieri, A., 47
Gambetti, P., 259, 264, 265
Gamson, J., 101, 104(2), 105(2)
Gangadharam, P. R., 304(24), 305
Gao, B., 85
Garban, F., 321, 322(4), 323(4), 325(4), 327(4)
Gardner, P. R., 9, 10, 12, 12(18), 13, 13(18), 18(24; 25), 19(20; 24; 25), 20, 20(24; 25), 21, 22, 22(18; 25; 26), 28, 149, 150(10), 200, 240, 338

Gargano, M., 193
Gariboldi, M. B., 241
Garland, S. A., 244(13; 20), 245, 245(20), 246
Garret, R. M., 108
Garrett, M. M., 244(15), 245
Gasnier, F., 107
Gateau-Roesch, O., 107
Geacintov, N. E., 102
Gelfand, E. W., 313
Gelinas, R. E., 191
Geller, B. L., 107, 108, 110(18), 113(9), 288
Geoghagen, N. S. M., 244(16), 245(16), 246
Gerasimov, A. M., 106, 107(3)
Gerstein, H. C., 337
Getzoff, E. D., 47, 49(11), 50, 51, 53(11), 54, 59(11; 26), 64, 90, 103, 218, 291
Ghanevati, M., 240
Ghosh, B., 114
Giard, J. C., 155
Giartosio, A., 49, 49(12), 50, 50(10), 53(10; 12; 16), 54(12; 16), 55(12), 56(12), 59(12; 16), 60(16), 61(16)
Gibson, C. M., 148
Gibson, R., 244(14), 245
Gibson, T. J., 82(15), 83
Gierse, J. K., 25
Giese, A., 258, 261(5)
Gilbert, K., 244(14), 245
Gill, D. M., 117
Gill, M. S., 223
Gill, S. R., 244(13; 15), 245
Gillespie, A. M., 192, 193, 194, 199(20)
Gilliland, G., 45
Gilmour, M. N., 200
Gilson, M. K., 46
Gingles, N., 148
Gitlin, J. D., 167
Giulivi, C., 282
Glassmith, L. L., 264
Glew, R. H., 112
Glicktein, H., 4
Glodek, A., 244(13), 245
Gnirke, A., 337
Gocayne, J. D., 244(13; 16), 245, 245(16), 246
Godinger, D., 240
Goeddel, D. V., 130, 139(6), 336
Goldek, A., 244(16), 245(16), 246
Goldstein, S., 24, 234, 235, 236
Gomes, C. M., 244(17), 245(17), 246, 248, 248(17), 254(31), 255(17; 31)

Gonzalez, A., 124, 245(21), 246, 247(21), 248(21)
Gonzalez, P. K., 334
Gonzalez-Flecha, B., 284, 285(17)
Goodfellow, B. J., 246, 250(25), 251(25), 253(25)
Goodman, M., 17, 200
Goodman, S. I., 19
Gordon, J. W., 192(11), 193
Gorman, C., 191, 296
Gorringe, A. R., 155
Gort, A. S., 157, 163(17)
Gossler, A., 192
Goto, H., 339
Goto, J. J., 115
Goubeaud, M., 91
Gow, A. J., 102
Goyal, A., 244(14), 245
Grabarse, W., 91
Graden, J. A., 115, 224, 243
Grady, R. W., 3
Graeff-Wohlleben, H., 145(5), 146
Graham, D. E., 244(13), 245
Grahame, D. A., 234, 239(5)
Gralla, E. B., 4, 5(23), 6(23), 10, 19, 21(16), 115, 168, 169, 243
Granger, D. N., 333
Grant, C. V., 90
Grant, J. W., 306
Gratton, E., 51, 54(20)
Graziano, J. H., 22
Greco, R., 161, 165
Green, M. J., 356
Greenawalt, J. W., 272, 273(11)
Greenwald, R. A., 92
Greenwood, N. N., 26
Gregory, E. M., 161
Greiner, R. A., 258, 261(4)
Griendling, K. K., 21
Grimsby, J., 274
Grisham, M. B., 101, 102, 104(2), 105(2; 12)
Grodkowski, J., 227, 228(19), 231(19), 232(19)
Groner, Y., 181, 182(3), 192, 197, 203, 241, 293, 334, 336(8)
Gross, R., 145(5), 146
Groves, J. T., 25, 26, 27(21b; 26), 28, 32
Gruer, M. J., 12
Gualco, G., 24, 34(6)
Guan, Y., 65, 66, 69, 69(21), 70(21), 71(21; 26), 72, 73(21; 42)

Guest, J. R., 5, 12
Guidot, D. M., 320
Guiso, N., 145(4; 5), 146
Gunther, M. R., 25
Guo, Q., 193
Gurney, M. E., 181, 182(4), 192(12), 193
Gutkind, J. S., 313(7), 314
Gutteridge, J. M. C., 131, 337
Guyonvarch, A., 145
Gwinn, M., 244(13; 15; 20), 245, 245(20), 246

## H

Haandrikman, A. J., 148
Habermehl, G. G., 314
Hafiz, F., 259, 264, 265(7)
Haft, D. H., 244(15), 245
Hagen, W. R., 248, 250(30), 251(30)
Hagstrom, R., 45
Hah, Y. C., 91, 92, 93(7–9), 95(7), 96(9), 98(19), 99(19), 100, 100(7; 19), 243
Haikawa, Y., 244(18; 19), 245(18; 19), 246
Hall, D. O., 106, 107(5)
Hallewell, R. A., 47, 49, 51, 53(9), 54, 54(9), 59(9; 26), 64, 68(13), 69
Halliwell, B., 25, 234, 235, 235(1), 337
Hamazaki, K., 346
Hambright, P., 26, 27(21), 28(21), 223, 224(2), 225(2), 226(2), 227, 227(2), 228(2; 19), 229(2), 230(2), 231, 231(2; 19), 232(2; 19), 233(2)
Hamer, D. H., 301
Hamilton, S. R., 333
Han, D., 271, 272, 274(10), 282
Han, H., 314
Hanevold, K., 114
Hankinson, O., 302
Hanley, M. E., 320
Hanna, M. C., 244(16), 245(16), 246
Hannun, Y. A., 313
Hardham, J. M., 244(20), 245(20), 246
Hardie, M. M., 157
Hardy, M. M., 235
Harford, J. B., 3
Harris, E. D., 259, 264(12)
Harris, S., 191
Harrison, D., 244(14), 245
Harrison, J., 37
Harrison, P. M., 5

Harrison, S. J., 275
Hartmann, H.-J., 116
Hartree, E. F., 107
Hashida, M., 350
Haskins, K., 226, 233(17a)
Hasset, D. J., 148, 153
Hässig, R., 258, 261(3)
Haswell, S. J., 259, 264, 265(7)
Hatanaka, H., 351, 354(19)
Hatch, B., 244(20), 245(20), 246
Hatchikian, E. C., 91
Hatmaker, T. L., 235, 236(10)
Haugland, R., 359
Hauptmann, N., 274
Hausinger, R. P., 253(39), 254
Hausladen, A., 10, 12(10), 13(10), 16, 17(10; 31), 22(10), 162, 165(34)
Hawkins, C., 5
Hayakawa, H., 115
Hayashi, H., 138
Hayashi, S., 214
Hazlett, K. R. O., 124, 245(23), 246, 247(23), 248(23), 250(23), 251(23), 254(23), 255(23), 256(23), 257(23)
Hearn, A. S., 63, 67(10), 68(33), 69, 72, 73(42)
Hearse, D. J., 241
Hedrick, J. L., 159
Heffetz, D., 314
Heffron, F., 164
Heidelberg, J., 244(15), 245
Heimler, I., 193
Helfand, S. L., 218
Hellner, L., 74
Hemsley, A., 108
Henderson, L., 350
Henry, L. E., 106, 107(5)
Hensley, K., 342
Hentati, A., 181, 182(4), 192(12), 193
Hentiques, A. O., 153
Herbert, S. K., 148, 153
Herms, J. W., 258, 261(5)
Hernandez-Saavedra, D., 25, 32(13), 34(13), 36(13), 37(13), 230
Herold, S., 102
Hérouart, D., 148(17), 151
Heslop, H. E., 140
Heydari, A. R., 112
Hibbs, J. B. J., 10, 16(19), 22(19)
Hibi, M., 313

Hickey, E. K., 244(13; 15; 20), 245, 245(20), 246
Hickey, M. J., 64, 66, 68(13), 69, 71(26)
Higashiyama, S., 131, 139
Higgins, C. F., 157
Higgins, D. G., 82(15), 83
Hill, H. A. O., 356
Hill, R. L., 49, 50(4)
Hill, W. G., 194
Hillen, M., 104
Hilliker, A. J., 214, 218, 291
Hiltermann, T. J. N., 320
Hino, Y., 244(18; 19), 245(18; 19), 246
Hinton, L. M., 337
Hirano, K., 339
Hirose, J., 115
Hirsch, O., 185
Hjalmarsson, K., 74, 75(2), 77
Ho, T. Y., 159
Ho, Y. S., 21, 130, 182, 193, 339
Hoang, L., 244(14), 245
Hochrein, E. C., 339
Hof, P. R., 192(11), 193
Hoffbrand, A. V., 140
Hoffman, E. K., 129, 129(2), 130, 193, 339
Hofmann, H., 102, 103(5), 106(5)
Hogg, N., 341
Holland, J. C., 340
Holm, R. H., 65
Holme, E., 74
Holtke, H. J., 206
Homma, T., 310
Honig, B., 45, 46, 46(17), 47
Hopkin, K. A., 80
Hoppel, C. L., 274
Hopwood, D. A., 91
Hopwood, L. E., 277
Hori, M., 336
Horikawa, H., 244(18; 19), 245(18; 19), 246
Horikoshi, K., 244(18; 19), 245(18; 19), 246
Hornsby, P., 241
Horst, K., 244(20), 245(20), 246
Horwitz, J., 24, 34(7)
Hoshida, S., 336
Hosoyama, A., 244(18; 19), 245(18; 19), 246
Houseweart, M. K., 213
Howard, A. J., 130
Howard, B. H., 296
Howell, J. K., 244(20), 245(20), 246
Hristova, M., 25
Hsi, L. C., 25

Hsieh, Y., 66, 67, 68(32), 69, 71(26)
Hsu, J. L., 67, 68(32)
Hu, Y., 124, 125(10), 245(22), 246, 247(22), 248(22)
Hu, Z., 91
Huang, M., 191
Huang, S., 130, 193, 211(18), 340
Huang, T. T., 19, 129(2), 130, 191, 192, 193, 194, 196, 199(20), 211, 340
Huang, Y. H., 124, 248, 250(29), 251(29), 253(29), 254(29)
Huang, Y. T., 155
Huber, H., 248, 250(32), 251(32), 255(32), 256(32)
Huber, R., 248
Hudson, A. J., 5
Huffman, K., 333
Hughes, M. N., 104, 105
Huie, R. E., 10, 24, 102, 104(3)
Hullihen, J., 272, 273(11)
Hunt, J. A., 25
Hunter, G. J., 70, 81, 83(10)
Hunter, T., 70, 81, 83(10)
Huntley, G. W., 192(11), 193
Hurst, M. A., 244(16), 245(16), 246
Hurt, J. B., 233
Hutz, R. J., 193
Huynh, B. H., 124, 243, 245(7; 23), 246, 247(23), 248, 248(7; 23), 250(23; 29), 251(23; 29), 253(7; 29; 39), 254, 254(7; 23; 29), 255(23), 256(23), 257(23)
Hyslop, P. A., 279, 342

I

Ianni, A., 313
Ichikawa, I., 310
Ichimori, K., 234
Iizuka, S., 131, 139
Ikebukuro, K., 70
Ikeburkuro, K., 81, 83(10)
Imai, T., 84
Imlay, I. A., 150
Imlay, J. A., 3, 3(17), 4, 4(14), 5, 5(14; 23), 6(23), 9, 10, 14(1), 21(15; 16), 147, 150(16), 151, 152, 152(21), 154(16), 157, 163(17), 232, 354, 357
Imlay, K. C., 51

Inaoka, T., 86
Iñarrea, P. J., 106, 271
Ingold, K. U., 240
Inoue, M., 346, 347, 350(7; 8; 10; 11), 351, 353, 354, 354(19; 20)
Iorndanova, I. K., 264
Irani, K., 314
Irvin, S. D., 168
Ischiropoulos, H., 24, 26, 27(21a), 32(4), 34(7), 37
Ishii, T., 108
Ishikawa, M., 131, 134, 138
Itzhaki, R. F., 117
Izokun-Etiobhio, B. O., 114

## J

Jackson, M. S., 219
Jackson-Lewis, V., 24, 34(7), 180, 182, 183, 184(9), 185(10), 186
Jacob, W. A., 320
Jacobs, A., 3
Jacobs, E. E., 273
Jacobsson, J., 74
Jacques, M., 4, 150
Jadot, G., 321
Jaenisch, R., 191
Jameson, G. B., 69, 80, 83(6), 87, 89
Jang, J.-H., 100
Jenkins, N. A., 192(10), 193
Jenney, F. E., 124, 125(10)
Jenney, F. E., Jr., 243, 244(6), 245(6; 22), 246, 247(22), 248(22), 251(6)
Jensen, L. T., 171, 172(11)
Jewett, S. L., 240, 264
Jiang, H., 186
Jiang, X., 312
Jin, N., 26, 27(21b)
Jiwani, N., 244(14), 245
Jochum, M., 139
Johnson, G. L., 313
Johnson, M. K., 124, 248, 250(29), 251(29), 253(29), 254(29)
Johnson-Schlitz, D. M., 216
Johnsson, B.-H., 70
Johsson, J., 130
Jones, A. D., 25
Jones, I. M., 259, 264, 265(7)
Jones, N. L., 337

Jones, P., 25
Jones, P. L., 341
Jonsson, B.-H., 66
Jonsson, J., 339
Jonsson, L. M., 339
Jonsson, P. A., 74
Jordán, J., 277
Jordan, M. C., 241, 335
Joseph, J., 229
Jourd'heuil, D., 101, 102, 104(2), 105(2; 12)
Jovanovic, T., 124, 245(23), 246, 247(23), 248(23), 250(23), 251(23), 254(23), 255(23), 256(23), 257(23)
Jun, A. S., 19
Jung, G., 293, 293(5), 294, 295(5), 298, 298(5), 299(14)

## K

Kachadourian, R., 27, 225, 226, 226(14), 233(17a)
Kadioglu, A., 148
Kagayama, A., 320
Kaine, B. P., 244(13; 16), 245, 245(16), 246
Kakhlon, O., 4
Kalyanaraman, B., 10, 229
Kaminski, P. M., 319
Kanayama, Y., 139
Kang, S.-O., 90, 91, 93(7; 8; 11), 95(7; 11), 96(11), 98(11), 99(11), 100, 100(7), 243
Karess, R. E., 213, 214(3)
Karin, M., 313, 313(8), 314
Karlsson, K., 74, 75, 75(8–10), 76(16–18), 77(17), 78(16)
Karnani, H., 320
Karsaksis, E. J., 307
Kasahara, E., 346
Kasarskis, E. J., 306, 310(5), 341
Kashiba, M., 346
Kato, K., 339
Kato, S., 213
Kawaguchi, T., 130, 133, 134, 139, 140(9)
Kawai, Y., 133
Kawamura, N., 139
Kawano, K., 131
Kawarabayasi, Y., 244(18; 19), 245(18; 19), 246
Kawase, M., 192
Kawata, S., 131
Kay, H. H., 74

Keagle, P., 244(14), 245
Keefer, L. K., 163
Keele, B. B. J., 9
Keil, A. D., 155, 157(4)
Kelley, J. M., 244(16), 245(16), 246
Kelly, K., 327
Kemler, R., 192
Keng, Y. F., 314
Kennedy, M. C., 10, 13, 254
Kennepohl, P., 65
Kerby, J. D., 32
Kerlavage, A. R., 244(13; 16), 245, 245(16), 246
Kerner, J., 274
Kessler, C., 206
Keston, A. S., 359
Ketchum, K. A., 244(13; 15; 20), 245, 245(20), 246
Kettle, A. J., 360
Keyer, K., 3, 4(14), 5(14), 10, 21(15), 147, 150
Khalak, H., 244(20), 245(20), 246
Khan, A. U., 102
Khan, M. F., 235, 242(21)
Khan, U., 187
Khelef, N., 145(4), 146
Kidani, Y., 115
Kieser, H. M., 91
Kieser, T., 91
Kijima, Y., 131
Kikuchi, H., 244(18; 19), 245(18; 19), 246
Killat, S., 145(5), 146
Kim, E.-J., 91, 92, 93(7; 9), 95(7), 96(9), 98(19), 99(19), 100, 100(7; 19), 243
Kim, H. P., 91, 93(9), 96(9), 130, 310
Kim, H. T., 294
Kim, J., 293, 293(5), 295(5), 298(5)
Kim, J.-S., 100
Kim, K.-S., 100
Kim, S., 101, 102, 104(2), 105(2; 12)
Kim, Y. C., 91, 148
Kim, Y. H., 293, 293(5), 294, 295, 295(5), 298, 298(5), 299(10; 14)
Kimland, M., 241
Kimmel, B. E., 337
King, S. B., 102
Kiningham, K. K., 306, 310, 310(5)
Kinoshita, N., 139
Kinouchi, H., 194
Kipnes, R. S., 333
Kirby, K., 218, 291
Kirby, W. M., 162

Kirkness, E. F., 244(13; 16), 245, 245(16), 246
Kissner, R., 24, 102
Kitano, M., 339
Klann, E., 74(13), 75
Klapper, I., 45
Klausner, R. D., 3
Klein, W. L., 21
Klenk, H.-P., 244(13; 16), 245, 245(16), 246
Klichko, V. I., 214, 291
Klomp, L. W., 167
Klotz, L.-O., 101
Klug-Roth, D., 224, 236, 237(25)
Kobayashi, K., 84, 89
Kobayashi, Y., 130, 139(7)
Koen, A. L., 17, 200
Koenigs, L. L., 320
Koepp, J., 226, 233(17a)
Kohda, H., 130, 140(9)
Kok, J., 148
Kolbanovskiy, A., 102
Kolkman, J. A., 248, 250(30), 251(30)
Kollman, P. A., 103
Kon, V., 310
Kondo, T., 129(2), 130, 193
Konno, M., 89
Konorev, E. A., 229
Kontula, K., 337
Koppenol, W. H., 24, 26, 27(21a), 102, 104, 115, 117, 122, 232, 235, 236(10), 271
Korn, R., 192
Korpela, H., 338
Kosman, D. J., 168, 168(7), 169
Koster, J. F., 337
Kostic, V., 182, 183, 184(9), 185(10), 187
Kosugi, H., 244(18; 19), 245(18; 19), 246
Kotamraju, S., 229
Kovacic, D., 102
Kowall, N. W., 129, 339
Kozlov, A. V., 3, 6
Kozy, H., 194, 199(20)
Krantz, M., 350
Krauss, P., 116
Krawiec, Z., 168
Krebs, C., 124, 245(23), 246, 247(23), 248(23), 250(23), 251(23), 254(23), 255(23), 256(23), 257(23)
Kredich, N. M., 151
Krems, B., 167
Kresge, A. J., 66
Kretzschmar, H., 258, 261(5)

Krishna, C. M., 238(30), 239
Krishna, M., 101, 102, 104(2), 105(2; 12)
Krishna, M. C., 234, 239(5)
Krishnan, A., 341
Krisnan, A., 307
Kroll, J. S., 59, 148, 155, 156, 157(4), 159, 160(15; 16), 161, 163(14)
Kronmal, G. S., 337
Kruck, T., 258, 261(5)
Kruk, H., 333
Kruuv, J., 49, 59(6)
Kudoh, Y., 244(18; 19), 245(18; 19), 246
Kumar, M., 91
Kumar, T. R., 193
Kunitomo, R., 351
Kuppusamy, P., 275
Kuramitsu, H., 152, 154(22), 156
Kuroki, T., 314
Kurtz, D. M., Jr., 229, 248, 255(27), 257(27)
Kushida, N., 244(18; 19), 245(18; 19), 246
Kuzuya, T., 139, 336
Kwan, T., 337
Kwon, Y. W., 181, 182(4), 192(12), 193
Kwong, L. K., 341, 342, 346(8)
Kyrpides, N. C., 244(13), 245

## L

Laessig, T., 155
Lah, M. S., 64, 67(17), 83
Lai, C. S., 277
Lai, S. S., 159
Laipis, P. J., 66
Lakka, T. A., 337
Lambeth, J. D., 21
Lampreia, J., 244(10), 245
Land, E. J., 39, 41(4), 357
Lander, E. S., 194
Lander, H. M., 314
Lanfranco, G., 333
Lange, D. J., 185
Langford, P. R., 59, 148, 155, 156, 157(4), 159, 160(15; 16), 161, 163(14)
Lania, A., 47
Lapinskas, P. J., 167
Larramendy, M., 3
Laski, F. A., 215
Lassegue, B., 21
Laudenbach, D. E., 148, 153

Lauer, P., 337
Lauey, M., 259, 264(13)
Lavelle, F., 65, 67(20)
Laver-Rudich, Z., 293
Lavie, V., 203
Lawrence, G. D., 224
Leatherbarrow, R. J., 290
Leber, B., 337
Lebovitz, R. M., 130, 193, 211(18), 340
Leclere, V., 91
Lee, C. C., 277
Lee, C. Y., 340
Lee, D., 107
Lee, F. J., 155
Lee, H., 91, 244(14), 245
Lee, H. J., 294
Lee, J., 25, 32
Lee, J. K., 91, 93(8), 100, 243
Lee, J.-W., 90, 91, 93(8; 11), 95(11), 96(11), 98(11), 99(11), 100, 243
Lee, M. K., 192(10), 193
Lee, N. H., 244(13), 245
Lee, P., 51
Lee, V. K., 337
Lee, Y., 350
Leenhouts, K. J., 148
Leff, J. A., 139, 319, 320
LeGall, J., 124, 243, 244(10; 17), 245, 245(7; 8; 17), 246, 247(8), 248, 248(7; 8; 17), 250, 250(8; 29), 251(8; 29; 35; 36), 253, 253(7; 29; 39; 40), 254, 254(7; 8; 29; 31), 255(17; 31)
Lehninger, A. L., 273
Leme Martins, E. A., 3
Lengfelder, E., 234, 235(2), 236
Lennon, P. J., 235
Lepock, J. A., 69
Lepock, J. R., 49, 49(11), 50, 53(9; 11), 54, 54(9), 59(6; 9; 11; 26), 65, 66, 69(21), 70(21), 71(21; 26), 72(27), 73(21)
Lerme, F., 107
Leroy, G., 244(9), 245
Levanon, D., 293
Leveque, V., 65, 66, 69(21), 70(21), 71(21), 72, 72(27), 73(21; 42)
Levin, E. D., 339
Levine, F., 235, 236(10)
Levrero, M., 313
Levy, A., 321, 322(4), 323(4), 325(4), 327(4)
Lewis, M. E., 206

AUTHOR INDEX 377

Lewis, R. V., 49
Li, R., 265
Li, S., 312
Li, Y., 19, 130, 192, 211, 340
Liang, L. P., 21, 226, 233(18)
Liao, C., 158, 159(21)
Liba, A., 4, 5(23), 6(23), 10, 21(16)
Libby, S. J., 153, 162, 163(33)
Liczmanski, L., 168
Lieman-Hurwitz, J., 203, 293
Lill, R., 171, 172(11)
Lima, M. J., 245(24), 246, 247(24), 248(24), 250(24), 251(24), 254(24), 255(24), 257(24)
Lin, A., 313, 313(8), 314
Lin, C. Y., 155
Lin, L.-N., 115, 116
Lind, J., 24
Lindahl, P. A., 91
Lindahl, U., 75
Lindenau, J., 114
Lindskog, S., 66, 70
Ling, X., 229
Linher, K. D., 244(15), 245
Linn, S., 3(17), 4, 150(16), 151, 154(16), 232
Lino, S., 134
Liochev, S. I., 4, 10, 21(14), 22, 27, 65, 103, 106, 107(8), 113(8), 149, 162, 165(34), 225, 229, 251, 355
Liou, L.-L., 19, 169
Lioua, G. G., 155
Lipman, D. J., 82(13), 83
Lippard, S. J., 116
Lipshitz, H. D., 214
Lipton, S. A., 105
Lithgow, G. J., 223
Litwiller, R., 124, 245(23), 246, 247(23), 248(23), 250(23), 251(23), 253(40), 254, 254(23), 255(23), 256(23), 257(23)
Litwinska, J., 168
Liu, D., 229
Liu, H. B., 115
Liu, J., 229
Liu, M. H., 26, 27(21b)
Liu, M. Y., 243, 244(17), 245(7; 17), 246, 248, 248(7; 17), 253(7), 254(7; 31), 255(17; 31)
Liu, T., 265
Liu, X. F., 167
Ljutakova, S. G., 106, 107(8), 113(8)
Lloyd, S. G., 250, 251(36)

Lo, Y. Y. C., 314
Loeb, D. B., 337
Loftus, B., 244(13), 245
Loktaeva, T. D., 106, 107(3)
Lombard, M., 123, 124, 124(2), 125(2), 127, 127(2), 152, 247, 248, 248(26), 251(26; 33), 254(26; 33), 255(26; 33), 257(26; 33)
Longo, D. L., 130, 139(7)
Longo, V. D., 19, 21, 169
Louisot, P., 107
Lovett, M., 181, 182(3), 192, 334
Lowe, D., 38
Loynds, B. M., 155, 159
Lu, N., 130, 193, 211(18), 340
Lucchesi, B. R., 235
Lucchi, S., 241
Luchter-Wasylewska, E., 107
Luckow, B., 294, 298(8)
Ludewig, G., 307, 341
Ludwick, N., 102, 105(12)
Ludwig, M. L., 64, 67(17), 83, 90, 90(6), 91
Lumm, W., 244(14), 245
Lydiate, D. J., 91
Lye, P. G., 24, 102
Lyman, R. F., 219
Lynch, M., 152, 154(22), 156
Lynch, R. E., 20
Lyons, T. J., 115

M

Macarthur, H., 235
Mach, H., 49
Mackay, T. F. C., 219
Mackensen, G. B., 226, 233(18)
MacMillan-Crow, L. A., 32
Madden, T. L., 82(13; 14), 83
Madlung, A., 258, 261(5)
Maeda, K., 346
Maeda, S., 108
Maellaro, E., 241
Majima, H., 306, 307, 310(5), 341
Makita, A., 131
Maklund, S. L., 114
Malek, J. A., 244(15), 245
Malfroy, B., 223, 333, 334
Malik, S., 223
Malinowski, D. P., 49

Maloriol, D., 155
Malvezzi-Campeggi, F., 51
Maniatis, T., 295, 296(9)
Manning, A. S., 241
Manson, J., 258, 261(5)
Manuel, S. M., 24
Manyak, M. J., 320
Mao, J.-I., 244(14), 245
Mao, S. J., 158, 159(21)
Mapa, F. A., 337
Marach, J. A., 240
Marcus, C. B., 333
Marcus, R. A., 224
Margoliash, E., 232, 271
Mariano, A. M., 243, 245(8), 247(8), 248(8), 250(8), 251(8), 254(8)
Maringanti, S., 8, 152
Marklund, G., 43, 74, 75(1), 161
Marklund, S. L., 43, 74, 75, 75(2; 3; 8; 9), 76, 76(16–18), 77, 77(3; 17), 78(16), 130, 161, 186, 265, 339
Marks, V., 134
Marky, L., 49, 50(7), 53(7), 54(7), 59(7)
Marla, S. S., 27(26), 28, 32
Marmocchi, F., 44, 45(9)
Marnett, L. J., 25
Maroney, M. J., 91, 93(11), 95(11), 96(11), 98(11), 99(11)
Marshall, P. A., 10, 24, 101
Martell, A. E., 121
Martin, E., 321
Martin, J. C., 37
Martin, L. A., 22
Martin, M. E., 84
Martin, R., 321, 323(6), 324(6)
Martin, W., 104
Martinez, G. R., 102
Martino, E., 359
Maruyama, T., 84
Mashino, T., 151
Maskos, Z., 235
Mason, A. B., 116
Mason, R. P., 25, 356, 359
Mason, T. M., 244(13), 245
Massano, G., 333
Masters, C., 108
Masuchi, Y., 244(18; 19), 245(18; 19), 246
Masuda, A., 130, 139(7)
Matak, D., 48, 59
Matias, P., 124, 245(21), 246, 247(21), 248(21)

Matsuda, Y., 131, 134, 138, 139
Matsumoto, T., 84
Matsumura, Y., 86
Matsuno, K., 351, 354, 354(19)
Matsushima, K., 130, 139(7)
Matzanke, B. F., 5
Matzuk, M. M., 130, 193, 211(18), 340
Maulik, N., 339
Maurer, P., 102
Mautner, G., 38
Mauze, S., 194
Mayfield, J. E., 155
Mazzetti, A. P., 50
McAdam, M. E., 65, 67(20)
McCammon, J. A., 44, 45, 46, 46(15), 47(23)
McClelland, E., 337
McClenaghan, M., 191
McClune, G. J., 62
McCord, J. M., 9, 25, 32(13), 34(13), 36(13), 37(13), 49, 64, 80, 85, 86, 92(22), 93, 108, 116, 139, 153, 162, 163(33), 200, 230, 241, 243, 256(1), 280, 331, 333, 335, 336(11), 340, 342, 354
McDonald, L., 244(13; 15; 20), 245, 246
McDonnell, P. J., 116
McDougall, S., 244(14), 245
McIntosh, L., 104
McKenna-Yasek, D., 185
McKenney, K., 244(13), 245
McLeod, M. P., 244(20), 245(20), 246
McNamara, J. O., 21
McNeil, L. K., 244(13), 245
McRee, D. E., 54, 59(26)
Meager, A., 140
Medrano, E., 283, 285(15)
Meese, E. U., 306
Mehlhorn, R. J., 240, 241
Mei, G., 51, 54(20; 22)
Meier, B., 84, 161, 314
Mello-Filho, A. C., 3
Melov, S., 19, 130, 211, 223, 340
Melsen, L. R., 153
Melvezzi-Campeggi, F., 51, 54(22)
Mena, M. A., 187
Menconi, M. J., 334
Mendes, P., 34
Meneghini, R., 3, 13, 22(27)
Meng, L.-J., 259
Merenyi, G., 24
Merkamm, M., 145

Merlino, G. T., 296
Merrick, J. M., 244(16), 245(16), 246
Meshnick, S. R., 241
Messner, K. R., 354, 357
Metzger, A. L., 64
Meyer, N. C., 337
Meyrick, B., 310
Michelson, A. M., 321
Middaugh, C. R., 49
Miki, T., 313(7), 314
Miller, A.-F., 65(24), 66, 67, 70(31), 90, 224
Miller, C. D., 148
Miller, D. M., 3
Miller, R. J., 277
Miller, W., 82(13), 83
Milman, N., 338
Minamino, T., 139
Minden, A., 313, 313(8), 314
Mintier, G. A., 337
Miranda, K. M., 102, 105(12)
Misko, T. P., 235
Misra, H. P., 93, 97, 109(20), 110, 283, 355, 358
Mitani, Y., 320
Mitchell, J. B., 101, 104(2), 105(2), 234, 239(5), 240
Mitchell, T. J., 148
Miyamoto, Y., 129
Miyauchi, Y., 351
Miyazawa, N., 132
Miyoshi, M., 346
Miziorko, H., 19
Mizui, T., 194
Mockett, R. J., 213, 214, 216, 219, 287
Moeller, N., 337
Mohazzab, K. M., 319
Mohsen, M., 240
Monaco, J., 188
Mondovi, B., 116
Montecucco, C., 275(22), 276, 283, 357
Monteiro, H. P., 301
Montero, G., 283, 285(14; 15)
Monti, E., 241
Moore, E. E., 139
Moore, F. A., 139
Moore, T., 337
Moore, W. F., 74
Moore-Jarrett, T., 310
Moran, C. P., Jr., 153
Morell, A., 350
Moreno, J. J., 26, 27(21a)

Morikang, E., 337
Morino, Y., 347, 350(8; 11)
Morita, A., 84
Morley, C. G. D., 3
Morrison, J. H., 192(11), 193
Morrow, J. D., 334
Moss, T. H., 71, 87, 89(32), 97
Mostert, L. J., 337
Moura, I., 124, 243, 244(10), 245, 245(7; 23; 24), 246, 247(23; 24), 248, 248(7; 23; 24), 250, 250(23–25; 29), 251(23; 25; 29; 35; 36), 253, 253(7; 25; 29; 39; 40), 254, 254(7; 23; 24; 29), 255(23; 24), 256(23), 257(23; 24)
Moura, J. J. G., 124, 243, 244(10), 245, 245(7; 23; 24), 246, 247(23; 24), 248, 248(7; 23; 24), 250, 250(23; 24), 251(23; 25; 29; 36), 253, 253(7; 25; 29; 40), 254, 254(7; 23; 24; 29), 255(23–25; 29), 256(23), 257(23; 24)
Moya, K. L., 258, 261(3)
Mulcahy, L. S., 313
Mullenbach, G. T., 51
Muller, G. I., 5
Münck, E., 91, 253(39), 254
Murakami, K., 129(2), 130, 192, 193
Murayama, K., 32, 34(32)
Murphy, M. E., 102, 103(9)
Murphy, P. L., 185
Musci, G., 49, 50(10), 53(10)
Musser, J. M., 200
Muthane, U., 186

# N

Nagai, Y., 244(18; 19), 245(18; 19), 246
Nagao, H., 115
Nagasawa, H. T., 102
Nagatomi, H., 351
Nagele, A., 234, 235(2)
Naini, A. B., 24, 34(7), 180, 183, 185, 185(10), 186
Nakahara, H., 346
Nakamura, Y., 244(18; 19), 245(18; 19), 246
Nakao, H., 139
Nakashima, Y., 133
Nakata, T., 131, 132, 138
Nakazawa, H., 234, 244(18; 19), 245(18; 19), 246

Nakazono, K., 354
Nam, J. H., 294
Namiki, M., 130, 138, 140(9)
Nardini, M., 59
Nasso, N. E., 153
Nathan, C., 153, 162, 163(33)
Natoli, G., 44, 45(9), 313
Nauser, T., 24, 102, 122
Naviliat, M., 24, 34(6)
Naylor, S., 124, 245(23; 24), 246, 247(23; 24), 248(23; 24), 250(23; 24), 251(23), 253(40), 254, 254(23; 24), 255(23; 24), 256(23), 257(23; 24)
Neese, F., 68(34), 69
Neiland, J. B., 3
Nelli, S., 104
Nelson, K. E., 244(13; 15), 245
Nelson, S. K., 335, 336(11)
Nelson, W. C., 244(15), 245
Nersissian, A., 115
Nesbitt, D. M., 277
Neta, P., 227, 228(19), 231(19), 232(19)
Neuburger, M., 22
Newell, D. G., 153
Nguyen, A. L., 5
Nguyen, D. D., 28, 240, 244(16), 245(16), 246
Nguyen, D.-D. H., 13, 21, 22(26)
Nguyen, D. H., 197
Nicholls, A., 46
Nicholls, D. G., 277
Nick, H. S., 61, 63, 65, 66, 67, 67(10), 68(32), 69, 69(21), 70(21), 71(21; 26), 72, 72(27), 73(21; 42), 130, 310
Niederhoffer, E. C., 65, 67(19), 73(19), 89
Niedzwiecki, A., 214
Nielson, S. O., 42, 61
Nilsen, S. P., 218
Nilsson, U. A., 240
Nims, R. W., 163
Nishida, C. R., 243
Nishida, T., 138
Nishikawa, M., 346, 350, 351
Nishikawa, N., 346
Nishiura, T., 134, 139
Nivière, V., 123, 124, 124(2), 125(2), 127, 127(2), 152, 247, 248, 248(26), 251(26; 33), 254(26; 33), 255(26; 33), 257(26; 33)
Niwa, T., 320

Noack, H., 114
Nobel, C. S., 241
Nobel, L. J., 340
Noble, L. J., 130, 211
Nohl, H., 361
Noij, M., 115
Noji, S., 133
Nölling, J., 244(14), 245
Nomura, N., 84
Norris, K. H., 241
Norris, S. J., 244(20), 245(20), 246
Northrup, S. H., 45, 46(15)
Nose, K., 314
Novogrodsky, A., 314
Nozik-Grayk, E., 74(12), 75
Nyman, L., 320
Nyyssonen, K., 337, 338

## O

Oberley, L. W., 131, 185, 200, 212(33), 265, 287, 288(3), 293, 312, 340, 341
Oberley, T. D., 182, 193, 341
Ochi, A., 346
Ochman, H., 200
Ochsner, U. A., 153, 155, 162, 163(33)
O'Connor, D., 67, 68(32), 69
Odomi, M., 320
O'Donnell, V. B., 226, 233, 233(16)
Offer, T., 240, 241
Ogasawara, N., 339
Ogino, T., 347, 350(8)
Ogiste, J. S., 314
Oguchi, A., 244(18; 19), 245(18; 19), 246
Ogura, K., 244(18; 19), 245(18; 19), 246
Ohama, E., 213
Ohba, M., 314
Ohfuku, Y., 244(18; 19), 245(18; 19), 246
Ohhira, M., 130, 138, 140(9)
Ohman, D. E., 148, 153
Ohmori, D., 84
Okada, S., 346
Oliva, C., 241
Oliveira, S., 244(17), 245(17), 246, 248(17), 255(17)
Oliver, S. G., 70
Olsen, G. J., 244(13; 16), 245, 245(16), 246
Olson, J. L., 130, 211, 340
Olson, M. O., 84

Olsson, L. I., 240
Omar, B. A., 241, 333, 335
Omar, R., 301, 304(25), 305
O'Neill, P., 38, 44, 45, 45(9), 47, 47(18), 48, 48(10), 59
Ono, M., 130, 134, 138, 140(9)
Ookawara, T., 129
Oppenheim, J. J., 130, 139(7)
Oraedu, A. C., 114
Orak, J. K., 114
Orr, W. C., 213, 214, 219, 291, 293(6), 294, 341
Orrenius, S., 241
Ose, D. E., 33
Oshino, N., 282
O'Sullivan, M. J., 134
Otsu, K., 336
Otsuka, K., 244(18), 245(18), 246
Otsuka, R., 244(19), 245(19), 246
Ott, W. C., 214
Ötting, F., 33, 256, 265, 287, 292(1)
Ottobrelli, A., 333
Oury, T. D., 74, 74(12; 14), 75, 182, 339
Overbeek, R., 244(13; 16), 245, 245(16), 246

P

Pacello, F., 165
Pacheco, I., 244(17), 245(17), 246, 248, 248(17), 254(31), 255(17; 31)
Packer, L., 277, 347
Padmaja, S., 10, 24, 102, 104(3)
Pallen, M. J., 156, 163(14)
Palmer, J. M., 106, 107(5)
Palmiter, R. D., 191
Pan, K.-M., 261
Pan, T., 265
Panchenko, L. F., 106, 107(3)
Panchuk-Voloshina, N., 359
Pantoliano, M. W., 55, 57(28), 116
Papazian, M. A., 80
Pappolla, M., 301, 304(25), 305
Pardo, C. A., 192(10), 193
Parge, H. E., 47, 64, 68(13)
Park, A., 346
Park, A.-M., 346
Park, K. H., 295, 299(10)
Park, S. F., 153
Parker, M. W., 64, 81, 83(8; 9), 90
Parkes, T. L., 214

Parks, D., 226, 233(16)
Parks, D. A., 333
Parsons, P. E., 139
Pastan, I., 296
Pasternack, R. F., 235
Patel, H. R., 313
Patel, M., 19, 21, 226, 229, 233(18)
Patierno, S. R., 320
Pattridge, K. A., 64, 67(17), 83, 90, 90(6), 91
Patwell, D., 244(14), 245
Pearlstein, R. D., 226, 233(18), 339
Pecoraro, V. L., 26
Pedersen, P. L., 272, 273(11)
Peeker, R., 74
Peeters-Joris, C., 106, 107(4), 113(4), 114(4)
Pegg, M., 108
Pelkonen, O., 320
Pesce, A., 38, 48, 59
Peter, R. M., 320
Peterson, J. D., 244(13; 15; 16; 20), 245, 245(16; 20), 246
Peterson, S., 244(13), 245
Petkau, A., 53(25), 54, 327
Petruzelli, R., 47
Petsko, G. A., 64, 83
Phillips, C. A., 244(15), 245
Phillips, J. P., 214, 218, 291
Phillis, R. W., 216
Piantadosi, C. A., 74(12), 75, 182
Pianzzola, M. J., 124, 152, 244(12), 245, 251(12)
Piermattei, D., 320
Piersma, S. R., 25
Pietrokovski, S., 244(14), 245
Piganelli, J. D., 226, 233(17a)
Ping, D., 341
Piras, C., 91
Poderoso, J. J., 281, 282(6), 284(6)
Polack, B., 153
Polchow, C. Y., 181, 182(4), 192(12), 193
Politi, L., 49
Polizio, F., 44, 48(10), 49(12), 50, 53(12), 54(12), 55(12), 56(12), 59, 59(12), 84, 115
Polticelli, F., 44, 45, 47, 47(18), 48, 48(10), 51, 59, 161
Ponka, P., 3
Ponthadavithi, S., 306
Popescu, C. V., 91
Popova, B., 310
Possel, H., 114

Pothier, B., 244(14), 245
Powers, T. B., 90
Poyart, C., 81, 82(11)
Prabhakar, S., 244(14), 245
Pratt, M. S., 244(15), 245
Preston, C. R., 216
Price, D. L., 192(10), 193
Prier, S. D., 153
Privalle, C. T., 235
Privalov, P. L., 51, 58(24)
Prusiner, S. B., 258, 259, 261
Pryor, W. A., 10, 26, 27(21a)
Przedborski, S., 24, 34(7), 180, 182, 183, 184(9), 185, 185(10), 186, 187
Pu, H., 181, 182(4), 192(12), 193
Puget, K., 321
Puntarulo, S., 284, 285(16)
Puppo, A., 148(17), 151
Purdy, D., 153

## Q

Qin, K., 258, 259, 261(5)
Quackenbush, J., 244(13), 245
Qui, D., 244(14), 245
Quijano, C., 23, 25, 32(13), 34(13), 36(13), 37(13), 230
Quintanilha, A. T., 277

## R

Rabani, J., 42, 61, 62, 224, 236, 237(25)
Rabinowitch, H. D., 235
Radeke, H. H., 314
Radi, R., 11(23), 12, 13, 13(23), 22(27), 23, 24, 25, 25(8), 27(27), 28, 28(15), 31(11), 32, 32(13), 33(11; 35), 34(6; 13), 36(8; 13), 37(13), 228, 229(20), 230
Radolf, J. D., 124, 245(23), 246, 247(23), 248(23), 250(23), 251(23), 254(23), 255(23), 256(23), 257(23)
Radyuk, S. N., 214, 291
Ragan, C. I., 271, 275(4)
Ragsdale, S. W., 91
Rahmandar, J. J., 214
Raikov, Z., 234
Raineri, I., 13, 18(25), 19(25), 20(25), 22(25), 191, 194, 199, 199(20), 338

Rajagopalan, K. V., 108
Rajendran, G., 188
Ramão, M. J., 248
Ramasarma, T., 3
Ramilo, C. A., 65, 69(21), 70(21), 71(21), 73(21)
Rankin, B. B., 112
Rao, G., 112
Raoult, D., 83
Rauckmann, E. J., 356
Raveau, M., 338
Ravenscroft, J., 223
Ravi, N., 124, 243, 245(7), 248, 248(7), 250(29), 251(29), 253(7; 29), 254(7; 29)
Ravindranath, S. D., 106, 107(6)
Rayport, S., 188
Reaume, A. G., 129, 129(2), 130, 193, 213, 339
Reaume, E. K., 193
Reddin, K. M., 155
Reddy, M. V., 304(24), 305
Redford, S. M., 49(11), 50, 53(11), 54, 59(11; 26)
Reenan, R. A., 218
Rees, D. C., 124, 125(10), 245(22), 246, 247(22), 248(22)
Reeve, J. N., 244(14), 245
Reich, C. I., 244(13; 16), 245, 245(16), 246
Reif, A., 105
Reisz-Porszasz, S., 302
Reiter, C., 24, 54
Reittie, J. E., 140
Ren, S., 22
Ren, X. D., 313
Renshaw, M. W., 313
Reola, L., 192
Repine, J. E., 139, 319, 320
Resch, K., 314
Rettie, A. E., 320
Reveillaud, I., 214
Rewers, K., 3
Reyes, H., 302
Reynafarje, B., 272, 273(11)
Reynolds, J. A. J., 33
Rho, H. M., 293, 293(5), 294, 295, 295(5), 298, 298(5), 299(10; 13; 14; 19–21; 23), 302
Ricciuti, F., 306
Rice, P., 244(14), 245
Richardson, A., 19, 112
Richardson, D. C., 50, 90, 103, 244(15; 20), 245, 245(20), 246

Richardson, D. L., 244(13), 245
Richardson, G. J., 24, 54
Richardson, J. S., 50, 90, 103
Richter, C., 275(22), 276, 283, 357
Riederer, P., 105
Riesz, P., 238(30), 239
Rigaut, J. P., 321, 323(6), 324(6)
Riley, D. P., 62, 223, 234, 235
Ringe, D., 64, 83
Rio, D. C., 215
Ripps, M. E., 192(11), 193
Rivers, W. J., 62, 235
Rizzolo, L. J., 64, 139
Robalinho, R. L., 13, 22(27)
Robb, F. T., 244(18; 19), 245(18; 19), 246
Roberts, K. M., 244(16; 20), 245(16; 20), 246
Roberts, L. J., 334
Roberts, P. B., 38, 72
Roberts, S., 3
Robertson, H. M., 216
Robinson, K. A., 342
Rocklin, A. M., 240, 264
Rodolfo, K., 258, 261(3)
Rodrigues-Pousada, C., 244(17), 245(17), 246, 248(17), 255(17)
Rodriguez, M., 11(23), 12, 13(23), 25
Roe, J. A., 49, 50(7), 53(7), 54(7), 59(7), 115
Roe, J.-H., 90, 91, 92, 93(7; 9), 95(7), 96(9), 98(19), 99(19), 100, 100(7; 19), 130, 310
Rogina, B., 218
Romão, M. J., 246, 248, 250(25), 251(25), 253(25), 254(31), 255(31)
Romero, N., 180
Rosano, C., 48, 59
Rosato, G., 51
Rosato, N., 51, 54(20; 22)
Rose, J. C., 199
Rosen, G. M., 235, 356
Rota, C., 359
Röthlisberger, U., 102
Rothman, R. J., 3
Rotilio, G., 38, 44, 45, 45(9), 47, 48, 48(10), 49, 49(12), 50, 51, 51(17), 53(12; 16), 54(12; 16), 55(12), 56(12), 59, 59(12; 16), 60(16), 61(16), 72, 84, 115, 116, 157, 158(18), 159(18), 161, 161(18), 165
Rotman, G., 181, 182(3), 192, 334
Rouati, D., 150
Rouault, T. A., 3
Rousson, R., 107
Rovin, B. H., 314
Rubbo, H., 24, 25(8), 36(8)
Rubin, G. M., 213, 214(3), 215
Ruddle, F. H., 306
Ruddy, D. A., 337
Ruiz-Ramirez, L., 27(27), 28
Rusakow, L. S., 197
Rusch, R., 51, 54(20)
Rush, J. D., 235, 236(10)
Rusnak, F., 124, 243, 245(23; 24), 246, 247(23; 24), 248(23; 24), 250(23–25), 251(23; 25), 253(25; 40), 254, 254(23; 24), 255(23; 24), 256(23), 257(23; 24)
Russanov, E. M., 106, 107(8), 113(8)
Russell, W. J., 241, 335
Russo, A., 234, 238(30), 239, 239(5), 240, 241
Ryan, U. S., 235
Rycroft, A. N., 156, 160(15)

S

Sadosky, A. B., 145(3), 146
Sadow, P. W., 244(13; 16), 245, 245(16), 246
Safer, H., 244(14), 245
Saidha, T., 107
Sailer, A., 258, 261(4)
Sakai, M., 244(18; 19), 245(18; 19), 246
Sakaki, J., 353, 354(20)
Sakata, N., 313
Sako, Y., 84
Sàles, N., 258, 261(3)
Salin, M. L., 64, 84, 139
Salonen, J. S., 320
Salonen, J. T., 337, 338
Salonen, R., 338
Salsman, S., 342
Salvemini, D., 223, 235
Salzberg, S., 244(20), 245(20), 246
Salzman, A. L., 10, 12(18), 13(18), 22, 22(18), 244(15), 245
Sambrook, J., 295, 296(9)
Samouilov, A., 235
Sampson, J. B., 24, 54, 115
Samson, G., 148, 153
Samulski, T. V., 226, 233(17b)
Samuni, A., 234, 238(30), 239, 239(5), 240, 241

Sanadi, D. R., 273
Sanchez, R. A., 284, 285(16)
Sanchez-Ruiz, J. M., 57
Sanders, J. W., 148
Sanders, S. P., 275
Sandler, S., 310
Sandström, J., 74, 75, 75(9; 10), 76(18), 77
Sandusky, M., 244(20), 245(20), 246
Sansone, A., 155, 156, 163(14)
Santos, R., 148(17), 151
Saraiva, L. M., 248, 250(32), 251(32), 255(32), 256(32)
Saran, M., 236
Sasaki, J., 351, 354, 354(19)
Sato, E., 346, 351
Sato, T., 354
Savini, I., 51, 54(20)
Sawada, M., 244(18; 19), 245(18; 19), 246
Sawamura, M., 108
Sawasdikosol, S., 188
Sawyer, D. T., 224
Scandurra, R., 49
Schaffer, A. A., 82(13), 83
Schasteen, C. S., 235
Schatz, G., 274
Scheuer, P. J., 138
Schiestl, R. H., 170
Schiff, J. A., 107
Schindler, U., 102, 103(5), 106(5)
Schinina, M. E., 59, 81, 83(9)
Schirmer, M. A., 70
Schmidt, H. H., 102, 103(5), 105, 106(5)
Scholler, D. M., 49, 50(7), 53(7), 54(7), 59(7)
Schork, N. J., 194
Schrempf, H., 91
Schulman, H. M., 3
Schulten, K., 45, 46(17)
Schulze-Osthoff, K., 320
Schulz-Schaeffer, W., 258, 261(5)
Schütz, G., 294, 298(8)
Schwartz, M. A., 313
Schwarz, H. A., 62, 63(11), 64, 93
Schweizer, H. P., 148, 153
Schwitzquebel, J. P., 106, 107(5)
Scott, J. L., 244(16), 245(16), 246
Scott, M. D., 241
Scott, R. W., 129, 129(2), 130, 193, 213, 339
Seidman, J. G., 203
Sekine, M., 244(18; 19), 245(18; 19), 246
Sekiya, C., 130, 134, 138, 140(9)

Selander, R. K., 200
Selle, S., 314
Sengelov, H., 338
Sentman, M. L., 339
Seo, H. G., 132
Seppanen, R., 338
Serfling, E., 192
Sergi, A., 45, 47(18)
Serroni, A., 3
Seto, N. O. L., 214
Sette, M., 44, 48(10), 59
Sezaki, H., 350
Sha, Z., 155
Shapiro, E. R., 71, 87, 89(32), 97
Sharma, M. L., 91, 93(11), 95(11), 96(11), 98(11), 99(11)
Sharma, P., 243, 245(8), 247(8), 248(8), 250(8), 251(8), 254(8)
Sharp, K., 45, 46, 46(17), 47
Sharpe, M. A., 102
Sheehan, B. J., 156, 160(15)
Shefner, J. M., 339
Shen, J., 44
Sheng, H., 226, 233(18), 339
Sherman, L., 293
Sherris, J. C., 162
Shewmake, K., 320
Shi, J., 21
Shi, W.-X., 188
Shi, Y. P., 192, 196
Shibanuma, M., 314
Shih, J. C., 274
Shiloh, M. U., 153, 162, 163(33)
Shima, S., 91
Shimer, G., 244(14), 245
Shimizu, T., 138
Shindyalov, I. N., 45
Shiraga, T., 320
Shizuyu, H., 244(18; 19), 245(18; 19), 246
Shuff, S. T., 235, 242(21)
Shuman, H. A., 145(3), 146
Shutenko, Z. S., 102, 103(5), 106(5)
Siddique, T., 115, 181, 182(4), 192(12), 193
Sieker, L. C., 248
Sies, H., 101, 102, 103(9), 241, 271, 281, 282(5), 283(5), 314, 342, 346
Sikkink, R., 124, 245(23), 246, 247(23), 248(23), 250(23), 251(23), 254(23), 255(23), 256(23), 257(23)
Sikorski, J. A., 235

Silva, G., 244(17), 245(17), 246, 248(17), 255(17)
Silva, N., Jr., 51, 54(20)
Silverman, D. N., 61, 63, 65, 66, 67, 67(10), 68(32), 69, 69(21), 70, 70(21), 71(21; 26), 72, 72(27), 73(21; 42)
Simonetti, S., 183, 185(10)
Simons, J. P., 191
Sines, J. J., 46
Singh, I., 114
Sisodia, S. S., 192(10), 193
Siwek, D. F., 129, 339
Sklar, L. A., 279, 342
Skogman, G., 74, 75(2)
Slater, A. F., 241
Slekar, K. H., 168(7), 169
Slot, J. W., 193
Slykehouse, T. O., 89
Smeal, T., 313
Smith, A. J., 159
Smith, C., 37
Smith, C. P., 91
Smith, D. R., 244(14), 245
Smith, H. O., 244(13; 15; 16; 20), 245, 245(16; 20), 246
Smith, M. R., 313
Smith, N. J., 338
Smith, R. M., 121
Smith, S., 181, 182(3), 192, 334
Snider, W. D., 129, 339
Snowden, A., 314
Sodergren, E., 244(20), 245(20), 246
Sohal, B. H., 216, 287
Sohal, R. S., 213, 214, 216, 219, 287, 293(6), 294, 341, 342, 346(8)
Solomon, E. I., 65, 68(34), 69
Soper, J. W., 272, 273(11)
Sorenson, J. R., 234, 235, 242(21)
Sorescu, D., 21
Sorkin, D. L., 67, 70(31)
Soubes, M., 124, 152, 244(12), 245, 251(12)
Souza, J. M., 24
Souza, R. F., 313, 320
Spadafora, R., 244(14), 245
Spasojevic, I., 25, 26, 27, 27(21), 28, 28(15; 21), 223, 224(2; 3), 225(2; 3), 226, 226(2), 227, 227(2), 228, 228(2; 3; 19; 33), 229(2; 3; 20), 230(2; 3), 231(2; 19), 232, 232(2; 19), 233(2; 17b)
Spear, N. H., 3, 24

Spechler, J., 320
Spitz, D. R., 185, 200, 212(33), 265, 287, 288(3)
Spottl, R., 236
Spradling, A. C., 213, 215, 216(11)
Spriggs, T., 244(13), 245
Squadrito, G. L., 10
Srinivasa, C., 4, 5(23), 6(23)
Srinivasan, C., 10, 21(16)
St. Clair, D. K., 182, 193, 199, 306, 307, 308(6), 310, 310(5), 340, 341
St. John, G., 153, 155
Stabel, T. J., 155
Stacey, D. W., 313
Stadtman, E. R., 240
Stahl, N., 261
Stallings, W. C., 64, 67(17), 83, 90, 90(6), 91
Stamler, J. S., 102
Stanbury, D. M., 237
Stanick, K., 361
Stein, A., 83
Stein, C. M., 334
Stein, O., 293
Steiner, H., 66
Steinman, H. M., 49, 49(11), 50, 50(4), 53(11), 59(11), 80, 86, 93, 145(3), 146, 153, 155, 157, 157(10), 159, 161(25)
Stella, L., 51
Stern, A. I., 107
Stetter, K. O., 248, 250(32), 251(32), 255(32), 256(32)
Stevens, R. D., 227, 228(19), 231, 231(19), 232(19)
Stewart, A. M., 244(15), 245
Stoddard, B. L., 64, 83
Stolk, J., 320
Stoppani, A. O. M., 271, 275(4), 359
Stowell, C., 350
Strain, J., 167
Striplin, S. P., 241, 335
Strohmeier Grot, A., 150, 152
Strome, R., 258, 261(5)
Strömkvist, M., 77
Strong, R. K., 83, 90(6), 91
Stroppolo, M. E., 38, 44, 48, 48(10), 51, 54(22), 59
Stroupe, M. E., 66, 72(27)
Sturtevant, J. M., 51, 57(23)
Sturtz, L. A., 167, 171, 172(11)
Su, Y. O., 26, 27(21b)

Sufit, R. L., 181, 182(4), 192(12), 193
Sugiyama, T., 132, 138
Suh, B., 92, 98(19), 99(19), 100, 100(19)
Suh, Y.-A., 21
Suliman, H. B., 74, 182
Sullivan, J. L., 338
Sulzer, D., 180, 187, 188, 189
Sun, J., 214
Sundaresan, M., 314
Supino, R., 241
Sutter, B., 115, 117, 122
Sutton, G. G., 244(13; 15; 16; 20), 245, 245(16; 20), 246
Suzuki, K., 129, 131, 132, 134, 138, 139, 197
Swaak, A. J. G., 337
Swallow, A. J., 357
Swanson, C. E., 240
Sy, M.-S., 259, 264, 265
Sykes, S. M., 244(13), 245
Sylvester, J. T., 275
Szabó, C., 10, 12(18), 13(18), 22(18)
Sztejnberg, L., 277, 355

## T

Taavitsainen, P., 320
Tabor, S., 249
Tada, M., 139
Tagawa, S., 84
Tainer, J. A., 47, 49(11), 50, 51, 53(11), 54, 59(11; 26), 64, 65, 66, 68(13), 69, 69(21), 70(21), 71(21; 26), 72, 72(27), 73(21; 42), 90, 103, 218, 291
Taka, H., 32, 34(32)
Takagi, H., 133
Takakura, Y., 350
Takamiya, M., 244(18; 19), 245(18; 19), 246
Takeda, M., 310
Takehara, Y., 346
Takeyasu, A., 130, 133, 140(9)
Tamaki, C., 350
Tamakura, F., 32, 34(32)
Tan, L. C., 314
Tanaka, T., 244(18; 19), 245(18; 19), 246
Tanaka, Y., 351, 353, 354(19; 20)
Tandler, B., 274
Tanhauser, S. M., 66
Taniguchi, N., 129, 130, 131, 132, 133, 134, 138, 139, 140(9), 197

Tanner, S. B., 334
Tardat, B., 4, 150
Tarpey, M. M., 24, 54
Tarui, S., 131
Tatar, M., 219
Tatsumi, H., 138
Tatusova, T. A., 82(14), 83
Tavares, P., 124, 243, 244(10), 245, 245(7), 248, 248(7), 250, 250(29), 251(29; 36), 253(7; 29; 40), 254, 254(7; 29)
Teixeira, M., 243, 244(17), 245(8; 17), 246, 247(8), 248, 248(8; 17), 250(8; 32), 251(8; 32), 254(8; 31), 255(17; 31; 32), 256(32)
Tekamp-Olson, P., 51
Templeton, D. J., 313
Tener, G. M., 214
Teng, K. K., 314
Terada, L. S., 313, 319, 320
Terada, N., 313
Terakawa, M., 320
Teramoto, H., 313(7), 314
Terranova, T., 271
Testerman, T., 155
Thacker, D., 320
Thauer, R. K., 91
Thomas, D. J., 153
Thomas, J. B., 153
Thomas, W., 337
Thompson, A., 124, 245(21), 246, 247(21), 248(21)
Thompson, J. A., 32, 82(15), 83, 115
Thompson, J. M., 105
Thompson, S. J., 320
Thomson, A. J., 12
Thorpe, A. N., 227, 228(19), 231(19), 232(19)
Thorpe, S., 350
Thummel, C. S., 214
Tibell, L., 74, 75, 75(2), 77
Timms, A. R., 5
Tischfield, J., 306
Tiscornia, B. C., 24, 34(6)
Tizio, S. C., 26, 27(21b)
Togasaki, D. M., 187
Tohyama, M., 133
Tokuma, Y., 320
Tolbert, C., 84
Tomb, J.-F., 244(13; 16), 245, 245(16), 246
Tomlinson, A. J., 253(40), 254
Torrie, B. H., 49, 59(6)

# AUTHOR INDEX 387

Touati, D., 4, 86, 123, 124, 124(2), 125(2), 127(2), 145, 148(17), 150, 151, 152, 162, 165, 244(12), 245, 247, 248, 248(26), 251(12; 26; 33), 254(26; 33), 255(26; 33), 257(26; 33)
Toussaint, B., 153
Tovar-Baraglia, A., 283, 285(14)
Tower, J., 214
Trager, W. F., 320
Trantwein, A. X., 5
Trent, J. M., 306
Trieu-Cuot, P., 81, 82(11)
Trifiletti, R. R., 24, 34(7)
Tsai, M., 37
Tsai-Wu, J. J., 155
Tsan, F., 310
Tsan, M. F., 340
Tsuchido, T., 86
Tsuchihashi, Z., 337
Tu, C. K., 63, 66, 67, 67(10), 68(32), 70, 71(26)
Tuinstra, R., 19
Tuminello, J. F., 4, 10, 12(11), 13(11), 17(11), 127, 149
Tuoati, D., 145(6), 146, 149(6), 150(6)
Tuomainen, T. P., 337
Tuomilehto, J., 338
Turck, M., 162
Turkaly, J., 274
Turkaly, P., 274
Turner, M., 140
Turrens, J. F., 273, 342
Tyler, D., 106, 107(7), 113(7)
Tyler, D. D., 271

## U

Uda, T., 133, 134
Ue, H., 89
Ugochukwo, E. N. A., 114
Ukeda, H., 108
Ullrich, V., 32
Upton-Rice, M., 339
Urcell, P. C., 130
Ursell, P. C., 211, 340
Utsumi, K., 346
Utterback, T., 244(13; 15; 16; 20), 245, 245(16; 20), 246
Uzzau, S., 155

## V

Vaganay, E., 107
Vaj, M., 333
Valdez, L., 280, 282, 285(10)
Valenti, P., 155, 161, 165
Valentine, J. S., 4, 5(23), 6(23), 10, 19, 21(16), 49, 50(7), 53(7), 54(7), 55, 57(28), 59(7), 115, 116, 169, 224, 243
Van Beeumen, J., 244(10), 245
Vance, C., 65(24), 66, 224
Van Damme, N., 244(10), 245
Van Der, Z. J., 105
Van der Kraaij, A. A. M., 337
van der Vliet, A., 25
Vandervoorde, A. M., 106, 107(4), 113(4), 114(4)
van der Woerd, M., 37
van Eijk, H. G., 337
Vanhaesebroeck, B., 320
Van Remmen, H., 19, 112
Vansteveninck, J., 105
Vanuffelen, B. E., 105
Vasquez-Torres, A., 153
Vásquez-Vivar, J., 10
Vassy, J., 321, 322(4), 323(4–6), 324(6), 325(4; 5), 327(4)
Vazquez-Torres, A., 162, 163(33)
Venema, G., 148
Venerini, F., 48
Vénien-Bryan, C., 259, 264
Venter, J. C., 244(13; 15; 16; 20), 245, 245(16; 20), 246
Verhagen, M. F. J. M., 248, 250(30), 251(30)
Verhoeven, A. J., 320
Verniquet, F., 22
Vicaire, R., 244(14), 245
Vichitbandha, S., 182, 193
Viezzoli, M. S., 47, 64
Vignais, P. M., 153
Vile, G. F., 301
Vincent, R., 193
Viola, K. L., 21
Visner, G. A., 130, 310
Vladimirov, Y. A., 3, 6
Vodovotz, Y., 101, 104(2), 105(2)
Voelcker, G., 116
Voelter, W., 116
Voetsch, W., 116
Vogel, H., 130, 193, 211(18), 340

Volbeda, A., 91
Volpe, C., 50, 53(16), 54(16), 59(16), 60(16), 61(16)
von Bohlen, A., 258, 261(5)
von Sonntag, C., 40
Voordouw, G., 124, 244(9; 11), 245
Voordouw, J. K., 124
Voorhorst, W. G. B., 248, 250(30), 251(30)
Vuillier, F., 24, 34(6)
Vujaškovic, Z., 226, 233(17b)
Vyakarnam, A., 140

## W

Walker, D. W., 223
Wallace, D. C., 19, 130, 211, 223, 340
Wallace, R., 191
Wallis, T. S., 156, 163(14)
Wan, X. S., 306, 307, 308(6), 341
Wang, H., 259
Wang, Y., 244(14), 245
Wang, Z., 226, 233(16), 235
Ward, J. M., 91
Ward, R. J., 3
Warner, D. S., 226, 233(18), 339
Watanabe, N., 347, 350(10; 11), 351, 353, 354, 354(19; 20)
Watson, S. P., 153
Weeraratna, A. T., 320
Weidman, J. F., 244(13; 16; 20), 245, 245(16; 20), 246
Weiner, P. K., 103
Weinstein, L., 80
Weinstock, G. M., 244(20), 245(20), 246
Weinstock, K. G., 244(16), 245(16), 246
Weisiger, R. A., 106, 107(1; 2), 272
Weiss, R. H., 62, 235
Weissig, H., 45
Weissmann, C., 258, 261(4)
Welsh, N., 310
Wen, J., 229
Weng, C., 158, 159, 159(21)
Weselake, R. J., 53(25), 54
Weser, U., 104, 116, 236
Westaway, D., 258, 259, 261(5)
Westbrook, J., 45
Westwick, J. K., 313
Wever, R., 25

White, C. W., 13, 18(25), 19(25), 20, 20(25), 21, 22(25; 26), 28, 197, 240, 338
White, J. E., 310, 340
White, O., 244(13; 15; 16; 20), 245, 245(16; 20), 246
Whitelaw, C. B., 191
Whittaker, J. W., 69, 70, 80, 83(6), 85, 86, 86(24; 25), 87, 87(24), 88(24; 33), 89, 89(24; 25; 33)
Whittaker, M. M., 69, 70, 80, 83(6), 85, 86, 86(24; 25), 87, 87(24), 88(24; 33), 89, 89(24; 25; 33)
Whittal, R. M., 259
Whittam, T. S., 200
Whyte, B., 170
Wierzbowski, J., 244(14), 245
Wigderson, M., 293
Wilcox, H. M., 129, 339
Wilczynska, A., 3
Wilhelm-Filho, D., 284, 285(17)
Wilks, K. E., 155, 157(4)
Williams, A. E., 161
Williams, E., 272, 274(10), 282
Williams, J. M., 5
Williams, M. D., 19
Williamson, R. A., 259
Willingham, I. R., 320
Willingham, M. C., 296
Williston, S., 115
Willson, J. M., 130
Wilmer, W. A., 314
Wilson, J. W., 145(3), 146, 310
Winge, D. R., 107, 108, 110(18), 113(9), 288
Wink, D. A., 101, 102, 104(2), 105(2; 12), 163
Winterbourn, C. C., 301, 360
Wiseman, T., 115
Woese, C. R., 244(13; 16); 245, 245(16), 246
Wolbert, R. B. G., 248, 250(30), 251(30)
Wolf, G., 114
Wolff, R. R., 337
Wolfgang, G., 90
Wolin, M. S., 319
Wong, B.-S., 259, 264, 265, 265(7)
Wong, G. H., 130, 139(6), 336
Wong, J. M. S., 314
Wong, P. C., 192(10), 193
Wong, Y., 44
Wood, P. M., 224
Woodgett, J. R., 313
Woodmansee, A. N., 3

Woodworth, R. C., 116
Wu, C. H., 155
Wunderlich, J. K., 250, 251(36)

## X

Xavier, A. V., 243, 244(17), 245(8; 17), 246, 247(8), 248, 248(8; 17), 250, 250(8), 251(8; 35), 253, 253(39), 254, 254(8; 31), 255(17; 31)
Xia, E., 112
Xia, J., 91
Xia, Y., 106
Xu, N., 313(7), 314
Xu, X., 21
Xu, Y., 153, 155, 162, 163(33), 199, 306, 307, 310, 310(5), 341

## Y

Yaginuma, Y., 138
Yakushiji, M., 138
Yamada, H., 339
Yamada, K., 102, 105(12)
Yamada, M., 115
Yamada, Y., 339
Yamakura, F., 84, 89
Yamamoto, S., 244(18; 19), 245(18; 19), 246
Yamano, S., 84
Yamaoka, K., 346
Yamashita, N., 336
Yamazaki, J., 244(18; 19), 245(18; 19), 246
Yamazato, A., 346
Yan, T., 312, 341
Yang, D.-S., 259
Yang, G., 25
Yang, Y., 259
Yasunami, M., 193
Yegorev, D. Y., 6
Yegorov, D. Y., 3
Yeh, A. P., 245(22), 246, 247(22), 248(22)
Yeh, C. C., 306, 307, 308(6), 310(5), 341
Yen, H. C., 182, 193
Yesilkaya, A., 32

Yesilkaya, H., 148
Yikilmaz, E., 90
Yim, M. B., 130, 240, 310, 314
Yim, Y.-I., 91, 93(8; 11), 95(11), 96(11), 98(11), 99(11), 243
Yip, C. M., 259
Yocum, C. F., 26
Yonezawa, T., 139
Yoo, H. Y., 293, 293(5), 295(5), 298, 298(5), 299(13; 14; 19–21; 23), 302
Yoshida, T., 65, 67(19), 73(19), 89, 339
Yoshimura, M. P., 130, 211, 340
Yoshioka, T., 310
Yoshizawa, T., 244(18; 19), 245(18; 19), 246
Yost, F. J., Jr., 97
Youn, H., 243
Youn, H.-D., 91, 93(7; 8), 95(7), 100(7), 243
Younes, M., 314
Young, R., 226, 233(17a)
Young, S. P., 3
Yu, C. H., 26, 27(21b)
Yu, J. C., 313(7), 314
Yu, L., 254
Yu, Z. X., 314

## Z

Zastawny, T. H., 19
Zecca, L., 105
Zhai, P., 181, 182(4), 192(12), 193
Zhang, H., 130, 193, 211(18), 312, 340
Zhang, J., 82(13), 83
Zhang, Z., 82(13), 83, 314
Zheng, W., 22
Zhong, W., 341
Zhou, L., 244(13; 16), 245, 245(16), 246
Zhou, M., 359
Zhu, L., 24, 32(4), 37
Zhuang, J., 334
Zhuang, Y., 24, 54
Zhuang, Y. X., 115
Zick, Y., 314
Zidian, J. L., 112
Zou, M., 32
Zweier, J. L., 106, 235, 275

# Subject Index

## A

Aconitase
  intracellular oxidative stress assay in superoxide dismutase transgenic mice
    activity staining of gel, 201–202
    cellulose gel electrophoresis, 201
    sample preparation, 200–201
  iron reactivity with superoxide, 11–13
  iron–sulfur cluster, 10–12, 21–23
  isoforms, 200
  kinetics of oxidative inactivation, 13
  nitric oxide inactivation, 13, 23
  superoxide assay
    activity assay, 15–16
    advantages and limitations, 22
    applications
      mitochondrial electron transfer inhibitors, 20
      oxidase effects, 21
      peptide studies, 21
      redox-cycling drug studies, 20
      superoxide diffusion through membranes, 19–20, 355
      superoxide dismutase effects on aconitase activity, 19
      superoxide dismutase mimetic effects on aconitase activity, 21
      tumor necrosis factor studies, 20
    calculations, 13–14
    fractionation of mitochondrial and cytosolic enzymes, 16–17
    inactive fraction measurement, 18–19
    rate constant determination
      inactivation, 17–18
      reactivation, 18
    sample preparation, 14–15
Aging, iron store relationship, 338–339
Amplex Red, see Hydrogen peroxide

## B

Brownian dynamics simulation, see Copper, zinc-superoxide dismutase

## C

Calcein, iron assay, 4
CD, see Circular dichroism
Circular dichroism
  manganese-superoxide dismutase, 87–88
  prion protein, 263–264
Confocal scanning laser microscopy, superoxide dismutase membrane-binding assay
  imaging, 322–323
  isolated cells, 324–325
  skin, 325
Copper, zinc-superoxide dismutase
  assays
    pulse radiolysis
      calculations, 42–43
      data collection, 41–42
      principles, 38–39
      solutions, 40–41
      superoxide generation, 39–40
    pyrogallol assay, 43–44
  bacterial functions, 155–156
  Brownian dynamics simulation of enzyme–substrate association rate
    diffusion-limited reaction, 44–45
    mutant enzyme analysis, 47–49
    pH-dependence studies, 48
    protein structure and electrostatic potential calculation, 45–46
    trajectory propagation and rate calculation, 46–47
  cell fractionation studies, 106–107
  differential scanning calorimetry
    data acquisition, 52–53
    intermediates in unfolding, 54
    melting temperatures of different enzymes, 53
    metals
      apoenzyme preparation, 55
      copper removal, 57
      effects on stability, 54–55
      zinc removal, 57
    mutation effects, 58–59, 61

pH dependence, 52
refolding, 53–54
sample preparation, 52
thermodynamic analysis of traces, 57–58
thermodynamic parameters, 51
disease mutations, 115
dismutation reaction, 38
extracellular enzyme, *see* Extracellular superoxide dismutase
fluorescence studies of unfolding, 51
human gene structure, 293
interconversion of nitric oxide and nitroxyl anion
  evidence, 102–104
  implications, 104–106
isothermal titration calorimetry of metal binding
  apoenzyme preparation with iminodiacetic acid chromatography, 116–118, 123
  controls, 119
  data analysis, 119–120
  pH dependence
    pH 5.0, 121–123
    pH 7.3, 120–121
  reconstitution, 122
  sample preparation, 118–119
knockout mouse phenotype, 129, 339
membrane-binding assays
  antiinflammatory function, 321
  cell isolation, 322
  cell type differences, 326
  confocal scanning laser microscopy
    imaging, 322–323
    isolated cells, 324–325
    skin, 325
  flow cytometry, 322, 324
  fluorescence labeling of enzyme, 321–322
  histologic studies, 323
  protective effects
    cytochrome $c$ assay, 323–324
    inosine triphosphate-induced clastogenic effects, 326
    phorbol myristate acetate-stimulated superoxide production, 325–326
    psoralin plus ultraviolet A-induced inflammatory skin reactions, 326
    superoxide-mediated clastogenesis, 326–327
metal affinity, 115
mitochondrial enzyme
  assay, 108
  inhibitor sensitivity, 109–110
  purification
    anion-exchange chromatography, 108
    marker enzyme assays, 107–108
    mitochondria isolation and digitonin treatment, 107
    purification table, 109
    sodium dodecyl sulfate effects in crude fractions, 110, 112–114
null mutants, *see* Superoxide dismutase
overexpression in transgenic systems, *see* Superoxide dismutase
rat tissue extract activities, 112–114
*Salmonella enterica* pathogenicity studies, *see* Superoxide dismutase
stability, 49–50
therapeutic targeting
  liver cell targeting
    galactose-conjugating enzyme, 350
    mannose-conjugating enzyme, 350
  low-pH tissue targeting
    co-poly(styrenemalic acid) conjugates, 347, 349–350
    fatty acid $N$-maleimide derivatives, 349–350
  rationale, 347
  renal proximal tubule cell targeting with hexamethylenediamine conjugate, 351
  vascular endothelial cell targeting with heparin-binding domain fusion protein, 351, 353–354
thermal inactivation assays, 50–51
transcriptional regulation and induction of *SOD1*
  chloramphenicol acetyltransferase assays of promoter strength
    activity assay, 296–297
    cell extract preparation, 296
    $\beta$-galactosidase as internal reference, 296
    transfection, 295–296
  *cis* element protein binding confirmation
    electrophoretic mobility shift assay, 297
    nuclear extract preparation, 297
  environmental factor induction
    chemical treatment of transfected cells, 301
    *cis* elements, 301–302, 304
    overview, 299, 301
    ginsenoside-Rb$_2$ induction, 299

modeling, 304–305
promoters from rat and human
  cloning, 294
  heterologous promoter construction, 298
  serial deletion mutant construction, 295
  structure, 293–294
  transcriptional regulation in normal state, 298–299
Cyclooxygenase, inhibitors, 320
Cytochrome $c$, superoxide assays
  acetylated cytochrome $c$ reduction, 283
  manganese-porphyrin, 232
  incubation conditions and detection, 357
  manganese-superoxide dismutase protective effects in cells, 323–324
  nickel-superoxide dismutase assay, 92–93
  precautions, 357–358
  principles, 356

# D

Desulfoferrodoxin, see Superoxide reductase
$o$-Diansidine, hydrogen peroxide fluorescence assay with horseradish peroxidase, 358–359
$2',7'$-Dichlorodihydrofluorescein diacetate, flow cytometry assay of oxidants, 318–319
Differential scanning calorimetry
  copper,zinc-superoxide dismutase
    data acquisition, 52–53
    intermediates in unfolding, 54
    melting temperatures of different enzymes, 53
    metals
      apoenzyme preparation, 55
      copper removal, 57
      effects on stability, 54–55
      zinc removal, 57
    mutation effects, 58–59, 61
    pH dependence, 52
    refolding, 53–54
    sample preparation, 52
    thermodynamic analysis of traces, 57–58
  thermodynamic parameters, 51
Dimethyl sulfoxide, superoxide stabilization, 62–63
DMSO, see Dimethyl sulfoxide
Drosophila transgenesis, see Superoxide dismutase
DSC, see Differential scanning calorimetry

# E

Electron paramagnetic resonance
  iron assay in cells
    calculations, 7
    cell growth, 6
    data acquisition, 7
    desferrioxamine mesylate chelation, 5–6
    performance, 7, 9
    reagents, 6
    sample preparation, 6–7
    theory, 5–6
  whole yeast superoxide dismutase null mutant analysis
    cell culture, 175–177
    data processing and calculations, 179–180
    low-temperature spectroscopy, 178–179
    media, 175–176
    sample preparation, 177–178
    signal features, 173–174
    superoxide stress indicator, 175
  manganese-superoxide dismutase, 89–90
  mitochondrial superoxide production assay, 275–277, 356
  nickel-superoxide dismutase, 97
  superoxide reductase, 254
Electrophoretic mobility shift assay, $SOD1$ regulators, 297–298
ELISA, see Enzyme-linked immunosorbent assay
EMSA, see Electrophoretic mobility shift assay
Enzyme-linked immunosorbent assay, manganese-superoxide dismutase
  additive experiments, 136
  applications, 141
  biotinylation of antibody, 132
  confounding factors, 136
  human serum levels
    adult respiratory distress syndrome, 139
    cancer, 138–140
    healthy controls, 136, 138
    liver disease, 138
    myocardial infarction, 139, 141
  incubation conditions, 133
  materials, 132
  monoclonal antibody preparation and assay, 133–134, 139
  polyclonal antibody preparation, 131–132
  reproducibility, 136

sample preparation
  serum, 134
  tissue, 134–135
serum dilution effects, 136
specificity, 136
stability analysis, 136
standard curve, 135
EPR, see Electron paramagnetic resonance
Extracellular superoxide dismutase
  distinguishing from copper,zinc-superoxide dismutase
    activity assay, 77
    antibody staining and separations, 79
    concanavalin A affinity chromatography, 77–78
    gel filtration, 77
    heparin affinity chromatography, 78
    sample preparation, 76
  heparin affinity, 75
  inhibitors, 75
  knockout mouse phenotype, 339
  overexpression in transgenic systems, see Superoxide dismutase
  polymorphism and disease, 339
  storage, 74–75
  tissue distribution, 74

# F

Flow cytometry
  oxidant detection using 2′,7′-dichlorodihydrofluorescein diacetate, 318–319
  superoxide dismutase membrane-binding assay, 322, 324

# G

Ginsenoside-Rb$_2$, *SOD1* induction, 299

# H

Hemochromatosis, superoxide-dependent pathology, 337
Hydroethine, mitochondrial superoxide production assay, 277, 279
Hydrogen peroxide
  assays
    Amplex Red fluorescence assay with horseradish peroxidase
      detection, 360
      incubation conditions, 359
      precautions, 361
      principles, 359
      standard curve, 359, 361
    *o*-diansidine fluorescence assay with horseradish peroxidase, 358–359
    horseradish peroxidase quantitative analysis assays, 284, 342, 358–359
    *p*-hydroxyphenylacetate fluorescence assay with horseradish peroxidase
      calculations, 344–345
      equipment, 343–344
      incubation conditions, 344
      interassay variation, 345–346
      principles, 342, 249
      reagents, 343
      sensitivity, 345
      standard curve, 344
      tissue-specific mitochondrial responses, 346
    mitochondrial assay as measure of superoxide production, 279–280
    parallel estimation with superoxide, 361
    production rates, 284–285
    scopoletin fluorescence assay with horseradish peroxidase, 359
    steady-state concentrations, 285, 287
  bacterial sensitivity and DNA damage assay, 149–151
  membrane diffusion, 283, 355
  organelle production, 281
  signal transduction, 280, 314
  T cell intracellular levels, 280–281
*p*-Hydroxyphenylacetate, *see* Hydrogen peroxide

# I

Inositol phosphate kinase, immunoassays for mitogen-activated protein kinase activation, 315–317
Iron
  binding proteins and superoxide damage, 337–338
  calcein assay, 4
  electron paramagnetic resonance assay in cells
    calculations, 7
    cell growth, 6
    data acquisition, 7

desferrioxamine mesylate chelation, 5–6
  performance, 7, 9
  reagents, 6
  sample preparation, 6–7
  theory, 5–6
  whole yeast superoxide dismutase null
    mutant iron content analysis
    cell culture, 175–177
    data processing and calculations,
      179–180
    low-temperature spectroscopy,
      178–179
    media, 175–176
    sample preparation, 177–178
    signal features, 173–174
    superoxide stress indicator, 175
intracellular pools, 3
Mšssbauer spectroscopy, 5
oxidative damage, 3–4
oxidative stress role
  overload exacerbation of superoxide-
    dependent pathology, 337–338
  stores and aging, 338–339
Isothermal titration calorimetry
  applications, 115–116
  copper,zinc-superoxide dismutase metal
    binding
  apoenzyme preparation with iminodiacetic
    acid chromatography, 116–118, 123
  controls, 119
  data analysis, 119–120
  pH dependence
    pH 5.0, 121–123
    pH 7.3, 120–121
  reconstitution, 122
  sample preparation, 118–119
ITC, see Isothermal titration calorimetry

## L

Lipid peroxidation, superoxide balance with
  superoxide dismutase, 335
Lucigenin, superoxide detection limitations, 355

## M

Manganese-porphyrin
  applications, 223
  brain homogenate antioxidant studies, 233
  cyclic voltammetry, 232

cytochrome $c$ assay of superoxide
  dismutation, 232
lipophilic compounds, 227
peroxynitrite reactions
  carbon dioxide effects, 30
  reactivity, 27–28
  stopped-flow studies, 29–30, 32
reaction rates with oxidants, 228–230
redox potentials, 224, 228
spectrophotometry of oxidation reactions, 29
superoxide dismutase-deficient *Escherichia
  coli* protection, 232–233
superoxide dismutase mimetic, 27
synthesis, 230–231
water-soluble compounds, 224–227
Manganese-superoxide dismutase
  catalytic efficiency, 64–65
  enzyme-linked immunosorbent assay
    additive experiments, 136
    applications, 141
    biotinylation of antibody, 132
    confounding factors, 136
    human serum levels
      adult respiratory distress syndrome, 139
      cancer, 138–140
      healthy controls, 136, 138
      liver disease, 138
      myocardial infarction, 139, 141
    incubation conditions, 133
    materials, 132
    monoclonal antibody preparation and assay,
      133–134, 139
    polyclonal antibody preparation, 131–132
    reproducibility, 136
    sample preparation
      serum, 134
      tissue, 134–135
    serum dilution effects, 136
    specificity, 136
    stability analysis, 136
    standard curve, 135
  human enzyme
    crystal structure, 68–70
    direct superoxide measurement in assays,
      61–64, 73
    product inhibition, 72–73
    site-directed mutagenesis, 70–72, 74
  induction of expression, 130, 139–140, 308,
    310, 312, 336–337
  knockout mouse phenotype, 130, 339–340

manganese redox properties, 26
membrane-binding assays
  antiinflammatory function, 321
  cell isolation, 322
  cell type differences, 326
  confocal scanning laser microscopy
    imaging, 322–323
    isolated cells, 324–325
    skin, 325
  flow cytometry, 322, 324
  fluorescence labeling of enzyme, 321–322
  histologic studies, 323
  protective effects
    cytochrome c assay, 323–324
    inosine triphosphate-induced clastogenic effects, 326
    phorbol myristate acetate-stimulated superoxide production, 325–326
    psoralin plus ultraviolet A-induced inflammatory skin reactions, 326
    superoxide-mediated clastogenesis, 326–327
mitochondrial localization, 106–107, 129
mutation and cancer, 340–341
null mutants, see Superoxide dismutase
overexpression in transgenic systems, see Superoxide dismutase
peroxynitrite reactions
  kinetic studies, 33–34
  metal center role, 32–33
  nitration of self and low-molecular-weight aromatics, 34, 36–37
  nitration site, 32
product inhibition, 67–68
prokaryotic enzymes
  absorption spectroscopy, 87, 89
  circular dichroism, 87–88
  cloning for recombinant expression, 84–86
  electron paramagnetic resonance, 89–90
  function, 80
  purification of recombinant enzymes, 86–87
  sequence homology, 81–83
  type identification from genomic data, 83–84
proton transfer in catalysis, 66–67
rat tissue extract activities, 112–113
redox properties, 65–66
structure, 64, 80, 131

transcriptional regulation of human gene
  cotransfection with transcription factor vectors, 308
  direct transcript detection versus reporter assays, 308, 312
  enhancer elements in induction, 310
  promoter
    footprinting studies, 306
    polymerase chain reaction amplification, 307
  transcription factor binding sites, 306, 312
  structure of gene, 306
  tumor necrosis factor-$\alpha$ response, 310, 312, 336
tumor expression, 131, 138–140
MAPK, see Mitogen-activated protein kinase
Mitogen-activated protein kinase
  activation assays
    inositol phosphate kinase immunoassays, 315–317
    oxidant detection using 2′,7′-dichlorodihydrofluorescein diacetate flow cytometry assay, 318–319
    oxidase assessment with inhibitors, 319–320
    phosphorylated protein Western blotting, 315
  cellular response specificity of signaling pathways, 313
  oxidative signaling in activation, 314
  types, 313
Mitoplast, see Superoxide
Mössbauer spectroscopy
  iron assay, 5
  superoxide reductase, 254–255

# N

NADPH oxidase, inhibitors, 320
NBT, see Nitroblue tetrazolium
Neelaredoxin, see Superoxide reductase
Nickel-superoxide dismutase
  assays
    activity staining in gels, 93
    column fractions, 93
    cytochrome c assay, 92–93
    nitroblue tetrazolium assay, 92
    superoxide absorption assay, 93

differential expression with iron enzyme in
    *Streptomyces,* 100
  discovery, 91
  inhibitors, 96
  kinetic parameters, 95
  nickel effects
    processing of enzyme, 99
    transcription, 99
  overexpression in scherichia coli, 99–100
  purification from *Streptomyces*
    cell culture, 91–92
    *Streptomyces coelicolor,* 94–95
    *Streptomyces griseus,* 94–95
    *Streptomyces seoulensis,* 94
  sequence homology, 91, 98–99, 101
  size, 95
  species distribution, 100
  spectroscopy
    electron paramagnetic resonance, 97
    ultraviolet–visible spectrum, 96–97
    X-ray absorption spectroscopy, 98
  stability, 95
Nitric oxide
  aconitase inactivation, 13, 23
  nitroxyl anion formation, *see* Nitroxyl anion
  peroxynitrite formation, 23–24
Nitroblue tetrazolium
  nickel-superoxide dismutase assay, 92
  superoxide detection limitations, 355
  superoxide dismutase assay in *Drosophila*
    homogenates
    calculations, 289–290
    distinguishing of isoform activity,
      291–292
    incubation conditions and absorbance
      measurement, 289
    materials, 288
    principles, 287–288
    sample preparation, 288–289
    validation, 290–291
Nitroxide radical, *see* Superoxide dismutase
  mimics
Nitroxyl anion
  copper,zinc-superoxide dismutase
    interconversion of nitric oxide and
      nitroxyl anion
    evidence, 102–104
    implications, 104–106
  formation from nitric oxide, 101–102
NO, *see* Nitric oxide

## O

Oxidative stress
  control
    modulation of superoxide dismutase
      activity, 336–337
    overview, 331–332, 346–347
    superoxide balance with superoxide
      dismutase, 334–336
  disease role
    malignancy, 340–341
    universality, 331, 333
    inflammation role, 333–334
  iron role
    overload exacerbation of superoxide-
      dependent pathology, 337–338
    stores and aging, 338–339
    mitochondrial injury, 332–333
  reactive oxygen species sources in cells,
    281–282, 342
  signal transduction, *see* Hydrogen peroxide;
    Superoxide

## P

*P* element, *see* Superoxide dismutase
Peroxynitrite
  diffusion, 24
  formation, 23–24
  Lewis acid reaction, 25–26
  manganese-porphyrin reactions
    carbon dioxide effects, 30
    reactivity, 27–28
    stopped-flow studies, 29–30, 32
  manganese-superoxide dismutase reaction
    kinetic studies, 33–34
    metal center role, 32–33
    nitration of self and low-molecular-weight
      aromatics, 34, 36–37
    nitration site, 32
  reactivity, 24–26
  reduction potentials, 24
Prion protein
  circular dichroism analysis, 263–264
  copper content analysis, 264
  diseases, 258
  isoforms, 258
  preparation
    native protein, 259
    recombinant protein

copper incorporation, 261–263
Escherichia coli growth and induction, 260
gene cloning, 260
histidine tag removal, 262–263
insoluble protein extraction, 260
nickel affinity chromatography, 261
soluble protein extraction, 260
signal peptides, 260
superoxide dismutase activity assays
gel activity assay, 265
overview, 259, 264–265
spectrophotometric assay, 265, 267
synapse localization, 258
PrP, see Prion protein
Pulse radiolysis
copper,zinc-superoxide dismutase assay
calculations, 42–43
data collection, 41–42
principles, 38–39
solutions, 40–41
superoxide generation, 39–40
direct superoxide measurement, 63–64
Pyrogallol, copper,zinc-superoxide dismutase assay, 43–44

## R

Rubredoxin, electron donor for superoxide reductase, 251–252
Reactive oxygen species, see Oxidative stress

## S

Scopoletin, hydrogen peroxide fluorescence assay with horseradish peroxidase, 359
SOD, see Superoxide dismutase
SOR, see Superoxide reductase
Superoxide
aconitase reaction and assay, see Aconitase
balance with superoxide dismutase, 334–336
cytochrome c assay
incubation conditions and detection, 357
precautions, 357–358
principles, 356
direct observation
pulse radiolysis, 63–64
stabilization for measurement, 62–63
stopped-flow spectrophotometry, 62–63
ultraviolet absorption, 61

intracellular levels, 9
lucigenin assay limitations, 355
membrane permeability, 19–20, 355
mitochondrial production
assays
acetylated cytochrome c reduction, 283
adenochrome formation, 282–283
electron paramagnetic resonance, 275–277, 356
hydroethine oxidation, 277, 279, 355–356
hydrogen peroxide assay, 279–280, 355
overview, 282, 355–356
intermembrane space detection approaches, 271–272
matrix space release, 271
mitoplast preparation
digitonin treatment, 273
functional integrity assessment with respiratory control ratio, 273–275
heart mitochondria isolation, 272
hypoosmotic treatment, 273
liver mitochondria isolation, 272
nitroblue tetrazolium assay limitations, 355
organelle production, 281
parallel estimation with hydrogen peroxide, 361
signal transduction, 280, 314
sources in cells, 281–282
targets in cells, 10
Superoxide dismutase
bacteria types and functional overview, 145, 155–156
balance with superoxide, 334–336
catalytic mechanism, 236–237, 243
copper,zinc enzyme, see Copper,zinc-superoxide dismutase
effects on aconitase activity, 19, 21
extracellular enzyme, see Extracellular superoxide dismutase
manganese enzyme, see Manganese-superoxide dismutase
meal coordination, 90
mimics, see Superoxide dismutase mimics
nickel enzyme, see Nickel-superoxide dismutase
nitroblue tetrazolium assay of Drosophila homogenates
calculations, 289–290

distinguishing of isoform activity, 291–292
incubation conditions and absorbance
   measurement, 289
materials, 288
principles, 287–288
sample preparation, 288–289
validation, 290–291
null mutants in bacteria
   compartmentalization problems, 146
   complementation assays
      metabolic defects, 152, 154
      superoxide-sensitive target identification,
         153–154
   cytoplasmic *sod* mutants
      aerobic growth and aerotolerance, 147
      DNA damage measured by hydrogen
         peroxide sensitivity, 149–151
      *Escherichia coli* as model system,
         146–147
      iron–sulfur cluster sensitivity, 149
      membrane damage, 151
      metabolic defects, 151
      paraquat sensitivity, 147–149
   generation, 145–146
   periplasmic copper,zinc-superoxide
      dismutase mutants, 151–152
   prospects for study, 154
null mutants in *Saccharomyces cerevisiae*
   advantages of system, 167
   amino acid auxotrophy, 168–169
   enzyme types, 167, 173
   handling of mutants, 167–168
   heat sensitivity, 170
   hyperoxia sensitivity, 170–171
   iron content analysis in whole yeast using
      electron paramagnetic resonance
      cell culture, 175–177
      data processing and calculations,
         179–180
      low-temperature spectroscopy, 178–179
      media, 175–176
      sample preparation, 177–178
      signal features, 173–174
      superoxide stress indicator, 175
   oxidant sensitivity, 168
   protein carbonyl assay, 171–172
   stationary-phase survival, 169, 173
overexpression in transgenic *Drosophila*
   constructs, 214
   life span studies

cohort variations, 219–220
insertional position effects, 218–219
*P* transposase insertion
   DNA preparation, 214–215
   microinjection, 215–216
   overview, 213–214
   transgene remobilization, 216–218
overexpression in transgenic mice
   aconitase assay of intracellular oxidative
      stress
      activity staining of gel, 201–202
      cellulose gel electrophoresis, 201
      sample preparation, 200–201
   amyotrophic lateral sclerosis model, 181,
      192, 213
   assay in fluids and tissue samples, 185–186,
      199–200
   cell culture
      cell death quantification and
         morphological characterization, 190
      fetal fibroblast derivation, 207–208
      fetal liver hematopoietic stem cell
         studies, 208–210
      immunostaining, 189
      preparation, 187–189
   commercial availability, 182
   complementation studies of knockout mice,
      210–213
   construction, 182, 191–193
   genotyping
      nondenaturing isoelectric focusing
         activity gel, 197–199
      nondenaturing polyacrylamide gel
         electrophoresis activity gel,
         184–185, 197–197
      polymerase chain reaction, 183–184,
         195–196
      Southern blot, 183, 195–196
   isoforms, 180–181
   *in situ* hybridization of
      copper,zinc-superoxide dismutase
      transcripts
      digoxigenin detection, 206
      fixation, 202–203
      hybridization, 205
      overview, 202
      prehybridization, 205
      radiation detection, 206
      riboprobe synthesis, 203–204
      sectioning of tissue, 203

tissue distribution analysis, 206
washing, 205
strain maintenance, 193–195
transgene copy number, 182–183
physiological function, 9
*Salmonella enterica* pathogenicity studies
activity staining in gels, 159–161
epithelial infection assays
intracellular survival assay, 166
invasion assay, 165
extraction of whole cells, 157
isoelectric focusing, 159
macrophage infection assay
bacterial uptake and killing, 164–165
macrophage isolation, 163–164
monolayer infection, 164
mouse serum preparation, 164
mouse infection and burden analysis, 166
periplasmic extract preparation, 157–158
polyacrylamide gel electrophoresis
separation of types, 158–159
solution activity assay, 161–162
strain susceptibility to superoxide
cytoplasmic superoxide sensitivity, 162
exogenous superoxide sensitivity, 163
Superoxide dismutase mimics, *see also*
Manganese-porphyrin; Prion protein;
Superoxide reductase
aconitase activity effects, 21
adverse effects, 240–241
catalytic mechanism, 236–237
characteristics, 235
dose response, 241–242
metal chelates, 235–236
nitroxide radicals
antioxidant mechanisms, 240
applications, 238–239
oxidative mode, 239–240
reductive mode, 239
redox requirements, 237–238
requirements for *in vivo* use, 237–238, 242
scavengers versus catalysts, 234
ternary complex, 242
therapeutic potential, 234
Superoxide reductase
active site structure, 247–248
assay
calculations, 127–129
direct measurement, 127

distinguishing from superoxide dismutase,
256–257
half-reactions, 126–127
catalytic reaction, 124, 243
classes, 124–125, 244
*Desulfoarculus baarsii* enzyme
kinetic parameters, 129
overexpression in *Escherichia coli*, 125
purification of recombinant enzyme,
125–126
electron paramagnetic resonance, 254
Mšssbauer spectroscopy, 254–255
prospects for study, 257–258
recombinant enzyme expression and
purification from *Escherichia coli*
advantages, 248–249
gene cloning, 249
*Treponema pallidum* neelaredoxin,
249–250
redox properties
redox potentials, 250–251
reduced and oxidized protein preparation
desulfoferrodoxin, 252–253
neelaredoxin, 252
rubredoxin as electron donor, 251–252
sequence homology between species,
245–247
species distribution, 124
superoxide dismutase activity, 255–256
ultraviolet–visible spectroscopy, 253–254

## T

TNF-$\alpha$, *see* Tumor necrosis factor-$\alpha$
Transgenic mouse, *see* Superoxide dismutase
Tumor necrosis factor-$\alpha$, manganese-superoxide
dismutase induction, 310, 312, 336

## W

Western blot, mitogen-activated protein kinase
phosphorylation assay, 315

## X

Xanthine oxidase, inhibitors, 319–320
X-ray absorption spectroscopy,
nickel-superoxide dismutase, 98

ISBN 0-12-182252-4